S0-BXD-956

MECHANISMS OF MIGRATION IN FISHES

NATO CONFERENCE SERIES

I Ecology
II Systems Science
III Human Factors
IV Marine Sciences
V Air–Sea Interactions
VI Materials Science

IV MARINE SCIENCES

Recent volumes in this series

Volume 4 Fjord Oceanography
edited by Howard J. Freeland, David M. Farmer, and Colin D. Levings

Volume 5 Bottom-interacting Ocean Acoustics
edited by William A. Kuperman and Finn B. Jensen

Volume 6 Marine Slides and Other Mass Movements
edited by Svend Saxov and J. K. Nieuwenhuis

Volume 7 The Role of Solar Ultraviolet Radiation in Marine Ecosystems
edited by John Calkins

Volume 8 Structure and Development of the Greenland–Scotland Ridge
edited by Martin H. P. Bott, Svend Saxov, Manik Talwani, and Jörn Thiede

Volume 9 Trace Metals in Sea Water
edited by C. S. Wong, Edward Boyle, Kenneth W. Bruland, J. D. Burton, and Edward D. Goldberg

Volume 10A Coastal Upwelling: Its Sediment Record
Responses of the Sedimentary Regime to Present Coastal Upwelling
edited by Erwin Suess and Jörn Thiede

Volume 10B Coastal Upwelling: Its Sediment Record
Sedimentary Records of Ancient Coastal Upwelling
edited by Jörn Thiede and Erwin Suess

Volume 11 Coastal Oceanography
edited by Herman G. Gade, Anton Edwards, and Harald Svendsen

Volume 12 Hydrothermal Processes at Seafloor Spreading Centers
edited by Peter A. Rona, Kurt Boström, Lucien Laubier, and Kenneth L. Smith, Jr.

Volume 13 Flows of Energy and Materials in Marine Ecosystems: Theory and Practice
edited by M. J. R. Fasham

Volume 14 Mechanisms of Migration in Fishes
edited by James D. McCleave, Geoffrey P. Arnold, Julian J. Dodson, and William H. Neill

[NATO Advanced Research Institute ...]

MECHANISMS OF MIGRATION IN FISHES

Edited by

James D. McCleave
University of Maine
Orono, Maine

Geoffrey P. Arnold
Ministry of Agriculture, Fisheries and Food
Lowestoft, Suffolk, England

Julian J. Dodson
Laval University
Quebec, Quebec, Canada

and

William H. Neill
Texas A&M University
College Station, Texas

Published in cooperation with NATO Scientific Affairs Division

PLENUM PRESS · NEW YORK AND LONDON

Library of Congress Cataloging in Publication Data

NATO Advanced Research Institute on Mechanisms of Migration in Fishes (1982: Acquafredda di Maratea, Italy)

Mechanisms of migration in fishes.

(NATO conference series. IV, Marine sciences; v. 14)
"Proceedings of a NATO Advanced Research Institute on Mechanisms of Migration in Fishes, held December 13–17, 1982, in Acquafredda di Maratea, Italy"—Verso CIP t.p.
"Published in cooperation with NATO Scientific Affairs Division."
Bibliography: p.
Includes index.
1. Fishes—Migration—Congresses. I. McCleave, James D. II. North Atlantic Treaty Organization. Scientific Affairs Division. III. Title. IV. Series.
QL639.5.N38 1982 597′.052′5 84-4754
ISBN 0-306-41676-X

Proceedings of a NATO Advanced Research Institute on Mechanisms of Migration in Fishes, held December 13–17, 1982, in Acquafredda di Maratea, Italy

This work relates to Department of the Navy Grant N00014-83-G-0013 issued by the Office of Naval Research. The United States Government has a royalty-free license throughout the world in all copyrightable material contained herein.

A Division of Plenum Publishing Corporation
233 Spring Street, New York, N.Y. 10013

Printed in the United States of America

PREFACE

The last major synthesis of our knowledge of fish migration and the underlying transport and guidance phenomena, both physical and biological, was "Fish Migration" published 16 years ago by F.R. Harden Jones (1968). That synthesis was based largely upon what could be gleaned by classical fishery-biology techniques, such as tagging and recapture studies, commercial fishing statistics, and netting and trapping studies. Despite the fact that Harden Jones also provided, with a good deal of thought and speculation, a theoretical basis for studying the various aspects of fish migration and migratory orientation, progress in this field has been, with a few exceptions, piecemeal and more disjointed than might have been expected.

Thus we welcomed the approach from the NATO Marine Sciences Programme Panel and the encouragement from F.R. Harden Jones to develop a proprosal for, and ultimately to organize, a NATO Advanced Research Institute (ARI) on mechanisms of fish migration. Substantial progress had been made with descriptive, analytical and predictive approaches to fish migration since the appearance of "Fish Migration." Both because of the progress and the often conflicting results of research, we felt that the time was again right and the effort justified to synthesize and to critically assess our knowledge. Our ultimate aim was to identify the gains and shortcomings and to develop testable hypotheses for the next decade or two.

While the above is an interesting intellectual excercise for basic science, a thorough knowledge of mechanisms of migration is equally important to the development of conservation and management policy and to the tactics and strategies for efficient commercial exploitation of migratory species. Successful development of new commercial uses of migratory species, such as ocean ranching of salmonids, will be assured only if migratory patterns and migratory responses to environmental factors are well known. Proper assessment and control, as well as purposeful manipulation, of environmental conditions depend upon this knowledge as well.

Our aims for the ARI were optimistically high. Whether they were met, and whether they are reflected, at least in part, in this volume, will remain to be seen. We can assure the reader that the 43 scientists--drawn from 10 countries and from biology, engineering, mathematics, oceanography, philosophy, and psychology--who met for five intense days in the charming isolation of Acquafredda di Maratea, Italy in December, 1982 were serious of purpose.

The ARI was organized into four sessions, three based upon major hydrographic regimes--open oceans, continental-shelf and estuarine waters, and rivers--and the fourth on special topics relevant to studies of fish migration--animal learning, methodology, physical oceanography, bioenergetics, and a review of relevant bird migration studies. (The special topics session led off our ARI, but here papers from this session follow other papers in order not to divert the reader from fish migration at the outset.) The papers in this volume are arranged under the four section headings; however, several papers might just as well have been placed in other sections. The first three sections begin with introductory papers by F.R. Harden Jones, W.C. Leggett, and T.G. Northcote, respectively, who served at the ARI as chairmen of the sessions dealing with the three hydrographic regimes. The volume concludes with a summary chapter of general recommendations arising from the final day of panel discussions.

The 43 participants presented 53 papers, 30 of which are published here. Many of the remaining contributions have been, or soon will be, published elsewhere.

We thank Mr. A. Guzzardi and the entire staff of the Villa del Mare Hotel for making our stay in Aquafredda di Maratea so delightful. The personal and friendly attention we received from the staff combined with the excellent conference facilities, accommodations and food (and spectacular views over the ocean) to contribute greatly to a successful ARI.

We thank the NATO Scientific Affairs Division, the US National Science Foundation (grant number BNS-8206546), and the US Office of Naval Research (grant number N00014-83-G-0013) for their generous financial support of the ARI. The senior editor is grateful to the Leopold Schepp Foundation, the Eppley Foundation for Research, the US National Science Foundation (grant numbers OCE-8208394, BNS-8122116) and the University of Maine for financial support during the organizing of the ARI and the preparation of this volume. We greatly appreciate assistance given by Edwana Albright (word processing), Suzanne Manlove (secretarial), and Barbara McCleave, Suzanne Manlove and Denise Brown (proofreading and indexing).

James D. McCleave
Geoffrey P. Arnold
Julian J. Dodson and
William H. Neill

CONTENTS

MIGRATION IN THE OPEN OCEAN

MIGRATION IN RIVERS

SPECIAL TOPICS

A VIEW FROM THE OCEAN

F. R. Harden Jones

Ministry of Agriculture, Fisheries and Food
Fisheries Laboratory
Lowestoft, Suffolk NR33 OHT England

ABSTRACT

Migration is defined as a coming and a going with the seasons on a regular basis and a distinction is made between local and more extensive movements in terms of fish-body lengths (10^3 L and 10^5 L, respectively), the former being likely to be associated with scalar and the latter with vector quantities. It is suggested that *clue* and *cue* be used, relative to migration, in the sense of information as to *where* and *when* respectively. The open ocean is considered as the sea area where oceanic currents are stronger than tidal currents and the importance to migration of the main circulatory features of the ocean, at different scales, is discussed. Boundaries (at the surface, between water masses, and near the bottom) are identified as sites where migrants might obtain directional clues.

Negative attitudes towards the publication of primary data are considered as a serious hindrance towards an advance in our knowledge and understanding of problems of homing and migration. While there is reason for concern over data that are known to have been collected, there is the further problem over data which may have been collected and about which nothing is known. There could be a need for a focal point for the exchange of information and the deposition of archival material.

To illustrate some of the more general points, aspects of the migratory behaviour of the southern bluefin tuna and the Pacific salmon are examined in detail. It is suggested that a proportion of the southern bluefin population may make a circumpolar migration to

---

match that of its distribution. The results of the open-ocean tagging experiments with sockeye, pink, and chum salmon show that there is an area of the North Pacific common to all three species from which there are no reported recoveries of tagged fish in coastal waters. The area is one in which there is no temperature stratification. It is estimated that up to 5% of the mature population could be at risk to being "lost at sea" within such a salmon "sink" or "hole."

The possible role of inertial clues in ocean migration is discussed. There are problems arising from the relatively low sensitivity of the vertebrate labyrinth to angular and linear accelerations, and from the equivalence of the effects of gravity and acceleration. But theoretical and experimental studies suggest that the use of inertial clues is not necessarily so unlikely as has been previously thought.

INTRODUCTION

When at sea, I like the view from the ocean. But waters in general, great or small, can be dangerous places: their very grandeur tempts us to play the braggart. I have in mind the scene in Shakespear's Henry IV Part I where Glendower boasts that, "I can call spirits from the vasty deep," and Hotspur cynically replies, "Why so can I, or so can any man; but will they come when you do call for them?" Perhaps we would all be wise to claim no more than can be sustained in argument and discussion. And in these encounters it should not be forgotten that our purpose is to move towards a better understanding of the problems of migration rather than to establish or maintain our individual positions. Furthermore, there are two important points to remember when making our contributions: the first is that any hypotheses should be simple, and the second is that we should at least know of, if not see, the original data claimed to support, or bearing in mind the cautions of van der Steen (1984--this volume), refute them.

SOME TERMS DEFINED

This section is entitled Open-ocean Migration. I use the word migration in the sense of a coming and a going with the seasons on a regular basis: for an animal which has a life-span of a year or more this definition would appear to meet our immediate needs.

If this definition is accepted, we can go further. Migratory movements involve a shift in the distribution of a section of the population--or of a stock if the population is so divided--and these movements are more than a shift of ground. I find it helpful to make a distinction between local and more extensive movements. If pressed to put in numbers, I would make the distinction in terms of fish lengths (L), with the local movements not exceeding 10^3 L and the more extensive movements usually associated with migration not less than 10^5 L. The difference is between a few kilometres on the one hand, and many tens, several hundreds, and occasionally a few thousands of kilometres on the other.

The differences in distance may reflect differences in behaviour, local movements being determined by reference to scalar quantities with the changes in distribution brought about by kineses and klino- or tropo-taxes, whereas the more extensive movements may involve vector quantities and the behavioural mechanisms klino-, tropo-, and telo-taxes.

Finally, there is one semantic cause which I would like to champion, and this concerns the meanings attached to the words *clue* and *cue*: and it is to our North American cousins that the point is made. The Shorter Oxford English Dictionary indicates that *clue* is a later form of *clew*; and a *clew* is a globular body, a ball, and especially a ball of thread or yarn. Such was the clew that Ariadne gave to Theseus when he was dealing with the Minotaur, and the thread from the clew allowed him to retrace his track through the labyrinth. So a *clue* has the meaning of an indication to follow, a key; and when dealing with migration I would suggest that we use *clue* in the sense of information as to *where*. At all events Ariadne did not give Theseus a *cue*, whose theatrical sense is that of a signal to an actor to enter or to begin his speech; and for migration I would suggest that *cue* be used in the sense of information as to *when*. To summarize: migratory behaviour may be triggered by cues; but when on migration the beast proceeds by reference to clues.

MIGRATION AND WATER CURRENTS

Many of the migratory movements of fishes can be related to water currents, and in the Epilogue to the book "Fish Migration" (Harden Jones 1968) I wrote that, "... one of the fundamental questions to be asked, and answered, is simply this: what are the movements of migrants relative to those of the water at the depth at which they are swimming?" I believe that the importance of this point is still generally accepted. In the January 1982 issue of the Underwater Telemetry Newsletter we are reminded that: "one aspect of the study of open water migration, one which is often ignored, however, is that the medium through which fish move is itself in motion. When the velocity of the local current approaches that of the fish, the swimming speed and direction of fish relative to ground may be very different to swimming behaviour relative to the surrounding body of water" (Smith et al. 1982, page 1); and Weihs (1984--this volume) has made us realize the importance of the relative motion between fish and medium in connection with studies on

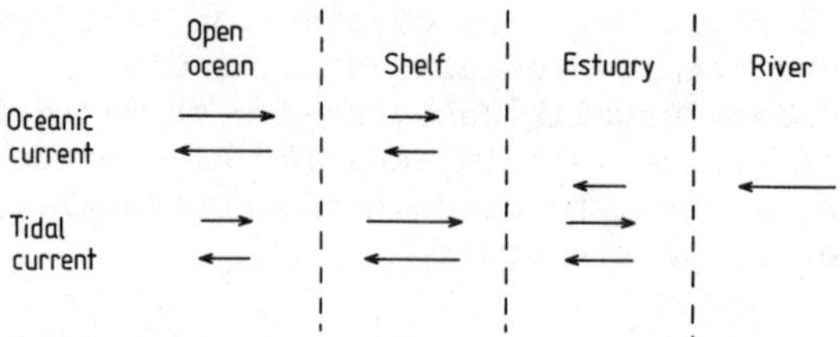

Figure 1. The relative importance of oceanic and tidal currents in the open ocean, continental-shelf waters, estuaries and rivers.

the energy cost of swimming and of migration.

A distinction is often made between open-ocean migration and that on the continental shelf, in the estuary, and up the river. If the four regimes are considered in relation to their water current patterns, the distinction between them would appear to be well founded. In the open ocean the oceanic currents are dominant and the effects of the tides are slight (Fig. 1). On the shelf the tidal currents are sometimes much stronger, although the effects of the ocean currents, usually expressed as a residual drift, may have an important effect on the distribution of plankton and pelagic nekton. In estuaries a downstream component is added to the tidal flow, while in the rivers there is only the flow to the sea.

In the open ocean the surface circulation is dominated by gyres, which are large-scale eddies whose dimensions are comparable to those of the basins within which they are contained. The distribution and migratory patterns of many stocks of fishes would appear to be closely related to these gyres, and what might be called the "doctrine of hydrographic containment" now enjoys a measure of respectability. But the ocean currents are neither as simple nor as consistent as the usual charts suggest. The effect of meanders, rings, and eddies (Fig. 2) on the abundance, distribution and movements of fish is something about which little is known. Within the basins occupied by the gyres there are water masses with differing temperatures and salinities. At the frontal zones between the water masses horizontal temperature and salinity gradients are much greater than on either side. The zones themselves may vary from a few hundred metres to several thousand metres wide--that is from 10^2 to 10^4 times the length of a fish--and the temperature gradients may range from greater than 0.01 C/m to 0.001 C/m. Fish collect in or near the frontal zones. One might like to think that they also travel along the frontal corridors, but the vector which would provide a directional clue is not immediately obvious. But if there are differences in the velocity at which the water masses are moving

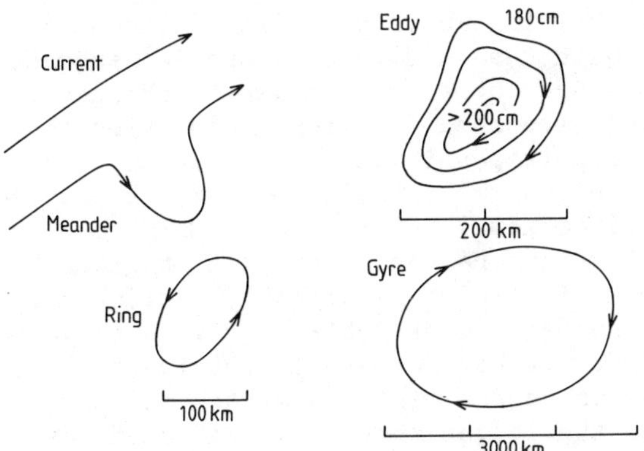

Figure 2. Some horizontal features of the open ocean of importance to the study of fish migration, following the definitions given by Pearce (1981). In the eddy the dynamic height contours are given in centimetres, and the anticyclonic flow is clockwise in the northern hemisphere (as shown) but anticlockwise in the southern hemisphere.

there will be a gradient of current speed across the front. Could there also be a significant geostrophic flow along one side of the front, similar to that which appear to be present on some shelf-sea fronts?

In the open ocean there are features of the vertical structure of the water column that may be important for fish on migration (Fig. 3). While the temperature structure is the most obvious--a warm

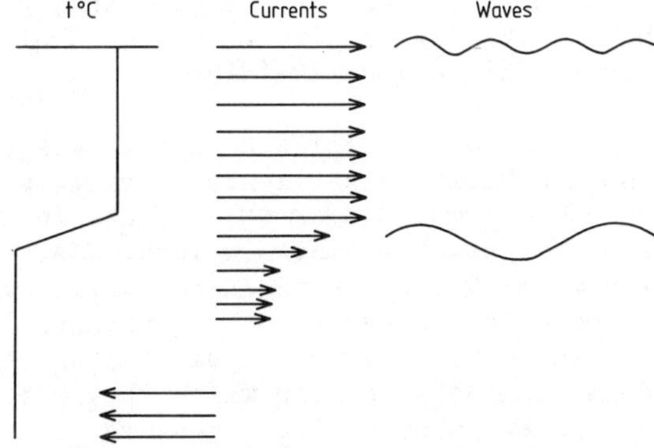

Figure 3. Some vertical features of the open ocean of importance to the study of fish migration: temperature structure; current structure; surface and internal gravity waves.

surface layer often lying above a clearly defined thermocline--tracking studies with tuna and related species suggest that their vertical movements are not so restricted by changes in temperature as had been previously supposed. Current velocities change with depth, and while the transport possibilities offered by undercurrents and countercurrents are generally well known, it is not clear if fish make use of the directional information likely to be available across the rheocline. Furthermore the possibility that fish use the directional information which is probably available from gravity waves at the sea surface (Cook 1984--this volume), or along internal boundaries, has not yet been seriously considered. If I were asked to identify sites where the migrants might find directional clues, it is the boundaries themselves that would be named: near the surface (waves from the sea and swell); in the water column between different water masses (internal waves and rheoclines); and, to include the shelf-seas, close to the bottom (flow lines over wave features of bottom topography) and into the boundary layer itself.

SOME PROBLEMS WITH MIGRATION DATA

I would like to comment on some problems relating to publication. Some authors appear to be reluctant or are unable to publish the original material on which their contributions are based. I came across a number of examples when doing my homework for "Fish Migration." For example Bowman's (1933) account of the tagging experiments with plaice (*Pleuronectes platessa*) carried out between 1923 and 1931 at the Shetland Islands was often cited as showing that migrating plaice swam against the prevailing or residual current. But the paper did not give the critical information--the time interval between the release and recapture of 99 tagged plaice--to support the conclusion. In 1960 I made some enquiries as to the availability of the original data, and fortunately the records had survived the war. The data did not support Bowman's conclusions: indeed, if they showed anything at all, they showed his hypothesis to be wrong (Harden Jones 1968, pages 179-181).

A second set of data to be published in part was that provided by Dence (1940, 1948) on homing experiments carried out with spawning dwarf white suckers (*Catastomus commersoni*) in the inlet streams of Wolf Lake, Huntington Forest, Newcomb, New York, USA. Suckers were marked in each inlet from 1938 to 1942 and recoveries were recorded from 1939 to 1946. Dence's two papers only gave details of the returns up to 1941. These results were clear and showed that many fish returned to spawn in the stream in which they were tagged while others strayed: the results should give a measure of homing success and straying rate respectively. I wrote and asked for a sight of the unpublished material. Professor Dence had retired and on receiving my letter returned to his laboratory to collect the data from a filing cabinet. He found that a student using the room had emptied the cabinet, and the papers had been destroyed. The reason that the data were not published in the first place was that the Editor of the Journal was unable, or unwilling, to spare pages for the tables.

I heard of a similar set of data relating to the homing behaviour of Pacific herring (*Clupea harengus pallasi*) when, in 1961, I visited the Fisheries Research Board of Canada's Biological Laboratory at Nanaimo, British Columbia. The experiments, carried out from 1936 to 1956, involved the tagging of over 500,000 fish from which there were over 15,000 recoveries. The detailed results had not been published but were made available to me by F.H.C. Taylor. They were summarized in "Fish Migration" (Harden Jones 1968, Table 18, page 142) as a unique set of data relating to homing and straying in herring on a scale which must have surprised all but the few who knew of the work.

My last example of a data problem concerns the incomplete publication of the leptocephalus, elver, and eel (*Anguilla* spp.) material collected by Johannes Schmidt and his colleagues. This has contributed to our failure to advance our understanding of the life history of the Atlantic eels during the last 60 years. In 1978 the United Kingdom's Natural Environment Research Council were asked to consider allocating ship's time for a proposed International Eel Expedition. In declining to help, the Council urged that existing collections should be properly examined before further sampling programmes were planned. While many of us were disappointed at the time, both the decision and the advice were sound; but it will have taken from 5 to 6 years for J. and I. Boëtius (at Charlottenlund, Denmark) with the statistical support of T. Harding (at Cambridge, UK) to make good the deficiency.

I suspect that there is something wrong with our attitudes to working up and publishing primary data; and when the hideous cost of collecting the material is considered, we should, perhaps, be doing much better. While there may be concern about the data that are known to have been collected, one cannot help wondering how much has been collected about which nothing is known. One contribution that we could all make to the common cause is to check our data inventories, our cupboards, and our cellars. If there is real difficulty in publishing the original material, we should at least make sure that its existence is known.

MATTERS OF SPECIAL INTEREST

Our knowledge of the currents and of the distribution and movements of fish in the open ocean is far from complete. To illustrate this point I am going to say something about the southern bluefin tuna and salmon in the Pacific Ocean.

The Southern Bluefin Tuna

I have recently been to Australia and have learned something of the work that has been carried out with the southern bluefin tuna, *Thunnus maccoyii*.

In 1980 the world catch of southern bluefin was 39,000 t, placing the species at the bottom of the list of the Top Ten Tuna. Two-thirds of the southern bluefin catch was taken by Japan and one-third by Australia. The fishery is valuable to both countries and there is concern as to the state of the population (Murphy and Majkowski 1981).

The southern bluefin is found across the three southern oceans between latitudes 25°S and 52°S. Between the 90°E meridian and the west coast of Australia its distribution extends north towards the equator to 10°S. Shingu (1978, his Fig. 14) summarizes the main features of the species' transoceanic distribution; longline catches of southern bluefin off Chile are figured by Shingu (1967, his Fig. 2) and off Uruguay by Hayashi et al. (1970, their Fig. 45). Within this largely zonal and virtually circumpolar distribution there are commercial fisheries between 8°W (to the west of South Africa) and 170°W (to the east of New Zealand). In Australian waters the domestic fishery is for young bluefin which are taken at the surface by line-and-pole or by purse seine (Majkowski et al. 1981). On the high seas Japanese pelagic longliners catch older tuna at greater depths on their feeding grounds in the Atlantic, Indian, and Pacific Oceans as far south as the West Wind Drift (Shingu 1970) and at water temperatures down to 5.5 C. This suggests that the tuna's southerly limit lies just to the north of the Antarctic Convergence. Figures 4 and 5 summarize the main features of the circulation and the distribution of the *T. maccoyii* in the southern oceans.

The southern bluefin is found at temperatures from 5-30 C and is remarkable in that it appears to have only one spawning area. This is in the eastern tropical Indian Ocean between the northwest coast of Australia and Java, within the latitudes 11-18°S and longitudes 102-121°E, where there is upwelling (for example, Ueyanagi 1969b, his Fig. 16). The evidence for a single spawning area is that 49 larvae identified as southern bluefin have been caught at 23 Indian Ocean stations listed by Ueyanagi (1969a, his Table 1) and, with one possible exception, no such larvae have been found elsewhere (Ueyanagi 1969b, his Appendix Fig. 9). Some of you might not find this evidence sufficient to be convincing; I certainly have my doubts.

Sund et al. (1981) summarized the currently accepted life history of the southern bluefin. Spawning in the Indian Ocean occurs from September to March. The subsequent drift of the larvae is not known. Some 1 year old bluefin (in their second year of life) are

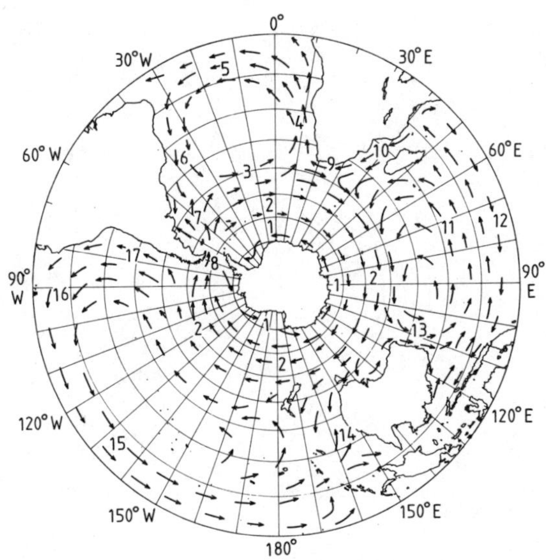

Figure 4. A simple presentation of the circulation of the southern oceans relevant to the distribution, life history, and biology of the southern bluefin tuna. The main surface currents are as follows: 1. East Wind Drift; 2. West Wind Drift. Atlantic Ocean: 3. South Atlantic Current; 4. Benguela Current; 5. South Equatorial Current; 6. Brazil Current; 7. Falkland Current; 8. Cape Horn Current. Indian Ocean: 9. Agulhas Current; 10. Mozambique Current; 11. South Equatorial Current; 12. Equatorial Counter-Current (in February and March); 13. West Australian Current. Pacific Ocean: 14. East Australian Current; 15. South Equatorial Current; 16. Peru Current; 17. Humboldt Current; 18. Cape Horn Current (through the Drake Passage into the Atlantic Ocean). The Antarctic Convergence is not shown. The Stereographic Polar projection used in Figures 4, 5, and 6 was provided by A.R. Tabor of the Marine Information and Advisory Service, Bidston Observatory, Birkenhead, Merseyside, England.

caught in Western Australia waters but most of the fish taken in the coastal fishery are 2-5 years old.

Recently the Australian fishery has extended further offshore to take some of the older fish which support the Japanese longline fleet in all three oceans. Southern bluefin mature as 7, 8, or 9 year olds. Their zonal distribution in the West Wind Drift is thought to contract and the maturing fish move from south to north between longitudes 90°E and 110°E, where they were caught by the Japanese as prespawning or spent fish on the so-called 'Oki' ground (20-30°S) and

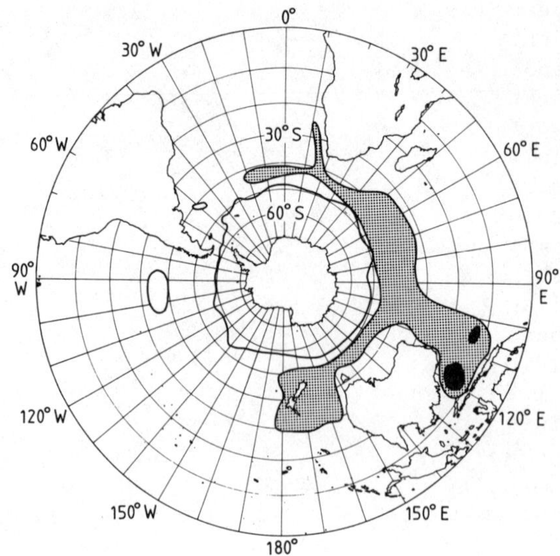

Figure 5. The southern bluefin tuna is reported as having a spawning area between northwest Australia and Indonesia (in black); it appears to have a circumpolar distribution and has been found off Chile and Uruguay (clear ellipses); and the population supports a widespread high seas Japanese longline fishery from off southwest Africa to New Zealand (stippled area). The position of the Antarctic Convergence is shown.

as spawning fish on the 'Oka' ground (10-20°S).

Some aspects of the bluefin tuna biology deserve further comment. It seems extraordinary that a fish so widely distributed across three oceans should be dependent on one spawning area, when it would seem to be biologically possible to have areas in all three oceans (off southwest Africa, in the southern Indian Ocean, in the Coral Sea, or to the northwest of New Zealand, and off the west coast of South America). If there is indeed only one spawning area, have others failed or are they yet to develop?

There are also problems with the drift of the larvae. Some must surely be carried to the west in the South Equatorial Current and so enter African waters, while others may complete the circuit of the gyral in the southern Indian Ocean and so enter Australian waters from the west. The drift of the satellite-tracked buoys reported by Creswell and Golding (1979) shows that the northern leg of the southern Indian Ocean gyral could be covered in under 1 year, which suggests that a 2.5-year period to complete the whole journey back to Australia might not be unreasonable. This would fit with the appearance of 2 year old fish in the Australian fishery. If some young bluefin do enter Australian waters from the west, the behavioural mechanisms involved might parallel those involved in the return of the planktonic phyllosoma larvae of the western rock lobster (*Panulirus cygnus*) (Phillips 1981): an understanding of one problem might help with an understanding of the other, and both could be relevant to the migration of eel larvae in the North Atlantic. An alternative route for bluefin to reach Australian waters would be in the intrusion of warm, low-salinity, tropical water which flows to the south along the Western Australian coast and then eastern into the Great Australian Bight (Fig. 6). This current, which has been known since at least 1949, has now been called the Leeuwin Current (Creswell and Golding 1980). The current is seasonal; the flow is strongest in April to June (that is in the southern autumn, at the end of the bluefin spawning season); and the temperature and salinity of the water suggests that part of the current may originate in the surface waters of the bluefin spawning area (see Wyrtki 1971, pages 462-465, Sections 16 and 17). Could the Leeuwin Current provide a transport system for young bluefin to West Australian waters and so round into the Bight? The buoy tracks given by Creswell and Golding (1980) suggest speeds up to 0.5 m/s, and while the pattern of

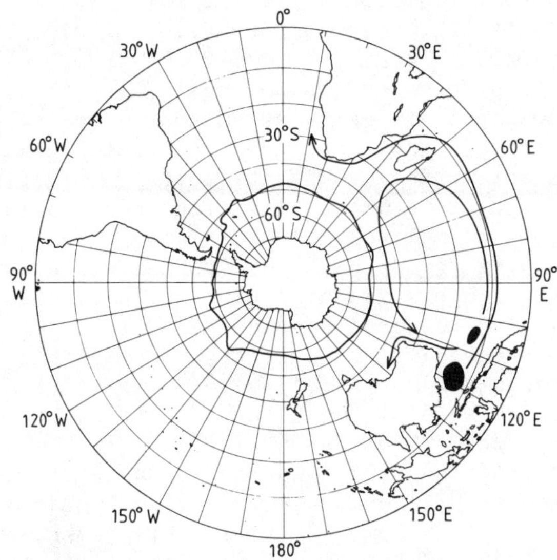

Figure 6. The larvae of the southern bluefin tuna could reach African and Australian waters in the anticlockwise circulation of the southern Indian Ocean gyral, or for Australian waters, more directly and more rapidly, in the seasonal south-going Leeuwin Current which turns into the Great Australian Bight. The position of the Antarctic Convergence is shown.

southerly movement could be interrupted by mesoscale cyclonic features to the west, it would be reasonable to suppose that juvenile bluefin entrained into the system at lower latitudes could reach the west coast of Australia as 1 year old fish.

If there is but a single spawning area in the eastern Indian Ocean, the question arises as to whether all the maturing fish are successful in making the return journey to spawn. I have been unable to find any details (length, age, sex, or maturity) of the southern bluefin caught off Chile, or Uruguay, and no fish have been tagged in these areas. Indeed the high-seas tagging is very limited. But from 1959 to 1973 some 47,000 bluefin were tagged in Australian waters. Nearly 7,000 recoveries were made from these experiments, and nearly 300 fish were returned by the Japanese longliners, the recapture zone extending from 2°W (off South Africa) to 180° (to the east of New Zealand). No Australian-tagged fish have been returned from the 'Oki' ground (as prespawners or as spent adults) or from the 'Oka' ground (as spawners), but the Japanese fishing effort in these areas has been greatly reduced over the period when the tagged fish might have been expected to reach maturity.

Some details of the Australian tagging experiments and the recoveries from them have appeared in annual or field reports and various papers. Looking at some of the results I was taken with the possibility that part of the adult population of southern bluefin might become entrained in the West Wind Drift and so make a circumpolar migration to match that of their distribution: in short, that the bluefin behaves, in part, as an underwater albatross.[1] The critical part of this journey would be the Drake Passage between South America and Antarctica: could the southern bluefin get through? Temperature would appear to be the limiting factor. The Southern Ocean Atlas (Gordon and Molinelli 1982, Plates 7 and 184) shows a layer of surface water round the Horn with temperatures over 7.5 C in February (above the temperature of 5 C at which bluefin are found in the West Wind Drift). So in an average year the bluefin should be able to pass from the Pacific to the Atlantic, and they might do so without difficulty in a warm year. There is no body of tagging results to support this idea, but a set of data picked out for special comment by Hynd (1969) is consistent with the hypothesis. Referring to southern bluefin tagged and released in Australian waters, he noted that, "It seems that all fish tagged between 1962 and 1965 were prevented in some way from moving westward (but not eastward) up till 1966: after this dispersion to the westward became possible..." (Hynd 1969, page 30). The point is brought out in his Figure 2 which shows, for the bluefin tagged and released in 1964,

---

[1] I found Stommel's (1957, page 179-181) account of the Antarctic circumpolar circulation interesting when considering this problem.

the position of the recaptures from 1964 to 1968. It was not until 1967 that fish were returned from the west, and the delay is consistent with the suggestion that it reflects the time taken by the fish to travel with the West Wind Drift from Australian waters to the Pacific, through Drake Passage into the Atlantic, and south of the Cape of Good Hope into the Indian Ocean.

Finally, there are the meridional movements that the bluefin make between the high-latitude feeding area and the low-latitude spawning area. These movements immediately recall those of whales, such as humpbacks (Lockyer and Brown 1981, their Fig. 5). The eastern area of the southern Indian Ocean is, somewhat exceptionally, an area with weak boundary currents and the north-going Western Australian current is not a particularly strong candidate for a transport system for the migrating tuna.

The biology and distribution of the bluefin tuna raise problems of homing, migration and guidance equal to those of any other species. But information on some important matters appears to be lacking. For example, I have not come across any information as to the depth at which the older tuna are caught on the high seas, which makes it difficult to relate changes in distribution to oceanographic features. Nevertheless, it is of interest to note that during the months when the bluefin are moving north between longitudes 90°E and 110°E, the depth of the maximum temperature gradient lies between 130 m and 150 m, and the maximum gradient varies between 0.35 C and 1.5 C per 10 m, the steeper gradient being to the north of latitude 18°S (Wyrtki 1971). The surface waves in this area predominantly come from between 80° and 130°, and while the majority of the wave periods are between 6 and 9 s, there are some waves of longer period (Hogben and Lumb 1967).

Salmon in the Pacific Ocean

In discussing the distribution and migration of the southern bluefin tuna, I raised the question as to whether the fish found off Chile and Uruguay would ever return to the Indian Ocean spawning ground; were they, for reproductive purposes, lost to the population? From time to time I have suggested that this might also occur for salmon of the genus *Oncorhynchus* in the North Pacific Ocean, the idea being fired by the relatively low proportion of the marked downstream migrants that return to their home stream and sustained by the equally low proportion of mature salmon tagged on the high seas that are recovered in coastal waters. But the heresy that some salmon are lost at sea has not been well received, particularly by those who believe that the beasts are endowed with very good guidance or navigation abilities and others who see in the suggestion some support for a high seas fishery in areas which could conflict with accepted management practice.

In what follows I will try to establish a prima facie case in support of the hypothesis that some salmon are lost at sea by showing that there are areas of the North Pacific from which no tagged salmon have been recorded as being recovered in coastal waters. The main surface currents of the North Pacific are shown in Figure 7. The original tagging data are published in Bulletins of the International North Pacific Fisheries Commission for sockeye salmon (*O. nerka*) by French et al. (1976), for chum salmon (*O. keta*) by Neave et al. (1976), and for pink salmon (*O. gorbuscha*) by Takagi et al. (1981). As can be seen from Table 1, the releases were made over the years 1956 to 1971, and the scale of the work is substantial. Fish selected for tagging were caught by purse seine or longline and were fitted with disc, spaghetti, or anchor tags. The data considered here relate to mature or maturing fish tagged on the high seas from April to August and the returns from the coastal waters and rivers in the year of tagging.

The data for the sockeye salmon, set out for sea areas of 2° of latitude and 5° of longitude, show the numbers tagged (Fig. 8), the numbers recovered in coastal waters (Fig. 9), and the returns as a proportion of the number tagged (Fig. 10). While large numbers of tagged sockeye were released in some areas, very few were released in others (Fig. 8); and in calculating the percentage returns (Fig. 10)

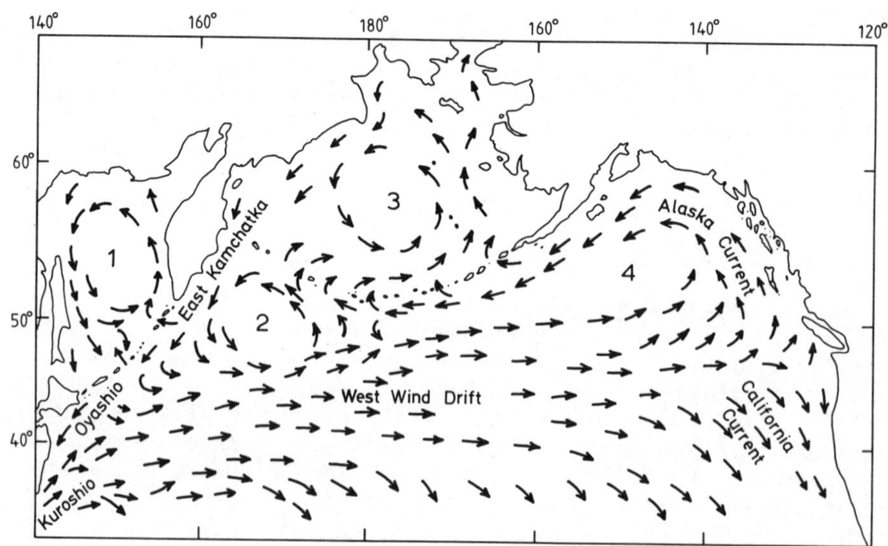

Figure 7. A simple presentation of the circulation of the North Pacific Ocean relevant to the return of salmon from the open ocean to coastal waters. 1. Sea of Okhotsk gyral; 2. Western subarctic gyral; 3. Bering Sea gyral; 4. Alaskan gyral.

Table 1. Numbers of mature or maturing salmon tagged in the North Pacific and subsequently recovered in coastal waters in the year of tagging. Data from American, Canadian and Japanese experiments carried out from 1956 to 1971 whose results have been published in the Bulletins of the International North Pacific Fisheries Commission, Vancouver, British Columbia, Canada.

Species of salmon	Number tagged	Recoveries in coastal waters		Author
		number	percent	
Sockeye	63,427	4,897	7.7	French et al. (1976)
Chum	101,685	1,391	1.4	Neave et al. (1976)
Pink	117,212	5,811	5.0	Takagi et al. (1981)

areas in which less than 40 tagged fish were released have been ignored. The proportion of returns is generally higher in areas closer to the coast (for example, the tagging areas bounded by latitudes 48°N to 52°N, and the tagging areas bounded by longitudes 145°W and 150°W), and there are areas from which no coastal returns

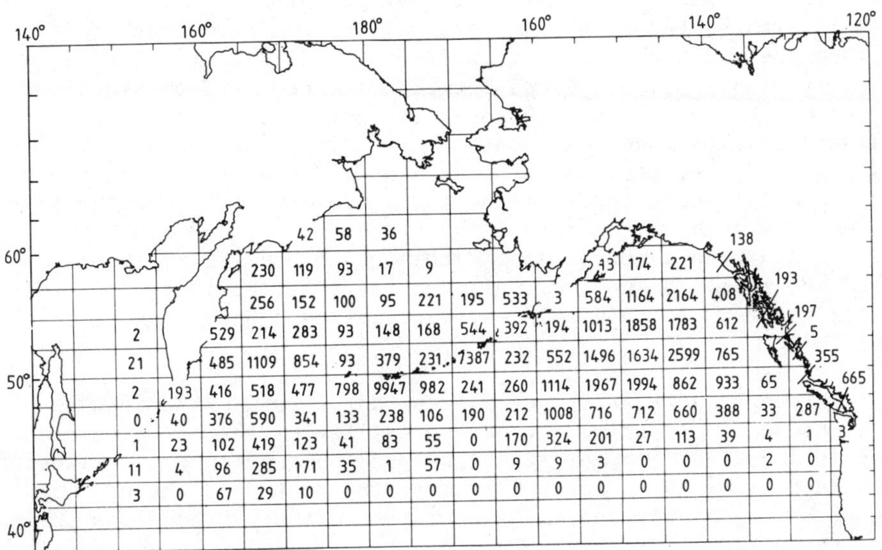

Figure 8. The numbers of mature or maturing sockeye salmon tagged in the North Pacific from 1956 to 1971 by rectangles of 2° latitude and 5° longitude. The total number of sockeye tagged was 63,427.

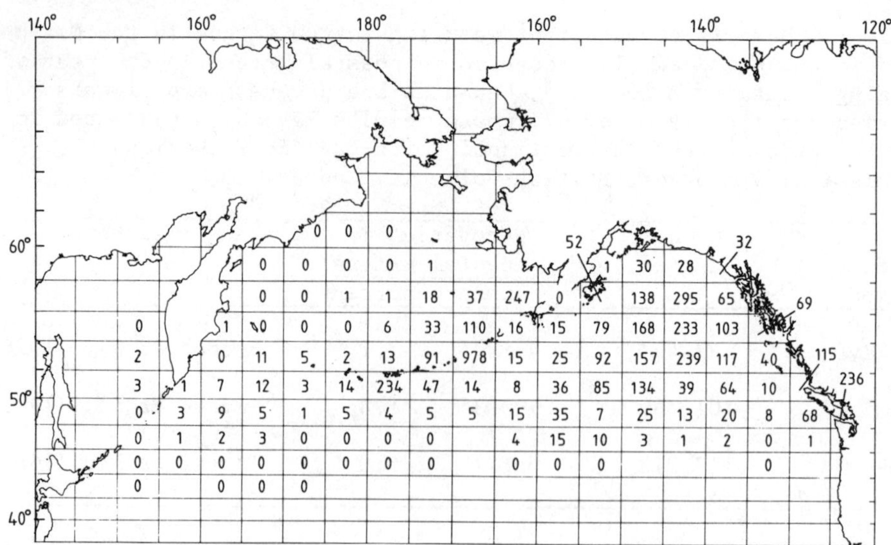

Figure 9. The number of sockeye salmon tagged in different rectangles that were recovered in coastal waters. The total number recovered was 4,897, an overall percentage of 7.7%.

have been recorded (Fig. 10). For sockeye salmon these areas form two main groups, one in the western part of the Bering Sea (Fig. 10) and the other in the middle of the North Pacific south of latitude 48°N (Fig. 11). Similar areas for pink and chum salmon are found only to the south of the Aleutian chain (Figs. 12, 13). The published results show that south of the Aleutians there are areas from which 1092 sockeye, 4207 pink, and 2287 chum salmon have been tagged and released; and from the total of 7586 fish no coastal recoveries have been recorded. The three "no-return" areas overlap (Fig. 14) to give a common area within which 3582 salmon have been tagged and released and from which no traveller has yet returned to coastal waters. At this point it would perhaps be wise to rest the case.

There are, however, three points that I would like to make by way of comment. Firstly, the northern limit of the common no-return areas at about 48°N appears to correspond to the boundary between the westward flow of the Alaskan Stream to the north and the eastward flow of the subarctic current system to the south. Sections on or close to the 180° meridian show a dip of the temperature and salinity isoclines in this area which reflects the sinking of surface water. Figure 20B of Favorite et al. (1976), here shown as Figure 15, indicates that the isotherms are running nearly vertically at latitudes 47°N to 45°N. The point I want to make is that the area

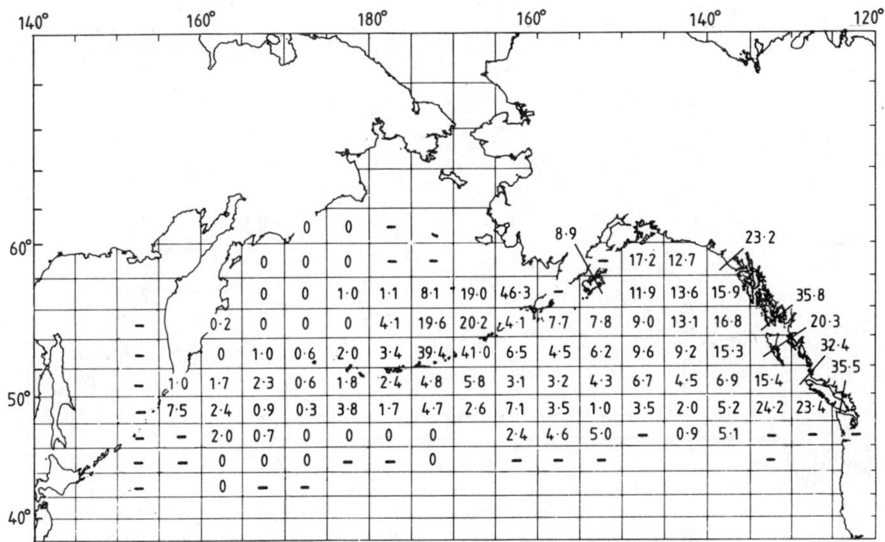

Figure 10. The number of sockeye salmon recovered in coastal waters expressed as a percentage of the number tagged in each rectangle. Rectangles in which no tagged fish were released are blank; rectangles in which less than 40 sockeye were released are indicated by a dash; rectangles from which no tagged fish were recovered are indicated by a zero.

within which salmon appear to get lost at sea does correspond to a distinct set of hydrographic conditions.

Secondly, it is argued against the suggestion that salmon get lost at sea that no species would have evolved and come to depend on a pattern of migratory behaviour which would allow a significant proportion of the stock to be lost for reproduction. What sort of losses are suggested? The tagging returns suggest that the overall proportion of salmon that return to coastal waters is not less than 1.4% for chum, 3.5% for pinks, and 7.7% for sockeye. But these are overall proportions and to obtain a meaningful estimate for the population, the tag returns from different areas should be weighted according to the local abundance of salmon. There can be little doubt that the highest proportion of tag returns come from those areas where salmon are most abundant; what proportion of the mature population would be contained within the areas of no-return? There are difficulties in estimating the abundance of salmon across the Pacific, if only because we do not know the efficiency of the various gears (purse seine, gill net and longline) that are used to catch them.

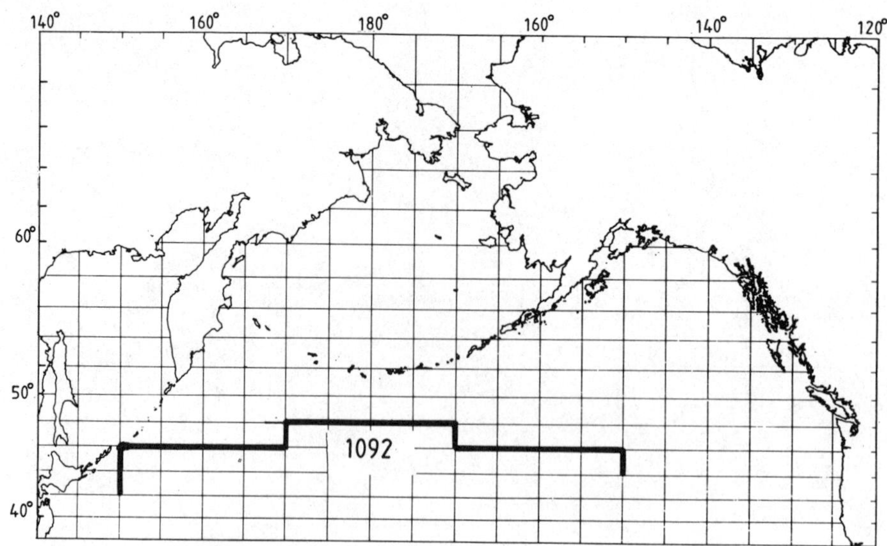

Figure 11. The area, south of the Aleutian chain, within which 1,092 sockeye salmon were tagged and from which no coastal-water recoveries have been reported.

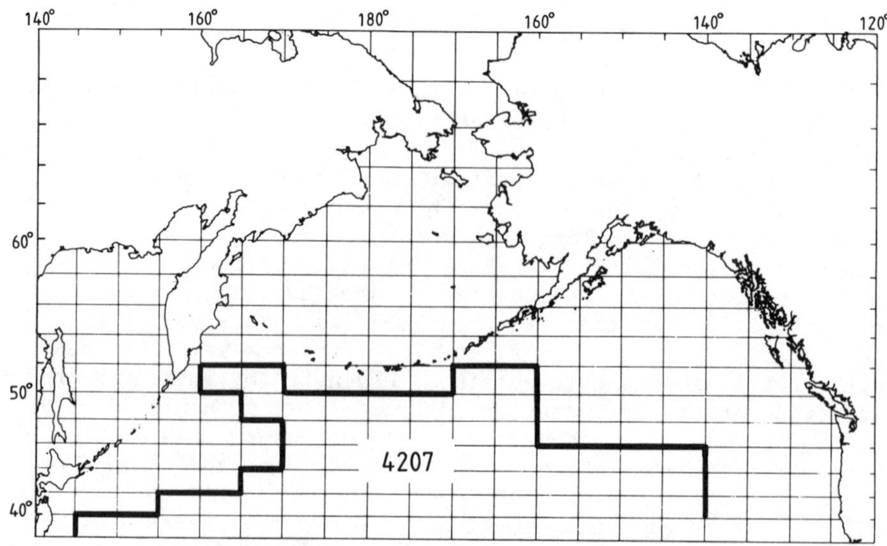

Figure 12. The area, south of the Aleutian chain, within which 4,207 pink salmon were tagged and from which no coastal-water recoveries have been reported.

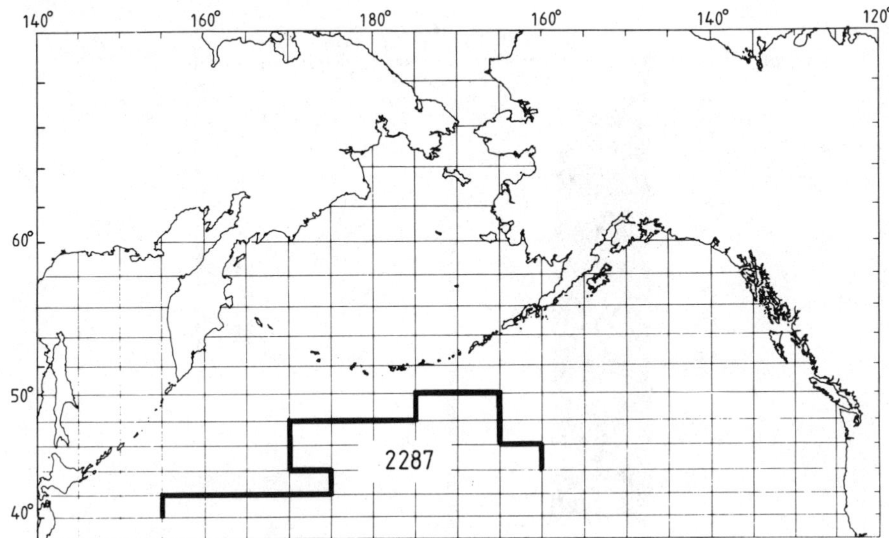

Figure 13. The area, south of the Aleutian chain, within which 2,287 chum salmon were tagged and from which no coastal-water recoveries have been reported.

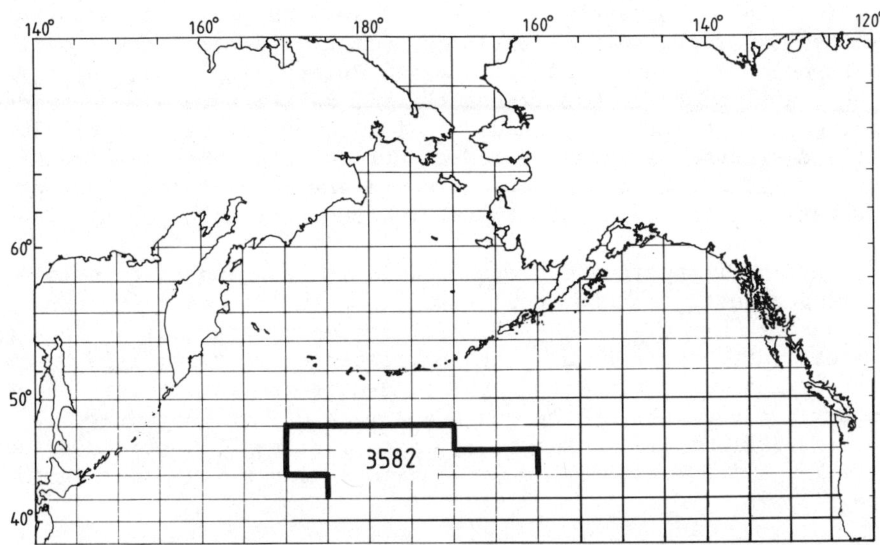

Figure 14. The common area, south of the Aleutian chain, within which a total of 3,582 sockeye, pink and chum salmon were tagged and from which no coastal-water recoveries have been reported.

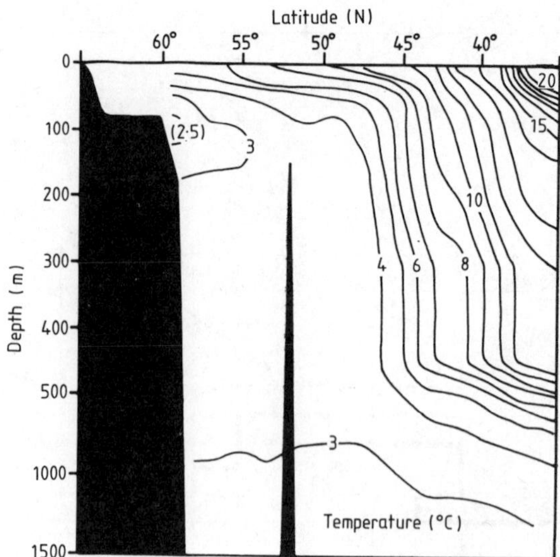

Figure 15. A vertical section along 175°W to show the long-term mean of temperature. Redrawn from Favorite et al.'s (1976) Figure 20B.

However, the levels of abundance in the no-return areas are low, and my estimate, taking into account the actual areas involved and the effort taken to catch the fish that were tagged, suggests that no more than 5% of the mature population would be lost; I admit that this is no more than a reasoned guess. But the important point may not be one of numbers, but rather that if salmon have a good guidance or navigation ability, why should it fail them in these particular areas? A hypothesis advanced to account for the successful return of salmon from one area of the open ocean might be reasonably expected to account for their failure to return from another.

Thirdly, if there are areas in which salmon appear to become lost in the North Pacific, could there be similar "salmon sinks" or "salmon holes" in the North Atlantic? We should perhaps be looking for areas where there is no vertical temperature stratification and where surface water sinks to intermediate or greater depths. Such an area certainly exists to the south and west of Cape Farewell, Greenland (Defant 1961, page 684; Grant 1968, page 40-45; McCartney and Talley 1982, their Fig. 6c). There is probably another in the Norwegian Sea.

A COMMENT ON INERTIAL SYSTEMS

Barlow (1964, 1966) has examined the possible role of inertial navigation in animal migration, with the vertebrate inner ear as the detector system. Kalmus (1964) took the view that the whole matter was "somewhat dubious," and until recently I shared this view. But the open ocean is an environment where such systems could come into their own and perhaps the suggestion merits further consideration.

In an attempt to avoid, or at least contain, semantic problems, *guidance* may be defined as the action of directing or controlling a vehicle or beast along a previously chosen path and *navigation* as the ability to proceed from one position to another position across unfamiliar ground. Following Draper et al. (1960), inertial guidance and inertial navigation are based on measurements made with respect to inertial space; and inertial space is the space or reference frame for which Newton's Laws of Motion are valid.

Biologists are not expected to expound on inertial systems and I am unable to astonish you. Murphy (1873)[1] gives a simple statement of the fundamental principles which, on account of its historical interest, will be quoted in full. "If a ball is freely suspended from the roof of a railway carriage, it will receive a shock sufficient to move it, when the carriage is set in motion: and the magnitude and direction of the shock thus given to the ball will depend on the magnitude and direction of the force with which the carriage begins to move. While the carriage is in uniform motion the ball will be relatively at rest; and every change in the velocity of the motion of the carriage, and of its direction, will give a shock of corresponding magnitude and direction to the ball. Now, it is conceivably quite possible, though such delicacy of mechanism is not to be hoped for, that a machine should be constructed, in connection with a chronometer, for registering the magnitude and direction of all these shocks, with the time at which each occurred; and from these data--the direction of the shock indicating the direction of the motion of the carriage, the magnitude of the shock indicating its velocity, and the interval of time between two shocks indicating the time during which the carriage has run without change of velocity or direction--from these data the position of the carriage, expressed in terms of distance and direction from the place from which it had set out, might be calculated at any moment."

Barlow's (1964, 1966) accounts show how Murphy's perceptive analogy has been realized in terms of linear and angular accelerometers, while technical accounts of inertial systems are given by O'Donnell (1958), Draper et al. (1960) and Fernandez and Macomber (1962).

---

[1] Joseph John Murphy, 1827-1894; a Belfast philosopher whose father was in the linen trade into which he followed. He wrote numerous memoirs and one of his papers in the Belfast Natural History and Philosophical Society is entitled "Ocean currents and their effect on climate."

The suggestion that the inertial detectors in the vertebrate labyrinth--the semicircular canals and otolith organs--could form the basis of a navigational system has been received with little enthusiasm, largely because the threshold levels ($0.1°/s^2$ for angular accelerations and 6 cm/s^2 for linear accelerations) are three to four orders of magnitude higher than could be accepted in a precision system. Barlow (1964, page 105) comments that the accepted thresholds, "... would not permit accurate navigation (e.g. errors of positions of, say, 5%) for travel times longer than the order of a few minutes." I have not found an argued case to substantiate this, and what passes for accurate navigation must of course be judged in relation to the needs of the animal.

In reviewing the evidence that birds might make use of inertial clues, Keeton (1974, page 79) suggested that, "we should look carefully, in future experiments, at the possibility that some birds may use inertial information as an aid to maintaining a straight course once they have chosen a bearing and are flying along it...this simpler use of inertial information makes less demand for accuracy on the vestibular system and is therefore more in line with the measured sensitivities..."

There is a further problem in accepting that animals could use inertial clues for navigation or guidance, and this is of a fundamental nature. An inertial system cannot readily resolve the ambiguity inherent in the effects of acceleration and gravity on the detectors.

For example consider the simple representation of an otolith

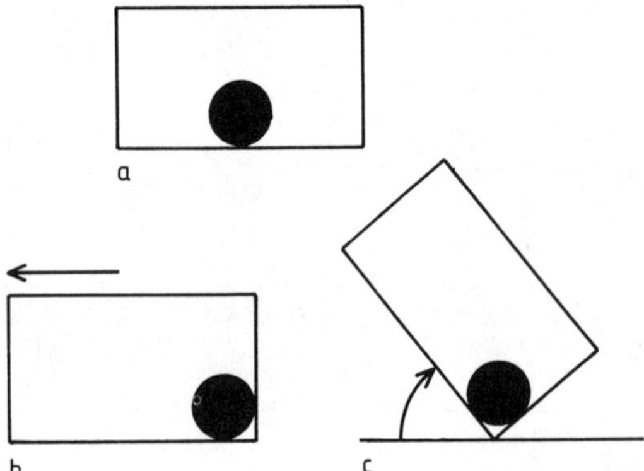

Figure 16. The effect of acceleration and gravity cannot be separated by a simple accelerometer, such as an otolith organ: the consequence of being accelerated forwards in b is identical to that of being tilted upwards in c.

shown in Figure 16. When the fish tilts its head up, the otolith is displaced posteriorly with a corresponding shearing force on the supporting sensory cells. But an identical effect is produced if the fish is accelerated forwards: effects of gravity and acceleration must be separated if the velocity and displacement of the fish are to be determined. In inertial navigation systems considerable engineering and signal processing skills have been applied to solving this very problem: can the vertebrates do the same?

Some observations of my own are relevant to this question. In 1953, in the Zoological Laboratory at Cambridge, I made some preliminary observations on the behaviour of blind cave fish (*Anoptichthys jordani*) accelerated along a bench in a closed aquarium tank or exposed to the oscillations of a rather primitive parallel swing. I was convinced that the fish made a 180° turn at some stage of the acceleration, and when I moved to the Lowestoft Laboratory I repeated the experiments with blind goldfish (*Carassius auratus*), which were accelerated along a special railway. These experiments showed that the fish turned through 180° when accelerated backwards (Harden Jones 1956). Baumgarten et al. (1971) have confirmed and extended these observations.

These results show that fish can make a distinction between the effects of gravity and acceleration. If the semicircular canals respond only to angular accelerations, the inner ear would appear to contain the detector system essential to discriminate between them. At the simplest level, a linear acceleration in the horizontal plane will only stimulate the otolith organs, but a pitching movement (head up or down) about the transverse axis will stimulate the otoliths and the anterior vertical canals. When one considers the well known interactions between the eyes and the ears--the ocular-vestibular responses--and that the reactions of fish to accelerations have hitherto been made with blind or blinded fish, it is certain much work remains to be done. Mayne (1974) discusses many of the problems; his work deserves study and should form an important part of the the backcloth to any discussion on guidance or navigation and as such is relevant to our consideration of open-ocean migration.

REFERENCES

Barlow, J.S. 1964. Inertial navigation as a basis for animal navigation. Journal of Theoretical Biology 6:76-117.

Barlow, J.S. 1966. Inertial navigation in relation to animal navigation. Journal of the Institute of Navigation 19:302-316.

Baumgarten, R.J. von, G. Baldrighi, J. Atema, and G.L. Shillinger. 1971. Behavioral responses to linear accelerations in blind goldfish. 1: The gravity reference response. Space Life Sciences 3:25-33.

Bowman, A. 1933. Plaice marking experiments in Shetland waters

1923-1931 (inclusive). Journal du Conseil Conseil Permanent International pour l'Exploration de la Mer 8:223-229.

Cook, P.H. 1984. Directional information from surface swell: some possibilities. Pages 79-101 *in* J.D. McCleave, G.P. Arnold, J.J. Dodson and W.H. Neill, editors. Mechanisms of migration in fishes. Plenum Press, New York, New York, USA.

Creswell, G.R., and T.J. Golding. 1979. Satellite-tracked buoy data report III. Indian Ocean 1977, Tasman Sea July-December 1977. Australia Commonwealth Scientific and Industrial Research Organization, Division of Fisheries and Oceanography Report 101.

Creswell, G.R., and T.J. Golding. 1980. Observations of a south-flowing current in the southeastern Indian Ocean. Deep-Sea Research 27A:449-466.

Defant, A. 1961. Physical oceanography volume 1. Pergamon Press, London, England.

Dence, W.A. 1940. Progress report on a study of the dwarf sucker (*Catastomus commersonnii utawana*). Roosevelt Wild Life Bulletin of the Roosevelt Wild Life Forest Experiment Station of the New York State College of Forestry at Syracuse University 7:221-233.

Dence, W.A. 1948. Life history, ecology and habits of the dwarf sucker, *Catastomus commersonnii utawana* Mather, at the Huntington Wildlife Station. Roosevelt Wild Life Bulletin of the Roosevelt Wild Life Forest Experiment Station of the New York State College of Forestry at Syracuse University 8:81-150.

Draper, C.S., W. Wrigley, and J. Hovorka. 1960. Inertial guidance. Pergamon Press, Oxford, England.

Favorite, F., A.J. Dodimead, and K. Nasu. 1976. Oceanography of the subarctic Pacific region, 1960-71. International North Pacific Fisheries Commission Bulletin 33.

Fernandez, M., and G.R. Macomber. 1962. Inertial guidance engineering. Prentice-Hall International, London, England.

French, R., H. Bilton, M. Osako, and A. Hartt. 1976. Distribution and origin of sockeye salmon (*Oncorhynchus nerka*) in offshore waters of the North Pacific Ocean. International North Pacific Fisheries Commission Bulletin 35.

Grant, A.B. 1968. Atlas of oceanographic sections. Atlantic Oceanographic Laboratory Report, Bedford Institute, Dartmouth, Nova Scotia, Canada. Report number AOL 68-5.

Gordon, A.L., and E.J. Molinelli. 1982. Thermohaline and chemical distributions and the Atlas data set. Southern Ocean Atlas. Columbia University Press, New York, New York, USA.

Harden Jones, F.R. 1956. An apparent reaction of fish to linear accelerations. Nature (London) 178:642-643.

Harden Jones, F.R. 1968. Fish migration. Arnold, London, England.

Hayashi, S., T. Koto, C. Shingu, S. Kume, and Y. Morita. 1970. Status of tuna fisheries resources in the Atlantic Ocean, 1956-1967. Bulletin Far Seas Fisheries Research Laboratory (Shimizu) S series 3:18-72.

Hogben, N., and F.E. Lumb. 1967. Ocean wave statistics. Her Majesty's Stationery Office, London, England.

Hynd, J.S. 1969. New evidence on southern bluefin stocks and migrations. Australian Fisheries 28(5):26-30.

Kalmus, H. 1964. Comparative physiology: navigation by animals. Annual Review of Physiology 26:109-130.

Keeton, W.T. 1974. The orientational and navigational basis of homing in birds. Pages 48-132 *in* D.S. Lehrman, J.S. Rosenblatt, R.A. Hinde and E. Shaw, editors. Advances in the study of behavior 5. Academic Press, New York, New York, USA.

Lockyer, C.H., and S.G. Brown. 1981. The migration of whales. Pages 105-137 *in* D.J. Aidley, editor. Animal migration. Society for Experimental Biology Seminar Series 13. Cambridge University Press, Cambridge, England.

Majkowski, J., K. Williams, and G.I. Murphy. 1981. Research identifies changing patterns in Australian tuna fishery. Australian Fisheries 40(2):5-10.

Mayne, R. 1974. A systems concept of the vestibular organs. Pages 493-580 *in* The handbook of sensory physiology 6: Vestibular system, part 2. Psychophysics, applied aspects and general interpretations. Springer-Verlag, Berlin, Federal Republic of Germany.

McCartney, M.S., and L.D. Talley. 1982. The subpolar mode water of the North Atlantic Ocean. Journal of Physical Oceanography 12:1169-1188.

Murphy, G.I., and J. Majkowski. 1981. State of the southern bluefin population: fully exploited. Australian Fisheries 40(11):20-29.

Murphy, J.J., 1873. Instinct: a mechanical analogy. Nature (London) 7:483.

Neave, F., T. Yonemori, and R.G. Bakkala. 1976. Distribution and origin of chum salmon in offshore waters of the North Pacific Ocean. International North Pacific Fisheries Commission Bulletin 35.

O'Donnell, C.F. 1958. Inertial navigation. Journal of the Franklin Institute 266:257-277 and 373-402.

Pearce, A. 1981. A brief introduction to descriptive physical oceanography. Australia Commonwealth Scientific and Industrial Research Organization, Division of Fisheries and Oceanography Report 132.

Phillips, B.F. 1981. The circulation of the southeastern Indian Ocean and the planktonic life of the western rock lobster. Oceanography and Marine Biology an Annual Review 19:11-39.

Shingu, C. 1967. Distribution and migration of the southern bluefin tuna. Nankai Regional Fisheries Research Laboratory Report 25:19-35.

Shingu, C. 1970. Studies relevant to distribution and migration of the southern bluefin tuna. Bulletin Far Seas Fisheries Research Laboratory (Shimizu) 3:57-113.

Shingu, C. 1978. Ecology and stock of southern bluefin tuna. Fisheries Studies of the Japanese Fisheries Research and Conservation Association 31. Translated by M.A. Hintze (1981) as an Australia Commonwealth Scientific and Industrial Research

Organization, Division of Fisheries and Oceanography Report 131.

Smith, G.W., A.D. Hawkins, and G.G. Urquhart. 1982. The open water behaviour of migrating fish: a current problem. Underwater Telemetry Newsletter 12(1):1-6.

Steen, W.J. van der. 1984. Methodological aspects of migration and orientation in fishes. Pages 421-444 *in* J.D. McCleave, G.P. Arnold. J.J. Dodson and W.H. Neill, editors. Mechanisms of migration in fishes. Plenum Press, New York, New York, USA.

Stommel, H. 1957. A survey of ocean current theory. Deep-Sea Research 4:149-184.

Sund, P.N., M. Blackburn, and F. Williams 1981. Tunas and their environment in the Pacific Ocean: a review. Oceanography and Marine Biology an Annual Review 19:443-512.

Takagi, K., K.V. Aro, A.C. Hartt, and D.B. Bell. 1981. Distribution and origin of pink salmon (*Oncorhynchus gorbuscha*) in offshore waters of the North Pacific Ocean. International North Pacific Fisheries Commission Bulletin 40.

Ueyanagi, S. 1969a. The spawning of the southern bluefin tuna (*Thunnus maccoyii*) as indicated by the occurrence of its larvae. Bulletin Far Seas Fisheries Research Laboratory (Shimizu) 1:1-14.

Ueyanagi, S. 1969b. Observations on the distribution of tuna larvae in the Indo-Pacific Ocean with emphasis on the delineation of the spawning areas of albacore, *Thunnus alalunga*. Bulletin Far Seas Fisheries Research Laboratory (Shimizu) 2:177-256.

Weihs, D. 1984. Bioenergetic considerations in fish migration. Pages 487-508 *in* J.D. McCleave, G.P. Arnold, J.J. Dodson and W.H. Neill, editors. Mechanisms of migration in fishes. Plenum Press, New York, New York, USA.

Wyrtki, K. 1971. Oceanographic atlas of the International Indian Ocean Expedition. National Science Foundation, Washington District of Columbia, USA.

ADVECTION, DIFFUSION, AND DRIFT MIGRATIONS OF LARVAL FISH

James H. Power

National Oceanic and Atmospheric Administration
National Marine Fisheries Service
Southwest Fisheries Center, La Jolla Laboratory
La Jolla, California 92038 USA

ABSTRACT

The advection-diffusion equation provides a conceptual framework for the study of oceanic drift migrations, i.e. the equation's parameters can be used to quantify and aid interpretation of drift of eggs and larval fishes. But first, relevant oceanic currents must be better characterized, and the capabilities of larvae to modify drift--through buoyancy, swimming, and orientation--need further study. Diffusion of larvae in the sea is not understood, and for the study of larval fish advection and diffusion it might be best to study special situations, e.g. discrete larval patch formation and dispersal.

INTRODUCTION

Displacement by water currents (drift) is a fact of life for all pelagic fish larvae. Some species have life histories that appear to include tactics for minimizing larval drift (Parrish et al. 1981; Iles and Sinclair 1982). Others, such as plaice (*Pleuronectes platessa*) and eels (*Anguilla* spp.), clearly use drift in their migrations. Historically, most studies of oceanic larval drift have been qualitative: the investigator examines a time series of larval distributions and usually finds that the influence of prevailing currents can be detected. This is satisfactory for documenting the occurrence of drift, but the future study of fish-migration mechanisms will require closer scrutiny of the components of drift.

My objective is to briefly describe these components and to use them in defining some unanswered questions concerning oceanic drift migrations. Drift probably has the greatest influence on eggs and larvae, although juvenile and adult fish may use drift in their migrations (Harden Jones 1968; Seckel 1972), and much of the following discussion can be applied to these later stages as well.

We cannot follow the movements of individual larvae and usually must study drift by observing the changes in the number of individuals at fixed locations (an Eulerian viewpoint). Because of this, it is useful to consider drift using the advection-diffusion equation familiar to physical oceanographers:

$$\frac{\partial F}{\partial t} + \frac{\partial}{\partial x}\left[uF - K_x \frac{\partial F}{\partial x}\right] + \frac{\partial}{\partial y}\left[vF - K_y \frac{\partial F}{\partial y}\right] + \frac{\partial}{\partial z}\left[wF - K_z \frac{\partial F}{\partial z}\right] = J,$$

where F is the number of fish per unit volume;
- u, v and w are velocities in the respective x, y, and z directions;
- K_x, K_y, and K_z are eddy diffusivity coefficients for the appropriate directions; and
- J is a sources-and-sinks term, expressing the fish's population dynamics.

This equation states that at a particular location the rate of change of fish numbers with respect to time is determined by the fish's population dynamics and the fluxes of fish to or from that location by means of advection (the velocity terms) and turbulent diffusion (the diffusivity terms). This equation usually has been applied to the transport of nonliving entities such as pollutants or salt, although its application to the transport of biota such as plankton is becoming more common.

The relative contribution of each term in the above equation must be known if the drift migration itself is to be understood, and each term may have both a biological and a physical component. For the physical process of transport the advective terms represent displacement due to the average currents while the diffusivity terms express the dispersal of fish by ocean turbulence. Both are important, although often only the advection terms are considered. If the velocity and diffusivity attributes of the water are known, then we can use the advection-diffusion equation to evaluate any possible biological components and determine how the larvae are interacting with the currents. Examples of this approach can be found in Talbot (1977) and Fortier and Leggett (1982). The advection-diffusion equation also provides the basis for the numerical simulation of drift (Wroblewski 1980; Ulanowicz et al 1982; Power 1982), and this can provide some insight into the overall problem of mechanisms of migration and the relative importance of each component.

ADVECTION

The study of a fish's drift migration requires good data concerning the water currents; satisfying this requirement forms a major obstacle to the study of larval drift. Ideally, current data with high spatial and temporal resolution should be taken at the same time the data on the fish's distribution are collected. The extent of most drift migrations ordinarily makes this objective impossible to achieve, and for this reason studies of oceanic drift migrations will continue to rely heavily upon data on currents collected for other purposes. The majority of the data available are from drifters, estimates of geostrophic flow, or ship's drift. Each of these presents inherent difficulties when used to interpret drift migrations.

Data from drift-bottle and drift-card experiments can give a rough indication of the currents encountered by the larvae but are of little value for the detailed study of drift. The fact that drift bottles must ultimately wash ashore for recovery, an event obviously to be circumvented by the fish, demonstrates that at some point the drift of the bottles and the larvae are not comparable. An alternative approach is for the investigator to track drogues released in concentrations of drifting larvae (e.g. Shelton and Hutchings 1982). Under suitable conditions, such as when adult spawning behavior creates an initial patch of eggs or larvae, this can be a powerful technique which should yield much information on larval drift in the near future.

Geostrophic flow estimates are computed from observations of the water's density distribution and are calculated by assuming that the Coriolis force and the forces arising from spatial gradients in water density are in balance. The computations are relative to the velocity at some selected depth, and if the velocity at that depth is not zero, the calculated current velocity is not a true "over-the-ground" speed. Geostrophic computations neglect currents driven by wind and other frictional forces, yet these currents may be responsible for the majority of the transport, if the larvae occupy the surface layer.

Current estimates from ship's drift are based upon the difference between the ship's dead-reckoning position and its true position after a given period of steaming and thus can give only an indication of the near-surface currents. Such data are notoriously "noisy," because of both the inherent surface-current variability and the numerous possible errors that can contaminate the ship's drift estimate.

Both ship's drift and geostrophic-flow calculations estimate the current between two geographic points, and this estimate may be seriously in error if there is significant spatial variation in the currents between those points. Additionally, one usually must summarize the ship's drift or geostrophic data over a time period or geographic area in order to obtain an adequately comprehensive picture of the currents. Such a summarization can mask the presence of spatial or seasonal aspects of the currents that may be important to the fish's drift.

The Antilles Current provides a good example of the dangers involved in trying to relate historical and published current data to a drift migration. This current, which may be of particular importance in the drift of eel leptocephali (Power 1982), appears in surface-current charts as a well-defined northwesterly flow along the Antilles/Bahamas arc. Two recent winter and summer near-synoptic surveys of the region (Gunn and Watts 1982) have indicated that the Antilles Current may be much more variable than one would suspect from examining summarizations of historical data, such as those that appear on current charts. In fact, the Antilles Current may not always exist, and in a survey done during the eel's 1973 spawning season the geostrophic flow estimates near the spawning region indicated the presence of a large eddy (Gunn and Watts 1982), which would entrain leptocephali rather than transport them into the Gulf Stream.

Thus one should be cautious about accepting the smoothed representations of currents presented in the literature as being relevant to the fish's drift. If a major current in a presumably well-traveled subtropical region has not yet been adequately characterized, it seems unlikely that the information can be any better for other areas where larval drift is important, such as the northeast Atlantic (Fraser 1958). When relying upon historical data it might be best to use only raw or partially processed data in order to evaluate deficiencies in the data relative to the problem at hand. The data should be examined for spatial or seasonal current features that may be important to the fish. It is unfortunate that in many cases investigators studying oceanic drift migrations may have to await (or undertake) the further accumulation of current data.

MODULATED DRIFT

Even if the currents are well known, larvae undergoing what are nominally drift migrations may nonetheless act in some way to modify their transport and add a biological component to the velocity vectors in the advection-diffusion equation. Because currents may vary substantially with depth, especially if they are produced by the wind or tides, an important way the fish may modify its drift is by controlling its vertical position. Sette (1943), Nelson et al. (1977), and Powles (1981) all provided examples in which fish larvae that were concentrated in surface waters were transported by Ekman (wind-induced) currents. Fortier and Leggett (1982) found that

Atlantic herring (*Clupea harengus*) and capelin (*Mallotus villosus*) larvae occupied different depth strata in the St. Lawrence estuary and thus achieved net downstream and upstream movement, respectively. Diurnal vertical migrations between opposing currents may in fact be how larvae avoid displacement from a geographic location (Marliave 1981). Tidal stream transport by larval or juvenile fish in estuaries, where the fish selectively occupies the water column to be carried on an appropriate tidal flow, has been well documented (Graham 1972; McCleave and Kleckner 1982). Tidal currents that larvae can exploit in modifying their drift may also be present well offshore of a coast, as for example on George's Bank (Loder 1980).

The relationship between a larva's drift and its vertical position means that future drift-migration research must incorporate discrete-depth sampling in order to determine what strata the larvae occupy. Such sampling must be done with sufficient frequency to resolve any periodic vertical migrations by the larvae or any ontogenetic changes in preferred depth of larvae. Because many larvae undergo diurnal vertical migrations, sampling must be done so as to clearly differentiate between any vertical migration and net avoidance. The possible interaction between vertical migrations and the tidal period means that sampling in tidal areas should be scheduled carefully and that this data should be examined for evidence of selective tidal transport. Finally, larvae at different ages may occupy different positions in the water column (Coombs et al. 1981), and these changes mean the larvae will undergo different drift regimes as they develop.

The dependence of a larva's drift on its vertical position in the water column means that it is also necessary to know how well larvae can maintain vertical position, especially if sampling has determined that the larvae do occupy a restricted depth range. A larva's vertical position is determined by its buoyancy or by active swimming. There have been investigations on the specific gravity of eggs and larvae of marine fishes (Blaxter 1969; Coombs 1981), but from the standpoint of egg and larval drift it would be worthwhile to examine the relationship between the depth of neutral buoyancy for the eggs and larvae and the currents present at that depth. Buoyancy regulation in clupeoid larvae (which have swim bladders) has been studied (Blaxter and Ehrlich 1974; Hunter and Sanchez 1976), but it is not known how well the larvae of other species (especially those lacking swim bladders) can actually select and maintain a preferred depth, especially in the face of currents or turbulence. Lack of ossified skeletons, tissues with high water content, and gel-filled subdermal spaces (Shelbourne 1956) all may serve a buoyancy function, but for many species we need to know what degree of control the larvae can exert over their buoyancy.

Apart from any intrinsic buoyancy, larvae may regulate their vertical position by swimming. It is also possible that some larval or postlarval fishes may modify their drift by horizontal swimming. This is a much less likely possibility, especially for smaller larvae which have difficulty overcoming the viscosity of the water, a necessity for significant sustained swimming (Weihs 1980). Because for larval fishes swimming can play an important role, both in maintaining vertical position and in possible horizontal movement, it is important for the study of fish migration to know how the fish's swimming performance develops over the course of its life history. It is often assumed that larvae do not add significantly to their movement by swimming, but this may be invalid for larger or metamorphosing larvae. Sette (1943) felt that postlarval Atlantic mackerel (*Scomber scombrus*) (fish greater than 10 mm) added a significant swimming component to their drift, and Nelson et al. (1977) felt that Atlantic menhaden (*Brevoortia tyrannus*) may use swimming in the last stages of movement towards their final estuarine habitat. Several investigators (Leis and Miller 1976; Leis 1982; Heufelder et al. 1982) have been unable to attribute the persistence of some observed larval fish distributions to passive mechanisms such as currents or eddies, and they speculate that active swimming may play a role. Bishai (1960) has documented some surprising swimming abilities in herring larvae, and it seems possible that eel leptocephali or glass eels may need to swim in the latter stages of their drift migration in order to enter coastal waters and approach land (Power 1982). Sustained (as opposed to burst) swimming in marine fish larvae has been studied only in a few species, such as plaice (Ryland 1963; Arnold 1969; Batty 1981) and herring (Rosenthal 1968). There is also an alternative role for swimming in drift migrations: the fish may swim not to augment drift, but rather to maintain drift by avoiding transport into an unfavorable water mass or current system.

If a migratory fish is to constructively modify its drift, it must use some sensory/orientation capability to detect when such behavior is appropriate. Larval phototaxis controlling vertical migrations is one obvious mechanism, but it would be interesting to know if (or when in their life cycle) larvae are capable of celestial or geomagnetic orientation. The presence of orientation and the timing of its development would also suggest how actively the fish may try to modify its drift. The detection of magnetite in yellowfin tuna (*Thunnus albacares*) (Walker 1984--this volume) raises the question as to when in the life cycle the magnetite is acquired. An even more critical lack of information concerns chemosensory abilities of larval fishes. Fishes are highly olfactory animals, and olfaction, apart from any role it may play in feeding, could provide important cues to larvae at fronts, convergences, and other boundaries to current systems by triggering behaviors that modify their drift. Adult eels have renowned olfactory capabilities, and leptocephali have a prominent olfactory rosette. Olfaction may guide the vertical movements of juvenile eels that are using tidal transport in an estuary (Creutzberg 1961), but it is unclear whether such a mechanism would be effective in offshore areas for eels or other species of larval fishes. Qasim et al. (1963) and Gibson (1982) have shown that even larvae without swim bladders are sensitive to the hydrostatic pressure changes associated with tidal fluctuations, and larvae might use such a sensitivity in order to time their vertical migrations appropriately.

DIFFUSION

The diffusion terms in the advection-diffusion equation express the dispersal of eggs and larvae by the turbulence of ocean currents. A fundamental question is whether eggs and larvae diffuse in the same way as passive particles. Presently, there are only a few quantitative studies of fish egg and larval diffusion (Hirano and Fujimoto 1970; Smith 1973; Talbot 1977; Hewitt 1982; Fortier and Leggett 1982). The extent of larval diffusion is important because it directly affects (in a statistical sense) the relative numbers of fish that arrive at the presumed goal of the drift migration. A low diffusion rate means that any given concentration of larvae (patch) will be advected by a subportion of the total current field, and the "appropriateness" of these currents in carrying the larvae to a suitable location becomes critical. Larvae found to be diffusing more slowly than passive particles implies that the larvae are maintaining some sort of discrete distribution through schooling or swarming. Hewitt (1981) observed that the "patchiness" of anchovy (*Engraulis mordax*) and jack mackerel (*Trachurus symmetricus*) was initially high due to spawning aggregations and then decreased as the larvae were dispersed. Later the patchiness began to increase again, and Hewitt attributes this to the onset of schooling behavior in the larvae. The probability that larvae in an incipient school encounter a particular and possibly necessary current (e.g. the coastal jet current described by Shelton and Hutchings 1982) may be less than that of passively diffusing larvae, just as the possibility that a predator encounters schooling prey is less.

The reverse of this hypothetical situation could also occur. If the larvae do not school, but in fact swim randomly, then they could disperse at a rate higher than that of a passive substance. In that case "biodiffusion" (Okubo 1980) must be considered along with the physical diffusion process. The net effect from a migration standpoint would be that the geographic range at the end of drift would be larger (lower "accuracy"), and that because of interannual current variations the consistency of the drift's end point also would vary (lower "precision").

One can think of other ways that the interaction between diffusion and larval behavior could become important during drift. For example, Talbot (1977) found that the dispersal of plaice larvae was largely due to vertical shear diffusion, which is the apparent horizontal diffusion arising from a vertical current shear. If larvae occupy a restricted depth stratum, they would be subject to a lesser portion of the current shear and could show an apparent horizontal dispersal that differs from that of larvae at other strata on those distributed throughout the water column.

The lack of information on diffusion of eggs and larvae in the ocean results from the extreme difficulty of observing diffusion quantitatively. One may be able to make inferences about diffusion by statistically evaluating quantitative ichthyoplankton samples (Smith 1973; Hewitt 1982), but such an approach examines the diffusive process indirectly and requires substantial sampling effort to be effective. Direct observation of diffusion will, as was mentioned in connection with drogue tracking, require the identification of systems where spawning behavior of adults or hydrographic conditions result in the formation of discrete larval patches that can be followed over time. Such situations will permit the repeated sampling necessary to determine the distribution of the fish, as well as that of a natural or introduced tracer for comparison. Complications are to be expected from advection, and the fact that the fish are not a conservative substance (i.e. the total concentration may change due to continued spawning or mortality). But in situations where one can assume that mortality is density independent, changes in the variance of the fish's distribution over time (from which one can make inferences about the diffusion) are not affected by mortality (Okubo 1980). The study of such "model systems" will materially advance our understanding of larval diffusion and consequently our understanding of larval drift migrations.

ACKNOWLEDGEMENTS

This paper was written while the author held a National Research Council Research Associateship. The National Research Council and NATO provided funds for the author to attend the NATO Advanced Research Institute on Mechanisms of Migration in Fishes, and this support is gratefully acknowledged. Dr. Roger Hewitt critically read the manuscript.

REFERENCES

Arnold, G.P. 1969. The orientation of plaice larvae (*Pleuronectes platessa* L.) in water currents. Journal of Experimental Biology 50:785-801.

Batty, R.S. 1981. Locomotion of plaice larvae. Pages 53-70 *in* M.H. Day, editor. Vertebrate locomotion. Academic Press, London, England.

Bishai, H.M. 1960. The effect of water currents on the survival and distribution of fish larvae. Journal du Conseil Conseil International pour l'Exploration de la Mer 25:134-146.

Blaxter, J.H.S. 1969. Development: Eggs and larvae. Pages 178-252 *in* W.S. Hoar and D.J. Randall, editors. Fish physiology, volume III. Academic Press, New York, New York, USA.

Blaxter, J.H.S., and K.F. Ehrlich. 1974. Changes in behavior during starvation of herring and plaice larvae. Pages 575-588 *in* J.H.S. Blaxter, editor. The early life history of fish. Springer-Verlag, New York, New York, USA.

Coombs, S.H. 1981. A density-gradient column for determining the specific gravity of fish eggs, with special reference to the eggs of the mackerel *Scomber scombrus*. Marine Biology 63:101-106.

Coombs, S.H., R.K. Pipe, and C.E. Mitchell. 1981. The vertical distribution of eggs and larvae of blue whiting (*Micromesistius poutassou*) and mackerel (*Scomber scombrus*) in the eastern North Atlantic and North Sea. Rapports et Procès-Verbaux des Réunions Conseil International pour l'Exploration de la Mer 178:188-195.

Creutzberg, F. 1961. On the orientation of migrating elvers (*Anguilla vulgaris* Turt.) in a tidal area. Netherlands Journal of Sea Research 1:257-338.

Fortier, L., and W.C. Leggett. 1982. Fickian transport and the dispersal of fish larvae in estuaries. Canadian Journal of Fisheries and Aquatic Sciences 39:1150-1163.

Fraser, J.H. 1958. The drift of the planktonic stages of fish in the northeast Atlantic and the possible significance to the stocks of commercial fish. International Commission for the Northwest Atlantic Fisheries Special Publication 1:289-310.

Gibson, R.N. 1982. The effect of hydrostatic pressure cycles on the activity of young plaice *Pleuronectes platessa*. Journal of the Marine Biological Association of the United Kingdom 62:621-635.

Graham, J.J. 1972. Retention of larval herring within the Sheepscot estuary of Maine. US National Marine Fisheries Service Fishery Bulletin 70:299-305.

Gunn, J.T., and D.R. Watts. 1982. On the currents and water masses north of the Antilles/Bahamas arc. Journal of Marine Research 40:1-18.

Harden Jones, F.R. 1968. Fish migration. Arnold, London, England.

Heufelder, G.R., D.J. Jude, and F.J. Tesar. 1982. Effects of upwelling on local abundance and distribution of larval alewife (*Alosa pseudoharengus*) in eastern Lake Michigan. Canadian Journal of Fisheries and Aquatic Sciences 39:1531-1537.

Hewitt, R.P. 1981. The value of pattern in the distribution of young fish. Rapports et Procès-Verbaux des Réunions Conseil International pour l'Exploration de la Mer 178:229-236.

Hewitt, R.P. 1982. Spatial pattern and survival of anchovy larvae: implications of adult reproductive strategy. Doctoral dissertation. University of California, San Diego, California, USA.

Hirano, T., and M. Fujimoto. 1970. Preliminary results of investigation of the Kuroshio functioning as a means of transportation and diffusion of fish eggs and larvae. Pages 405-416 *in* J.C. Marr, editor. The Kuroshio. East-West Center Press, Honolulu, Hawaii, USA.

Hunter, J.R., and C. Sanchez. 1976. Diel changes in swim bladder inflation of the larvae of the northern anchovy, *Engraulis mordax*. US National Marine Fisheries Service Fishery Bulletin 74:847-855.

Iles, T.D., and M. Sinclair. 1982. Atlantic herring: stock discreteness and abundance. Science (Washington DC) 215:627-633.

Leis, J.M. 1982. Nearshore distributional gradients of larval fish (15 taxa) and planktonic crustaceans (6 taxa) in Hawaii. Marine Biology 72:89-97.

Leis, J.M., and J.M. Miller. 1976. Offshore distributional patterns of Hawaiian fish larvae. Marine Biology 36:359-367.

Loder, J.W. 1980. Topographic rectification of tidal currents on the sides of Georges Bank. Journal of Physical Oceanography 10:1399-1416.

Marliave, J.B. 1981. Vertical migrations and larval settlement in *Gilbertidia sigalutes*, F. Cottidae. Rapports et Procès-Verbaux des Réunions Conseil International pour l'Exploration de la Mer 178:349-351.

McCleave, J.D., and R.C. Kleckner. 1982. Selective tidal stream transport in the estuarine migration of glass eels of the American eel (*Anguilla rostrata*). Journal du Conseil Conseil International pour l'Exploration de la Mer 40:262-271.

Nelson, W.R., M.C. Ingham, and W.E. Schaaf. 1977. Larval transport and year-class strength of Atlantic menhaden, *Brevoortia tyrannus*. US National Marine Fisheries Service Fishery Bulletin 75:23-41.

Okubo, A. 1980. Diffusion and ecological problems: mathematical models. Springer-Verlag, Berlin, Federal Republic of Germany.

Parrish, R.H., C.S. Nelson, and A. Bakun. 1981. Transport mechanisms and reproductive success of fishes in the California Current. Biological Oceanography 1:175-203.

Power, J.H. 1982. A numerical method for simulating plankton transport and numerical simulation of the North Atlantic Ocean drift of *Anguilla* leptocephali. Doctoral dissertation. University of Maine, Orono, Maine, USA.

Powles, H. 1981. Distribution and movements of neustonic young estuarine dependent (*Mugil* spp., *Pomotomus saltatrix*) and estuarine independent (*Coryphaena* spp.) fishes off the southeastern United States. Rapports et Procès-Verbaux des Réunions Conseil International pour l'Exploration de la Mer 178:207-209.

Qasim, S.Z., A.L. Rice, and E.W. Knight-Jones. 1963. Sensitivity to pressure changes in teleosts lacking swim bladders. Journal of the Marine Biological Association of India 5:289-293.

Rosenthal, H. 1968. Schwimmverhalten und Schwimmgeschwindigkeit bei Larven des Herings *Clupea harengus*. Helgoländer wissenschaftliche Meeresuntersuchungen 18:453-486.

Ryland, J.S. 1963. The swimming speeds of plaice larvae. Journal of Experimental Biology 40:285-299.

Seckel, G.R. 1972. Hawaiian-caught skipjack tuna and their physical environment. US National Marine Fisheries Service Fishery Bulletin 72:763-787.

Sette, O.E. 1943. Biology of the Atlantic mackerel (*Scomber scombrus*) of North America. US Fish and Wildlife Service Fishery Bulletin 50:149-237.

Shelbourne, J.E. 1956. The effect of water conservation on the structure of marine fish embryos and larvae. Journal of the Marine Biological Association of the United Kingdom 35:275-286.

Shelton, P.A., and L. Hutchings. 1982. Transport of anchovy, *Engraulis capensis* Gilchrist, eggs and early larvae by a frontal jet current. Journal du Conseil Conseil International pour l'Exploration de la Mer 40:185-198.

Smith, P.E. 1973. The mortality and dispersal of sardine eggs and larvae. Rapports et Procès-Verbaux des Réunions Conseil International pour l'Exploration de la Mer 164:282-292.

Talbot, J.W. 1977. The dispersal of plaice eggs and larvae in the Southern Bight of the North Sea. Journal du Conseil Conseil International pour l'Exploration de la Mer 37:221-248.

Ulanowicz, R.E., J.M. Lindsay, W.C. Caplins, and T.T. Polar. 1982. Simulating the lateral transport of ichthyoplankton in the Potomac estuary. Estuaries 5:57-67.

Walker, M.M. 1984. Magnetic sensitivity and its possible physical basis in the yellowfin tuna, *Thunnus albacares*. Pages 125-141 *in* J.D. McCleave, G.P. Arnold, J.J. Dodson and W.H. Neill, editors. Mechanisms of migration in fishes. Plenum Press, New York, New York, USA.

Weihs, D. 1980. Energetic significance of changes in swimming modes during growth of larval anchovy, *Engraulis mordax*. US National Marine Fisheries Service Fishery Bulletin 77:597-604.

Wroblewski, J.S. 1980. A simulation of the distribution of *Arcatia clausi* during Oregon upwelling, August 1973. Journal of Plankton Research 2:43-68.

DRIFT OF LARVAL FISHES IN THE OCEAN: RESULTS AND PROBLEMS FROM PREVIOUS STUDIES AND A PROPOSED FIELD EXPERIMENT

Hans-Christian John

Taxonomische Arbeitsgruppe der Biologischen Anstalt Helgoland
Zoologisches Institut und Museum
Martin-Luther-King-Platz 3
2000 Hamburg 13, Federal Republic of Germany

ABSTRACT

Literature concerning the drift of fish larvae is fairly abundant. Because it is impossible to trace directly the drift routes and fate of individual larvae, evidence for direction and duration of drift of planktonic stages presented in previous publications is necessarily scarce. This paper discusses the possible methods to obtain indirect evidence and recommends combined investigations of reproduction of adults, taxonomy of larvae, larval distribution and behaviour in situ as well as data of the relevant hydrographical features. Examples for the success of such a research concept are presented.

INTRODUCTION

Marine fish larvae, as part of the plankton, are distributed by currents. Partly due to this fact the overall distributions of adults are generally wider than their areas of reproduction. On the other hand currents may cause temporary or definite loss of off-spring, thus reducing the reproductive potential of a population. Larvae of truly oceanic species are analogous to holoplanktic invertebrates and are negatively affected by currents to a much lesser degree than neritic species as long as they stay within a physiologically adequate environment (Ekman 1953). In the open ocean horizontal changes in environmental parameters generally occur slowly and on a large geographical scale. This may be exemplified by the spawning of Exocoetidae (flying fishes) in their northeasternmost zone of distribution. As already demonstrated by Bruun (1935) these animals generally occur in waters warmer than 21 C. John (1979) documented the presence of exocoetid fry (besides other warm water species) during summer in areas of the subtropical NE Atlantic Ocean, where, due to lower temperatures, adults and fry under normal conditions are missing during winter. I speculated that southwestward drift with the Canary Current was essential for survival of the larvae (Fig. 1).

Neritic species are believed to have in general a higher tolerance against changes in temperature and salinity than oceanic taxa (Johnson and Brinton 1963). However, they are analogous to meroplankton and are more likely carried to areas unfavourable for metamorphosis. This must not necessarily be detrimental, if the larvae (or the older stages) have a chance to return to their places of birth later, because drift may remove larvae from the range of predators (Johannes 1978) or reduce interspecific and intraspecific

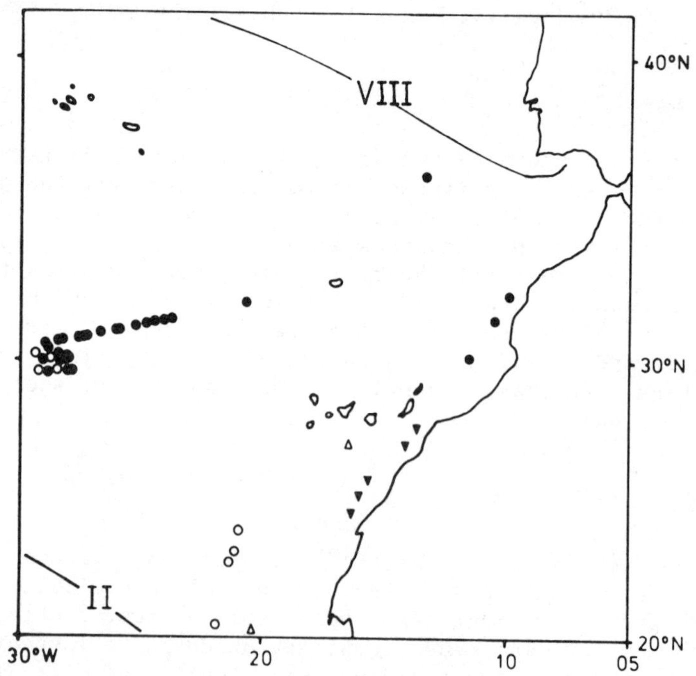

Figure 1. The 21°C surface isotherm in the coldest (II, February) and warmest (VIII, August) month off northwest Africa, and seasonal catches of exocoetid larvae. Open circles = spring; solid circles = summer; open triangles = autumn; solid triangles = winter. Positive winter catches were exclusively due to a positive temperature anomaly in winter 1970.

competition (Marty 1965). If the recirculation would take longer than the normal planktonic lifespan, at least for some invertebrates the planktonic phase may be prolonged under such circumstances (e.g. Wilson 1952, 1968; Thorson 1961). The occurrence of meroplanktonic larvae of neritic benthic invertebrates in the open ocean was reported, for example, by Johnson (1960), Lutjeharms and Heydorn (1981) and Scheltema (1971 review, 1972, 1973).

PROBLEMS IN THE LITERATURE

The influence of currents on the distribution of fish larvae or recruitment of parental stocks is frequently mentioned in the literature. Some of these papers are listed in Bishai (1960) and in Laevastu and Hayes (1981) or cited by Harden Jones (1968) and Marty (1965), but a special review of this subject seems to be lacking. Much of the existing information is a theoretical consideration of the relation between removal of larvae from the spawning grounds and resulting yearclass strength but does not deal with the passive migration of fish larvae. When dealing with the direction, distance or speed of passive drift, the publications generally concern well investigated, commercially-important species restricted to coastal areas where the hydrography is well known.

A rather common problem in these studies seems to be how to obtain evidence for passive drift of larvae. Evidence may be gained by repeatedly sampling identical areas and tracing the distributions of larvae in space and time (e.g. Sette 1943; Saville 1965). The possibility to investigate the (at least relative) age composition in larval distributions yields good proof for the direction of drift (e.g. Schmidt 1925; Sette 1943). Another way to prove the existence of drift and determine its direction is the relation of overall distribution of fry to the overall distribution of adults and the location of their spawning places (e.g. Marty, 1965). However, this method has its limits in tracing the speed and true directions of drift and may be misleading in case the spawning grounds are incompletely known. This becomes even more important when the information on larval drift is derived from the single fact that larvae were caught at a place "where they did not belong" and this fact is related to knowledge or assumption of the currents in the area (e.g. Scheltema 1971, 1972, 1973 and Taning 1937).

Due to both these mentioned difficulties and the distribution of research effort, information on drift of oceanic fish larvae of neritic fishes in the open ocean seems scarce. Nafpaktitis (1968) concluded from the limited regional distribution of individuals with maturing gonads in one species of lanternfish that larvae drifting within one year from the Eastern towards the Western Atlantic Ocean recruited the western "sterile expatriated population." However, recent results by Karnella and Gibbs (1977) demonstrated maturation in the Western Atlantic too. Fraser (1958) doubted that larvae of any other fish species than *Anguilla anguilla* might use the North Atlantic Drift for a non-stop transatlantic drift. This probably holds true for coastal fishes, but for oceanic species the possibility should not yet be excluded. Parin (1960) related the distribution of saury (*Cololabis saira*) larvae to the surface current system. Hamann et al. (1981) and John et al. (1980) described the drift mechanisms of larvae of one oceanic and one neritic species, respectively, off Mauritania. Dooley (1972) discussed the possible drift pattern of fishes associated with *Sargassum* weed. John (1975), Krefft (1976) and Schmidt (1925) documented the mid-ocean occurrence of larvae of fishes whose adults are neritic for most of their life span. The facts that only two of the mentioned papers deal with truly oceanic species and that only three publications provide evidence for the discussed drift pathways stress the importance of further research in the open ocean.

EVIDENCE FOR LARVAL DRIFT

The direct study of plankton drift in the ocean by tagging and recovering is, of course, an illusion. So any estimation must be made indirectly, needing both detailed knowledge on the biology of the species considered and exact knowledge of the hydrography in the species' range of occurrence. The general circulation systems at the surface of the oceans are generally well known; Figure 2, taken from Angel (1979), is one example of many similar charts. Much more detailed information on monthly or quarterly mean direction, speed and variability of surface currents is available from the hydrographic departments of several nations (pilot charts, e.g. DHI 1967), but these sources seem to be little used by marine biologists, Harden Jones (1968) making one of the few exceptions.

For species with a prevailing two dimensional distribution at the sea surface such data from pilot charts allow reasonable estimations of the effect of currents on the distribution of fish larvae. For one species of saury the qualitative aspects have been described by Parin (1960). Another, more quantitative analysis has been done of the distribution and regional abundance of fry of the flying fish, *Exocoetus volitans* (Fig. 3; from John 1983). As mentioned, reproduction occurs only at temperatures above 21 C and the fry are strictly neustonic during daytime and slightly more dispersed near the surface at night. The Atlantic system of surface currents causes a broader latitudinal belt of warm water and exocoetid distribution in the west. Furthermore, equatorward currents in the east cause high gradients of abundance near the limits of distribution, whilst poleward directed currents in the west cause dispersion and an extended zone of moderate to low abundance.

Another example suggests (or, in one case, proves) drift with

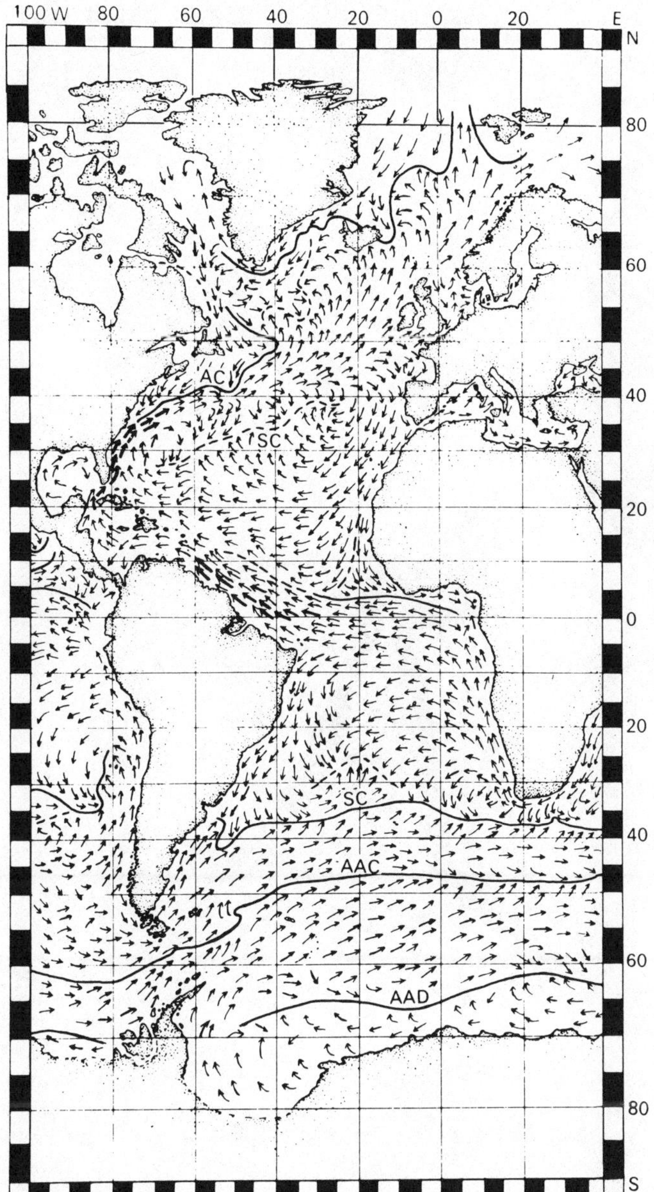

Figure 2. Surface circulation of the Atlantic Ocean (from Angel 1979).

currents, but does not allow quantitative analysis due to the scarcity of material. Adults of *Dactylopterus volitans* are demersal, neritic fishes occurring along both warm coasts of the Atlantic at

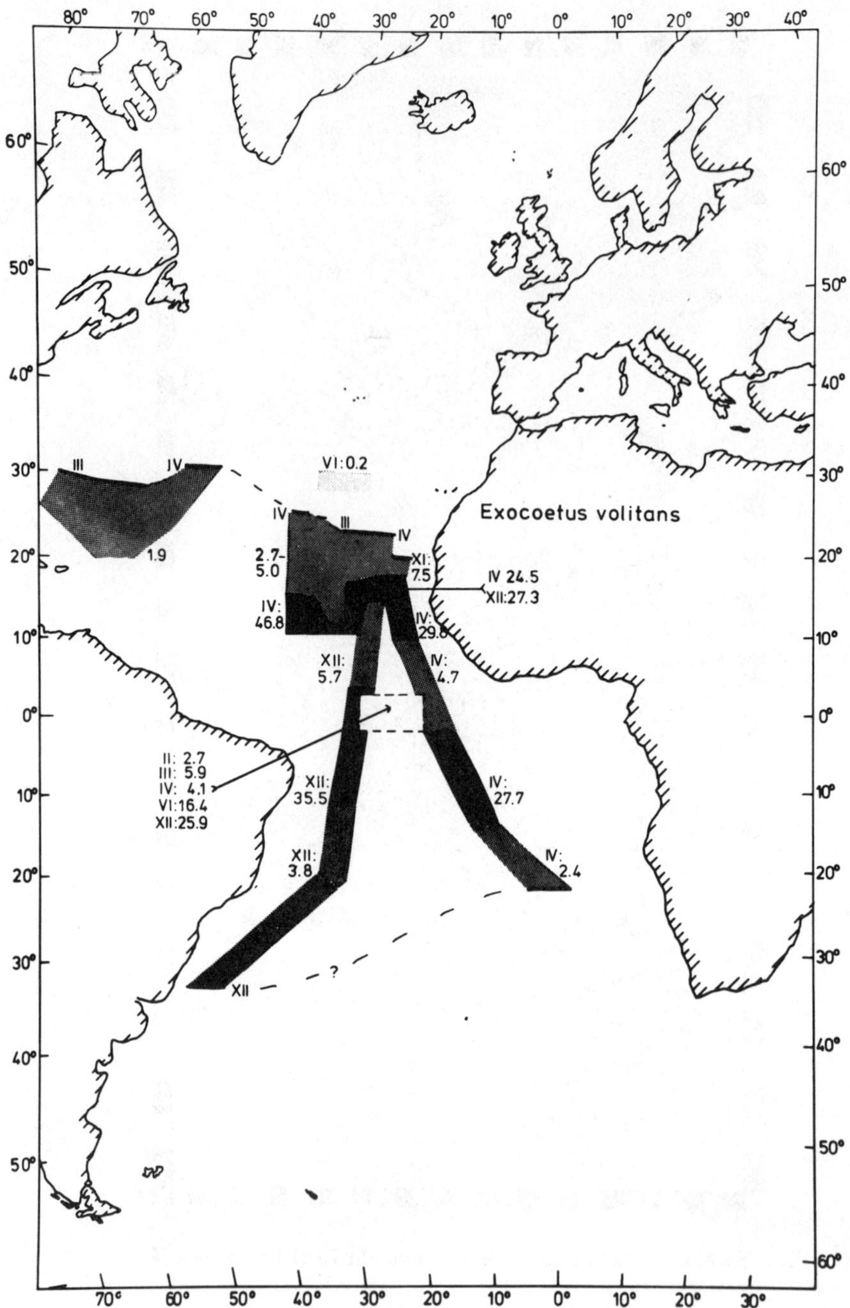

Figure 3. Distribution and abundance of fry of *Exocoetus volitans* (from John 1983). Roman numerals = months of sampling; arabic numerals = numbers of fry per 1000 m^2 water.

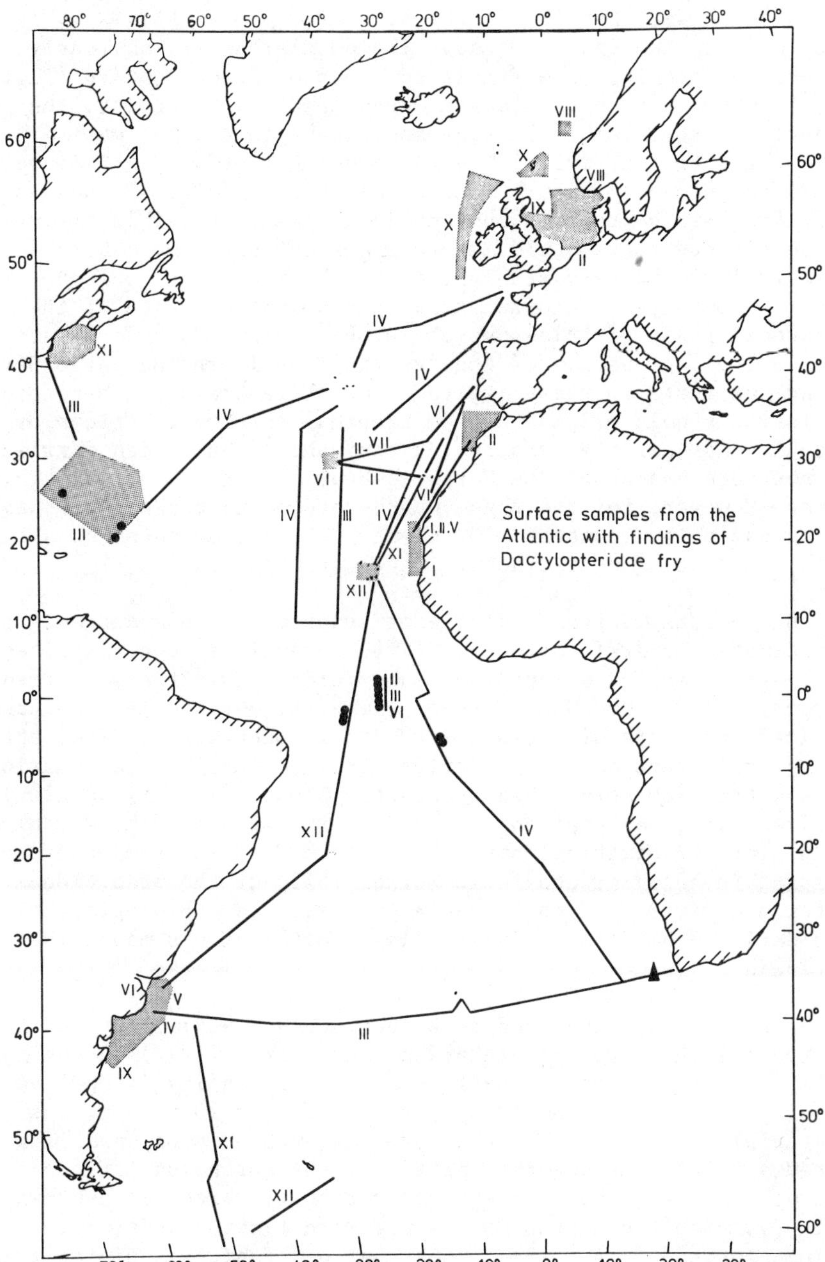

Figure 4. Catches of neustonic postlarvae of family Dactylopteridae. Solid lines = transects; shaded areas = station grids sampled; Roman numerals = months of sampling; solid triangle = Indian Ocean specimen of *Dactyloptena orientalis*; solid dots = *Dactylopterus volitans*.

depths shallower than 100 m, whilst *Dactyloptena orientalis* is restricted to the Indian Ocean. The postlarvae are obviously strictly neustonic. The single specimen of *D. orientalis* (Fig. 4) was, together with other Indian Ocean species, swept into the Atlantic by the Agulhas Current and caught in a water mass detrimental for survival. The 141 specimens of *D. volitans* were mainly caught in the South Equatorial Current (Fig. 4), suggesting drift from Africa towards the American coast. Possible arrival of the postlarvae there would mean a transport of 2,500 nautical miles. For this fast and very stable current the time span would be in the order of 100 days. Unfortunately, the normal duration of the planktonic phase of this species is not known. Much less satisfactory is the explanation of the few findings in the Sargasso Sea. Surface currents in these latitudes point westwards. Therefore, the postlarvae should originate from tropical northwest African or Cape Verde Islands parental stocks and thus should have been carried by the northern branch of the North Equatorial Current at least 2,400 miles. However, for the three months preceding the catch, this branch has only a constancy of about 50% and mean speed of about 0.5 knots, suggesting a distinctly longer transport.

Biologists generally use information on surface currents as well to estimate the drift of deeper living organisms (e.g. Scheltema 1971). This may be allowed for some vertically extended current systems, such as the Gulf Stream. However, our knowledge of the vertical structure of oceanic currents is much more limited than that for coastal areas and often derived from geostrophic calculations rather than from actual measurements. These calculations always involve several assumptions. The pattern calculated by Worthington (1976) for the North Atlantic Circulation differs from other recent geostrophic calculations (e.g. Wegner 1973) or the traditional patterns derived from ships set and drift. To a biologist, considering Schmidt's (1925) results, Worthington's model is less convincing.

As generally suggested by Harden Jones (1968) or stated for neustonic fish larvae by Richardson and Pearcy (1977) we can confirm that information from the surface may be misleading if applied to deeper layers, occasionally only as shallow as 30 m (e.g. the Equatorial Undercurrent or the undercurrents of some upwelling systems). This may be exemplified for the northwest African coastal area by comparing the mean surface currents, which run southward (Fig. 2), with the depths of the poleward flowing undercurrent at various latitudes (Fig. 5). This must be of biological importance off Mauritania. Hamann et al. (1981) presented an example for fish larvae carried northward by this undercurrent (Figs. 6, 10); further examples are at hand. The southward drift of surface-living fish larvae in the same area has also been demonstrated (Fig. 7). *Belone svetovidovi* deposits "demersal" eggs on seaweed in shallow areas, so the majority of the larvae caught should have originated from the bay

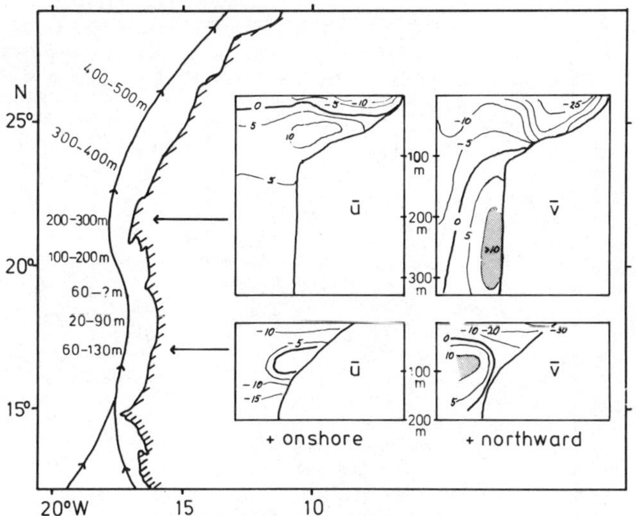

Figure 5. Depth of the core of the undercurrent along the northwest African coast at various latitudes (after Mittelstaedt 1978, and Hamann et al. 1981). Insets show cross-shelf current structures at two selected transects (different longitudinal scales of 80 km (above) and 100 km (below)). Velocities are given in cm/sec.

south of Cape Blanc and the few northern findings perhaps from Rio de Oro. Both the patterns of abundance and the regional length distributions point to drift with the Ekman layer.

Even for long-known, huge systems like the Gulf Stream-North Atlantic Drift the traditional information should be used with caution. Worthington's (1976) view has been mentioned. The existence of mesoscale variations like meanders or core rings is proven. Though Wiebe et al. (1976) stated that the decay of the species assemblage in such a core ring is faster than the change of its physical properties, the relatively short duration of the planktonic phase of many species might result in a delay of drift or even in a transport back towards the place of origin for parts of the population. Bakun et al. (1979), Harden Jones (1968) and Johannes (1978) assumed that species might time or locate spawning to take advantage of circulation systems, which might 1) keep the larvae near their birth place, 2) allow larvae to be swept to nurseries, or 3) allow larvae to be temporarily removed from the areas of heavy predation. The latter two theories demand a later transport or migration back to the spawning grounds.

Even if the horizontal and vertical structures of a current system were stable and well known, biologists often do not possess

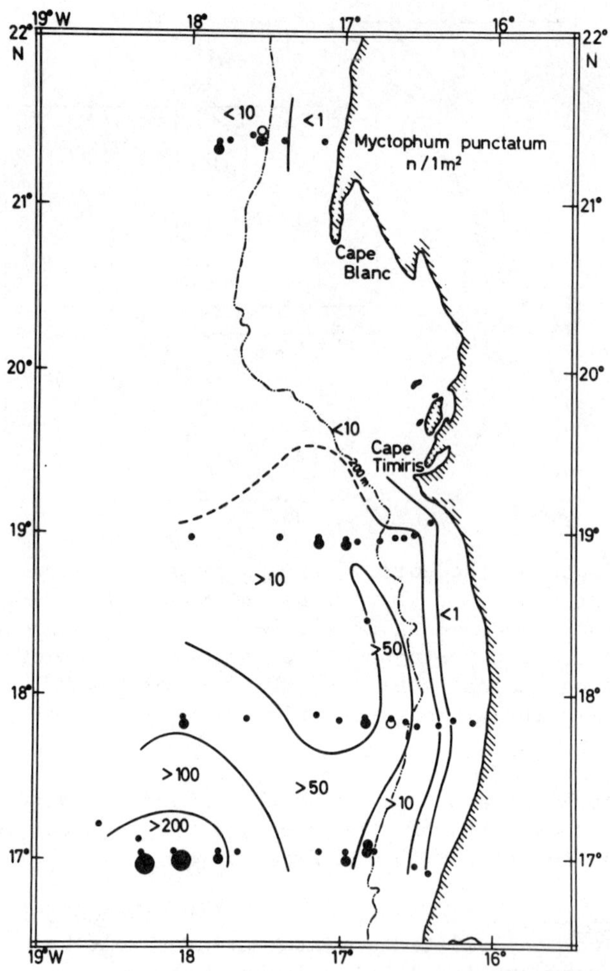

Figure 6. Abundance of *Myctophum punctatum* larvae (from Hamann et al. 1981.). Small closed circles = sampling stations; large closed circles = presence of adults; open circles = adults absent, though expected.

the biological information necessary to calculate the probable transport. Identification of many fish larvae is not yet possible to the required level (e.g. species level in small *Anguilla* or population level in North Sea herring). Knowledge of the vertical distribution of plankton is still scarce. As already reviewed by Friedrich (1965) the vertical distribution of a single species may show diurnal, seasonal, ontogenetic and geographical changes or may differ between sexes. Diurnal changes in vertical distribution of fish larvae are rather common, and behaviour of the same species may change during ontogeny too (John 1973). *Macrorhamphosus scolopax* is

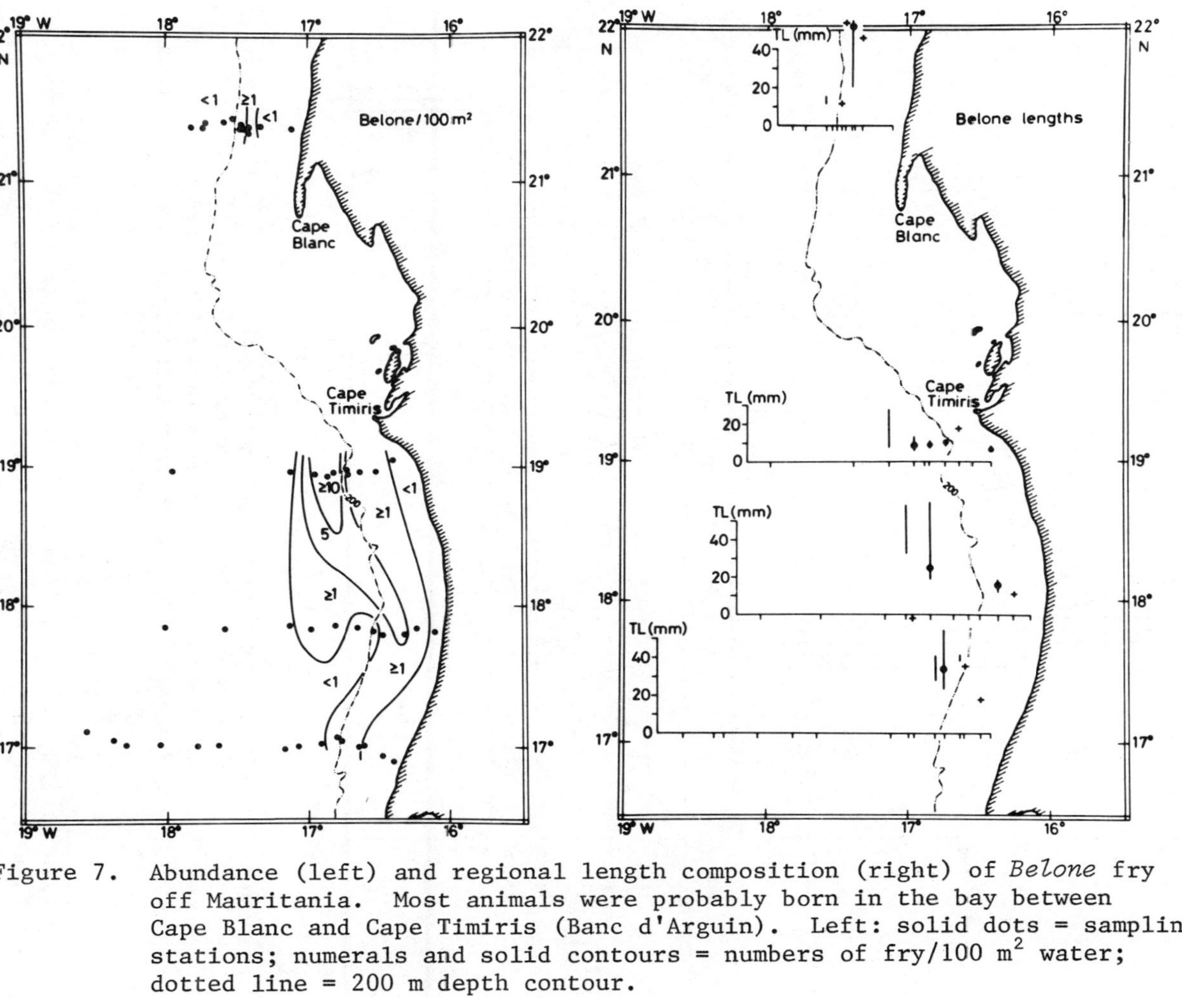

Figure 7. Abundance (left) and regional length composition (right) of *Belone* fry off Mauritania. Most animals were probably born in the bay between Cape Blanc and Cape Timiris (Banc d'Arguin). Left: solid dots = sampling stations; numerals and solid contours = numbers of fry/100 m^2 water; dotted line = 200 m depth contour.

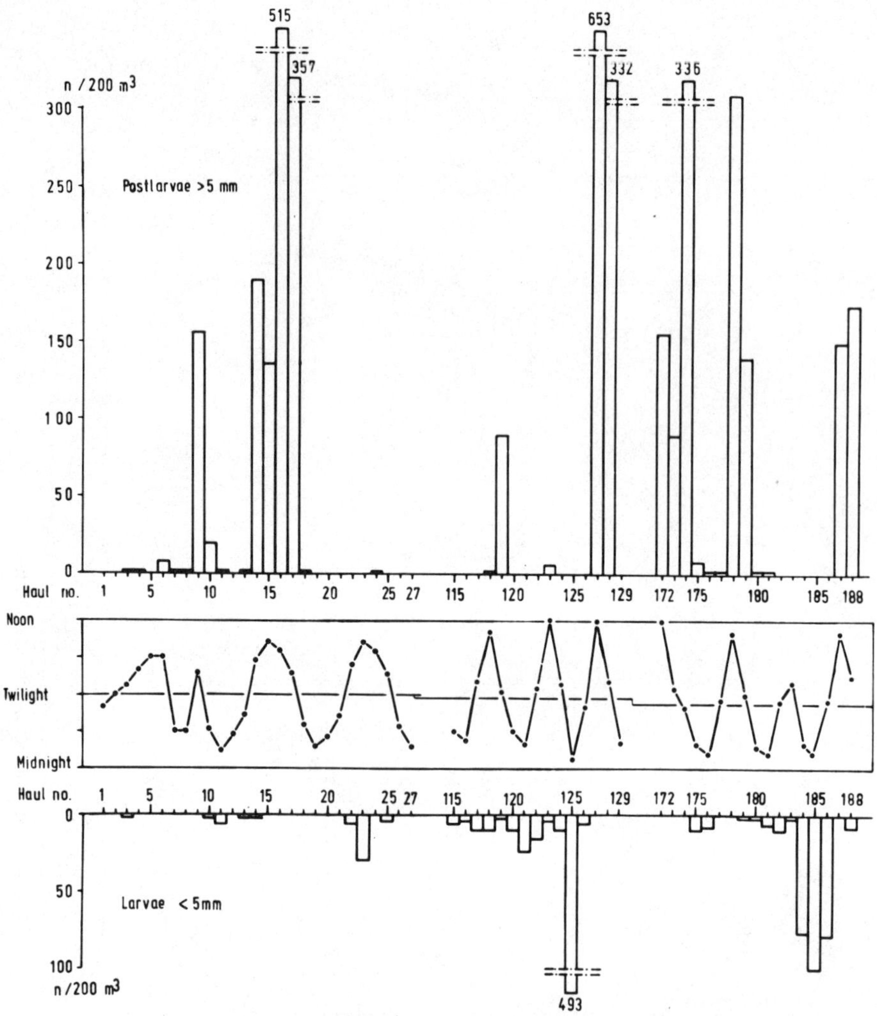

Figure 8. Alternating diurnal surface occurrence of postlarvae (top) and larvae (bottom) of *Macrorhamphosus scolopax* (after John 1973). Middle diagram shows time scale of sampling: upper half corresponds to daytime samples and lower part to nighttime samples in the layer 0-25 cm.

an example (Fig. 8). The small larvae correspond to the "normal" type, avoiding the surface layer during the day and showing a relatively small extent of diurnal vertical migration (less than 50 m). The postlarvae represent a rather rare, though not unique, daytime-surface occurrence and a vertical range of migration down to 200 m at night (Badcock and Merrett 1976). Vertical distribution of one and the same developmental stage of the same taxon may differ

Table 1. Mean vertical distribution of *Myctophum punctatum* larvae in two different areas.

Off Mauritania		Mediterranean Sea[a]		
Depth, m	Percentage of larvae caught	Depth, m	Percentage of hauls positive	Number of larvae per haul
0-30	40.1	30	21.8	3.3
30-60	47.8	30-70	10.5	3.6
60-90	9.5	71-150	45.2	8.5
90-120	1.3	150	35.6	1.6
120-150	1.2			

[a] Data from the Mediterranean Sea were calculated from original data of Taning (1918), assuming a wire length to depth ratio of 2.3:1.

between areas due to different environmental parameters, among which light seems to be most important. This is illustrated by the differing vertical distributions of *Myctophum punctatum* off Mauritania and in the Mediterranean Sea; in the latter the larvae occur much deeper (Table 1). Ichthyoplankton may reach much deeper than the previously assumed limit of 140 m depth (Badcock and Merrett 1976; Gorbunova 1977; John unpublished data). As already suggested by Pearcy and Meyers (1974), one might even speculate that vertical distribution is a direct response to currents in order to keep a population in its horizontal range.

Nevertheless, the classical work by Schmidt (1925) provides an example that occasionally fish larval drift can be estimated without previous knowledge of many details. However, this worked only because *Anguilla* has a limited spawning place and season, its planktonic phase is much longer than the spawning season, and for the time scale involved the North Atlantic Drift is rather constant. It would otherwise not have been possible to separate age classes by length.

Mostly we have to deal with species in which the planktonic phase is much shorter, sometimes even shorter than the spawning season (but knowledge of the duration of the planktonic phase of oceanic fishes is scarce). Several breeding grounds or seasons or prolonged batch-spawning would render Schmidt's method meaningless. We have some examples at hand in which length distributions in space and time indicate the direction of drift (Figs. 7,9; also Sette 1943) or even the mean velocity of drift by applying an age-length

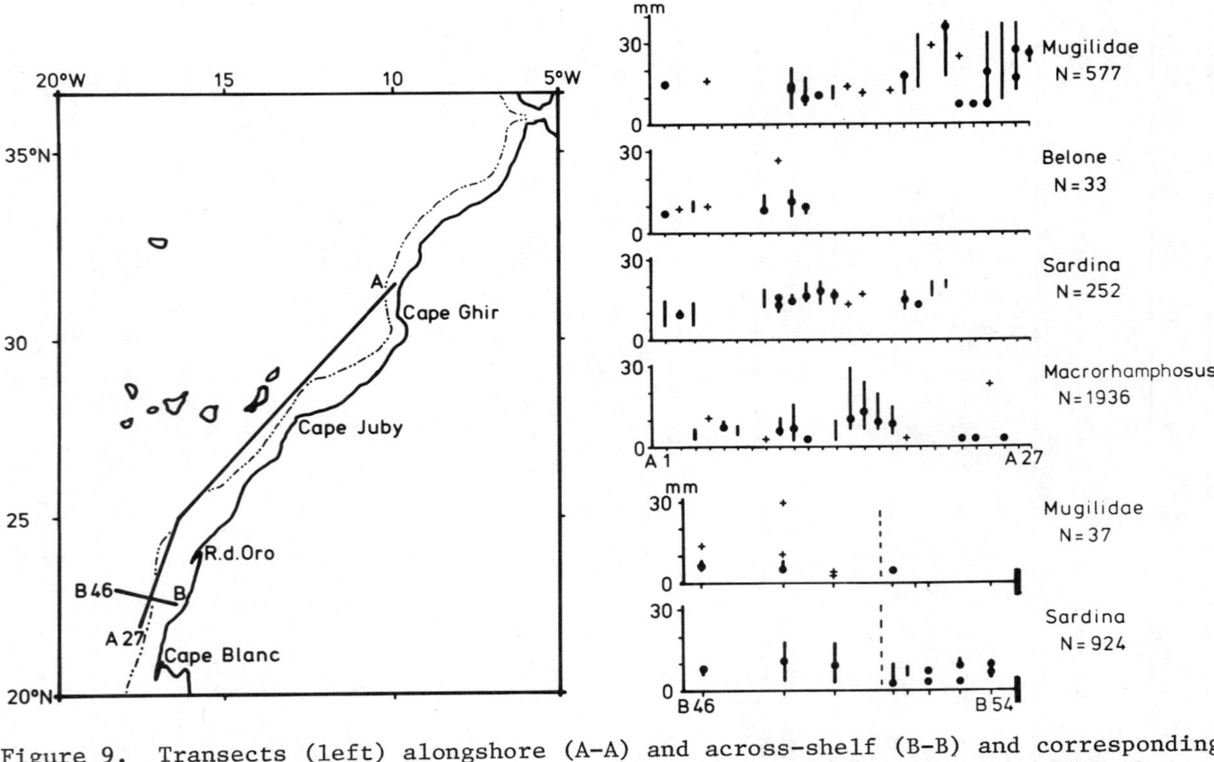

Figure 9. Transects (left) alongshore (A-A) and across-shelf (B-B) and corresponding length compositions (right) of fish larvae. Length increases with the direction of drift of the Ekman layer, except when vertical migration reaches below this layer (as in *Macrorhamphosus scolopax*).

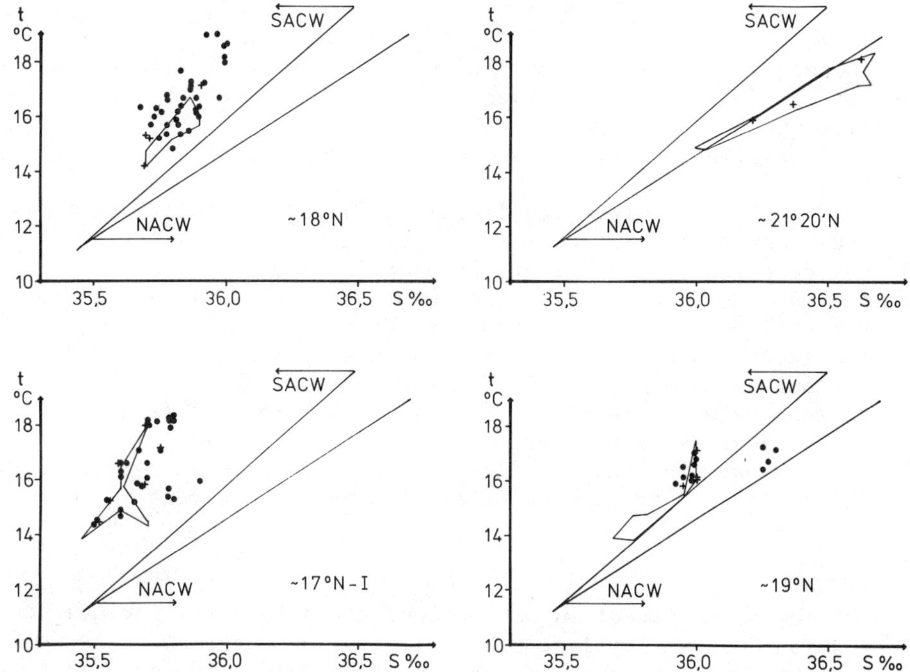

Figure 10. Water mass characteristics associated with positive (= closed circles) and negative (envelopes) catches of *Myctophum punctatum* larvae. Larvae were carried by South Atlantic Central Water Mass (SACW). NACW = North Atlantic Central Water Mass. (Water mass definitions after Tomczak 1978.)

key (John et al. 1980), but many more examples in which the length distributions proved to be worthless. On the other hand modern plankton sampling methods allow quantitative analysis of plankton abundance, and abundance patterns indeed yield another indication of the direction and distance from the spawning ground (in case this is limited) and may provide an indirect key to relative age (see Fig. 6 as an example). *Myctophum punctatum* is an oceanic species here found in its southernmost range. Adults were most abundant off the slope at 17°N and 89% were near spawning (maturity stages V-VII). The quantitative distribution of larvae suggests drift along the slope towards north. This could not be corroborated by length distributions, because the larvae seem to grow very slowly, but the assumed drift is in agreement with water mass analysis and current meter data (see Figs. 5,10; also Hamann et al. 1981).

ASSESSMENT OF LARVAL DRIFT

Consequently for the assessment of the drift of fish larvae as many of the following requirements as possible should be satisfied or be investigated:

1) the spawning area(s) and season(s) of the species considered are known;
2) all planktonic stages are identifiable with certainty to the necessary taxonomic level;
3) the duration of the planktonic phase is known;
4) regional abundance, limits of occurrence and regional length composition of each planktonic stage are known;
5) the short-scale vertical distributions of all stages and in all regions and seasons involved are known;

6a) current velocity is measured in situ over the entire period, depth range and area involved; or

6b) it is known that the stability of the hydrographic system allows the applicability of data other than in situ current measurements.

Of course satisfaction of the conditions in this list demands cooperation between physical and biological oceanographers as well as high efforts by gear, ship and manpower employed. In the past this effort would have been unreasonably high; it would still be unreason-reasonable to apply such an expensive research concept to a species so widely distributed and with such a long planktonic phase as *Anguilla anguilla*. For certain other species with more limited distribution and a planktonic phase not longer than three months modern methods are at hand. Multiple-opening-closing plankton nets allow nearly simultaneous samples with the desired vertical resolution and desired sample size in a minimum of ship time. Commonly-used arrays of current meters can at least partly be replaced by satellite-tracked drift buoys or current profilers. In case current measurements are not available Johnson and Brinton (1963) demonstrated the value of water mass analysis to trace the movement of plankton, whilst indicator species would yield a higher margin of error. (Figure 10 exemplifies the value of water mass analysis for the investigation of larval drift.) The use of geostrophic calculations of currents for the analysis of drift of fish larvae has been questioned above, but this is a personal view. Bakun et al. (1982) recommended this method, and Krauss and Meincke (1982) proved that such calculation can agree with simultaneous actual measurements. Taxonomic problems in studies of fish larvae generally diminish when more and better material becomes available. Lastly, electronic processing of data makes feasible the collecting of such a huge lot of essential information. Modelling of plankton drift shows convincing results when a reliable hydrographic data base is at hand and biological data allow a test of the model (e.g. Brockmann 1979; Wroblewski 1982).

For smaller areas the success of this or a similar research concept has already been demonstrated (Hamann et al. 1981; Peterson et al. 1979). Off Mauritania (during a survey under the leadership of E. Mittelstaedt and H. Weikert) the different distributional patterns of fish larvae proved to depend on the interrelationship of spawning place, vertical distribution of planktonic stages and currents in the respective depths. All this information was gathered during the same survey. For several species the existence of faunal boundaries in or near the investigated area could be explained by the hydrographical structure (John unpublished data).

For the adjacent area north of Cape Blanc hydrographic information of desired quality is available from earlier cruises, but data on fish larvae are far from satisfying, except for strictly neustonic taxa. John (1979) speculated on drift of larvae in this region and discussed mechanisms whereby adult stocks may be maintained in spite of removal of offspring. The theories are:

1) active migration of adults to the spawning places (e.g. Fig. 1);
2) foreign recruitment by offspring born farther upstream (probably effective for the Mauritanian sardine, John et al. 1980);
3) hydrographically-isolated nurseries with little or no net transport (probably effective for *Pagellus acarne*); and
4) diurnal vertical migration between opposite currents, resulting in reduced net transport (probably effective for *Macrorhamphosus scolopax* in its southernmost range).

To test these theories a field experiment was carried out under the direction of H. Weikert during January to March 1983. An array of 36 current meters was moored between 16° and 32°N. Besides sampling the entire area, the area between 76° and 27°N was densely covered by means of CTD casts, Nansen bottle casts for chemistry, and intense plankton sampling with high vertical and horizontal resolution. Information on species composition and spawning habits of adult fishes and on the taxonomy of many of their larvae is rather good for the area (John unpublished data). If, upon analysis of data, the hydrographic conditions turn out to be broadly the same as those encountered between 1970 and 1977, this strategy should allow evaluation of the proposed theories as far as planktonic stages are concerned.

REFERENCES

Angel, M.W. 1979. Zoogeography of the Atlantic Ocean. Pages 168-190 *in* S. Van der Spoel and A.C. Pierrot-Bults, editors. Zoogeography and diversity in plankton. Bunge, Utrecht, The Netherlands.

Badcock, J., and N.R. Merrett. 1976. Midwater fishes in the Eastern North Atlantic I. Vertical distribution and associated biology in 30°N, 23°W, with developmental notes on certain myctophids.

Progress in Oceanography 7:3-58.

Bakun, A., C.S. Nelson, and R.H. Parrish. 1979. Determination of surface drift patterns affecting fish stocks in the California Current upwelling region. UNESCO, Intergovernmental Oceanographic Commission, Workshop Report 17:7-21.

Bakun, A., J. Beyer, D. Pauly, J.G. Pope, and G.D. Sharp. 1982. Ocean sciences in relation to living resources. Canadian Journal of Fisheries and Aquatic Sciences 39:1059-1070.

Bishai, H.M. 1960. The effect of water currents on the survival and distribution of fish larvae. Journal du Conseil Conseil International pour l'Exploration de la Mer 25:134-146.

Brockmann, C. 1979. A numerical upwelling model and its application to a biological problem. Meeresforschung 27:137-146.

Bruun, A. 1935. Flying-fishes (Exocoetidae) of the Atlantic, systematic and biological studies. Dana Report Carlsberg Foundation 6.

DHI. 1967. Monatskarten für den Nordatlantischen Ozean. Deutsches Hydrographisches Institut, Hamburg, Federal Republic of Germany.

Dooley, J.K. 1972. Fishes associated with the pelagic Sargassum complex, with a discussion of the Sargassum community. Contributions in Marine Science 16:1-32.

Ekman, S. 1953. Zoogeography of the sea. Sidgwick and Jackson, London, England.

Fraser, J.H. 1958. The drift of the planktonic stages of fish in the Northeast Atlantic and its possible significance to the stocks of commercial fish. International Commission for the Northwest Atlantic Fisheries Special Publication 1:289-310.

Friedrich, H. 1965. Meeresbiologie. Borntraeger, Berlin, Federal Republic of Germany.

Gorbunova, N.N. 1977. Vertical distribution of the fish larvae in the eastern equatorial Pacific Ocean. Polskie Archivum Hydrobiologii Supplement 24:377-386.

Hamann, I., H.-Ch. John, and E. Mittelstaedt. 1981. Hydrography and its effect on fish larvae in the Mauritanian upwelling area. Deep-Sea Research 28A. Part A. Oceanographic Research Papers 561-575.

Harden Jones, F.R. 1968. Fish migration. Arnold, London, England.

Johannes, R.E. 1978. Reproductive strategies of coastal marine fishes in the tropics. Environmental Biology of Fishes 3:65-84.

John, H.-Ch. 1973. Oberflächennahes Ichthyoplankton der Kanarenstrom-Region. Meteor Forschungsergebnisse Reihe D-Biologie 15:36-50.

John, H.-Ch. 1975. Untersuchungen am oberflächennahen Ichthyoplankton des mittleren und südlichen Atlantischen Ozeans. Doctoral dissertation. Kiel University, Kiel, Federal Republic of Germany.

John, H.-Ch. 1979. Regional and seasonal differences in ichthyoneuston off Northwest Africa. Meteor Forschungsergebnisse Reihe D-Biologie 29:30-47.

John, H.-Ch. 1983. Quantitative distribution of fry of beloniform fishes in the Atlantic Ocean. Meteor Forschungsergebnisse Reihe D-Biologie 36:21-33.

John, H.-Ch., U.J. Böhde, and W. Nellen. 1980. *Sardina pilchardus* larvae in their southernmost range. Archiv für Fischereiwissenschaft 31:67-85.

Johnson, M.W. 1960. Production and distribution of the spiny lobster *Panulirus interruptus* (Randall), with records of *P. gracilis* Streets. Bulletin of the Scripps Institution of Oceanography of the University of California 7:413-461.

Johnson, M.W., and E. Brinton. 1963. Biological species, water masses and currents. Pages 381-414 *in* M.N. Hill, editor. The Sea, volume 2. Wiley, Chichester, England.

Karnella, C., and R.H. Gibbs. 1977. The lanternfish *Lobianchia dofleini:* An example of the importance of life-history information in prediction of oceanic sound scattering. Pages 361-379 *in* N.R. Andersen and B.J. Zahuranec, editors. Oceanic sound scattering prediction. Plenum Press, New York, New York, USA.

Krauss, W., and J. Meincke. 1982. Drifting buoy trajectories in the North Atlantic Current. Nature (London) 296:737-740.

Krefft, G. 1976. Results of the research cruises of FRV "Walther Herwig" to South America. 41. Fishes of the order Beryciformes from the western South Atlantic. Archiv für Fischereiwissenschaft 26:65-86.

Laevastu, T., and M.L. Hayes. 1981. Fisheries Oceanography and Ecology. Fishing News Books, Farnham, England.

Lutjeharms, J.R.E., and A.E.F. Heydorn. 1981. The rock-lobster, *Jasus tristani*, on Vema Seamount: Drifting buoys suggest a possible recruiting mechanism. Deep-Sea Research 28A. Part A. Oceanographic Research Papers 631-636.

Marty, J.J. 1965. Drift migrations and their significance to the biology of food fishes of the North Atlantic. International Commission for the Northwest Atlantic Fisheries Special Publication 6:355-361.

Mittelstaedt, E. 1978. Physical oceanography of coastal upwelling regions with special reference to Northwest Africa. International Council for the Exploration of the Sea. Symposium on the Canary Current: Upwelling and living resources. Number 40.

Nafpaktitis, B.G. 1968. Taxonomy and distribution of the lanternfishes, genera *Lobianchia* and *Diaphus*, in the North Atlantic. Dana Report Carlsberg Foundation 73.

Parin, N.V. 1960. The range of the saury (*Cololabis saira* Brev.-Scomberesocidae, Pisces) and effects of oceanographic features on its distribution. Doklady Akademii Nauk SSSR 130:649-652. English translation by L. Penny, Ichthyological Laboratory, United States National Museum.

Pearcy, W.G., and S.S. Myers. 1974. Larval fishes of Yaquina Bay, Oregon: A nursery ground for marine fishes? US National Marine Fisheries Service Fishery Bulletin 72:201-213.

Peterson, W.T., C.B. Miller, and A. Hutchinson. 1979. Zonation and maintenance of copepod populations in the Oregon zone. Deep-Sea Research 26A. Part A. Oceanographic Research Papers 467-494.

Richardson, S.L., and W.G. Pearcy. 1977. Coastal and oceanic fish larvae in an area of upwelling off Yaquina Bay, Oregon. US National Marine Fisheries Service Fishery Bulletin 75:125-145.

Saville, A. 1965. Factors controlling dispersal of the pelagic stages of fish and their influence on survival. International Commission for the Northwest Atlantic Fisheries Special Publication 6:335-348.

Scheltema, R.S. 1971. The dispersal of the larvae of shoal-water benthic invertebrate species over long distances by ocean currents. Pages 7-28 *in* D.J. Crisp, editor. Fourth European Marine Biological Symposium. University Press, Cambridge, England.

Scheltema, R.S. 1972. Eastward and westward dispersal across the tropical Atlantic Ocean of larvae belonging to the genus (Prosobranchia, Mesogastropoda, Bursidae). Internationale Revue der Gesamten Hydrobiologie 57:863-873.

Scheltema, R.S. 1973. Dispersal of the protozoan *Folliculina simplex* Dons (Ciliophora, Heterotricha) throughout the North Atlantic Ocean on the shells of gastropod veliger larvae. Journal of Marine Research 31:11-20.

Schmidt, J. 1925. The breeding places of the eel. Smithsonian Institution Annual Report (1924):279-316.

Sette, O.E. 1943. Biology of the Atlantic mackerel (*Scomber scombrus*) of North America. Part I: Early life history, including the growth, drift, and mortality of the egg and larval populations. US National Marine Fisheries Service Fishery Bulletin 50:149-237.

Tåning, A.V. 1918. Mediterranean Scopelidae (*Saurus, Aulopus Chlorophthalmus* and *Myctophum*). Report on the Danish Oceanographical Expeditions 1908-10 to the Mediterranean and Adjacent Seas 2 A 7, Biology.

Tåning, A.V. 1937. Some features in the migration of cod. Journal du Conseil Conseil International pour l'Exploration de la Mer 12: 1-35.

Thorson, G. 1961. Length of pelagic larval life in marine bottom invertebrates as related to larval transport by ocean currents. Pages 455-474 *in* M. Sears, editor. Oceanography. Publications of the American Association for the Advancement of Science 67.

Tomczak, M. 1978. Distribution of water masses at the surface as derived from T-S diagram analysis from surface observations in the CINECA area. International Council for the Exploration of the Sea. Symposium on the Canary Current: Upwelling and living resources. Number 45.

Wegner, G. 1973. Geostrophische Oberflächenströmung im nördlichen Nordatlantischen Ozean im Internationalen Geophysikalischen Jahr 1957/58. Berichte der Deutschen Wissenschaftlichen Kommission fur Meeresforschung 22:411-426.

Wiebe, P.H., E.M. Hulburt, E.J. Carpenter, A.E. Jahn, G.P. Knapp, S.H. Boyd, P.B. Ortner, and J.L. Cox. 1976. Gulf Stream cold core rings: large-scale interaction sites for open ocean plankton communities. Deep-Sea Research 23:695-710.

Wilson, D.P. 1952. The influence of the nature of the substratum on the metamorphosis of the larvae of marine animals. Annales de l'Institut Oceanographique 27:49-156.

Wilson, D.P. 1968. Some aspects of the development of eggs and larvae of *Sabellaria alveolata* (L.). Journal of the Marine Biological Association of the United Kingdom 48:367-386.

Worthington, L.V. 1976. On the North Atlantic circulation. Johns Hopkins Oceanographic Studies 6.

Wroblewski, J.S. 1982. Interaction of currents and vertical migration in maintaining *Calanus marshallae* in the Oregon upwelling zone - a simulation. Deep-Sea Research 29A. Part A. Oceanographic Research Papers 665-686.

BEHAVIORAL ENVIROREGULATION'S ROLE IN FISH MIGRATION

William H. Neill

Department of Wildlife and Fisheries Sciences
Texas A&M University
College Station, Texas 77843 USA

ABSTRACT

Migration in fishes logically had its evolutionary origins in the "migration" of required or preferred environments. The extent to which fishes now have come to rely on sign stimuli to guide their migrations must be measured against antecedent responses that are strictly enviroregulatory. Separation of the two kinds of mechanisms will require imaginative application both of emergent modeling techniques and new technologies for data gathering.

MIGRATION VIA BEHAVIORAL ENVIROREGULATION

The adaptive value of migration is for fishes, as for other animals, optimization of surrounding environment. Were the distribution of environment and fishes' environmental needs constant through time, there would be no benefit in moving from place to place. However, the distribution of environment shifts through time in both patterned and random ways; even the fishes themselves alter the spatial distribution of environment, primarily by consuming oxygen and food, attracting predators, and producing wastes. Moreover, the set of environmental conditions that is required or preferred by individual fish changes--from day to night, from summer to winter, and from the time of hatching to adulthood. Thus, the spatial distributions of many fishes change temporally in more or less regular, cyclical patterns that may be termed "migrations."

One must envision an early stage in the evolution of every migratory pattern in which moment-to-moment enviroregulation comprises not only the "goal" of migration but also the set of necessary and sufficient mechanisms for achieving that goal. But, because some important elements of environment not only are scarce but also are distributed in persistent or quasipredictable ways (e.g., suitable spawning grounds), most if not all migratory fishes have come to rely partly on sign stimuli to bias their movements in favor of the appropriate heading. Still, even the remarkable spawning migrations of the Pacific salmons (*Oncorhynchus* spp.) must be strongly influenced by the homing fish's efforts to continuously optimize immediate environment. Among many observations that could be cited in support of this contention is that of Neave (1964): Sockeye salmon (*O. nerka*) homing to the Fraser River from the southern Gulf of Alaska apparently swing somewhat to the north before making a landfall, thus avoiding an area of too-warm water (> 14 or 15 C). In a similar vein Laevastu and Hela (1970) cite an unpublished claim of Michitaka Uda that Pacific salmon in the Bering Sea migrate along a path characterized by abundant food associated with the frontal zone of the Oyashio and North Pacific Drift currents: the location of the front and, thus, the salmon's migratory route vary from year to year in response to changing climatic conditions.

Like the Pacific salmons, the American shad (*Alosa sapidissima*) is an anadromous fish that returns to its natal stream to spawn. Leggett and Whitney (1972) proposed that shad approach their spawning grounds (Atlantic drainages of North America) along paths bounded by the 13 and 18 C sea-surface isotherms; the result is a seasonal progression of spawning, from the southern limit of the range in early spring to the northern limit in late summer. On the basis of additional information, Neves and Depres (1979) have argued that the "migratory corridor" of homing shad is better defined by sea-bottom temperatures between 3 and 15 C, with most fish moving over bottoms between 7 and 13 C.

To more nomadic fishes such as tunas (Scombridae) "home" is not so much a particular bit of geographic space as it is a restricted (but, as the fish grown and mature, perhaps changing) set of preferred physical, chemical, and biotic conditions. Were it not that the requisite set of environmental conditions shifts geographically in regular fashion, owing to the annual cycle of weather, the movements of these fishes might not form the repeatable patterns that we recognize as migrations. Laurs and Lynn (1977) have described the seasonal migration of one tuna, the albacore (*Thunnus alalunga*), into Pacific coastal waters of North America: As seasonal warming during May and June raises surface temperatures off the coast of California to the albacore's preferred level (16-19 C), the fish approach the coast via the Transition Zone between the cool, low-salinity waters of the Pacific Subarctic to the north and the warmer, more saline North Pacific Central waters to the south. The Transition Zone is normally bounded by sharp temperature and salinity gradients, which albacore seem to regard as barriers or demarcations. During a year (1974) when frontal structure was weakly developed, albacore were relatively widely distributed and their migration poorly defined.

For another tuna, the skipjack (*Katsuwonus pelamis*), Barkley et al. (1978) have suggested that the regularly observed westward migration of larger fish from the eastern tropical Pacific (Williams 1970) is a size-dependent change in temperature-oxygen requirements. As the fish grow they must occupy progressively cooler water if they are to avoid metabolic overheating of core tissues (Neill et al. 1976) and so, beginning at a weight of 4-5 kg, must vacate increasingly larger parts of the eastern tropical Pacific in which required temperatures lie at depths below that at which dissolved oxygen is adequate. (Given that this is one potentially clear-cut case of fish migration mandated by ontogenetically changing environmental requirements, it is ironic that simple passive drift in the North Equatorial Current seems to constitute a sufficient mechanism for achieving the observed migration [Seckel 1972].)

MODELING ENVIROREGULATION

The preceding paragraphs were intended to suggest the important role of enviroregulation as a set of mechanisms whereby fishes achieve their migrations. Harden Jones (1968) and Laevastu and Hela (1970), among others, have emphasized the effects of environment on fish movements, but the case for enviroregulation as the mechanism of fish migration has been made most explicitly by Balchen (1976, 1979). He argues that the migratory movements of fish are the outcome of a continuous "process of maximizing 'comfort'," which I take to be the equivalent of enviroregulation. The fish's behavioral response to the joint distribution of environment (apparently including sign stimuli) is always conditional on its physiological state, which in turn has been dictated by prior environmental experience. This, Balchen believes, leads to changing environmental preferences and consequent movement in a direction that tends to increase the comfort state. Balchen (1979) has extended his concept as a mathematical model in which it is assumed that "...the fish searches in the field of comfort in such a way as to move toward the optimum with a speed that is proportional to the gradient of the comfort [hyperspace] with respect to [spatial] location..."

A less encyclopedic but conceptually similar model of fish movements in environmental fields has been built by DeAngelis (1978; see also DeAngelis and Yeh 1984--this volume). DeAngelis' model simulates the movements of fish in two-dimensional space relative to the distribution of multiple environmental variables, for each of which (independently) the fish has a preferred level. The simulated fish swims a stochastic sequence of "steps" biased by the distribution of environment; the result is a migratory track probabilistically consistent with enviroregulation.

Both Balchen's (1979) and DeAngelis' (1978) models beg the issue of exactly how fishes are able to translate their perceptions of environmental distribution relative to the preferred state into appropriate changes of swimming velocity (speed and direction). If one accepts that fishes migrate by enviroregulation (= maximizing comfort), the next question must be, "How do fishes enviroregulate?" Put another way, "How does a fish organize its behavior so that it is most likely to achieve its joint environmental preferendum?"

This question remains largely unanswered. One approach that can suggest answers is diagnostic modeling. For example, I have used computer simulation to explore several models of fish thermoregulatory behavior (Neill 1979). The model that best mimicked the distribution of real fish in one-dimensional temperature gradients invoked "klinokinetic avoidance" behavior--an increase in the probability of changing direction whenever recent experience implies worsening environmental (thermal, in this case) conditions. (A similar model appears consistent with the movements of motile bacteria in chemical gradients: *Salmonella typhimurium* "tumbles" more frequently while ascending gradients of repellents and descending gradients of attractants than while moving in the reverse directions [Tsang et al. 1973].)

Such computer-modeling exercises serve a valuable (and cost-effective) screening function in that they facilitate identification of sufficient mechanisms, but the likelihood that sufficient mechanisms are also necessary can be established only through careful testing in the laboratory and field. Because the relevant responses of fishes are likely to be subtle and to contain a large stochastic component, the full power of modern technology must be brought to bear. This technology--much of which is less than 10 years old--includes automated process control and data-capture systems; satellite imaging and other remote-sensing techniques for synoptic mapping of environment; high-resolution sonar; and microprocessor-based, ultraminiature, multichannel transmitters that can be put in or on free-swimming fish.

RESOLVING COMPONENT MECHANISMS OF MIGRATION

How can models and hardware be used together to better understand the mechanisms of fish migration? Harden Jones (1981, 1984--this volume), in describing emergence of the inertial-guidance hypothesis, has provided an exemplary answer to this question. The generic answer is that movements of migrating fish can be partitioned into components attributable to accepted mechanisms, so that one can focus more closely on what is less certain and even unsuspected.

Models can be used to forecast active movements and drift of migrating fish in the context of observed environmental situations. Fish movements monitored with the aid of technology then can be compared with model predictions. If the researcher is both fortunate and clever, a new and more interesting model may arise from the pattern of residual vectors. In particular, analysis of movements divested of their supposed enviroregulatory and drift components should facilitate the search for unequivocal evidence of bicoordinate navigation in fishes.

REFERENCES

Balchen, J.G. 1976. Principles of migration in fishes. Foundation for Scientific and Industrial Research of The Norwegian Institute of Technology (SINTEF), Report STF48-A76045, Trondheim, Norway.

Balchen, J.G. 1979. Modeling, prediction, and control of fish behavior. Pages 99-146 *in* C.T. Leondes, editor. Control and dynamic systems, volume 15. Academic Press, New York, New York, USA.

Barkley, R.A., W.H. Neill, and R.M. Gooding. 1978. Skipjack tuna, *Katsuwonus pelamis*, habitat based on temperature and oxygen requirements. US National Marine Fisheries Service Fishery Bulletin 76:653-662.

DeAngelis, D.L. 1978. A model for the movement and distribution of fish in a body of water. Oak Ridge National Laboratory, Report ORNL/TM-6310, Oak Ridge, Tennessee, USA.

DeAngelis, D.L., and G.T. Yeh. 1984. An introduction to modeling fish migratory behavior. Pages 445-469 *in* J.D. McCleave, G.P. Arnold, J.J. Dodson, and W.H. Neill, editors. Mechanisms of migration in fishes. Plenum Press, New York, New York, USA.

Harden Jones, F.R. 1968. Fish migration. Arnold, London, England.

Harden Jones, F.R. 1981. Fish migration: strategy and tactics. Pages 139-165 *in* D.J. Aidley, editor. Animal migration. Cambridge University Press, Cambridge, England.

Harden Jones, F.R. 1984. Could fish use inertial clues when on migration? Pages 67-78 *in* J.D. McCleave, G.P. Arnold, J.J. Dodson, and W.H. Neill, editors. Mechanisms of migration in fishes. Plenum Press, New York, New York, USA.

Laevastu, T., and I. Hela, editors. 1970. Fisheries oceanography. Fishing News (Books) Limited, London, England.

Laurs, R.M., and R.J. Lynn. 1977. Seasonal migration of North Pacific albacore, *Thunnus alalunga*, into North American coastal waters: Distribution, relative abundance, and association with Transition Zone waters. US National Marine Fisheries Service Fishery Bulletin 75:795-822.

Leggett, W.C., and R.R. Whitney. 1972. Water temperature and the migrations of American shad. US National Marine Fisheries Service Fishery Bulletin 70:659-691.

Neave, F. 1964. Ocean migrations of Pacific salmon. Journal of the Fisheries Research Board of Canada 21:1227-1244.

Neill, W.H. 1979. Mechanisms of fish distribution in heterothermal environments. American Zoologist 19:305-317.

Neill, W.H., R.K.C. Chang, and A.E. Dizon. 1976. Magnitude and ecological implications of thermal inertia in skipjack tuna, *Katsuwonus pelamis* (Linnaeus). Environmental Biology of Fishes 1:61-80.

Neves, R.J., and L. Depres. 1979. The oceanic migration of American shad, *Alosa sapidissima*, along the Atlantic coast. US National Marine Fisheries Service Fishery Bulletin 77:199-212.

Seckel, G.R. 1972. Hawaiian-caught skipjack tuna and their physical environment. US National Marine Fisheries Service Fishery Bulletin 70:763-787.

Tsang, N., R. Macnab, and D.E. Koshland, Jr. 1973. Common mechanism for repellents and attractants in bacterial chemotaxis. Science (Washington DC) 181:60-63.

Williams, F. 1970. Sea surface temperature and the distribution and apparent abundance of skipjack (*Katsuwonus pelamis*) in the eastern Pacific Ocean, 1951-1968. Inter-American Tropical Tuna Commission Bulletin 15:229-281.

COULD FISH USE INERTIAL CLUES WHEN ON MIGRATION?

F. R. Harden Jones

Ministry of Agriculture, Fisheries and Food
Fisheries Laboratory
Lowestoft, Suffolk NR33 OHT England

ABSTRACT

Plaice (*Pleuronectes platessa*) fitted with acoustic transponding, compass tags, released and tracked at sea with sector-scanning sonar, show a remarkably consistent heading when in midwater and at night under conditions when normal visual and tactile clues are absent. The results from the tracking studies are consistent with the hypothesis that the heading maintained in midwater may be selected by reference to topographical clues before the fish leaves the bottom. The suggestion that the fish could maintain a chosen heading in midwater by reference to intertial clues using the labyrinth as a detector is examined in the light of the known sensitivity of the semicircular canals to angular accelerations. A simple random walk model is used to calculate the extent to which a fish would be carried off course, when subjected to turning movements accompanying undetected angular accelerations of varying duration. It is shown that the spread of headings increases slowly with time for turning movements of high acceleration and short duration, but more rapidly with turning movements of low acceleration and long duration. If a fish were to restrict undetectable, and therefore uncorrectable, turns to a duration of less than 2 s by introducing movements of its own, the probability of the fish being within ±45° of its original heading after 3 h would be 0.46. It is suggested that the labyrinth could provide a limited facility for an inertial guidance system, which could be useful if the heading could be updated and corrected by reference to external clues at regular intervals. For plaice this could be provided by regular movements

---

down into the bottom during the period spent in midwater. But this pattern of behaviour could not account for the consistent headings maintained by Atlantic salmon (*Salmo salar*) which keep close to the surface, and under these circumstances the question arises as to whether the fish is responding to inertial information provided by the orbital wave motion derived from the sea or swell.

INTRODUCTION

At the Lowestoft Laboratory we have been using a high resolution sector-scanning sonar to track individual fish fitted with acoustic transponding tags. Observations have been made in the open sea on plaice, cod, sole, dogfish, eels and salmon. It has been shown that plaice (*Pleuronectes platessa*) (Greer Walker et al. 1978) use the tide when moving from one position to another, following a pattern of behaviour now called "selective tidal stream transport." Fishing experiments (Harden Jones et al. 1979) have shown that plaice behaved in a similar way when on migration between their feeding and spawning areas in the Southern Bight of the North Sea. But although we may have found a mechanism to account for the migration of this species in coastal waters, there are still many problems to be solved both in the sea and in the laboratory (Harden Jones 1980). For example, a fish must join the appropriate tide at the start and leave the system at the end of the journey: timing is of the essence. We are building a large annular tank in which the flow can be reversed to study some of these problems under controlled conditions.

Tidal stream transport could save energy and relieve the migrant of problems of guidance and navigation. Nevertheless, vector analysis of some of the plaice tracks (Greer Walker et al. 1978, their Figs. 5, 11 and 17) showed surprisingly steady headings when plaice were in midwater, both by night and by day. To study the matter further, we asked our colleagues at the Fisheries Laboratory to make a compass tag which would provide on-line information on the heading of a free-swimming fish.

THE COMPASS TAG

Trials of the new compass tag have been described by Mitson et al. (1982). The compass reference is provided by a magnet attached to a slotted disc, which is mounted on a jewelled pivot. A circular array of eight infra-red emittors and detectors is used to determine the heading of the slot, and so that of the fish, which can be assigned to one or other of eight 45° sectors. When this tag is interrogated at 300 kHz, it responds with an easily recognisable signal from which the depth of the fish and the range and bearing from the ship can be determined in the usual way using the sector-scanning sonar. The heading of the fish is indicated by a second signal, the delay between the two identifying the relevant sector. The two signals appear on the B-scan sonar display and, from selected channels of the receiver beam, on an Alden paper recorder. The heading of the fish can also be read directly from a digital decoder.

So far we have tracked seven plaice fitted with compass tags and our colleague, M. Greer Walker, has tracked two Atlantic salmon (*Salmo salar*). Some preliminary results from one of the plaice tracks have been published (Harden Jones 1981; Harden Jones and Arnold 1982).

We have been particularly interested in the heading of plaice in midwater at night, when it would be unlikely that the fish could receive visual or tactile clues. In the tracks that have been analysed in detail a midwater plaice generally headed northeast or southeast. The heading was maintained within one 45° sector for up to 15 min and within two adjacent 45° sectors for over 3 h. These results may underestimate the ability of plaice to hold to a steady course as there were instances when the compass signals were alternating, over a 5 to 10 min period, between two adjacent sectors, suggesting that the fish was keeping to a narrow range of bearings towards the edge of either sector.

The consistency with which the fish held course was quite unexpected, and the results raise two questions: the first, what is the reference used to select the heading that the fish adopts when in midwater; and the second, how is the course adopted subsequently maintained?

THE CHOICE OF THE ORIGINAL COURSE

Several observations have suggested a possible answer to the first question. The picture-forming qualities of the sector-scanner display give a clear plan view of the alignment of sandwaves (down to wavelengths of 1 m) on the bottom. Fish have been tracked while they swam very close to the seabed near the top of the wave crests before "taking-off" into midwater. The directional movement was maintained for 5 to 10 min, and after leaving the bottom the fish kept to the same sector heading for several minutes. A comparison between the fish leaving the bottom and an aircraft leaving a runway was almost irresistable. This led to the suggestion that the heading maintained by plaice in midwater might be related to that held on the bottom shortly before "take-off." The results that have been analysed are consistent with this suggestion.

Topographical features could provide the bottom reference. Sandwaves, whose crests are aligned at 90° to the tidal axis, are a possible clue. The movements of plaice on the bottom have already been noted as being related to features of bottom topography and to sandwaves in particular (Greer Walker et al. 1978, page 82, plaice 9). In the area where the compass-tag tracks were made, the tidal axis was along the line northeast to southwest. A fish swimming across or along the sandwaves might therefore be expected to maintain a northeast to southwest or a northwest to southeast course, respectively, after leaving the bottom for midwater. These possibilities correspond, in part at least, to the northeast and southeast courses actually observed.

MAINTAINING A HEADING WHEN IN MIDWATER

We do not know how a plaice maintains its heading to within ±45° for 3 to 4 h in midwater in the absence of visual and tactile clues. There are several possibilities. The fish could maintain its heading by reference to the earth's magnetic field, by reference to the electrical field induced by the flow of water through the earth's magnetic field, and by reference to mechanical clues associated with the flow of water.

But here I am going to follow Keeton's (1974) suggestion--on which I have already commented (Harden Jones 1984--this volume)--and examine the possibility that a heading originally chosen, perhaps by reference to a topographical feature, could be maintained in midwater using inertial clues. To give a specific example, consider a plaice swimming close to the bottom on a heading of 090° which was maintained by eye as the fish followed the crest of a sandwave. From time to time the visual reference might also be reinforced by tactile clues. If the fish left the bottom and so lost the visual reference and tactile clues, for how long would the fish be expected to maintain the original heading if it were dependent on inertial guidance using the semicircular canals as the angular sensors?

Clearly the fish will be able to maintain its original heading if it corrects for any turning movement in the horizontal plane around its vertical (the dorsal-ventral or Z) axis. Such turns could arise as a consequence of its own swimming movements or its displacement by the medium. There will be some turns which the fish detects and for which it can correct, and others which may be so slow or of such short duration that they are undetected and uncorrected. It is these subliminal turning movements which could take the fish away from the original heading. What are the thresholds for the detection of angular displacements in the horizontal plane?

The plan, properties, and performance of the cupula-endolymph system of the semicircular canals are similar in all vertebrates. This much seems clear in electrophysiological studies on elasmobranchs (Groen et al. 1952) and clinical studies on man (for example, Hulk and Jongkees 1948). The results obtained with fish-labyrinth preparations have been used to interpret those obtained with human subjects (for example, the nature of secondary after-sensations, Groen et al. 1952, page 342). In man the minimum sustained angular acceleration that gives rise to a sense of rotation is about $0.5°/s^2$ (Groen and Jongkees 1948), and it will be assumed that the threshold for a fish is similar. The physical properties of the cupula-endolymph system--which functions as a highly-damped torsion pendulum--ensure that angular accelerations must be of a certain minimum duration before they are detected, and the product of the angular acceleration α in degrees/s^2 and the time t in seconds for detecting a rotation is constant. In man the critical value of αt, the so-called Mulder product, lies between 2 and 3. Figure 1 shows the form of the threshold curve in man, the lower accelerations having to be sustained for a longer period before giving rise to a sense of rotation. The reason that sustained angular accelerations less than $0.5°/s^2$ are not perceived is probably because the extent of the cupula deflection does not lead to a sufficient change in the resting discharge of the appropriate nerve fibres. Let me assure you, from personal experience, that the threshold does exist. I have myself sat in the special revolving chair designed by Byford et al. (1952) and been brought, by a sustained subliminal acceleration, to a substantial angular velocity without any sensation of rotation.

If Figure 1 is taken to represent the threshold curve for a fish, the magnitude of the angles through which it could be turned without detection can be calculated for a range of angular accelerations from $0.25°/s^2$ to $16.0°/s^2$ (Table 1).

The results can be used to estimate the extent to which a fish

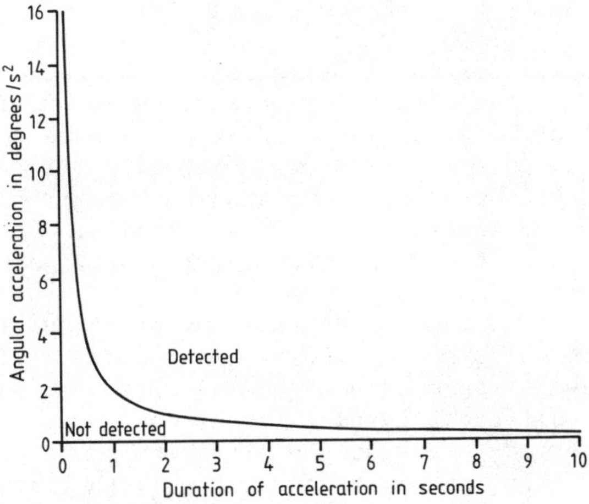

Figure 1. The detection of a sense of rotation in man and its dependence on the magnitude and duration of the angular acceleration.

Table 1. Calculation of the angle θ turned through from rest when a body is subjected to a constant angular acceleration $\alpha°/s^2$ for t seconds so that $\alpha t = 2$ and $\theta° = \frac{1}{2}\alpha t^2$.

Acceleration, $\alpha°/s^2$	16	8	4	2	1	0.5	0.25
Duration, s	0.125	0.25	0.5	1.0	2.0	4.0	8.0
Angle turned through, θ	0.125°	0.25°	0.5°	1.0°	2.0°	4.0°	8.0°

might deviate from its original heading using a simple model in which the fish turns, or is turned, at a particular acceleration until the Mulder product reaches the value 2. The turning movement then stops, and being undetected, is uncorrected. In the model--deliberately choosing the worst possible case--the fish immediately starts, or is subjected to, another turning movement of a similar acceleration and duration; and so on. If the probability of turning is equal for left and right turns, the standard deviation of the spread of turns about the original heading after a given time interval t will be

$$\sigma = \sqrt{pqk} \ ,$$

where p and q are the probabilities of turning left and right, respectively (here $p = q = 0.5$), and k the number of turns in the time interval t. In terms of degrees, the standard deviation will be $\sigma\theta$, where θ is the number of degrees per turn.

Table 2 summarises the calculations for σ where the angular acceleration is $1.0°/s^2$ and the time extends to 60 min. Figure 2 shows the value of σ, in degrees, for the range of accelerations given in Table 1: σ increases slowly with time for turning movements of high acceleration and short duration but more rapidly with turning movements of low acceleration and long duration. It is the latter movements which will carry the fish off its original heading.

The point can be made in another way by calculating the probability with which an individual fish would be within a given number of degrees of an original heading after various time intervals when exposed to different turning movements. A set of calculations for a fish subjected to a sequence of random turns of angular acceleration of $1°/s^2$ and duration of 2 s, with error limits of ±22.5° and ±45.0° (corresponding to containment within one and two 45° compass sectors, respectively), is shown in Table 3. The results, together with those for other turning movements, are shown in Figure 3.

Table 2. Calculation of the standard deviation σ for the distribution of headings about an original direction for a body subjected to a series of equal turns of $\theta°$, with equal probabilities $p = q = 0.5$ of turning left or right, and at a particular turning frequency k for different periods. In the example chosen $\alpha = 1°/s^2$, $t = 2s$, $\theta = 2°$ (see Table 1), $k = 30/min$ and $\sigma = \sqrt{npq}$, where n = number of turns in a given period. In a normal distribution $\pm 1\sigma$ will include 68% of the population, $\pm 2\sigma$ will include 95% of the population, and $\pm 3\sigma$ will include 99% of the population.

Period, min	Number of turns	σ for turns	σ $\theta°$ for degrees
5	150	6	12
10	300	8	16
20	600	12	24
40	1200	17	34
60	1800	21	42

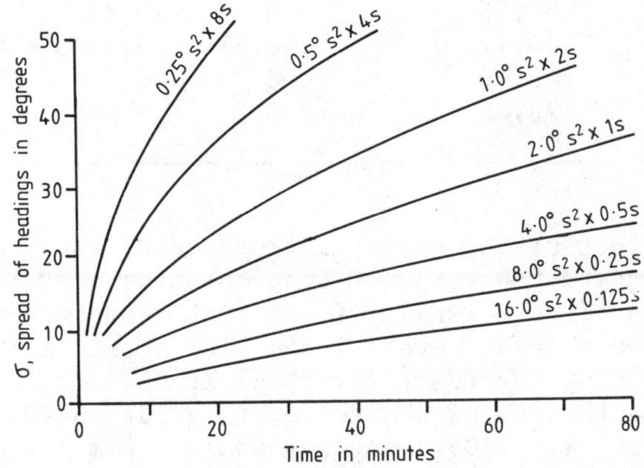

Figure 2. The expected increase in the spread of headings (σ in degrees) with time about an original course when a fish is subjected to undetected random turning movements of a given magnitude and duration.

DISCUSSION

A guidance system based on the performance of the semicircular canals would be at risk to substantial errors introduced by turns of low acceleration which would pass undetected and uncorrected. The lower the angular acceleration, the longer could the turn continue until the critical Mulder product is reached, and the greater will be the angle turned through to the left or to the right. If the

Table 3. The probability with which a fish might be expected to remain within ±22.5° or ±45° of an original course (corresponding to containment within one or two 45° compass sectors) when subjected to undetected random turning movements of 2° at a frequency of 30 turns/min. The unit variance, in degrees, is calculated as shown in Table 2, and the probabilities associated with the quotients 22.5/σ and 45/σ are taken from Table II1 in Fisher and Yates (1963, page 45).

		Probability of remaining within			
Time, min	σ, degrees	one 45° sector		two 45° sectors	
		22.5/σ	P	45/σ	P
5	12.25	1.84	0.93	3.67	1.00
10	17.32	1.30	0.81	2.60	1.00
20	24.50	0.92	0.64	1.84	0.93
40	36.64	0.61	0.46	1.23	0.78
60	42.40	0.53	0.40	1.06	0.71
90	51.96	0.43	0.33	0.87	0.62
100	54.77	0.41	0.32	0.82	0.59
120	60.00	0.38	0.30	0.75	0.55
180	73.48	0.31	0.24	0.61	0.46

cumulative effect of the errors introduced by these movements is to be reduced, undetected turns of long duration must be curtailed. For errors introduced as a result of the fish's own movements, this curtailment could be achieved if the fish itself made regular and detectable turning movements for which it could correct. Such turning movements have been observed in goldfish (*Carassius auratus*) (Kleerekoper et al. 1970), and they would break the continuity of the more persistent turns which will carry the fish off its original heading. For example, if a fish were to restrict undetectable turns to a duration of less than 2 s by introducing movements of its own, at the end of 3 h the probability of the fish being within ±45° of its original heading would be 0.46 (Table 3). It may therefore be possible for the labyrinth to provide a limited facility for an inertial guidance system, which could be useful if the heading could be updated and corrected by reference to external clues at regular intervals. In the case of our plaice this could be by regular movements down onto the bottom during the period spent in midwater.

Movements introduced by the fish would not break the continuity of the slow turns to which it could be subjected by displacement of the water in which it was swimming: this source of error would remain. The extent to which the fish might be carried off course by

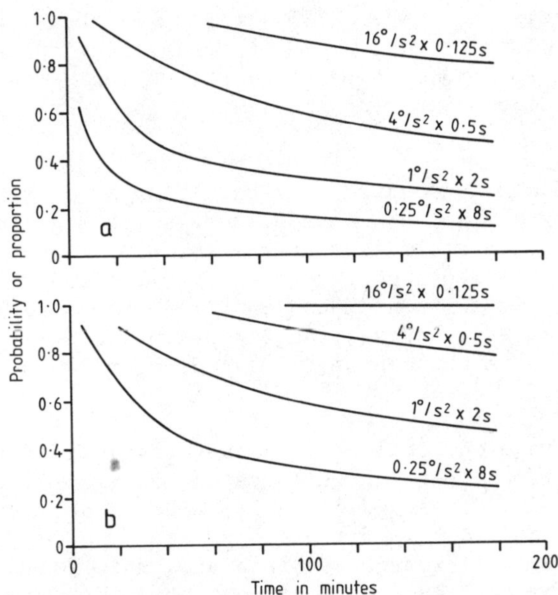

Figure 3. The probability with which a fish might be expected to remain (a) within ±22.5° or (b) ±45° of an original course when subjected to undetected random turning movements of a given magnitude and duration.

such turns will then depend on the turbulent movements of the water at the appropriate scale. It might be unreasonable to have assumed that such movements would be of frequent occurrence or that they would cease when the Mulder product equalled two, just before they would be detected and corrected. If this is true, the difficulties to be faced when a fish tries to maintain a straight course may have been exaggerated.

There are few data to confirm or refute the suggestion that vertebrates have, or make use of, an inertial guidance system. Kleerekoper et al. (1969, 1970) drew attention to the fact that goldfish maintained an approximate balance between the sum of the angles turned to the left or right over periods up to 60 h, which suggests that there must be some form of summation in the central nervous system. Both Keeton (1974, page 79) and Schmidt-Koenig (1979, pages 138-141) refer to evidence that birds continue to fly along straight lines under conditions which would appear to exclude external reference points. The original data are those of Drury and Nisbet (1964) and are based on radar observations on songbird migrants in New England. An important question is whether there is any radar evidence to show that birds above fog and below cloud, at night, can maintain their ground track following a change of wind direction. Drury and Nisbet (1964, page 107) themselves give the answer: "an important test would be to investigate whether birds can remain orientated *between* a layer of ground fog and a layer of high clouds, but ground-based weather stations do not record such situations."

So far as man is concerned, I have always found Schaeffer's (1928) results convincing: blindfolded subjects go round in circles and cannot maintain a straight course for any length of time. I have made similar experiments on myself and my wife and children: muffled and blindfolded we all went round in circles of 8 to 10 m diameter. My first attempt was, however, an exception and I managed to sustain a more or less straight course (±10°) for over 200 m before walking into a hedge.

At this stage one could perhaps be forgiven for echoing Kalmus' (1964) doubts about inertial guidance and navigation were it not for Mittelstaedt and Mittelstaedt's (1980) experiments with gerbils (*Meriones unguiculatus*). If the babies are displaced from their nest, the mother will search for them and, when found, will carry them back to the nest on a more or less straight course. The return appears to be made to the original position of the nest and can be accomplished in complete darkness, and the mother will ignore the real nest if it is displaced close by. If, when the mother is recovering the young, she is rotated at a subliminal acceleration and deceleration before returning, the nest is missed by an angle corresponding to that through which she was turned, whereas an appropriate correction is made for a brisker movement. These observations suggest that inertial clues are normally used in returning to the nest. If confirmed, these results would demonstrate the use of inertial clues in homing, even if on a limited scale: the capacity to use such information would appear to be present.

Finally, I wish to refer to Sergeant's (1964, page 441) remarks on the northward movement of harp seals (*Pagophilus groenlandicus*) in Canadian waters. He reported that a colleague had seen "...schools of adult and immature seals in late April off the Labrador coast swimming northwards on a cloudy night. This was observed from a sealing ship, and other ships in the area reported the same direction of movement. The seals maintained their direction under loose icefields." These observations recall the unpublished results obtained with Atlantic salmon fitted with compass tags and tracked by my colleague, M. Greer Walker. As with the seals, these fish were close to the surface and maintained a consistent direction by day and by night, when celestial clues could probably be excluded. Could the sea surface be the reference point? Recalling the use of swell patterns by the Polynesian navigators (Lewis 1978), could both seals and salmon maintain course in a similar way, by using the inertial information provided by the orbital wave motion which would be available, online, in the surface waters to depths depending on the wave period and wave amplitude? We have had some earlier interest in the effect of sea and swell on fish (Harden Jones and Scholes 1980) but only in relation to problems of catch, weather and motion sickness. Our colleague, P. Cook, has examined the possibility that fish might use inertial information derived from sea and swell (Cook 1984--this volume).

REFERENCES

Byford, G.H., C.S. Hallpike, and J.D. Hood. 1952. The design, construction and performance of a new type of revolving chair. Acta oto-laryngologica 42:511-538.

Cook, P.H. 1984. Directional information from surface swell: some possibilities. Pages 79-101 *in* J.D. McCleave, G.P. Arnold, J.J. Dodson and W.H. Neill, editors. Mechanisms of migration in fishes. Plenum Press, New York, New York, USA.

Drury, W.H., and I.C.T. Nisbet. 1964. Radar studies of orientation of songbird migrants in southeastern New England. Bird-Banding 35:69-119.

Fisher, R.A., and F. Yates. 1963. Statistical Tables for Biological, Agricultural and Medical Research. 6th edition, Oliver and Boyd, Edinburgh, Scotland.

Greer Walker, M., F.R. Harden Jones, and G.P. Arnold. 1978. The movement of plaice (*Pleuronectes platessa* L.) tracked in the open sea. Journal du Conseil Conseil International pour l'Exploration de la Mer 38:58-86.

Groen, J.J., and L.B.W. Jongkees. 1948. The threshold of angular acceleration perception. Journal of Physiology 107:1-7.

Groen, J.J., O. Lowenstein, and A.J.H. Vendrik. 1952. The mechanical analysis of the responses from the end-organs of the horizontal semicircular canal in the isolated elasmobranch labyrinth. Journal of Physiology 117:329-346.

Harden Jones, F.R. 1980. The migration of plaice (*Pleuronectes platessa*) in relation to the environment. Pages 383-399 *in* J.E. Bardach, J.J. Magnuson, R.C. May and J.M. Reinhart, editors. Fish behavior and its use in the capture and culture of fishes. International Center for Living Aquatic Resources Management, Manila, Philippines.

Harden Jones, F.R. 1981. Fish migration: strategy and tactics. Pages 139-165 *in* D.J. Aidley, editor. Animal migration. Society for Experimental Biologial Seminar Series 13: Cambridge University Press, Cambridge, England.

Harden Jones, F.R. 1984. A view from the ocean. Pages 1-26 *in* J.D. McCleave, G.P. Arnold, J.J. Dodson and W.H. Neill, editors. Mechanisms of migration in fishes. Plenum Press, New York, New York, USA.

Harden Jones, F.R., and G.P. Arnold. 1982. Acoustic telemetry and the marine fisheries. Pages 75-93 *in* C.L. Cheeseman and R.B. Mitson, editors. Telemetric studies of vertebrates. Symposia of the Zoological Society of London 49. Academic Press, London, England.

Harden Jones, F.R., and P. Scholes. 1980. Wind and the catch of a Lowestoft trawler. Journal du Conseil Conseil International pour l'Exploration de la Mer 39:53-69.

Harden Jones, F.R., G.P. Arnold, M. Greer Walker, and P. Scholes. 1979. Selective tidal stream transport and the migration of plaice (*Pleuronectes platessa* L.) in the southern North Sea. Journal du Conseil Conseil International pour l'Exploration de la Mer 38:331-337.

Hulk, J., and L.B.W. Jongkees. 1948. The turning test with small regulable stimuli. II. The normal cupulogram. Journal of Laryngology and Otology 62:70-75.

Kalmus, H. 1964. Comparative physiology: navigation by animals. Annual Review of Physiology 26:109-130.

Keeton, W.T. 1974. The orientational and navigational basis of homing in birds. Pages 48-132 *in* D.S. Lehrman, J.S. Rosenblatt, R.A. Hinde and E. Shaw, editors. Advances in the study of behavior 5. Academic Press, New York, New York, USA.

Kleerekoper, H., A.M. Timms, G.F. Westlake, F.B. Davy, T. Malar, and V.M. Anderson. 1969. Inertial guidance system in the orientation of the goldfish (*Carassius auratus*). Nature (London) 223:501-502.

Kleerekoper, H., A.M. Timms, G.F. Westlake, F.B. Davy, T. Malar, and V.M. Anderson. 1970. An analysis of locomotor behavior of goldfish (*Carassius auratus*). Animal Behaviour 18:317-330.

Lewis, D. 1978. The voyaging stars, secrets of the Pacific Island navigators. Collins, Sydney, Australia.

Mitson, R.B., T.J. Storeton-West, and N.D. Pearson. 1982. Trials of an acoustic transponding fish tag compass. Biotelemetry and Patient Monitoring 9:69-79.

Mittelstaedt, M-L., and H. Mittelstaedt. 1980. Homing by path integration in a mammal. Naturwissenschaften 67:566-567.

Schaeffer, A.A. 1928. Spiral movements in man. Journal of Morphology 45:293-398.

Schmidt-Koenig, K. 1979. Avian orientation and navigation. Academic Press, New York, New York, USA.

Sergeant, D.E. 1964. Migrations of harp seals *Pagophilus groenlandicus* (Erxleben) in the Northwest Atlantic. Journal of the Fisheries Research Board of Canada 22:433-464.

DIRECTIONAL INFORMATION FROM SURFACE SWELL: SOME POSSIBILITIES

P. H. Cook[1]

Ministry of Agriculture, Fisheries and Food
Fisheries Laboratory
Lowestoft, Suffolk NR33 OHT England

ABSTRACT

Calculations using first order wave theory show that the accelerative forces produced below a single repeating linear wave with a height of 2 m and a period of 10 s have values that might reasonably be detected by the labyrinth of a typical open ocean migrant. These forces are examined in relation to both a passive and a swimming fish, and it is shown that the direction of the swell could be detected by changes in the phase relationships of the horizontal and vertical components of acceleration acting on the fish.

On the assumption that a typical open ocean migrant is able to maintain a consistent orientation in relation to the direction of the swell, the movements of adult Atlantic salmon (*Salmo salar*) and smolts are projected from an analysis of swell and current patterns in the North Atlantic Ocean. The calculations assume that fish travel in the direction of swell propagation and that swimming speed is expressed as the number of rectangles of 5° latitude by 5° longitude crossed in 28 days. Adult salmon are assumed to swim at rates of four and six rectangles per month and smolts at a rate of one rectangle per month. The effect of currents is ignored for adult salmon but is reflected in the resultant direction of smolt movement.

---

1 Natural Environment Research Council postgraduate research student based at Fisheries Laboratory, Lowestoft. Present address Institute of Oceanographic Sciences, Wormley, Godalming, Surrey GU8 5UB England.

Errors in the calculation affecting the rate of transit are shown to be primarily due to the decrease in zonal distances between the centres of adjacent rectangles with increasing latitude.

Migration paths are predicted for adult salmon leaving southwest Greenland and for smolts leaving northeast Scotland at the beginning of each calendar month. Eastward movement of adult salmon towards Britain and western Europe is predicted during summer and autumn. Westward movement towards Canada is confined to the spring, while adult fish leaving in winter should move to the north towards the Arctic. A northward movement of smolts into the Norwegian Sea is predicted throughout the year except during the spring. Subsequent movement through the Denmark Strait is shown to be possible after a period of 10 months.

Future work needed to elaborate the details of a possible mechanism is discussed. Proposals are made for: 1) further theoretical studies of the accelerative forces acting on the fish; 2) experimental studies to determine the threshold of response to linear accelerations; 3) tracking work in the open sea to describe the orientation of fish to swell waves; and 4) computer studies to develop predictive statistical models of fish distributions using ocean swell and current data.

INTRODUCTION

Swell is a dominant feature of the surface of the world ocean. Arising as sea waves generated by local wind systems, swell waves radiate out from their source and can travel hundreds of kilometres before breaking on a shore remote from their site of origin. The key characteristic of swell waves is that their magnitude and direction can be predicted on a seasonal basis for different parts of the world. Indeed the consistency in the patterns produced by the swell was used for many hundreds of years by the Polynesian Islanders as a crude but effective form of navigation (Lewis 1978).

The passage of a swell wave induces motion in the water particles beneath the wave. In the ocean this motion dominates the upper 30 m and has been well described in standard textbooks on wave and fluid mechanics. The essential features of the subsurface motion are shown schematically in Figure 1. The water particles describe closed circular orbits in the vertical plane perpendicular to the wave crest. The diameter of the orbits decreases exponentially with depth and the time taken to complete one revolution is equal to the wave period.

For fish migrating in the surface waters of the open ocean (for example, salmon, tuna, herring) the forces induced by these motions might provide the fish with on-line directional information. This

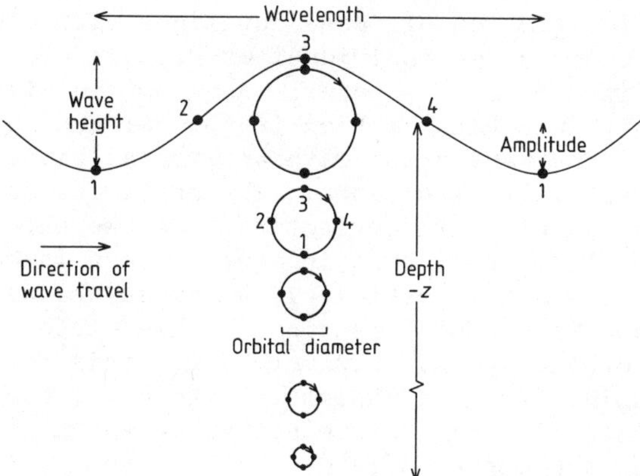

Figure 1. The orbital motions of water particles under a swell wave.

paper examines the physical stimuli available beneath surface swell waves and discusses how directional information might be obtained from them and incorporated into a guidance mechanism for migrating fish.

THEORY

Water Motion Under A Wave

To a first approximation, a surface swell wave can be described by a velocity potential ($\emptyset$) derived from first order potential flow wave theory as:

$$\emptyset = \frac{ag}{\sigma} \cdot \frac{\cosh k\ (h+z)}{\cosh kh} \cdot \cos\ (kx - \sigma t), \qquad (1)$$

where a is the wave amplitude, g is the acceleration due to gravity (9.81 m s^{-1}), σ is the wave frequency ($\sigma = 2\pi/T$, where T is the wave period), k is the wave number ($k = 2\pi/L$, where L is the wavelength), h is the depth of water, z is the depth (negative) below the surface and ($kx - \sigma t$) is the phase angle. Deriving equation (1) with respect to x and z components, the horizontal (U) and vertical (W) velocities of a water particle defined at any point in time and space under the wave are given by:

$$U = -\frac{\partial\emptyset}{\partial x} = \frac{agk}{\sigma} \cdot \frac{\cosh k(h+z)}{\cosh kh} \cdot \sin\ (kx - \sigma t); \qquad (2)$$

and

$$W = -\frac{\partial\emptyset}{\partial z} = \frac{agk}{\sigma} \cdot \frac{\sinh k(h+z)}{\cosh kh} \cdot \cos\ (kx - \sigma t). \qquad (3)$$

Equations (2) and (3) represent the motions of the fluid under the wave. The integration of these equations with time defines the successive positions of an individual water particle in space in relation to a fixed reference point. As shown in Figure 1 the particles exhibit an orbital motion in the vertical plane perpendicular to the wave crest with an orbital diameter that decreases progressively with depth. The particles do not rotate, however, but maintain a fixed orientation as the wave passes. The time for one complete revolution is equal to the wave period. The velocity of any one water particle is given as the vector sum of equations (2) and (3). The direction of this vector is always at a tangent to the orbital path. However, since this direction is constantly changing from one instant to the next, the particle experiences a rate of change of velocity, or an acceleration, directed towards the centre of the orbit. This acceleration is the centripetal force. Figure 2 illustrates the vectors involved. The horizontal (U') and vertical (W') components of this acceleration can be defined as:

$$U' = -agk \cdot \frac{\cosh k(h+z)}{\cosh kh} \cdot \cos(kx - \sigma t); \tag{4}$$

and

$$W' = -agk \cdot \frac{\sinh k(h+z)}{\cosh kh} \cdot \sin(kx - \sigma t). \tag{5}$$

Over a single wave period the horizontal and vertical components of velocity and acceleration fluctuate with a sinusoidal pattern as

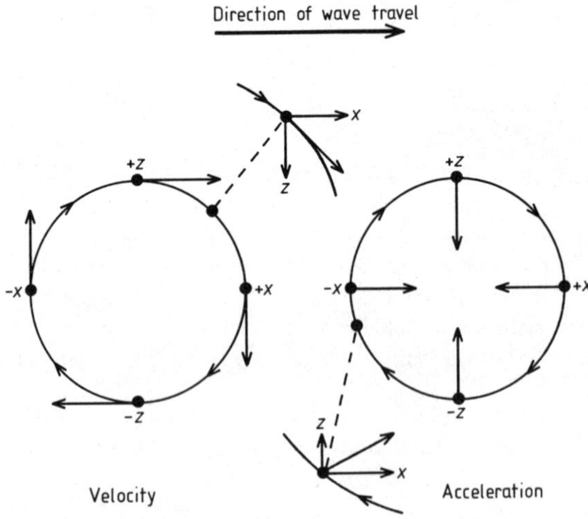

Figure 2. Velocity and acceleration vectors experienced by a water particle at different positions on the orbital path. The vectors can be represented by horizontal (x) and vertical (z) components.

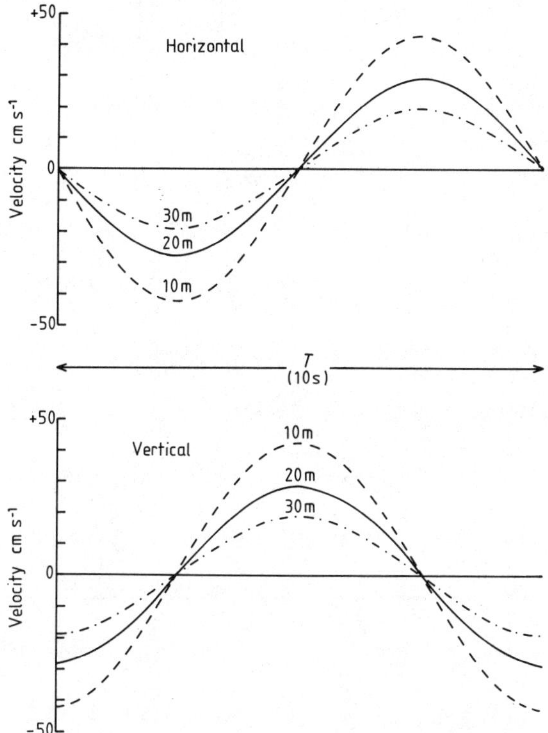

Figure 3. Fluctuations in the vertical and horizontal components of the velocity of a water particle under a wave with a height of 2 m and a period of 10 s at mean depths of 10, 20 and 30 m below the surface.

shown for a series of depths in Figures 3 and 4. There is a 90° phase lag between the two components in each case, with the vertical component reaching a maximal value 90° ahead of the horizontal one. The form of this fluctuation is maintained at different depths under the wave but the amplitude decreases progressively with depth. Different wave heights and periods do not alter the form of these periodic variations, but they do influence the depth to which the wave-induced motion penetrates the surface layer of the ocean. Figure 5 shows how the values of speed and acceleration decay with depth in relation to wave height.

The Effect of Wave-Induced Water Motions on the Fish

Within a wave field a fish sufficiently close to the surface will experience a motion induced by the surface waves. The exact nature of this motion is unknown, but it is likely to be a function of the size and swimming speed of the fish and the height and period

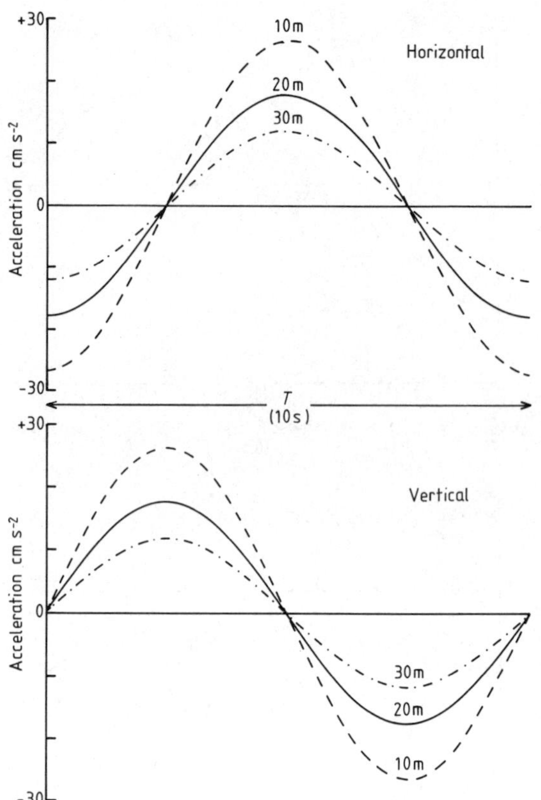

Figure 4. Fluctuations in the vertical and horizontal components of the acceleration of a water particle under a wave with a height of 2 m and a period of 10 s at mean depths of 10, 20 and 30 m below the surface.

of the wave. In the ocean a typical swell wave can have a height of 2 m and a period of 10 s, giving (from first order theory) a wavelength of 156 m and wave celerity of 15 m s^{-1}. A typical open ocean migrant, such as the Atlantic salmon (*Salmo salar*), with a body length of 1 m can maintain a sustained cruising speed of 2 m s^{-1}, equivalent to 12% of the wave speed.

If it is assumed that the wave is sufficiently large compared with the fish, then to a first approximation a passive fish will behave like a water particle. The fish will be translocated in a circular orbit with a tangential velocity equal to that of the water particle as illustrated in Figure 6. Equations (2) to (5) may then be used to describe the movements of the fish and the forces acting on the fish.

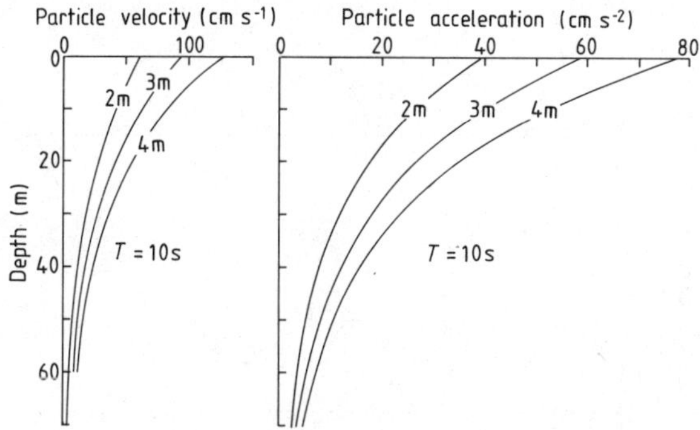

Figure 5. Variations in the velocity and acceleration of a water particle with depth under waves with a period of 10 s and heights of 2, 3 and 4 m.

For a fish moving at significant speed relative to the wave these equations are no longer valid and they must be modified to include a term for the fish's motion. Considering only horizontal movement, the phase angle for a fish swimming at a speed F and at an angle θ to the wave crest becomes $K\sigma t$, where $K = (F\sin\theta/C-1)$ and C is the wave celerity.

Equations (2) to (5) then become:

$$U = \frac{agk}{\sigma} \cdot \frac{\cosh k(h+z)}{\cosh kh} \cdot \sin(K\sigma t) + F; \tag{6}$$

$$W = -\frac{agk}{\sigma} \cdot \frac{\sinh k(h+z)}{\cosh kh} \cdot \cos(K\sigma t); \tag{7}$$

$$U' = Kagk \cdot \frac{\cosh k(h+z)}{\cosh kh} \cdot \cos(K\sigma t); \text{ and} \tag{8}$$

$$W' = Kagk \cdot \frac{\sinh k(h+z)}{\cosh kh} \cdot \sin(K\sigma t). \tag{9}$$

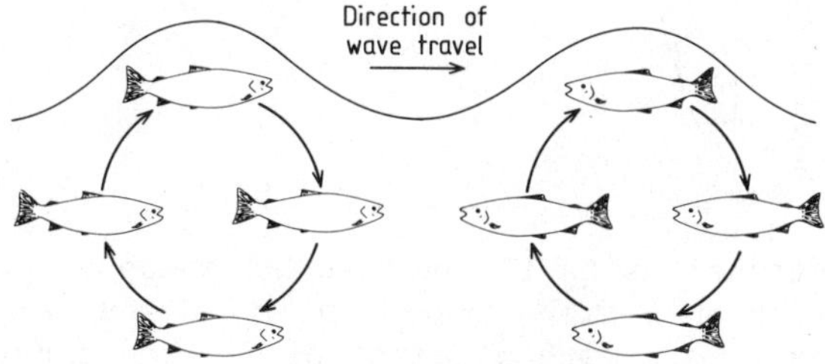

Figure 6. Schematic representation of the motion of a passive fish under a swell wave when heading both with and against the direction of wave propagation.

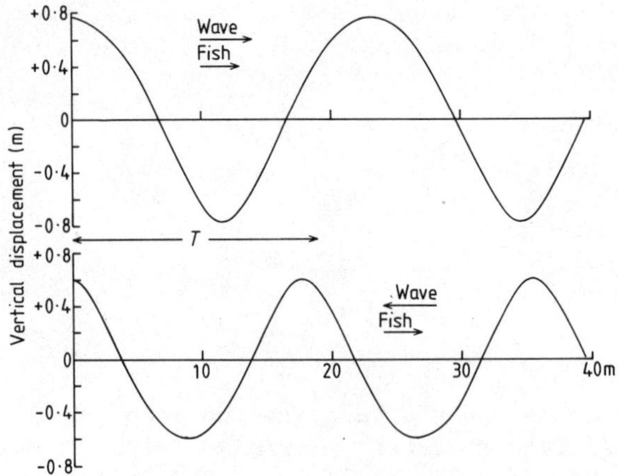

Figure 7. The vertical displacement about a mean depth of 10 m of a fish swimming at a speed of 2 m s^{-1} under a repeating linear wave with a height of 2 m and a period of 10 s. In the upper graph the fish is swimming in the direction of wave propagation; in the lower graph it is swimming against the direction of wave propagation.

The displacement of a fish swimming under a wave can be determined from these equations. Figure 7 shows the vertical displacement of a fish about a mean position for two extreme cases, that when the fish swims in the direction of wave propagation and that when it swims in the opposite direction. In both cases swimming

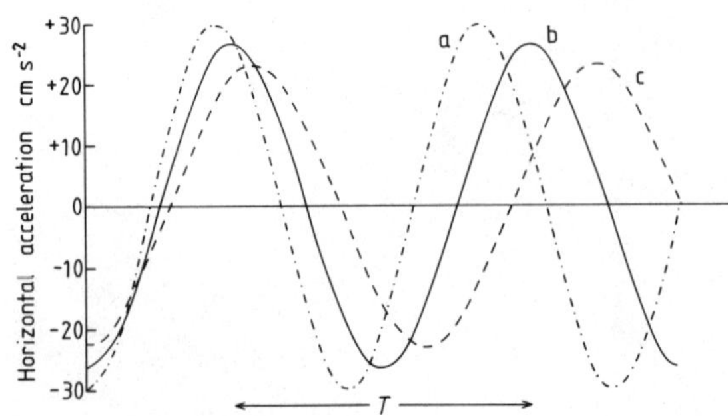

Figure 8. Fluctuations in the horizontal component of acceleration acting on a fish at a depth of 10 m under a repeating linear wave with a height of 2 m and a period of 10 s: (a) fish swimming against the direction of wave propagation at a speed of 2 m s^{-1}; (b) passive fish; (c) fish swimming in the direction of wave propagation at a speed of 2 m s^{-1}.

results in the fish following a distorted sinusoidal path. The amplitude and period of the displacement increase when the fish swims with the wave and decrease when it swims against the direction of wave propagation.

The accelerative force experienced by the fish is also distorted by its movement. Swimming with the wave decreases the magnitude of the force, whereas swimming in the opposite direction increases it (Fig. 8). Swimming in the direction of wave propagation increases the apparent period of the wave; swimming in the opposite direction decreases it.

The Detection of Wave-Induced Motion

The ability of a fish to directly detect wave-induced motion depends on its ability to detect the centrifugal force. This force is equal and opposite to the centripetal force, which directs the fish in a circular orbit beneath the wave. The centrifugal force is a linear acceleration and like other such accelerations should be capable of being detected by the otoliths. Thresholds for the detection of linear acceleration have been calculated for man in both horizontal and vertical planes (Jongkees and Groen 1946; Walsh 1962, 1974; Benson et al. 1975) and range from 6-29 cm s^{-2}. Although Harden Jones (1957) tested blind goldfish (*Carassius auratus*) under varying horizontal accelerations with little response, it is not unreasonable to assume that fish may have a threshold of around 10 cm s^{-2} in the vertical plane, comparable to that of man. With such a threshold value a fish could detect the passage of a typical swell wave at any depth from the surface down to 30 m (Fig. 9). The maximum depth of detection does not appear to be significantly affected by the direction in which the fish swims.

To gain directional information the fish must be able to detect the direction in which it is being translocated, as illustrated in Figure 6. This could be achieved if the fish were to determine the phase difference in the peak values of the horizontal and vertical components of acceleration. When the fish is heading in the same direction as that in which the wave is travelling, the vertical component precedes the horizontal component of acceleration by 90°; when the fish heads against the direction of wave propagation this phase difference is reversed. A further indicator of heading could also be given by a comparison of the periods of rotation of the centrifugal force vector (or the fluctuations in one of its components) experienced during swimming with those experienced when at rest. No direct evidence exists to suggest that fish can in fact discriminate in this manner, but it is not unreasonable to suppose that some such mechanism may exist.

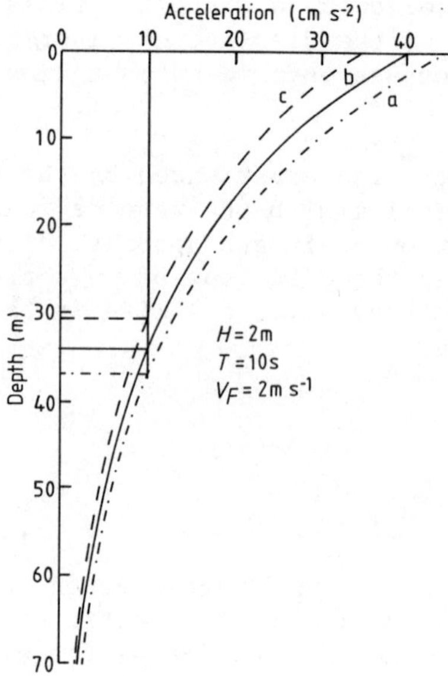

Figure 9. The variation with depth in the peak accelerations experienced by a fish under a wave with a height of 2 m and a period of 10 s: (a) fish swimming against the direction of wave propagation at a speed of 2 m s^{-1}; (b) passive fish; (c) fish swimming in the direction of wave propagation at a speed of 2 m s^{-1}. The maximal depth of detection is shown in each case assuming a detection threshold of 10 cm s^{-2}.

APPLICATION

If fish were able to detect the direction of the swell in the open ocean and to maintain a consistent heading in relation to it, it should be possible to predict migration routes from the swell patterns observed at the appropriate times of the year. This possibility has been explored in relation to the open ocean migrations of the Atlantic salmon.

Migrations of Salmon, Current and Swell Patterns in the North Atlantic Ocean

The movements of salmon in the North Atlantic Ocean have been well documented. Adult salmon tagged on the feeding grounds off southwest Greenland in August have been recaptured around the coastline of western Europe and eastern Canada from February to July of the following year (Fig. 12). Smolts tagged in the British Isles have appeared in Norwegian, Faroes, West Greenland and Nova Scotian waters up to two years later (Fig. 13).

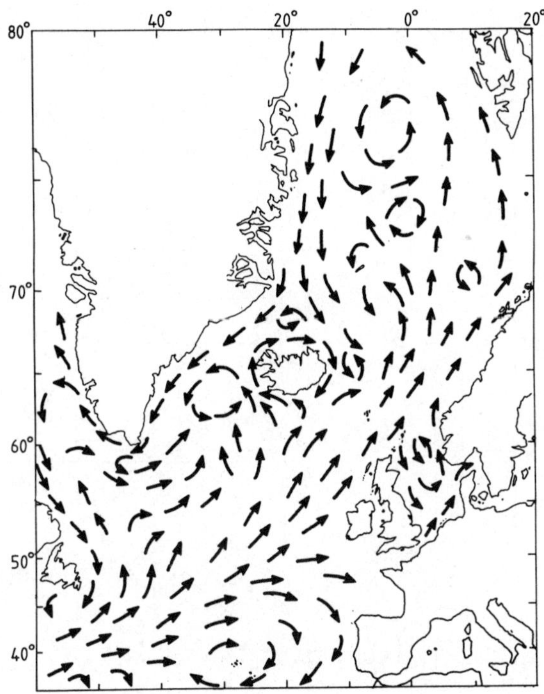

Figure 10. The pattern of the surface water currents in the North Atlantic based on Figures 4, 28 and 44 of Harden Jones (1968).

The surface current pattern of the northern North Atlantic is shown in Figure 10. Water flowing down from Labrador and eastward across the Atlantic is carried in a northeasterly direction by the North Atlantic Drift into the Norwegian Sea and on towards Spitzbergen and the Arctic. Return currents from the Arctic flow south towards Iceland, through the Denmark Strait and westwards around the southern tip of Greenland. Cyclonic gyres exist to the southwest of Iceland in the North Atlantic and to the northeast of Iceland in the Norwegian, Icelandic and Greenland Seas. Typical current speeds range from high values of 3.5 $m\ s^{-1}$ in parts of the Gulf Stream to low values of 0.1 $m\ s^{-1}$ off southwest Iceland. In the Northeast Atlantic speeds are typically 0.15 $m\ s^{-1}$ (Hermann and Thomsen 1946).

The swell direction for the North Atlantic is given in two US Naval Oceanographic Atlases (Anonymous 1958, 1963) for rectangles of 5° latitude by 5° longitude for individual months and by half-cardinal compass points. Above a latitude of 70°N the same data are given in some areas for rectangles of 5° latitude by 10° longitude. The observed mean swell patterns derived from these

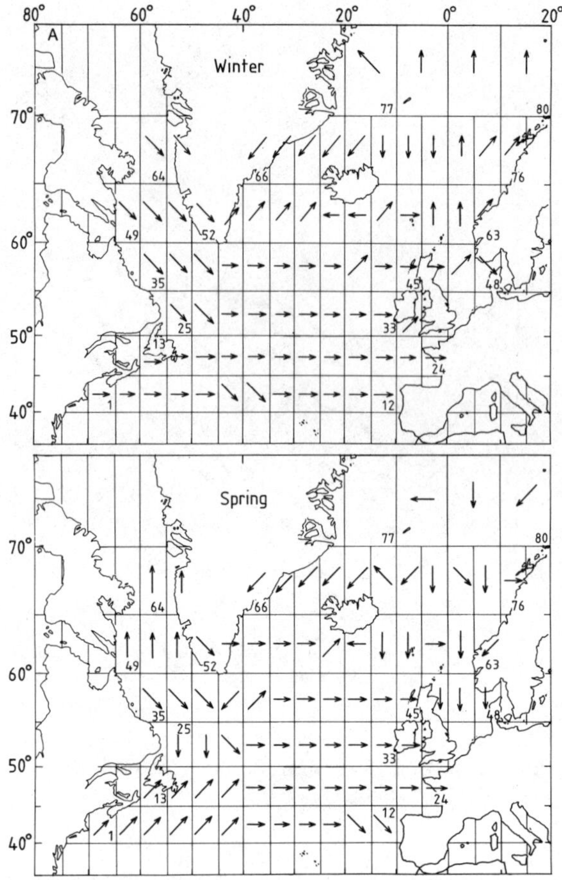

Figure 11. Seasonal mean swell directions in the North Atlantic for rectangles of 5° latitude x 5° longitude based on the data given in the US Naval Oceanographic Atlases (Anonymous 1958, 1963). The data are given for rectangles of 5° latitude x 10° longitude for some areas above a latitude of 70°N. The rectangles are numbered sequentially from west to east and south to north, as indicated. (a) winter, spring; (b) summer, autumn.

atlases show little variation throughout the year (Fig. 11), and indeed between latitudes 40°N and 60°N there is an almost constant eastward heading.

A series of calculations has been performed using these data to investigate how swell patterns might influence the migrations of both smolts and adult salmon in the open ocean.

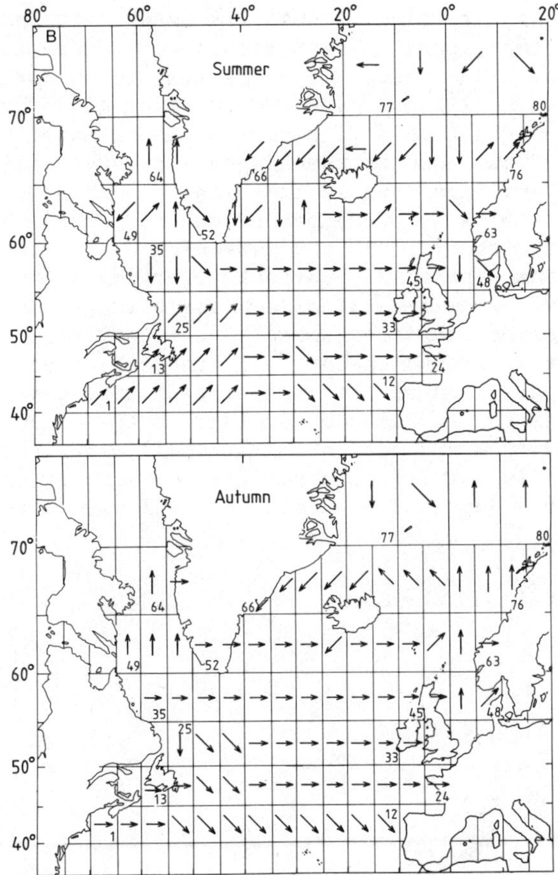

Figure 11 continued.

The Calculations

The calculations are designed to determine the paths which migrating salmon might follow in the open ocean and the transit times associated with them. It is assumed that in all cases the fish swims in the surface water (depth < 30 m) normal to the crest of the swell wave and in the direction of wave propagation.

Swimming speed is expressed as the number of rectangles of 5° latitude by 5° longitude crossed in 28 days. Smolts are assumed to travel at a rate of one rectangle per month and adult salmon at rates of four and six rectangles per month. For adult salmon these rates correspond to swimming speeds of 0.6 to 1.3 m s^{-1}, speeds that can be sustained by a 1-m long fish for an indefinite period. No account has been taken of the effect of the ocean current, so that, for adult salmon, speed and direction through the water correspond to speed and direction over the ground. The direction of movement of an adult fish is thus determined purely by reference to the predicted mean direction of the swell in a given month for a given rectangle. But this assumption is not valid for smolts, whose swimming speeds are very much less than those of the adults and very similar to the speed of 0.15 m s^{-1} typical of the North Atlantic currents. The direction of movement of smolts is therefore calculated as the product of the predicted swell direction and the direction of the current interpolated from Figure 12. Current direction is rounded to the nearest cardinal or half-cardinal compass point to match the swell data. It is assumed that the speed of the current is accommodated within the overall ground speed of one rectangle per month because the speed and direction of the current cannot be determined with sufficient precision to allow a complete vector calculation. The drift resulting from the motion of the swell waves themselves (Stokes' drift) is sufficiently small to be ignored for both smolts and adults. At a depth of 4 m below a swell wave with a height of 2

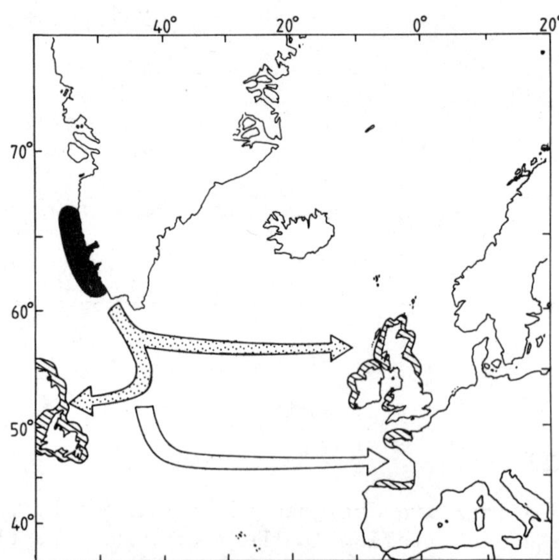

Figure 12. Predicted migration routes for adult Atlantic salmon leaving the feeding grounds off southwest Greenland (Rectangle 52, Fig. 11) based on the data given in Tables 3 and 4. The stippled arrows indicate the predicted routes of salmon swimming at a rate of four rectangles (5° latitude x 5° longitude) per month. The open arrow indicates the predicted route of salmon swimming at a rate of six rectangles per month. The recapture areas for salmon tagged off southwest Greenland are indicated by cross hatching.

m and a period of 10 s it has a speed of approximately 0.03 m s^{-1}.

The assumption that the swimming speed can be expressed as a constant number of rectangles crossed per month introduces an error into the calculation. Distances between the centres of adjacent rectangles on the same meridian are constant but meridional distances between centres decrease with increasing latitude (Table 1). Thus the distance crossed in unit time decreases as a function of latitude and the ground speed varies correspondingly (Table 2). The meridional speed varies by a factor of two between latitudes 45°N-50°N and 70°N-75°N, and there is a corresponding but smaller difference in the equivalent diagonal speeds.

Migration paths have been calculated for adult salmon leaving southwest Greenland and for smolts leaving northeast Scotland at the beginning of each calendar month and the results are shown in Tables 3 to 5. The rectangles of 5° latitude by 5° longitude are numbered sequentially in Figure 11.

Predicted Movements of Adult Salmon in the North Atlantic Ocean

It is clear from Tables 3 and 4 that the migration of adult salmon from southwest Greenland (rectangle 52, Fig. 11) should be directed towards Britain and western Europe throughout most of the year (Fig. 12), reflecting the consistent easterly movement of the swell over much of the North Atlantic (Fig. 11). Indeed during late summer and autumn, when salmon are thought to begin their spawning migration from Greenland, only an easterly migration would appear to be feasible. Salmon leaving rectangle 52 between July and October and migrating at a rate of four rectangles per month would arrive at rectangle 45 on the west coast of Scotland after 2 months. Salmon leaving between August and October, but migrating at the higher rate of six rectangles per month, would arrive at the same destination but within 1.5 months. Fish migrating at the faster rate and leaving southwest Greenland in June should take a more southerly route (Fig. 12) and arrive off the Bay of Biscay after 2 months. Those leaving in July should take a more northerly route passing to the east of Iceland before entering the northern North Sea via the Faroes. The effect of the higher rate of migration is thus to make a significant alteration to the destination of salmon leaving southwest Greenland in June and July.

In winter, however, the direction of movement should be very different. Fish beginning their migration in December would appear to be caught about 1 month later in a 'swell gyre' off southwest Iceland (rectangles 40-42 and 56-58), while fish leaving southwest Greenland in January would appear to move northward between Greenland and Iceland and continue on towards the Arctic.

Westward movement of salmon towards Canada would appear to be

Table 1. Distances between the centres of adjacent rectangles of 5° latitude x 5° longitude for 5° intervals of latitude between 45°N and 75°N. The distances are calculated from the equation:

$$D = R \cos^{-1}[\cos(x_2-x_1) \cos(y_1) \cos(y_2) + \sin(y_1) \sin(y_2)],$$

where D is the distance in kilometres, R is the radius of the earth taken as 6378.16 km, y_1 and y_2 are the latitudes and x_1 and x_2 the longitudes at the centres of the adjacent rectangles expressed in radians.

Direction of measurement	Distances (km)										
	45–50°N		50–55°N		55–60°N		60–65°N		65–70°N		70–75°N
Zonal	376		339		299		257		213		167
Meridional		557		557		557		557		557	
Diagonal		661		641		622		604		588	

Table 2. Ground speeds (km day^{-1}) between the centres of adjacent rectangles of 5° latitude x 5° longitude for 5° intervals of latitude between 45°N and 75°N assuming a transit rate of one rectangle per calendar month.

Direction of measurement	Ground speed (km day^{-1})										
	45–50°N	—	50–55°N	—	55–60°N	—	60–65°N	—	65–70°N	—	70–75°N
Zonal	13		12		11		9		8		6
Meridional		20		20		20		20		20	
Diagonal		24		23		22		22		21	

Table 3. Rectangles of 5° latitude x 5° longitude traversed by adult salmon leaving southwest Greenland (Rectangle 52, Fig. 11) at the start of each calendar month and swimming normal to the direction of swell propagation at a rate of four rectangles per month. The calculations were completed for eight successive 7-day periods or until the fish reached a rectangle containing a coastline or reached latitude 75°N. The rectangles are numbered in Figure 11.

Month of starting	Numbers of sequential rectangles crossed in each of eight successive 7-day periods									Approximate track length (km)	Mean swimming speed ($m\ s^{-1}$)
	0	1	2	3	4	5	6	7	8		
January	52	38	39	55	68	69	77	-	-	2948	0.81
February	52	38	28	29	30	31	32	33	-	2958	0.70
March	52	38	39	40	41	31	32	33	-	2838	0.67
April	52	38	26	25	-	-	-	-	-	1602	0.88
May	52	-	-	-	-	-	-	-	-	-	-
June	52	38	26	15	27	39	40	41	57	4342	0.90
July	52	38	39	40	41	42	43	44	45	2715	0.56
August	52	38	39	40	41	42	43	44	45	2715	0.56
September	52	38	39	40	41	42	43	44	45	2715	0.56
October	52	38	39	40	41	42	43	44	45	2715	0.56
November	52	38	28	29	30	31	32	33	-	2958	0.70
December	52	38	39	40	41	57	56	40	41	3319	0.69

Table 4. Rectangles of 5° latitude x 5° longitude traversed by adult salmon leaving southwest Greenland (Rectangle 52, Fig. 11) at the start of each calendar month and swimming normal to the direction of swell propagation at a rate of six rectangles per month. The calculations were completed for twelve successive 4.7-day periods or until the fish reached a rectangle containing a coastline or reached latitude 75°N. The rectangles are numbered in Figure 11.

Month of starting	Numbers of sequential rectangles crossed in each of twelve successive 4.7-day periods													Approximate track length (km)	Mean swimming speed ($m\ s^{-1}$)
	0	1	2	3	4	5	6	7	8	9	10	11	12		
January	52	38	39	55	68	55	68	69	77	-	-	-	-	2948	0.91
February	52	38	28	29	30	31	32	33	-	-	-	-	-	2958	1.04
March	52	38	39	40	41	42	43	44	45	-	-	-	-	2715	0.84
April	52	38	26	25	-	-	-	-	-	-	-	-	-	1602	1.32
May	52	-	-	-	-	-	-	-	-	-	-	-	-	-	-
June	52	38	26	15	27	17	18	19	20	21	22	23	24	5774	1.19
July	52	38	39	40	41	57	58	59	72	60	61	62	-	4330	0.97
August	52	38	39	40	41	42	43	44	45	-	-	-	-	2715	0.84
September	52	38	39	40	41	42	43	44	45	-	-	-	-	2715	0.84
October	52	38	39	40	41	42	43	44	45	-	-	-	-	2715	0.84
November	52	38	28	29	30	31	32	33	-	-	-	-	-	2958	1.04
December	52	38	39	40	41	42	58	57	56	40	41	42	58	4796	0.99

Table 5. Rectangles of 5° latitude x 5° longitude (5° latitude x 10° longitude above 70°N), traversed by smolts leaving northeast Scotland (Rectangle 46, Fig. 11) at the start of each calendar month and swimming normal to the direction of swell propagation at a rate of one rectangle per month. The calculations were completed for ten successive 28-day periods or until the fish reached a rectangle containing a coastline or reached latitude 75°N. The rectangles are numbered in Figure 11.

Month of starting	Numbers of sequential rectangles crossed in each of ten successive 28-day periods											Approximate track length (km)	Mean swimming speed ($m\ s^{-1}$)
	0	1	2	3	4	5	6	7	8	9	10		
January	46	62	74	74	74	78	-	-	-	-	-	1767	0.15
February	46	62	74	74	74	78	-	-	-	-	-	1767	0.15
March	46	60	72	72	60	-	-	-	-	-	-	1736	0.18
April	46	-	-	-	-	-	-	-	-	-	-	-	-
May	46	-	-	-	-	-	-	-	-	-	-	-	-
June	46	-	-	-	-	-	-	-	-	-	-	-	-
July	46	62	63	76	80	-	-	-	-	-	-	2040	0.21
August	46	62	74	79	78	77	77	70	69	68	55	3991	0.16
September	46	62	75	79	-	-	-	-	-	-	-	1783	0.25
October	46	62	74	79	78	77	70	69	68	55	39	4613	0.19
November	46	62	74	79	78	-	-	-	-	-	-	2070	0.21
December	46	60	61	62	74	74	78	-	-	-	-	2281	0.16

restricted to the spring. A fish leaving southwest Greenland in April would arrive in rectangle 25 off Nova Scotia within 1 month. During May the swell direction in rectangle 52 is northeasterly such that any offshore migration would have to be in the opposite direction to that of swell propagation.

Predicted Movements of Smolts in the North Atlantic Ocean

The predicted movement of the smolts appears to be significantly affected by the current pattern in addition to the swell pattern. Fish leaving rectangle 46 (Fig. 11) on the northeast coast of Scotland between November and March would move to the north and into the Norwegian Sea (Fig. 13) and become entrained in the gyre centred on 70°N, 0° longitude and bounded by the North Atlantic, West Spitzbergen, East Greenland and East Icelandic currents (Harden Jones 1968, his Fig. 28). During the spring, however, the swell direction is to the south and smolts leaving the coast in these months should move into the North Sea (Table 5). But in summer and early autumn after Atlantic salmon smolts have actually entered the sea the swell pattern is directed to the north again. In July a smolt leaving rectangle 46 should cross the northern North Sea and move up the

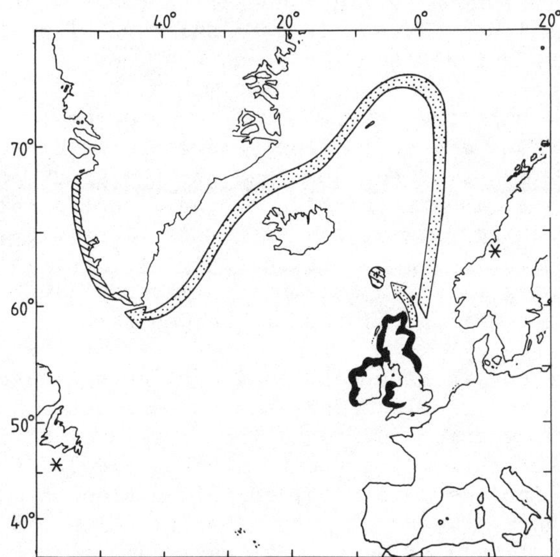

Figure 13. Predicted migration routes for salmon smolts leaving northeast Scotland (Rectangle 46, Fig. 11) and travelling at a rate of one rectangle (5° latitude x 5° longitude) per month based on the data given in Table 5. The recapture area for smolts tagged in the British Isles is indicated by cross hatching; areas from which occasional recaptures are reported are indicated by asterisks.

Norwegian coast arriving off the Lofoten Islands in November. Between August and October the northward movement should occur further to the west in rectangles 62, 74 and 79 taking the fish up to latitudes of 70-75°N. Thereafter movement should be to the west at these high latitudes as far as rectangle 77, thence southwest through the Denmark Strait to rectangles 68 and 55 after a total time of approximately 10 months.

FUTURE WORK

It would appear from the above calculations that wave-induced motion could provide information of use to migrating fish. Much work remains to be done, however, to elaborate the details of a possible mechanism.

Further theoretical work is required firstly to take into account the shape and the mass of the fish--both have been ignored in the present calculations--and secondly to consider the effects of the varying accelerations generated beneath an ocean wave train. The present calculations relate solely to the case of a simple repeating linear wave with a fixed amplitude and wavelenth.

Laboratory experiments are required to determine whether the accelerative forces are sufficiently large to be detected. Seagoing experiments are required to show that ocean migrants can maintain a consistent orientation in relation to the direction of the swell. Some recent tracking work at the Fisheries Laboratory, Lowestoft (M. Greer Walker personal communication) suggests that salmon in the North Sea swimming close to the surface may move in the direction of swell propagation during periods of bad weather.

Finally, there is considerable scope for the development of computer simulation models comparable to that described by Arnold and Cook (1984--this volume) for migration by selective tidal stream transport on the continental shelf. Such a model for the open ocean could usefully include more detailed information on the ocean currents--meanders, rings and gyres (Harden Jones 1984--this volume, his Fig. 2) are likely to be significant--and the consequences of statistical variations in the observed swell patterns. The model should include a piecewise linear interpolation to obviate the constraints imposed by the presentation of the swell data on a rectangular grid. It could also usefully be extended to permit statistical predictions of fish distributions likely to arise from the adoption of this mode of migration. The ability to make such predictions could be of significance to the management of high seas fisheries.

REFERENCES

Anonymous. 1958. Oceanographic Atlas of the Polar Seas. Part II: Arctic. US Navy Hydrographic Office Publication number 705.

Anonymous. 1963. Oceanographic Atlas of the North Atlantic Ocean. Section IV: Sea and Swell. US Naval Oceanographic Office Special Publication number 700.

Arnold, G.P., and P.H. Cook. 1984. Fish migration by selective tidal stream transport: first results with a computer simulation model for the European continental shelf. Pages 227-261 *in* J.D. McCleave, G.P. Arnold, J.J. Dodson and W.H. Neill, editors. Mechanisms of migration in fishes. Plenum Press, New York, New York, USA.

Benson, A.J., E. Diaz, and P. Farrugia. 1975. The perception of body orientation relative to a rotating linear acceleration vector. Pages 264-274 *in* H. Schone, editor. Mechanisms of spatial perception and orientation as related to gravity. Fortschritte der Zoologie 23(1).

Harden Jones, F.R. 1957. Rotation experiments with blind goldfish. Journal of Experimental Biology 34:259-275.

Harden Jones, F.R. 1968. Fish migration. Arnold, London, England.

Harden Jones, F.R. 1984. A view from the ocean. Pages 1-26 *in* J.D. McCleave, G.P. Arnold, J.J. Dodson and W.H. Neill, editors. Mechanisms of migration in fishes. Plenum Press, New York, New York, USA

Hermann, F., and H. Thomsen. 1946. Drift-bottle experiments in the northern North Atlantic. Meddelelser fra Kommissionen for Danmarks Fiskeri og Havundersøgelser. Serie Hydrografi 3(4).

Jongkees, L.B.W., and J.J. Groen. 1946. The nature of the vestibular stimulus. Journal of Laryngology and Otology 61:529-541.

Lewis, D. 1978. The voyaging stars, secrets of the Pacific Island navigators. Collins, Sydney, Australia.

Walsh, E.G. 1962. The perception of rhythmically repeated linear motion in the horizontal plane. British Journal of Psychology 53:439-445.

Walsh, E.G. 1964. The perception of rhythmically repeated linear motions in the vertical plane. Quarterly Journal of Experimental Psychology 49:58-65.

INFLUENCE OF STOCK ORIGIN ON SALMON MIGRATORY BEHAVIOR

Ernest L. Brannon

School of Fisheries
University of Washington
Seattle, Washington 98105 USA

ABSTRACT

Information on the spatial and temporal distribution of Pacific salmon stocks indicates that marine migratory patterns are not random events. Emigration appears to follow an innately directed pattern of movement, while homing is directed by cues acquired during emigration.

INTRODUCTION

In the general commentary on homing of the salmonid fishes one finds differences of opinion as to the degree of orientation and precision of their migration. In the opinion of Saila and Shappy (1963) Pacific salmon (genus *Oncorhynchus*) need only show a slight bias in their oceanic migration to eventually reach the continental coast, and henceforth random choice will eventually lead sufficient numbers to within olfactory detection of their home stream to account for return percentages that characterize marine survival. In contrast when it is demonstrated that juveniles of some species of Pacific salmon are innately oriented during limnetic freshwater migrations (Groot 1965; Brannon 1972; Quinn 1980; Clarke 1981) and use celestial and magnetic cues to negotiate complex migration routes (Quinn 1980; Groot 1982; Quinn and Brannon 1982), it is difficult to believe that adult salmon are not similarly oriented and make use of the same mechanisms in migrations at sea.

This paper presents information indicating that homing in salmon is a rather precisely directed movement in space and time and thus is best described as an oriented migration. Selected examples of different species will be used as evidence of such directed behavior.

SPATIAL DISTRIBUTION IN HOMING BEHAVIOR

The five species of salmon in the Eastern Pacific Ocean distribute differently in both the freshwater and marine environment (Miller and Brannon 1982). Sockeye, pink and chum salmon (*O. nerka, O. gorbuscha, O. keta*) were generally limited only by the spawning area available in their streams, which gave them a large numerical abundance and directed their evolution toward distant migrations to the nutrient-rich waters of the north. Development of an oriented migration appears necessary for these species to have returned to their home stream from distances in excess of 2000 km. Chinook and coho salmon (*O. tshawytscha, O. kisutch*), limited by the productivity of their freshwater rearing habitat, were less abundant and often showed local or intermediate distribution patterns. However, even in these species patterns of movement indicate an oriented migration.

An example of this is with two coho stocks from a major tributary of the Columbia River that show different patterns of marine distribution (T. Rasch personal communication). Based on long-term marine tag return data, Cowlitz River stock makes a predominant contribution to the fishery off the Washington Coast, with a much reduced interception in Oregon and a small catch in British Columbia. Coho from the Toutle River, a tributary to the Cowlitz, distribute much stronger to the south with some caught in Washington, but the majority caught off the coast of Oregon and the remainder in California (Fig. 1). Although both stocks come from the same river system, they segregate in different patterns during marine residence. Transplantation of Cowlitz River coho to other Columbia River hatcheries as a management tool to increase the coho catch off Washington has occurred without noticeably changing the marine distribution of the stock.

Once in the influence of the Columbia River, transplanted groups of Cowlitz stock return to their new home stream rather than the Cowlitz River, which conforms to the evidence that homing in freshwater is directed by odor cues imprinted on as emigrants (Hasler et al. 1978). However, regardless of the change in their freshwater homing migration, coho originating from the Cowlitz River stock continue to migrate in marine waters along the ancestral routes of their kin.

Further evidence on the nature of the marine distribution is provided by a transplant study of chinook salmon. Stock from the Elwha River, which empties into the Straits of Juan de Fuca, were spawned and their eggs transferred 150 km east for incubation and subsequent release at the University of Washington hatchery along with University stock and a hybrid group made from crossing the

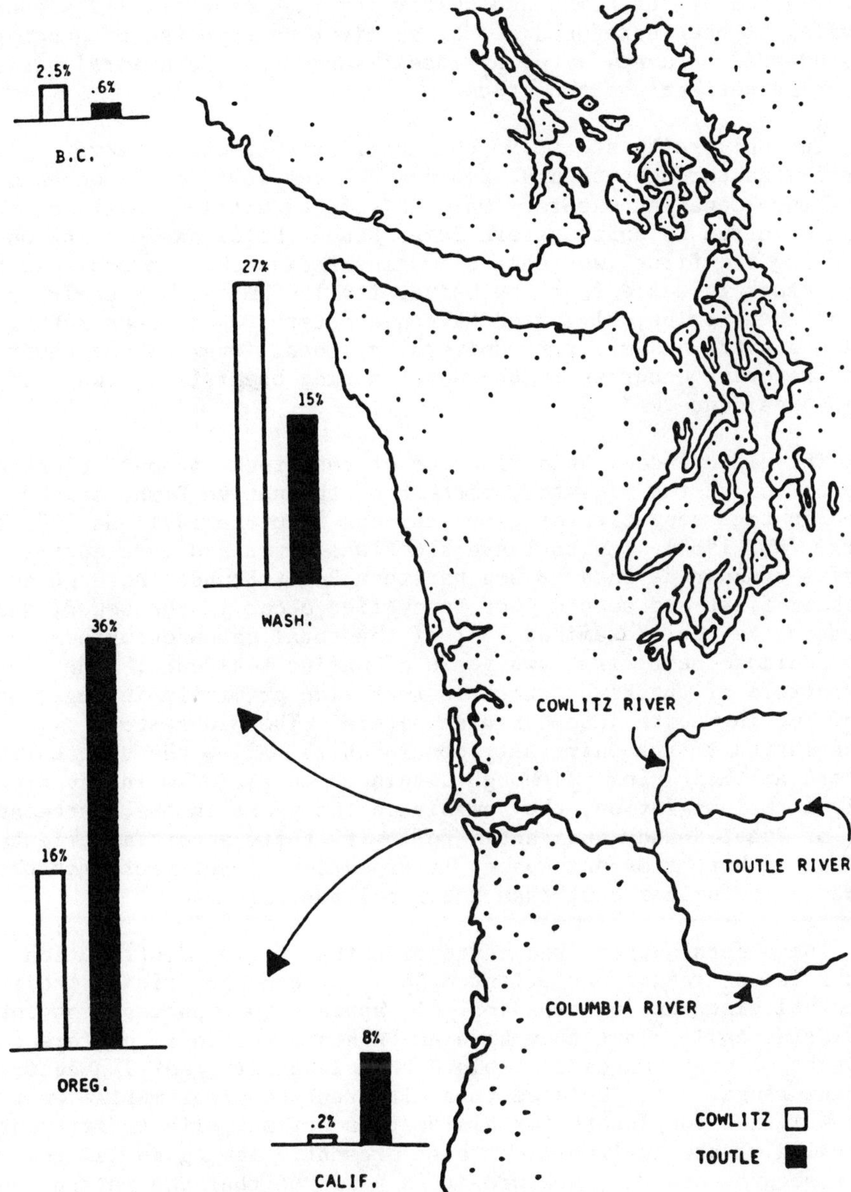

Figure 1. Distribution of marine recoveries of tagged Cowlitz River and Toutle River coho salmon.

stocks (E.L. Brannon and W.K. Hershberger unpublished data). Although the introduced stock showed poorest survival, followed by the hybrid, based on tag recoveries, differences in marine distribution of the Elwha and University fish occurred. The poorer survival of the introduced stock, as might be expected of genotypes less adapted to a new site, by itself suggests a behavioral basis for variation between the two stocks.

The University stock, as characteristic of other years, distributed throughout Puget Sound and moved west to the ocean or north into Canadian waters. Only 18% of the total recovery of the stock occurred in southeastern Puget Sound (below dashed line on Fig. 2) during the first two years of marine residence. In contrast the Elwha stock released from the University 150 km east of their natural marine entry point, clustered in southeastern Puget Sound rather than follow a pattern similar to University stock. Over 45% of their total recovery occurred in the sound during their first two years of marine residence.

The difference in distribution of the stocks is made clear upon examination of the migratory pattern of the native Elwha stock. Based on tag recoveries of Elwha chinook from the 1971 and 1973 brood years[1] (WDF 1983), smolts leave the Elwha River and move east in the Straits of Juan de Fuca to southeastern Puget Sound. As high as 70% of their first and second year recoveries occur in the sound, and based on the years examined, 36% of the total catch occurred in the sound during their first two years of marine residence. Therefore, the pattern of the Elwha stock is to reside primarily in Puget Sound before moving north into Canadian waters. The Elwha stock transplanted to the University continued to follow the same marine pattern as their kin. Although displaced east 150 km in advance of their normal migration, they persisted two years in the southeastern part of Puget Sound, apparently following their ancestral orientation pattern in both time and space, irrespective of the fact that they could go no further east than their release point.

These data support the contention that marine distribution of stocks is not random but oriented in a pattern specific to their ancestral distribution behavior. It appears that once salmon enter the marine environment they have an innate directional preference or sequence of preferences that place them in their general ancestral feeding range. If displaced to a different point of marine entry, they will not compensate for dislocation but may orient differently depending on the environmental cues present. Homing to new release sites occurs, however, because it is believed that the return route to the point of marine entry is guided by clues learned or imprinted on as emigrating smolts.

---

[1] 1983 Pacific Coast data base of coded-wire tagged fisheries recoveries of salmon. Washington State Department of Fisheries, 115 General Administration Building, Olympia, Washington 98504 USA.

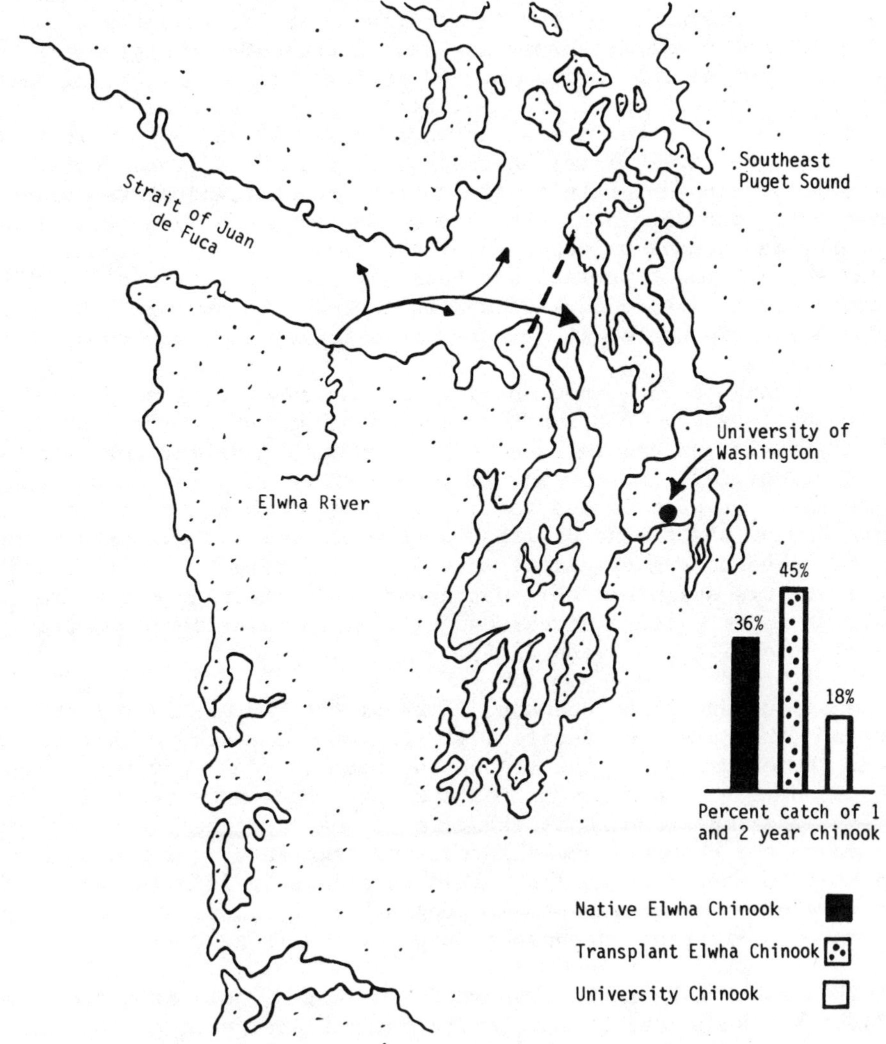

Figure 2. Distribution of native Elwha chinook, and Southeast Puget Sound Catch of Elwha native, Elwha transplant and University chinook during the first two years of marine residence expressed as percent of total harvest for all years.

TEMPORAL DISTRIBUTION IN HOMING BEHAVIOR

Evidence for oriented migrations is even more available in run timing. Precise timing in homing behavior is most easily demonstrated in Fraser River sockeye salmon. Spawning time is a genetic trait selected by emergence timing of the fry, given the incubation temperature of the natal stream, and coordinates the timing of the homeward journey at sea. Because fry emergence is timed to occur at the optimum period in the spring, coinciding with the productivity bloom of the specific nursery lake, sockeye returning to cold streams must spawn earlier and those in warmer streams progressively later to compensate for the differences in embryonic development rate in the corresponding temperature regimes. Consequently the 45 stocks that are monitored annually on the Fraser River for management purposes enter the trunk stream at separate times over a 6-month period, but peak fry emergence from all stocks is condensed to within days of one another in the spring with some stocks having incubated 16 or more weeks longer than others.

This phenomenon is not specific to sockeye but is responsible for the differences in spawning times among spring, summer, and fall chinook, early run and late run coho, and timing differences within the other species also. Other factors such as location at sea and distance to travel will add their influence on when the journey begins and on the timing of freshwater entrance, but ultimately they must be keyed to the synchrony of emergence timing. Moreover, when stocks are transplanted they continue to show their ancestral return time, and under culture situations timing can be manipulated through genetic selection (Feldmann 1974).

Examination of the spawning times of the Fraser River sockeye shows a rather precise specificity from year to year. Within most stocks the annual variation around the mean in peak spawning is less than ±10 days. As one moves from the lower Fraser to the more distant areas upstream, spawning time becomes progressively earlier, related to the change in mean incubation temperature of the spawning streams from warmer to cooler. This relationship is more easily demonstrated by examining stocks geographically close but separated by markedly different incubation temperatures (Fig. 3).

Chilliwack Lake sockeye spawn by the end of July as one of the earliest stocks in the Fraser. With optimum emergence timing around the first of May and with cold incubation water originating from the snow pack, selection has placed spawning very early to acquire sufficient time to gain the necessary temperature units for emergence. In contrast only 30 km from Chilliwack Lake the latest Fraser River stock is found in Cultus Lake spawning on a spring-fed beach irrigated at a constant 8 C. At that incubation temperature spawning occurs at the first of December to synchronize emergence timing with the spring plankton bloom.

Similar differences in timing occur between lake inlet and outlet spawners. Stocks utilizing tributaries or lake inlet streams, such as those shown in Figure 3, must spawn 3-7 weeks sooner than their respective lake outlet stocks because of cooler incubation

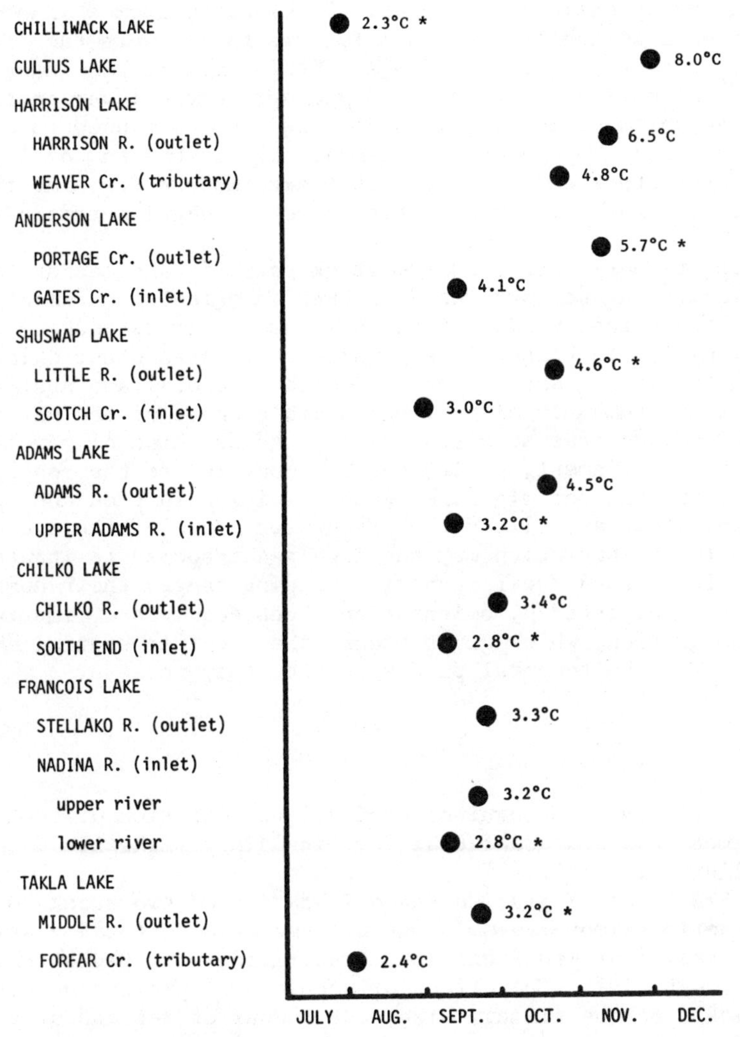

Figure 3. Spawning times and mean incubation temperatures in spawning streams of sockeye salmon stocks returning to the Fraser River system.

temperatures. However, emergence timing of the fry within each inlet-outlet system overlaps and is synchronized with the productivity cycle of the specific nursery lake.

Stocks can be isolated in time even within a single stream by temperature. An example is Nadina River flowing from Nadina Lake to Francois Lake in which two stocks of sockeye are separated in spawning by approximately 2 weeks. Originating from a lake the river is quite warm in August, but cool fall weather lowers the river temperature as the river flows over its 25 km length. The early stock spawns during the first of September in the lower portion of the stream which cools first. As the weather continues to cool, the upper river next to Nadina Lake is suitable at the end of September for the late stock to spawn, and thus accommodate the same fry emergence timing but from slightly warmer incubation temperatures.

These temporal patterns are stock specific and cannot tolerate the variability that anything less than an oriented movement would cause in the spawning time of the numerous discrete sockeye populations in the Fraser River system as well as other salmon stocks along the Pacific coast. Because similar temporal and spatial regularity is found in all salmonids and because of the persistence of stocks to continue ancestral timing and patterns of movement in the marine environment, their behavior conforms to the concept that homing is an oriented migration with precision in time and space in the marine phase as well as the freshwater phase of the cycle. Although smolt emigration may be largely a response to innate influences to reach their ancestral feeding range, their homing migration is directed by orientation clues acquired or imprinted upon during emigration, which gives stocks the specific spatial accuracy necessary for the temporal predisposition they possess.

REFERENCES

Brannon, E.L. 1972. Mechanisms controlling migration of sockeye salmon fry. International Pacific Salmon Fisheries Commission Bulletin 21:1-86.

Clarke, L.A. 1981. Migration and orientation of two stocks of Atlantic salmon (*Salmo salar* L.) smolts. Doctoral dissertation. University of New Brunswick, Fredericton, New Brunswick, Canada.

Feldmann, C.L. 1974. The effect of accelerated growth and early release on the timing, size, and number of returns of coho salmon (*Oncorhynchus kisutch*). Master's thesis. University of Washington, Seattle, Washington, USA.

Groot, C. 1965. On the orientation of young salmon (*Oncorhynchus nerka*) during their seaward migration out of lakes. Behaviour Supplement 14:1-198.

Groot, C. 1982. Modifications on a theme - A perspective on migratory behavior of Pacific salmon. Pages 1-21 *in* E.L. Brannon and E.O. Salo, editors. Proceedings of the Salmon and Trout Migratory Behavior Symposium. University of Washington, Seattle, Washington, USA.

Hasler, A.D., A.T. Scholz, and R.M. Horrall. 1978. Olfactory imprinting and homing in salmon. American Scientist 66:347-355.

Miller, R.J., and E.L. Brannon. 1982. The origin and development of life history patterns in Pacific salmonids. Pages 296-309 *in*

E.L. Brannon and E.O. Salo, editors. Proceedings of the Salmon and Trout Migratory Behavior Symposium. University of Washington, Seattle, Washington, USA.

Quinn, T.P. 1980. Evidence for celestial and magnetic compass orientation in lake migrating sockeye salmon fry. Journal of Comparative Physiology A. Sensory, Neural and Behavioral Physiology 137A:243-248.

Quinn, T.P., and E.L. Brannon. 1982. The use of celestial and magnetic cues by orienting sockeye salmon smolts. Journal of Comparative Physiology A. Sensory, Neural, and Behavioral Physiology 147A:547-552.

Saila, S.B., and R.A. Shappy. 1963. Random movement and orientation in salmon migration. Journal du Conseil Conseil International pour l'Exploration de la Mer 28:153-166.

AN EXPERIMENTAL APPROACH TO FISH COMPASS AND MAP ORIENTATION

Thomas P. Quinn

Department of Fisheries and Oceans
Fisheries Research Branch
Pacific Biological Station
Nanaimo, British Columbia V9R 5K6 Canada

ABSTRACT

A review of field studies indicates that the homeward migrations of maturing Pacific salmon from the open ocean are well timed and oriented. It is hypothesized that salmon can determine their location relative to home and that this map sense is based on the inclination and declination of the earth's magnetic field. In order to test this hypothesis experimental techniques which successfully documented compass orientation in small fishes may be modified to allow adult salmon to display directional orientation. The existence of a map sense could be evaluated by testing salmon after displacement. Subtle, systematic changes in the magnetic field around the arena simulating displacement could provide data with which to evaluate the role of the magnetic field in the map sense.

INTRODUCTION

Traditionally, animal migrations have been studied by observing seasonal abundance patterns and by tagging, banding or marking individuals for subsequent recapture. While valuable for determining general migration routes and timing, these techniques reveal only circumstantial or correlative information on the mechanisms guiding migrations. Kramer's (1949) discovery that caged migratory birds display restless perch-hopping behavior (Zugunruhe) in the compass direction appropriate for seasonal migration provided a powerful tool for the study of migratory orientation mechanisms. Since then experimental arenas have been used to successfully demonstrate compass orientation in a wide variety of animals.

Using various arenas and techniques researchers have established the existence of time-compensating sun orientation in fishes and have determined some of the characteristics of the orientation (Hasler et al. 1958; Hasler and Schwassmann 1960; Schwassman and Hasler 1964; Groot 1965; Goodyear 1970, 1973; Loyacano et al. 1977; Goodyear and Bennett 1979). Recent experiments have shown that juvenile sockeye salmon (*Oncorhynchus nerka*) complement their sun compass with a magnetic compass, which they use at night and during heavy overcast (Quinn 1980, 1982a; Quinn and Brannon 1982).

It would probably be possible to establish sun or magnetic compass orientation in many species of fishes using experimental arenas. However, the hypothesis that migrating animals possess a map as well as a compass (Kramer 1957) is of much greater significance. Research with homing pigeons implies the existence of a map sense or sense of location relative to home, but the physical basis of the map remains in question (see reviews by Schmidt-Koenig 1979; Gould 1982).

The oceanic homing migrations of adult Pacific salmon may cover thousands of kilometers and constitute one of the classic examples of animal migration. This paper assumes that the majority of surviving adult salmon home to spawn in their natal stream and that final home stream selection is based primarily if not exclusively on odors (Hasler et al. 1978). However, the distinction between open ocean orientation and home stream selection in salmon was emphasized by Hasler (1956), and this paper also assumes that these migratory phases are essentially separate, though movement in coastal waters may be guided by both oceanic and riverine mechanisms. Six lines of evidence indicate that salmon movements in the open ocean are directed, not random, and together they argue for the existence of some map sense.

1) Research gillnetting and purse seining have shown that salmon move in particular directions at sea. These directions vary with season, species, location and age class, but are quite predictable (Johnsen 1964; Larkins 1964; Dunn 1969).

2) Fish tagged at a given location at sea disperse to a wide range of home streams (Neave 1964). Indeed, several species represented by many populations may be caught in a single net haul (Royce et al. 1968). Thus, despite common environmental cues, different populations migrate to their respective home streams.

3) Tagging and sampling studies indicate relatively rapid movement to precise coastal areas from broadly defined feeding areas (Neave 1964; Royce et al. 1968; Straty 1975). Members of a population do not travel as a group, but may be found over 3700 km in the east-west axis, experiencing different environmental conditions. They return home with little error, though the homeward directions vary greatly depending on where the maturing salmon are feeding.

4) Individuals in a population converge on the mouth of their home stream with remarkable temporal precision as well. The timing of sockeye salmon runs to rivers in the Fraser River and Bristol Bay

systems is difficult to explain without postulating some form of active orientation (Killick 1955; Gilhousen 1960; Burgner 1980). Between 1956 and 1976, 80% of the Bristol Bay sockeye runs (averaging 17 million salmon) passed a narrow fishery on the average in a 13-day period (Burgner 1980). If the individual populations that comprise the Bristol Bay fishery could be distinguished at sea, the timing of each population might prove to be even more precise than the pooled catch statistics indicate.

5) The rate at which salmon move between mark and recapture sites also indicates that they are actively orienting. Maturing sockeye approaching coastal waters cover 46-56 km/day over many days (Hartt 1966). Ultrasonic tracking of salmon in coastal waters indicates that they swim at about 2 km/hour (Madison et al. 1972; Groot et al. 1975; Stasko et al. 1976), slightly exceeding their most efficient cruising speed (Brett 1965). In order to travel 48 km/day, a salmon swimming 2 km/hour would have to maintain perfect orientation. Even slight decreases in orientation require substantial increases in swimming speed to cover 48 km/day, and the 5 km/hour fatigue speed of a typical adult sockeye (Brett 1983) sets the lower limit on orientation, given the observed travel times (Fig. 1).

6) Experimental studies have demonstrated that open-water movements of sockeye fry and smolts in lakes are aided by celestial and magnetic compass orientation abilities (Groot 1965; Brannon 1972; Quinn 1980; Quinn and Brannon 1982). Evidence has also been presented for compass orientation in adult pink salmon (Churmasov and Stepanov 1977; Stepanov et al. 1979). These studies demonstrate that salmon can select an appropriate compass direction in the absence of rheotactic, thermal, olfactory or salinity cues.

There are many proximate factors influencing the distribution of salmon at sea. Relative densities of food, predators and competitors are undoubtedly important, as are oceanographic features such as currents, salinity and temperature. When these features are abnormally distributed, the spatial and temporal pattern of salmon migrations may be affected. This should not be construed as evidence that these factors actually guide the migrations. While behavioral

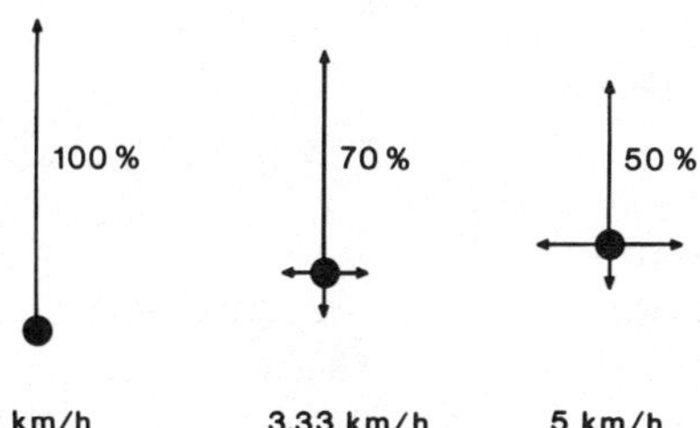

Figure 1. The relationship between degree of directional orientation and swimming speed required to travel 48 km/day.

responses to environmental parameters may influence distribution and migration, active orientation is the most straightforward explanation for the documented movement patterns of salmon. These movement patterns seem best explained by the hypothesis that the salmon possess a map, a compass and a calendar. As was succinctly stated by Larkin (1975), "It thus becomes necessary to postulate that salmon have a bi-coordinate system of navigation that enables them to know where they are and where they are to go (and when to leave in order to get there on time)." This paper outlines a hypothesis first proposed by Quinn (1982b) for the physical basis of map navigation in salmon and describes critical experiments to test the hypothesis. In the final analysis, testability is an essential component of a good hypothesis. While strong indirect evidence indicates that homing pigeons perform true navigation, the sensory basis of the map has been difficult to pin down (see Gould 1982; Walcott 1982).

THE HYPOTHESIS

It is hypothesized that salmon detect the magnetic field's inclination (or dip) and declination (angle between the geographic and magnetic poles) in a system modified from that of Viguier (1882). Inclination isolines are analogous to lines of latitude, roughly concentric around the magnetic poles (Fig. 2). However, because the

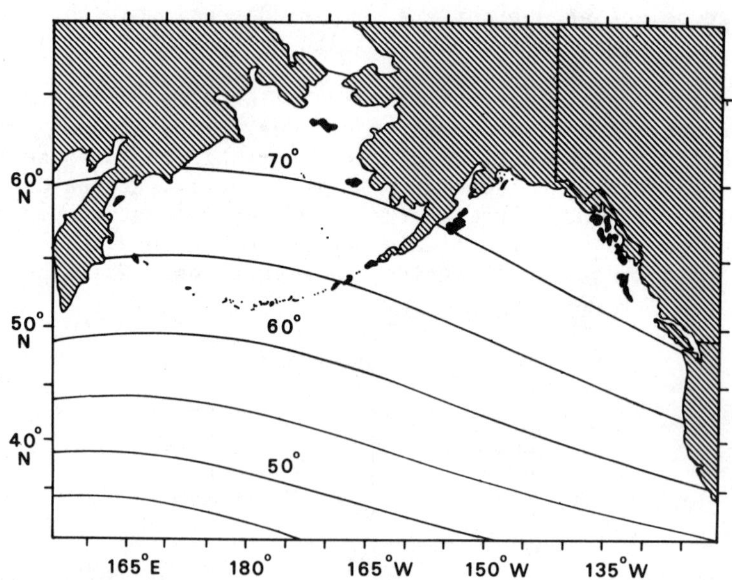

Figure 2. The inclination of the earth's magnetic field in 5° increments in the North Pacific Ocean. Adapted from Quinn (1982).

earth's field has approximately 10% nondipole components, the inclination isolines are not quite parallel. Perception of the magnetic field's vertical component has been demonstrated in birds (Wiltschko and Wiltschko 1972), though the level of sensitivity to inclination differences has not been determined. Experiments with sockeye salmon fry indicated that they use primarily the field's horizontal component and not the vertical component for compass orientation (Quinn et al. 1981). However, the relatively short movements of fry do not necessitate a sense of inclination, and the results do not preclude such sensitivity in adults.

If the geographic and magnetic poles coincided, and the magnetic field was entirely dipole in nature, the field would provide no way to determine position along an east-west axis. However, the north magnetic pole was estimated at 76.2° N, 101° W in 1970, approximately 1500 km from the geographic pole, and the south magnetic pole was estimated at 66° S, 139° E, about 2600 km from the geographic pole. East-west movement at northern latitudes exposes an animal to significant changes in the field's declination (Fig. 3). A sense of declination requires detection of the magnetic field's horizontal component and determination of geographic north. Studies of salmon compass orientation indicate that they can determine both the magnetic and geographic (rotational) poles. The angle between them at any location approximates the field's declination. Error is

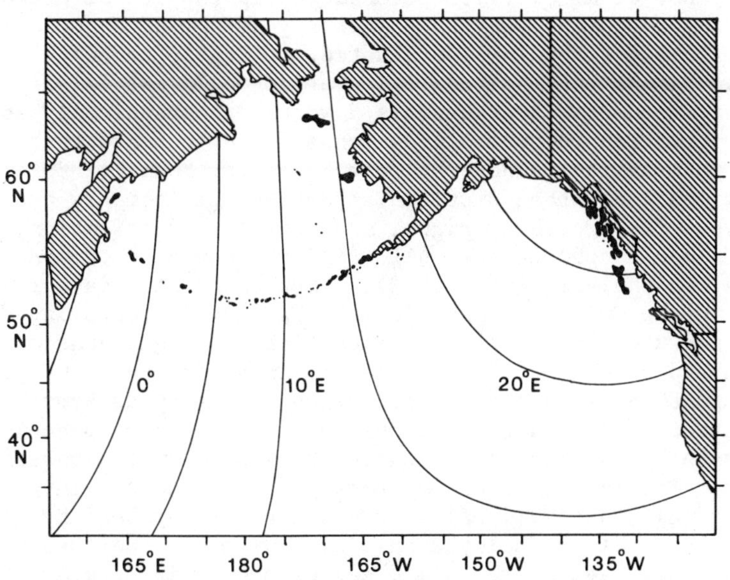

Figure 3. The declination of the earth's magnetic field in 5° increments in the North Pacific Ocean. Adapted from Quinn (1982).

caused by spatial and secular variations in the field of various time periods, some of which were discussed by Quinn et al. (1981).

TESTING THE HYPOTHESIS

The "map navigation hypothesis" states that salmon have a sense of their position relative to home. One competing hypothesis is that they have seasonally changing compass directional preferences, but no map. This may be termed the "sequential compass hypothesis." Evidence indicates that a pattern of changing compass preferences aids migrations of sockeye salmon smolts in lakes (Groot 1965) and long-distance migrations of garden warblers (*Sylvia borin*) (Gwinner and Wiltschko 1978, 1980). As a third alternative salmon may have little or no directional sense but rather attempt to optimize physiologically important environmental parameters such as temperature, salinity and depth (Leggett 1977). By testing salmon in large experimental arenas, it may be possible to falsify or support these hypotheses.

The first priority is to design an arena in which salmon will display meaningful migratory behavior. Most fish orientation arenas have been round, but multi-armed tanks have been used as well. Round tanks encourage circling behavior, unless they are very large or have baffles of some sort. A hexagonal maze, 5 m across, provided encouraging results with adult pink salmon (Churmasov and Stepanov 1977; Stepanov et al. 1979). Because subtle aspects of arena design or recording techniques may substantially affect data quality, a variety of designs should be evaluated. Owing to technical problems failure to demonstrate compass orientation would not falsify the map or sequential compass hypotheses. However, geographically appropriate orientation in controlled arenas would indicate that physiological optimization is not the sole basis for homing. Experimental manipulations could then distinguish between the compass and map hypotheses. The probable origin, and therefore destination, of sockeye salmon caught in the Gulf of Alaska can be determined by scale pattern analysis (Cook 1982). Fish of Bristol Bay origin are abundant in the Gulf of Alaska in late spring, and maturing fish could be captured and selected for testing. Using arenas on the Gulf of Alaska and on one of the Aleutian Islands, it should be possible to distinguish between the map and sequential compass hypotheses. The map hypothesis predicts orientation altered by displacement; the compass hypothesis predicts orientation altered by the passage of time. Specifically, Bristol Bay sockeye salmon tested in arenas on the Gulf of Alaska should maintain a southwesterly heading if they possess a map, but gradually change clockwise to a northeasterly heading if they have a programmed set of sequential compass preferences. Displacement to the Aleutians should not influence sequential compass orientation, but if salmon have a sense of location they should change to northeasterly bearings (Fig. 4).

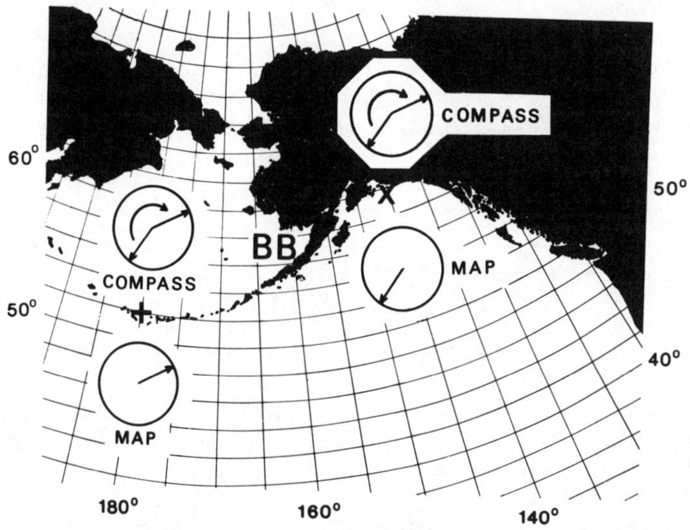

Figure 4. Predicted directional preferences of maturing Bristol Bay (BB) sockeye salmon, caught in the Gulf of Alaska at 55°N, 140°W and tested in arenas at 60°N, 145°W(X) and 52°N, 175°W(+), based on map and sequential compass hypotheses.

If the results of the displacement tests indicated that the salmon changed to a compass preference appropriate for the new location, the specific map hypothesis proposed in this paper could be tested. The cube-surface coil (Rubens 1945) generates a uniform magnetic field inside the coil when connected to a DC power supply. Changes in the horizontal and vertical components of the field generated by the coil could be used to change the declination and inclination and thus simulate displacement. For example, a capture site in the Gulf of Alaska, 55° N 140° W, has a declination and inclination of about 25° E and 70° down, respectively. The Adak Island area, 52° N 175° W (over 2000 km away), has about 10° E declination and 62.5° downward inclination. If salmon tested on the coast of the Gulf of Alaska in the simulated Adak Island magnetic field reversed their orientation, the magnetic map hypothesis would be supported. If the orientation of the salmon changed in proportion to the change in the horizontal field (about 15°), then a magnetic compass, not a map, would be indicated.

The question of how fish migrate long distances through open water is perhaps the most intriguing problem in fish migration research. While field studies have provided essential descriptive information on migratory patterns, the mechanisms underlying the migrations will probably remain unknown until controlled experiments are conducted to evaluate specific hypotheses with testable predictions.

ACKNOWLEDGMENTS

I thank the following individuals for valuable discussions and comments: E. Brannon, R. Burgner, C. Groot, C. Harris, A. Hartt, L. Law, R. Merrill, J. Phillips and J. Waldvogel.

REFERENCES

Brannon, E.L. 1972. Mechanisms controlling migration of sockeye salmon fry. International Pacific Salmon Fisheries Commission Bulletin 21:1-86.

Brett, J.R. 1965. The swimming energetics of salmon. Scientific American 213(2):80-85.

Brett, J.R. 1983. Life energetics of sockeye salmon, *Oncorhynchus nerka*. Pages 29-63 *in* W.P. Aspey and S.I. Luslick, editors. Behavioral energetics: the cost of survival in vertebrates. Ohio State University Press, Columbus, Ohio, USA.

Burgner, R.L. 1980. Some features of the ocean migration and timing of Pacific salmon. Pages 153-164 *in* W.J. McNeil and D.C. Himsworth, editors. Salmonid ecosystems of the North Pacific. Oregon State University Press, Corvallis, Oregon, USA.

Churmasov, A.V., and A.S. Stepanov. 1977. Sun orientation and guideposts of the humpback salmon. Soviet Journal of Marine Biology 3(5):55-63.

Cook, R.C. 1982. Stock identification of sockeye salmon (*Oncorhynchus nerka*) with scale pattern recognition. Canadian Journal of Fisheries and Aquatic Sciences 39:611-617.

Dunn, J.R. 1969. Direction of movement of salmon in the North Pacific Ocean, Bering Sea, and Gulf of Alaska as indicated by surface gillnet catches, 1962-1965. International North Pacific Fisheries Commission Bulletin 26:27-55.

Gilhousen, P. 1960. Migratory behavior of adult Fraser River sockeye. International Pacific Salmon Fisheries Commission Progress Report 7:1-78.

Goodyear, C.P. 1970. Terrestrial and aquatic orientation in the starhead topminnow, *Fundulus notti*. Science (Washington DC) 168:603-605.

Goodyear, C.P. 1973. Learned orientation in the predator avoidance behavior of mosquitofish, *Gambusia affinis*. Behaviour 45: 191-224.

Goodyear, C.P., and D.H. Bennett. 1979. Sun compass orientation of immature bluegill. Transactions of the American Fisheries Society 108:555-559.

Gould, J.L. 1982. The map sense of pigeons. Nature (London) 296: 205-211.

Groot, C. 1965. On the orientation of young sockeye salmon (*Oncorhynchus nerka*) during their seaward migration out of lakes. Behaviour, Supplement 14:1-198.

Groot, C., K. Simpson, I. Todd, P.D. Murray, and G.A. Buxton. 1975.

Movements of sockeye salmon (*Oncorhynchus nerka*) in the Skeena River estuary as revealed by ultrasonic tracking. Journal of the Fisheries Research Board of Canada 32:233-242.

Gwinner, E., and W. Wiltschko. 1978. Endogenously controlled changes in migratory direction of the garden warbler, *Sylvia borin*. Journal of Comparative Physiology A. Sensory, Neural, and Behavioral Physiology 125A:267-273.

Gwinner, E., and W. Wiltschko. 1980. Circannual changes in migratory orientation of the garden warbler, *Sylvia borin*. Behavioral Ecology and Sociobiology 7:73-78.

Hartt, A.C. 1966. Migrations of salmon in the North Pacific Ocean and Bering Sea as determined by seining and tagging, 1959-1960. International North Pacific Fisheries Commission Bulletin 19: 1-141.

Hasler, A.D. 1956. Influence of environmental reference points on learned orientation in fish (*Phoxinus*). Zeitschrift für Vergleichende Physiologie 38:303-310.

Hasler, A.D., R.M. Horrall, W.J. Wisby, and W. Braemer. 1958. Sun-orientation and homing in fishes. Limnology and Oceanography 3: 353-361.

Hasler, A.D., A.T. Scholz, and R.M. Horrall. 1978. Olfactory imprinting and homing in salmon. American Scientist 66:347-355.

Hasler, A.D., and H.O. Schwassmann. 1960. Sun-orientation of fish at different latitudes. Cold Spring Harbor Symposia on Quantitative Biology 25:429-441.

Johnsen, R.C. 1964. Direction of movement of salmon in the North Pacific Ocean and Bering Sea as indicated by surface gillnet catches, 1959-1960. International North Pacific Fisheries Commission Bulletin 14:33-48.

Killick, S.R. 1955. The chronological order of Fraser River sockeye salmon during migration, spawning and death. International Pacific Salmon Fisheries Commission Bulletin 7:1-95.

Kramer, G. 1949. Über Richtungstendenzen bei der nächtlichen Zugunruhe gekäfigter Vogel. Pages 269-283 *in* E. Mayr and E. Schuz, editors. Ornithologie als biologische Wissenschaft. Heidelberg, Federal Republic of Germany.

Kramer, G. 1957. Experiments of bird orientation and their interpretation. Ibis 99:196-227.

Larkin, P.A. 1975. Some major problems for further study on Pacific salmon. International North Pacific Fisheries Commission Bulletin 32:3-9.

Larkins, H.A. 1964. Direction of movement of salmon in the North Pacific Ocean, Bering Sea and Gulf of Alaska as indicated by surface gillnet catches, 1961. International North Pacific Fisheries Commission Bulletin 14:49-58.

Leggett, W.C. 1977. The ecology of fish migrations. Annual Review of Ecology and Systematics 8:285-308.

Loyacano, H.A., J.A. Chappell, and S.A. Gauthreaux. 1977. Sun-compass orientation in juvenile largemouth bass, *Micropterus salmoides*. Transactions of the American Fisheries Society 106:77-79.

Madison, D.M., R.M. Horrall, A.B. Stasko, and A.D. Hasler. 1972. Migratory movements of adult sockeye salmon (*Oncorhynchus nerka*) in coastal British Columbia as revealed by ultrasonic tracking. Journal of the Fisheries Research Board of Canada 29:1025-1033.

Neave, F. 1964. Ocean migrations of Pacific salmon. Journal of the Fisheries Research Board of Canada 21:1227-1244.

Quinn, T.P. 1980. Evidence for celestial and magnetic compass orientation in lake migrating sockeye salmon fry. Journal of Comparative Physiology A. Sensory, Neural, and Behavioral Physiology 137:243-248.

Quinn, T.P. 1982a. Intra-specific differences in sockeye salmon fry compass orientation mechanisms. Pages 79-85 *in* E.L. Brannon and E.O. Salo, editors. Salmon and trout migratory behavior symposium. School of Fisheries, University of Washington, Seattle, Washington, USA.

Quinn, T.P. 1982b. A model for salmon navigation on the high seas. Pages 229-237 *in* E.L. Brannon and E.O. Salo, editors. Salmon and trout migratory behavior symposium. School of Fisheries, University of Washington, Seattle, Washington, USA.

Quinn, T.P., R.T. Merrill, and E.L. Brannon. 1981. Magnetic field detection in sockeye salmon. Journal of Experimental Zoology 217:137-142.

Quinn, T.P., and E.L. Brannon. 1982. The use of celestial and magnetic cues by orienting sockeye salmon smolts. Journal of Comparative Physiology A. Sensory, Neural, and Behavioral Physiology 147:547-552.

Royce, W.F., L.S. Smith, and A.C. Hartt. 1968. Models of oceanic migrations of Pacific salmon and comments on guidance mechanisms. US National Marine Fisheries Service Fishery Bulletin 66:441-462.

Rubens, S.M. 1945. Cube-surface coil for producing a uniform magnetic field. Review of Scientific Instruments 16:243-245.

Schmidt-Koenig, K. 1979. Avian orientation and navigation. Academic Press, London, England.

Schwassmann, H.O., and A.D. Hasler. 1964. The role of the sun's altitude in sun orientation of fish. Physiological Zoology 37: 163-178.

Stasko, A.B., R.M. Horrall, and A.D. Hasler. 1976. Coastal movements of adult Fraser River sockeye salmon (*Oncorhynchus nerka*) observed by ultrasonic tracking. Transactions of the American Fisheries Society 105:64-71.

Stepanov, A.S., A.V. Churmasov, and S.A. Cherkashin. 1979. Migration direction finding by pink salmon according to the sun. Soviet Journal of Marine Biology 5(2):92-99.

Straty, R.R. 1975. Migratory routes of adult sockeye salmon (*Oncorhynchus nerka*), in the Eastern Bering Sea and Bristol Bay. National Oceanic and Atmospheric Administration Technical Report, US National Marine Fisheries Service Special Scientific Report Fisheries Series 690:1-32.

Viguier, C. 1882. Le sens d'orientation et ses organes chez les animaux et chez l'homme. Revue de Philosophie 14:1-36.

Walcott, C. 1982. Is there evidence for a magnetic map in homing pigeons? Pages 99-108 *in* F. Papi and H.G. Wallraff, editors. Avian navigation. Springer-Verlag, Berlin, Federal Republic of Germany.

Wiltschko, W., and R. Wiltschko. 1972. Magnetic compass of European robins. Science (Washington DC) 176:62-64.

MAGNETIC SENSITIVITY AND ITS POSSIBLE PHYSICAL BASIS IN THE YELLOWFIN TUNA, *THUNNUS ALBACARES*

Michael M. Walker

National Oceanic and Atmospheric Administration
National Marine Fisheries Service
Southwest Fisheries Center, Honolulu Laboratory
Honolulu, Hawaii 96812 USA and
Department of Zoology
University of Hawaii
Honolulu, Hawaii 96822 USA

ABSTRACT

Many animals are known to orient to magnetic fields. However, two central problems in the study of magnetic sensitivity have been the almost complete failure of magnetic field conditioning experiments and the lack of evidence for a feasible transduction mechanism. In the studies reported here yellowfin tuna learned to discriminate between two Earth-strength magnetic fields in a discrete-trials/ fixed-interval conditioning procedure. Magnetometry experiments, diffraction spectra and electron microscope analyses demonstrated single-domain crystals of the ferromagnetic mineral magnetite in the head of this species. The crystals are concentrated in tissue contained within a sinus formed by the ethmoid bones of the skull. Theoretical analyses show that the crystals would be suitable for use in magnetoreception if linked to the nervous system. The physical properties of the crystals would determine the operation of magnetoreceptor organelles and constrain the capacities of the magnetic sense. Tests of these constraints in appropriately designed conditioning experiments will provide powerful tests of the ferromagnetic magnetoreception hypothesis.

INTRODUCTION

For pelagic fishes migration represents a substantial investment of energy (Sharp and Dotson 1977). These energetic costs imply intense selection on the sensory mechanisms that guide migration. With the exception of the upstream migration of salmon (Hasler et al. 1978) the known capacities of the sensory systems of fishes are not sufficient to explain their navigational achievements (Tesch 1980). We must therefore either reexamine the capacities of the known sensory systems or attempt to find new systems that could provide fish with the necessary sensory abilities (Kreithen 1978). This paper takes the second of these approaches, investigating the responses of yellowfin tuna to magnetic fields and a possible transduction mechanism for the magnetic sense of this and other migratory fishes.

Many animals from different taxa are known to respond to one or more features of the geomagnetic field (Keeton 1971, 1972; Wiltschko 1972; Walcott and Green 1974; Lindauer and Martin 1972; Martin and Lindauer 1977; Quinn 1980; Wiltschko et al. 1981). These responses fall into two general categories--responses to magnetic field direction and to magnetic field intensity. Although the experimental results are repeatable, they are primarily based on unconditioned responses and tell us little about the sensory mechanism and its capacities.

The central problems in the study of magnetic sensitivity in animals have been the almost universal failure of magnetic field conditioning experiments and the lack of evidence for any of the hypothesized magnetic field transduction mechanisms. Conditioning experiments have either failed or been unrepeatable (Reille 1968; Kreithen and Keeton 1974; Beaugrand 1976; Bookman 1977, 1978). Where conditioning was obtained (Kalmijn 1978) subsequent psychophysical analyses of the capacities of the sense were either not done or not reported. Numerous magnetic field transduction mechanisms have been suggested (e.g., Kalmijn 1974; Leask 1977; Jungerman and Rosenblum 1980), but many are unacceptable because they do not explain the general responses to magnetic fields by animals or because magnetoreception is known to occur when the special conditions required by the hypotheses are not met (e.g., Phillips and Adler 1978; Quinn et al. 1981).

The hypothesis that the basis of the magnetic sense is single-domain crystals of magnetite produced by animals has attracted much attention in recent years (Gould et al. 1978; Walcott et al. 1979; J.L. Kirschvink, M.M. Walker, A.E. Dizon, and K.A. Peterson unpublished). This hypothesis is appealing because it can explain the general responses of animals to magnetic fields (Yorke 1979, 1981; Kirschvink and Gould 1981) and lends itself to behavioral testing (Kalmijn 1981; Kirschvink 1981). It also provides us with the basis for a search for receptors which could mediate the behavioral responses.

In this paper I report the use of an orthodox behavioral

conditioning paradigm (Woodard and Bitterman 1974) to train yellowfin tuna to discriminate between two different earth-strength magnetic fields. I also demonstrate the presence of single-domain magnetite crystals in the skull of this species. Based on the theoretical analysis of J.L. Kirschvink and M.M. Walker (unpublished) I then propose behavioral experiments to test for constraints on magnetoreceptor operation resulting from the physical properties of the magnetite crystals. Pending demonstration of a functional link between the crystals and the nervous system these experiments will provide the best test of the ferromagnetic magnetoreception hypothesis.

CONDITIONING EXPERIMENTS

Different magnetic field stimuli can be delivered only successively and not simultaneously. Therefore, in magnetic field conditioning experiments the subject cannot make a simultaneous comparison of stimuli. This limits the choice of conditioning procedures to those which are effective using singly presented discriminative stimuli. The approach I chose was to define a single response, to reward that response under one set of magnetic field conditions and not under another, and to compare the readiness with which the response was expressed under the two conditions (Bitterman 1966). The measure of behavior compared between the stimulus conditions was the rate of performance of a conditioned response. The primary advantage of rate as a measure of discrimination is its sensitivity; it can vary widely and rapidly in response to changes in experimental conditions and can accommodate short-term variability in behavior (Kling 1971). This approach seemed likely to test efficiently the ability of yellowfin tuna to discriminate between different magnetic fields.

These experiments were conducted at the Kewalo Research Facility of the National Marine Fisheries Service, Honolulu Laboratory. The fish used were juvenile yellowfin tuna (40-50 cm fork length) tested individually in one of two cylindrical test tanks (6 m diameter, 0.75 m depth). The experimental tanks contained no metal and each had 100 turns of number 18 AWG copper wire wrapped around its perimeter. A 1-ampere direct current passing through these wires added a vertical magnetic field to the background field. This field was nonuniform, adding from 10 microTesla (μT) in the center to 50 μT at the edge of each tank. The response apparatus was a 60- x 30-cm pipe frame lowered into the tank during trial periods and retracted during intertrial periods. The frame, the magnetic field and a semi-automatic feeder mounted at the side of the tank (Jemison et al. 1982) were operated by mechanical and electrical linkages from the experimental control room. The control room was physically isolated from the experimental tank and was darkened during experiments. The fish were observed through small viewing ports and their responses recorded manually.

The differences between the two magnetic fields used in the discrimination experiments were as follows. The local Hawaiian field was uniform throughout the tanks. That is, inclination, declination and total field intensity were the same at any point in the area occupied by the fish. The altered field introduced significant radially oriented gradients of both intensity and inclination within the tanks. These experiments therefore provided the fish with two very different magnetic fields as discriminative stimuli. The fish could conceivably monitor differences in magnetic field inclination, intensity or the gradients in these two parameters to make the discrimination.

After being allowed to acclimate to the experimental tanks fish were trained to swim through the pipe frame for a food reward (a piece of cut smelt, Osmeridae) delivered from the feeder. In discrimination testing a trial began with simultaneous presentation of the pipe frame and either the positively (S+) or negatively (S-) reinforced stimulus. All responses by the fish within a 30-second trial period were counted. In S+ trials the fish was rewarded with a piece of food following the first response after 30 seconds. In S- trials a 10-second penalty timer started at the end of the trial period. The timer was reset by each subsequent response by the fish until either the penalty time elapsed or a total of 30 seconds of penalty had accumulated. Response to S- was thus penalized by extending the trial without any possibility of the fish obtaining food. After reinforcement had been given the pipe frame was retracted for a variable intertrial interval (mean 90 seconds) after which another trial sequence began.

The fish were given 30 trial training sessions once daily. In any trial session the S+ and S- were presented 15 times, with no more than three S+ or S- trials in succession. Testing was balanced by training two fish with the normal Hawaiian field and two with the altered field as S+. Statistical treatment of the data was by analysis of variance.

Discrimination between the two magnetic field conditions occurred after two 30 trial sessions (Fig. 1). During the first two sessions there was no separation of response rates but from the third day all four fish maintained different rates of responding during S+ and S- trials. An analysis of variance comparing S+ and S- response rates before and after 60 trials for all four fish yielded an $F(1,3)$ stimuli 7.61, $P = 0.07$, and an $F(1,3)$ stimuli by blocks = 102.55, $P = 0.002$. All other comparisons within the analysis did not approach significance.

Control trials were performed with one fish by interrupting the current to the coil; normal procedures were then followed. The

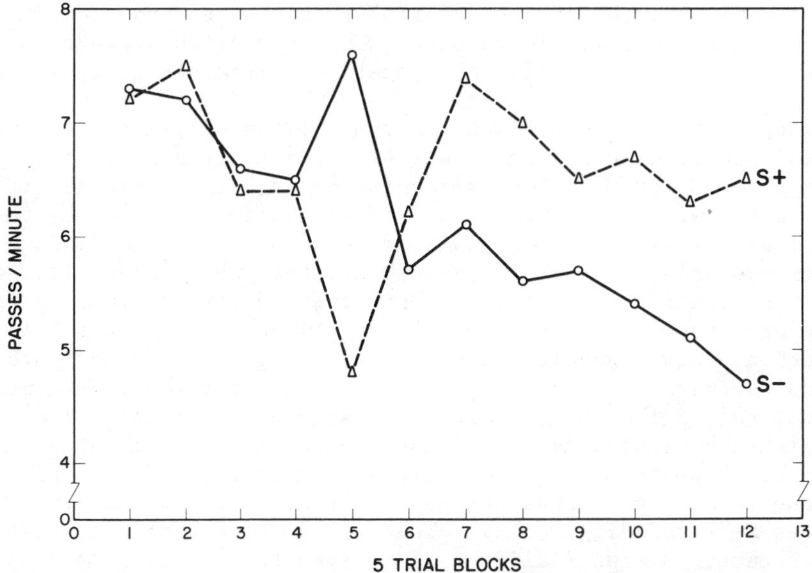

Figure 1. Magnetic field discrimination learning in yellowfin tuna. Each point is the average of five S+ or S- trials for the four fish tested. (Walker et al. 1982; (C) 1982 IEEE.)

response rates during S+ and S- trials fluctuated randomly during this period (Fig. 2). When the power supply was reconnected to the coil the fish was again able to make the discrimination, although

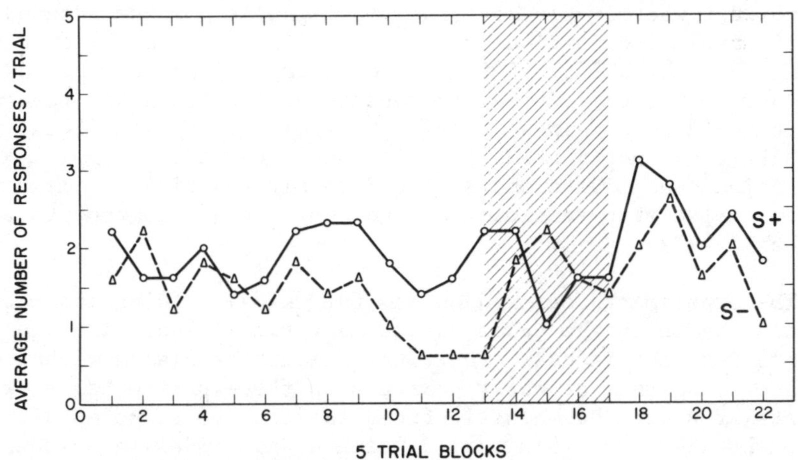

Figure 2. Magnetic field discrimination learning in yellowfin tuna. Control tests shaded. (Walker et al. 1982; (C) 1982 IEEE.)

less strongly than before the control trials were conducted. Evidently the magnetic fields used in the experiment were necessary for anticipation of positive and negative reinforcement by the fish.

These experiments demonstrate the ability of yellowfin tuna to discriminate between different earth-strength magnetic fields. Although the discrimination is consistent in all individuals it is consistently weak. This results from limitations imposed by the fish themselves. Tunas swim continuously and can only be conditioned to produce responses involving movement of their whole body (Iversen 1967; Nakamura 1968). The fish therefore did not attain high rates of performance of the response used to compensate for this limitation. Consequently the scope for change in response rate was low and the variation in responding was high compared to the mean response rate. The upper limit of 30 seconds of penalty time was established because some fish jumped out of the experimental tanks during development of the procedure. This made the "cost" of response to S- low so that the penalty time could only be expected to suppress S- responding weakly. That the fish were able to discriminate between the two fields at all, given these limitations on the procedure, suggests that their magnetic sense is well developed and underlines the sensitivity of response rate as a measure of discrimination.

The conditioning technique used here should be applicable to a range of studies on magnetic sensitivity. Other tunas (skipjack tuna, *Katsuwonus pelamis*, and kawakawa, *Euthynnus affinis*) have shown evidence of magnetic sensitivity in preliminary conditioning experiments (Walker unpublished data). These studies should be continued and extended to other species. The technique can also be used to test whether fishes possess a magnetic compass responding to magnetic field inclination or polarity. This can be achieved by testing for the ability of fish to discriminate between fields in which horizontal or vertical components of the field are reversed. A polarity sensitive compass should discriminate between fields with horizontal, but not vertical, components reversed. An axial or inclination sensitive compass should permit the fish to discriminate between fields with vertical, but not horizontal, components reversed (Wiltschko 1972).

The experiments carried out so far reveal nothing about the mechanism by which the tunas detect magnetic fields. I saw no evidence that the induced electrical fields associated with the presence or absence of water currents in the experimental tanks or the rate at which the magnetic field was changed affected discrimination by the fish. Similarly, Quinn et al. (1981) concluded that sockeye salmon fry and smolts must be able to detect magnetic fields in the absence of water flow in both fresh and salt water. From this behavioral evidence it seems unlikely that these teleost fishes detect magnetic fields by electrical induction in spite of the known electric field sensitivity of Atlantic salmon, *Salmo salar* (Rommel and McCleave 1973). It therefore seems important to identify and test other hypotheses for magnetic field transduction mechanisms in teleost fishes.

PHYSICAL BASIS FOR MAGNETIC FIELD TRANSDUCTION

Magnetometry Experiments

Magnetic material has been detected in a variety of domain states in tissues of a number of different animals (Gould et al. 1978; Walcott et al. 1979; Presti and Pettigrew 1980; Mather and Baker 1981; Zoeger et al. 1981; Jones and MacFadden 1982). In the yellowfin tuna I sought to distinguish magnetic material that might be used in magnetoreception from other deposits by identifying particles in a suitable domain state that could be consistently found in the same place in different individuals. All the tissues that could be identified and extracted from the body of the yellowfin tuna using glass microtome knives were tested for saturation isothermal remanent magnetization (sIRM) in a superconducting magnetometer. A feature of these studies is the small amounts of material involved and the ease with which contamination can influence the results. Procedures for avoiding entry of contaminants and tests for their presence in sample tissues have been developed and described in detail elsewhere (Kirschvink 1983).

There is a battery of paleomagnetic techniques available for the study of magnetic minerals. These techniques require some special adaptations for use in the study of biomagnetism. The magnetic fields associated with crystals of magnetic minerals in animals are very small and could not be detected in the presence of a background field. In a null field environment and at room temperature the crystal moments would be randomly oriented and not detectable. To detect the crystals it is necessary to freeze the sample tissue and prevent the orientation of the crystals from changing under the influence of thermal agitation (Kirschvink 1983). It is then possible to realign the moments of the crystals with a strong inducing magnetic field pulse (>0.3 T). The crystal moments then sum to produce the sIRM (also known as the moment of the sample) and become detectable in a superconducting magnetometer, if they are sufficiently concentrated in a tissue (1 100 nm crystal per 30,000 cells in a 9 mm^3 sample (Kirschvink 1983)).

A superconducting magnetometer detects weak magnetic fields in the samples and also magnetic field noise within the magnetometer itself. These measurements are combined to produce a signal to noise ratio for a tissue sample. Intensity of magnetization indicates the concentration of magnetic material within a sample and is estimated by dividing its moment by its volume (Kirschvink 1983). In this study to be considered magnetic a tissue had to show high values for both these measures compared to other tissues taken from the same fish.

Magnetic crystals tend to have their moments aligned in an "easy" direction, usually the long axis of the crystals (Kirschvink and Gould 1981). The microscopic coercivity of the crystals is the magnetic field required to cause their moments to reverse direction. This coercivity is dependent on the size, shape and domain state of the crystals. Progressive alternating field (AF) demagnetization of a magnetic sample subjects the aligned crystal moments to decaying sinusoidal magnetic fields oriented in one axis. The orientation of the moments of magnetic crystals with coercivities less than the maximum alternating field applied will be randomized by this procedure. The contribution of the randomized crystal moments to the IRM will thus be removed and can be detected in the magnetometer as a reduction in the moment of a sample. The size of the randomizing field required to decrease the IRM to half its saturation value gives an estimate of the median coercivity of the crystals present in a sample. From this we can determine the likely source of the sIRM and the size, shape and domain state of the magnetic crystals.

The magnetometry experiments demonstrated that magnetic material is concentrated only in the head of the yellowfin tuna. Eighteen samples of tissues and organs extracted from a representative fish were not magnetic. A few samples showed either high signal to noise

Table 1. Values of saturation isothermal remanent magnetization in various tissues of a representative yellowfin tuna. Signal-to-noise values are the ratio of the measured value of the tissue to the mean of the noise level before and after the sample was measured. (Walker et al. 1982; (C) 1982 IEEE.)

Tissue	Signal-to-noise ratio	Magnetic intensity (picoTesla)
Left pectoral fin	1.5	45.3
Peduncle tendon	2.5	35.8
Gill	4.6	20.3
Eye	10.1	4.1
Whole parethmoid	15.4	39.4
Left rear parethmoid	11.0	47.4
Right rear parethmoid	18.5	255.3

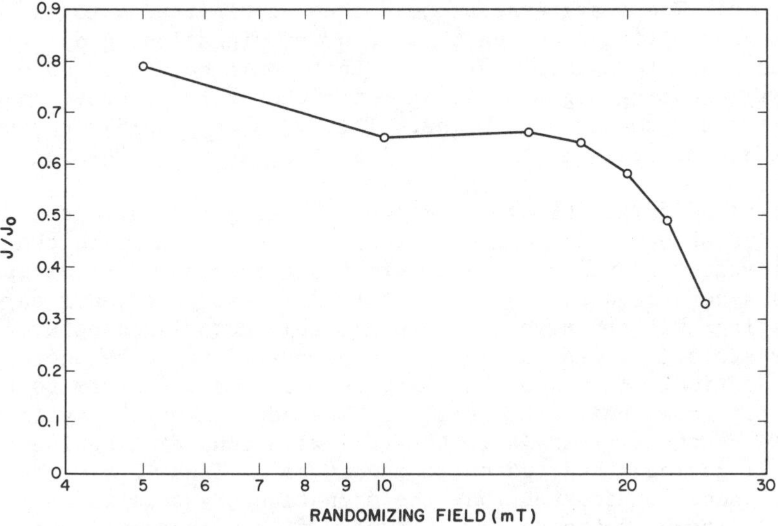

Figure 3. Progressive alternating field demagnetization of the ethmoids of three yellowfin tuna.

ratios or intensities of magnetization (Table 1). These I judged to be nonmagnetic and excluded from consideration. Only samples from the ethmoid bones of the fish's skull exhibited high values for both measures of magnetization (Table 1) in that and all other individuals tested. Progressive AF demagnetization of the ethmoid bones of three yellowfin tuna gave an estimate of median coercivity for the magnetic particles of 22.5 mT (Fig. 3). This is consistent with the presence of a dispersion of single-domain magnetite crystals with particle lengths and diameters of approximately 50 and 40 nm (McElhinny 1973). The narrow range of particle sizes is unlike the pattern observed for geologic and synthetic magnetite crystals, which commonly exhibit log-normal distributions (J.L. Kirschvink personal communication).

Identification of the Magnetic Material

Although the magnetometry experiments indicate that the ethmoids of the yellowfin tuna contain ferromagnetic particles they do not uniquely identify the mineral involved. Extraction of the crystals themselves is necessary to do this and to confirm the estimates of their size and shape. Using glass knives I removed the tissue from within the ethmoids of five yellowfin tuna, ground the tissue in a glass tissue grinder, extracted released fats with ether and digested the remaining cellular material in several changes of millipore-filtered 5% hypochlorite solution (commercial bleach). After centrifuging and washing the digested material small black aggregates of crystals could be attracted to a magnet held to the side of the test tube. These aggregates could then be pipetted onto slides for analyses of diffraction spectra. A brief digestion in buffered EDTA was necessary to obtain aggregates that could be dispersed completely by an alternating magnetic field after they were pipetted onto carbon-coated copper mesh grids. This made it possible to examine individual crystals in the transmission electron microscope (TEM).

X-ray diffraction of the magnetic aggregates extracted from the ethmoid tissue of the yellowfin tuna uniquely identified the crystals as magnetite. An electron microprobe analysis showed that the crystals contained no measurable titanium, manganese, nor chromium, indicating that the crystals were probably not of geologic origin. The crystals averaged 45 x 38 nm and appeared to be hexagonal in cross section. They are thus single-domains and conform to the sizes estimated from their coercivities. The nonoctahedral crystal habit adopted by these crystals in the yellowfin tuna distinguishes them from all geologic and synthetic magnetites. Thus we have evidence from a number of sources that the magnetite crystals found in the ethmoid tissue of the yellowfin tuna must have been produced by the fish themselves. The fish do this with close control over the size, shape and composition of the crystals. They therefore control the physical properties of the crystals and so control their operation in the hypothetical magnetoreceptors considered next.

THE FERROMAGNETIC MAGNETORECEPTION HYPOTHESIS

The basis for magnetite-based magnetoreception is the torque exerted on single-domain magnetite crystals by the geomagnetic field. This torque is described by the relation $\tau = \mu \times B$ where τ is the torque, μ is the moment of a magnetite crystal and B is the intensity of the external magnetic field. The torque will cause the crystals to align with the external field. At physiological temperatures in a live fish the crystals will be subject to thermal agitation, which will cause their vector direction to wander randomly about the applied field direction. The mean alignment of the crystals will be in the direction of the external field so that a compass receptor system needs only to monitor the orientation of up to 1,000 crystals or groups of crystals to detect magnetic field direction accurately (Yorke 1979). The variance of the orientation of the crystals about the external field will depend on its intensity (Kirschvink and Gould 1981; Yorke 1981). Monitoring the variance will thus provide the physical basis for a response to magnetic field intensity (Gould 1980, 1982; Moore 1980; Walcott 1980).

The above analysis yields three testable behavioral predictions (J.L. Kirschvink and M.M.Walker unpublished). The first is that the magnetic sense organ should be made up of separate, independent compass and intensity receptors with different magnetic moments. The second is that the accuracy of the compass response should be

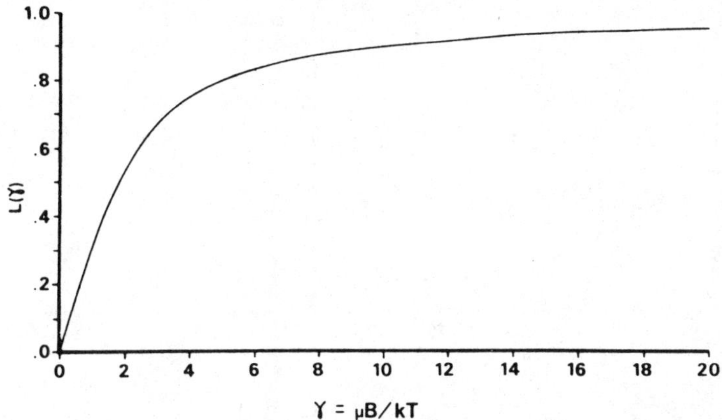

Figure 4. The Langevin function plotted against γ, the ratio of magnetic to thermal energies for single domain magnetite crystals. The accuracy of alignment of the crystals (L(γ)) increases rapidly up to γ values of about six and increases asymptotically thereafter. The accuracy of behavioral responses to magnetic field direction mediated by magnetite-based magnetoreceptors should at first increase rapidly with external field intensity. Beyond a certain point further increase in external field intensity should not lead to greater compass accuracy. (Modified from Kirschvink 1981.)

quantitatively defined by the Langevin function (Fig. 4; Kirschvink 1981), i.e., the accuracy of the compass response should increase asymptotically with external field strength. The third prediction is that the threshold sensitivity to changes in magnetic field intensity will be defined by the first derivative of the equation for the root mean square deviation of the crystals' alignment in the intensity receptor organelle. Plotted against external field strength the threshold sensitivity should increase to a maximum at about 50 μT (= 0.5 Gauss) and decline monotonically thereafter (Fig. 5; J.L. Kirschvink and M.M. Walker unpublished).

These predictions are testable with currently available behavioral conditioning procedures. The first prediction is testable using the conditioning paradigm reported here. A factorial design is suggested in which experiments test for separate responses to magnetic field direction and intensity. The tests require fish to discriminate between magnetic fields with different vector directions but the same total intensity and between fields with the same direction but different intensities. The second and third predictions are testable with suitable modifications of the conditioning procedure used for testing sun-compass orientation in

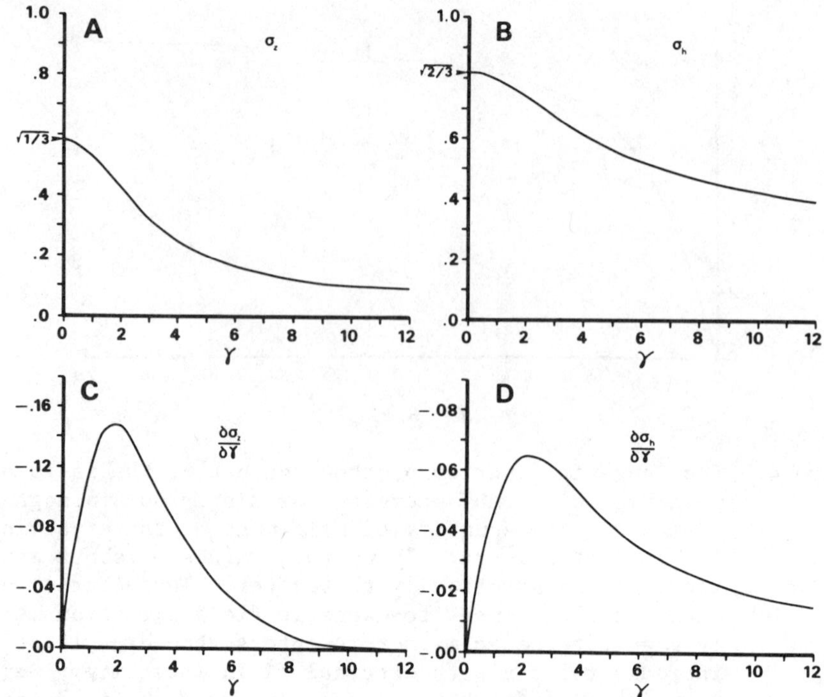

Figure 5. A,B. Plots of the variance of the Langevin function parallel and perpendicular to external field direction against γ, the ratio of magnetic to thermal energies of single domain magnetite crystals. C,D. Plots of the first derivative of the Langevin variance parallel and perpendicular to external field direction. The Langevin variance declines with increasing γ, and, for magnetite-based magnetoreceptors, will be dependent on external field intensity. It follows that receptors monitoring some component of this variance will be most sensitive to changes in intensity when the change in variance with intensity is a maximum. C,D show that the maximum rates of change of variance with intensity occur at about $\gamma = 2$. Behaviorally measured thresholds to magnetic field intensity changes should follow the form of these plots. Threshold sensitivity should increase rapidly with external field intensity to a maximum and decline thereafter. (J.L. Kirschvink and M.M. Walker, unpublished data.)

white bass by Hasler et al. (1958). The accuracy of a directional response should increase asymptotically with magnetic field strength and should conform to the Langevin function (Fig. 4) (Kalmijn 1981; Kirschvink 1981). The same conditioning approach can be adapted to test for the smallest changes in magnetic field intensity that the fish can detect. Fish should be required to produce or withhold a single response with accuracies approaching 90-100%. Threshold sensitivity is then measured by decreasing the difference between positively and negatively reinforced stimuli using appropriate procedures Engen 1971). This threshold sensitivity to magnetic field intensity changes should change with field strength in the manner shown in Fig. 5 (J.L. Kirschvink and M.M. Walker unpublished).

CONCLUDING REMARKS

This and other recent studies have demonstrated repeatable behavioral responses to magnetic fields by migratory fishes (Quinn 1980; Quinn et al. 1981; Quinn and Brannon 1982). Theoretical analyses based on the physical properties of single-domain magnetite discovered in yellowfin tuna and other animals easily explain the separate magnetic compass and intensity responses of vertebrates and invertebrates (Yorke 1979, 1981; Kirschvink 1981; Kirschvink and Gould 1981). J.L. Kirschvink and M.M. Walker (unpublished) extend these analyses and make testable predictions on the nature and organization of ferromagnetic magnetoreceptor organelles and the constraints they place on the capacities of the magnetic sense.

Gross dissection of the anterior region of the skull of yellowfin tuna reveals the supraophthalmic nerve. This trunk carries the anterior acousticolateralis nerve and ramifies in the region of the ethmoid bones. The hair cells associated with the acousticolateralis nerve system could provide ideal mechanoreceptors for use in magnetoreception with the magnetite crystals. Preliminary histological studies suggest the presence of nerves in a thin band of tissue in the midline immediately beneath the dermethmoid bone (Walker unpublished data). Thus for the first time there is found together in a species a repeatable behavioral response to magnetic fields and the necessary physical and neural components of a ferromagnetic sense organ. Proving the links between these components of a magnetic sense should be an important goal for future research.

It now seems opportune to speculate on the possible utility of the magnetic sense of pelagic fishes for navigation. Acoustic tracking experiments have shown that swordfish *Xiphias gladius* can maintain a compass heading for periods of up to several days (Carey and Robison 1981). Skipjack tuna make diurnal horizontal movements without retracing their path onto and off shallow banks where they feed during daylight hours (Yuen 1970). These movements suggest that these fishes are able to determine their position and set a compass course toward a goal. The knowledge that pelagic fishes respond to magnetic fields suggests the possibility that they use the geomagnetic field to guide their migrations. Field experiments to test this hypothesis should not only show whether fish do use the geomagnetic field in navigation but may also indicate how the fish use the sense. This should contribute greatly to our understanding of the mechanisms of migration in fishes.

ACKNOWLEDGMENTS

A.E. Dizon, J.T. Corwin, J.L. Kirschvink and staff and graduate students at the National Marine Fisheries Service, Kewalo Research Facility have contributed to this paper through critical discussions and comments on an earlier draft. Research funds from the National Marine Fisheries Service, and travel funds from the University of Hawaii and this Institute are gratefully acknowledged.

REFERENCES

Beaugrand, J.P. 1976. An attempt to confirm magnetic sensitivity in the pigeon, *Columba livia*. Journal of Comparative Physiology A Sensory, Neural and Behavioral Physiology 110A:343-355.

Bitterman, M.E. 1966. Animal learning. Pages 451-484 *in* J.B. Sidowski, editor. Experimental methods and instrumentation in psychology. McGraw-Hill, New York, New York, USA.

Bookman, M.A. 1977. Sensitivity of the homing pigeon to an Earth-strength magnetic field. Nature (London) 267:340-342.

Bookman, M.A. 1978. Sensitivity of the homing pigeon to an Earth-strength magnetic field. Pages 127-134 *in* K. Schmidt-Koenig and W.T. Keeton, editors. Animal migration, navigation, and homing. Springer-Verlag, Heidelberg, Federal Republic of Germany.

Carey, F.G., and B.H. Robison. 1981. Daily patterns in the activities of swordfish, *Xiphias gladius*, observed by acoustic telemetry. U.S. National Marine Fisheries Serice Fishery Bulletin 79: 277-292.

Engen, T. 1971. Psychophysics. I. Discrimination and detection. Pages 11-46 *in* J.W. Kling and L.A. Riggs, editors. Woodworth & Schlosberg's experimental psychology. Holt, Rinehart and Winston, New York, New York, USA.

Gould, J.L. 1980. The case for magnetic sensitivity in birds and bees (such as it is). American Scientist 68:256-267.

Gould, J.L. 1982. The map sense of pigeons. Nature (London) 296: 205-211.

Gould, J.L., J.L. Kirschvink, and K.S. Deffeyes. 1978. Bees have magnetic remanence. Science (Washington DC) 201:1026-1028.

Hasler, A.D., R.M. Horrall, W.J. Wisby, and W. Braemer. 1958. Sun-orientation and homing in fishes. Limnology and Oceanography 3: 353-361.

Hasler, A.D., A.T. Scholz, and R.M. Horrall. 1978. Olfactory imprinting and homing in salmon. American Scientist 66:347-355.

Iversen, R.T.B. 1967. Response of yellowfin tuna (*Thunnus albacares*) to underwater sound. Marine Bio-Acoustics 2:105-121. Pergamon Press, Oxford, England.

Jemison, H.A. III, A.E. Dizon, and M.M. Walker. 1982. An automatic feeder for liquids and wet or dry solids. Behavior Research Methods & Instrumentation 14:54-55.

Jones, D.S., and B.J. MacFadden. 1982. Induced magnetization in the monarch butterfly, *Danaus plexippus* (Insecta, Lepidoptera). Journal of Experimental Biology 96:1-9.

Jungerman, R.L., and B. Rosenblum. 1980. Magnetic induction for the sensing of magnetic fields by animals--An analysis. Journal of Theoretical Biology 87:25-32.

Kalmijn, A.J. 1974. The detection of electric fields from inanimate and animate sources other than electric organs. Pages 147-200 *in* A. Fessard, editor. Handbook of sensory physiology. Springer-Verlag, New York, New York, USA.

Kalmijn, A.J. 1978. Experimental evidence of geomagnetic orientation in elasmobranch fishes. Pages 347-353 *in* K. Schmidt-Koenig and W.T. Keeton, editors. Animal migration, navigation, and homing. Springer-Verlag, Heidelberg, Federal Republic of Germany.

Kalmijn, A.J. 1981. Biophysics of geomagnetic field detection. IEEE Transactions of Magnetics 17:1113-1124.

Keeton, W.T. 1971. Magnets interfere with pigeon homing. Proceedings of the National Academy of Sciences 68:102-106.

Keeton, W.T. 1972. Effects of magnets on pigeon homing. Pages 579-594 *in* S.R. Galler, K. Schmidt-Koenig, G.J. Jacobs, and R.E. Belleville, editors. Animal orientation and navigation. National Aeronautics and Space Administration, Special Publication 262, Washington, District of Columbia, USA.

Kirschvink, J.L. 1981. The horizontal magnetic dance of the honeybee is compatible with a single-domain ferromagnetic magnetoreceptor. BioSystems 14:193-203.

Kirschvink, J.L. 1983. Biogenic ferrimagnetism: A new biomagnetism. Pages 501-532 *in* S.J. Williamson, G.-L. Romani, L. Kaufman, and I. Modena, editors. Biomagnetism: An interdisciplinary approach. Plenum Press, New York, New York, USA.

Kirschvink, J.L., and J.L. Gould. 1981. Biogenic magnetite as a basis for magnetic field detection in animals. BioSystems 13:181-201.

Kling, J.W. 1971. Learning. Introductory survey. Pages 551-563 *in* J.W. Kling and L.A. Riggs, editors. Woodworth & Schlosberg's Experimental psychology. Holt, Rinehart and Winston, New York, New York, USA.

Kreithen, M.L. 1978. Sensory mechanisms for animal orientation--can any new ones be discovered? Pages 25-34 *in* K. Schmidt-Koenig and W.T. Keeton, editors. Animal migration, navigation, and homing. Springer-Verlag, Heidelberg, Federal Republic of Germany.

Kreithen, M.L., and W.T. Keeton. 1974. Attempts to condition homing pigeons to magnetic stimuli. Journal of Comparative Physiology 91:355-362.

Leask, M.J.M. 1977. A physicochemical mechanism for magnetic field detection by migratory birds and homing pigeons. Nature (London) 267:144-145.

Lindauer, M., and H. Martin. 1972. Magnetic effect on dancing bees. Pages 559-567 *in* S.R. Galler, K. Schmidt-Koenig, G.J. Jacobs, and R.E. Belleville, editors. Animal orientation and navigation. National Aeronautics and Space Administration, Special Publication 262, Washington, District of Columbia, USA.

Mather, J.G., and R.R. Baker. 1981. Magnetic sense of direction in woodmice for route-based navigation. Nature (London) 291: 152-155.

McElhinny, M.W. 1973. Palaeomagnetism and plate tectonics. Cambridge University Press, London, England.

Martin, H., and M. Lindauer. 1977. The effect of the Earth's magnetic field on gravity orientation in the honey bee (*Apis mellifica*). Journal of Comparative Physiology A. Sensory, Neural and Behavioral Physiology 12A:145-187.

Moore, B.R. 1980. Is the homing pigeon's map geomagnetic? Nature (London) 285:69-70.

Nakamura, E.L. 1968. Visual acuity of two tunas, *Katouwonus pelamis* and *Euthynnus affinis*. Copeia 1968:41-49.

Phillips, J.B., and K. Adler. 1978. Directional and discriminatory responses of salamanders to weak magnetic fields. Pages 325-333 *in* K. Schmidt-Koenig and W.T. Keeton, editors. Animal migration, navigation, and homing. Springer-Verlag, Heidelberg, Federal Republic of Germany.

Presti, D., and J.D. Pettigrew. 1980. Ferromagnetic coupling to muscle receptors as a basis for geomagnetic field sensitivity in animals. Nature (London) 285:99-101.

Quinn, T.P. 1980. Evidence for celestial and magnetic compass orientation in lake migrating sockeye salmon fry. Journal of Comparative Physiology A. Sensory, Neural and Behavioral Physiology 137A:243-248.

Quinn, T.P., and E.L. Brannon. 1982. The use of celestial and magnetic cues by orienting sockeye salmon smolts. Journal of Comparative Physiology A. Sensory, Neural, and Behavioral Physiology 147A:547-552.

Quinn, T.P., R.T. Merrill, and E.L. Brannon. 1981. Magnetic field detection in sockeye salmon. Journal of Experimental Zoology 217:137-142.

Reille, A. 1968. Essai de mise en évidence d'une sensibilitié du pigeon au champ magnétique à l'aide d'un conditionnement nociceptif. Journal of Physiologie (Paris) 60:85-92.

Rommel, S.A. Jr., and J.D. McCleave. 1973. Sensitivity of American eels (*Anguilla rostrata*) and Atlantic salmon (*Salmo salar*) to weak electric and magnetic fields. Journal of the Fisheries Research Board of Canada 30:657-663.

Sharp, G.D., and R.C. Dotson. 1977. Energy for migration in albacore, *Thunnus alalunga*. US National Marine Fisheries Service Fishery Bulletin 75:447-450.

Tesch, F.-W. 1980. Migratory performance and environmental evidence of orientation. Pages 589-612 *in* M.A. Ali, editor. Environmental physiology of fishes. Plenum Press, New York, New York, USA.

Walcott, C. 1980. Magnetic orientation in homing pigeons. IEEE Transactions on Magnetics 16:1008-1013.

Walcott, C., J.L. Gould, and J.L. Kirschvink. 1979. Pigeons have magnets. Science (Washington DC) 205:1027-1029.

Walcott, C., and R.P. Green. 1974. Orientation of homing pigeons altered by a change in the direction of an applied magnetic field. Science (Washington, DC) 184:180-182.

Walker, M.M., A.E. Dizon, and J.L. Kirschvink. 1982. Geomagnetic field detection by yellowfin tuna. Pages 755-758 *in* Oceans 82 Conference Record: Industry, Government, Eduction - Partners in Progress. Institute of Electrical and Electronics Engineers, New York, New York, USA.

Wiltschko, W. 1972. The influence of magnetic total intensity and inclination on directions preferred by migrating European robins (*Erithacus rubecula*). Pages 569-577 *in* S.R. Galler, K. Schmidt-Koenig, G.J. Jacobs, and R.E. Belleville, editors. Animal orientation and navigation. National Aeronautics and Space Administration Special Publication 262. Washington, District of Columbia, USA.

Wiltschko, R., D. Nohr, and W. Wiltschko. 1981. Pigeons with a deficient sun compass use the magnetic compass. Science (Washington, DC) 214:343-345.

Woodard, W.T., and M.E. Bitterman. 1974. A discrete-trials/fixed-interval method of discrimination training. Behavior Research Methods & Instrumentation 6:389-392.

Yorke, E.D. 1979. A possible magnetic transducer in birds. Journal of Theoretical Biology 77:101-105.

Yorke, E.D. 1981. Sensitivity of pigeons to small magnetic field variations. Journal of Theoretical Biology 89:533-537.

Yuen, H.S.H. 1970. Behavior of skipjack tuna, *Katsuwonus pelamis*, as determined by tracking with ultrasonic devices. Journal of the Fisheries Research Board of Canada 27:2071-2079.

Zoeger, J., J.R. Dunn, and M. Fuller. 1981. Magnetic material in the head of the common Pacific dolphin. Science (Washington DC) 213:892-894.

ARTIFICIAL MATURATION AS A TECHNIQUE FOR INVESTIGATING ADAPTATIONS FOR MIGRATION IN THE EUROPEAN EEL, *ANGUILLA ANGUILLA* (L.)

N. W. Pankhurst

Department of Zoology
University of Alberta
Edmonton, Alberta T6G 2E9 Canada

ABSTRACT

In the European eel, *Anguilla anguilla*, artificial maturation using hormone treatment offers an opportunity to study eels of advanced sexual maturity, which are not otherwise available to researchers. Maturation and the accompanying spawning migration impose a radical change in environment and a cessation in feeding, both of which require certain morphological and physiological changes if successful migration and spawning are to be achieved. In artificially matured and some naturally maturing, early migratory stage eels, modifications in vision, olfaction, integumental structure and colour, and skeletal muscle have been identified. These can generally be related to specific demands of the oceanic environment or spawning. Some changes occur in hormone-treated eels which may be artefacts of the hormone treatment itself. Improvements in the methods of artificial maturation which reduce the likelihood of appearance of unnatural side-effects are suggested and directions for further research are discussed.

INTRODUCTION

In migratory fish species in which the location of migratory routes, spawning grounds or occurrence of a particular life-history stage is unknown or inaccessible, biological descriptions must remain incomplete. When the area of study involves features which change with sexual maturation or gonadal recrudescence, one approach to alleviate this problem is to mimic the changes in the laboratory, by treatment of selected individuals with gonadotrophic or steroid hormones. The success and validity of this approach depends on the availability of suitably receptive sexual stages and the appropriate hormone(s), the provision of required physical cues and conditions, and some indication that the observed changes in artificially matured or hormone treated fish are a reasonable approximation of the events occurring during natural maturation.

The European eel is a species for which the study of sexual maturation and migration in the adults is often frustrated by the difficulty of obtaining migrating fish past a certain level of sexual maturity. Once the seaward-migrating, sexually maturing eels leave the rivers and estuaries, they are effectively lost to researchers. Small numbers of migrating eels have been taken over the continental shelves of Britain and Europe (Tesch 1977), but often in nonideal conditions with the result that their informational value has been limited. Artificial maturation does, in this case, provide an avenue for investigating the changes occurring with sexual maturation. In the absence of precise information on the marine phase of the migration, these changes can be used as the basis for speculation on the probable conditions of, and problems posed by, the migratory environment. In the lower reaches of rivers where migrating eels are still readily available, the eels are sufficiently sexually mature that gonadotrophic hormones can be used to maintain sexual development in the laboratory. A number of hormones have been used for inducing sexual development, including human chorionic gonadotropin (hCG) and carp and salmon pituitary extracts (Boëtius et al. 1962; Fontaine 1964; Boëtius and Boëtius 1967, 1980; Nose 1971; Bieniarz and Epler 1975; Villani and Lumare 1975; Kokhnenko et al. 1978; Pankhurst 1982a,b,c). The majority of the earlier studies have been concerned solely with inducing sexual maturation and not with the somatic changes which accompany maturation. The use of heterologous hormones in these experiments is not ideal, and the question can be raised as to whether the observed maturational changes are artefacts of hormone treatment, which do not occur naturally in maturing fish. Examination of untreated, seaward-migrating fish does provide some basis for evaluating the changes in hormone-injected eels. However, the changes found in artificially matured eels should be treated with some caution and can only be verified ultimately by capture of wild sexually-mature individuals.

The success of artificial maturation techniques depends largely upon selection of eels which have already reached a certain stage of maturity. One criterion which is commonly used for the identification of migrating (maturing) eels is the colour of the integument (Sinha and Jones 1975). This is not always reliable as skin colour is known to be influenced by a number of factors in addition to maturity phase (Pankhurst and Lythgoe 1982). A more reliable indication of the onset of sexual maturation is given by the increase in eye size which occurs in maturing eels; an index based on the external surface area of the eye and body length (Table 1) has proved useful in estimating stage of gonadal development in intact eels (Pankhurst 1982a).

TABLE 1. The relationship between eye size, body length and sexual maturity in *A. anguilla*.

Physical Appearance	Eye Index (I)[1]	Body Length (cm)	Sex[2,3]	Maturity Phase	Gonad Size	Migration Status
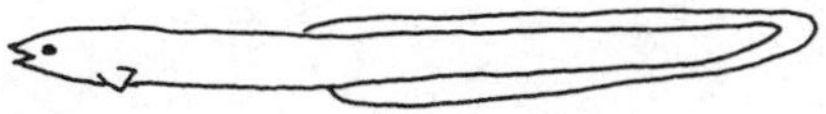	≤ 6.5	0-49.99	male or female	sexually immature adult	Stage 1 testes[4], oocyte diameter ≤ 0.10 mm	non-migratory
		≥ 50	female	sexually immature adult	Oocyte diameter ≤ 0.10 mm	non-migratory
	> 6.5	0-49.99	male	sexually maturing adult	Testes Stage 2-7[4]	migratory
		≥ 50	female	sexually maturing adult	Oocyte diameter 0.10-1.00 mm	migratory

[1] $I = \left[\frac{\left(\frac{A+B}{4}\right)}{L}\right]^2 \pi \times 100$, A = horizontal diameter of the eye, B = vertical diameter and L = total body length (mm) (Pankhurst 1982a).

[2] Sinha and Jones (1975).

[3] Tesch (1977).

[4] Boëtius and Boëtius (1967).

The sexually immature phase of post-larval life is primarily a growth phase and occurs while the eel is resident in fresh water. During this phase the eel lives in water which is often turbid and stained with the products of vegetable decay, and conditions for vision can be very poor (Lythgoe 1979). Instead there is heavy reliance on olfactory cues, particularly in prey location (Teichmann 1959; Tesch 1977). The end of this phase is signalled by the onset of sexual maturation and the beginning of the marine migration to the spawning grounds in the western North Atlantic Ocean (Tesch 1977). Here the eel encounters a very different environment. The waters of the open ocean are clear and blue with light being predominantly downwelling (Jerlov 1976), and photic conditions for vision are good (Lythgoe 1979). The gut of sexually maturing eels atrophies, and feeding ceases once migration begins (Tesch 1977) with the result that the type of olfactory information required, or available, may be quite different during the migratory phase. In addition the change in the photic environment is likely to dictate strategies for camouflage and predator avoidance different from those used in fresh water. The cessation of feeding prior to migration means that the metabolic demands of migration and spawning must be met from a stored reserve, accrued before migration begins. An examination of changes in sensory function, the integument and muscle in naturally maturing and artificially matured eels has been conducted to investigate some of the ways by which the demands of migration might be met (Pankhurst 1982a,b,c; Pankhurst and Lythgoe 1982, 1983).

SENSORY SYSTEMS

In line with the change in quality of sensory information received by migrating eels, modifications do occur in sensory systems, one of the most notable being in the visual system. In conjunction with a large increase in eye size there are various changes in retinal structure (Carlisle and Denton 1959; Stramke 1972; Pankhurst 1982a; Pankhurst and Lythgoe 1983). These include an increase in the total number of rods, the outer segments of which become slightly wider and longer, and decreases in cone density with cone degeneration occurring in more mature eels. Numbers of inner nuclear layer cells and ganglion cells also decrease. These changes are generally complete at or soon after the beginning of downstream migration. Figure 1 summarises the major structural changes. Similar changes do occur in the eyes of other teleosts with growth (Blaxter and Jones 1967; Blaxter and Staines 1970; Johns and Easter 1977; Boehlert 1979; Sandy and Blaxter 1980); however, in eels the modifications are more closely related to sexual maturity. The structural changes are accompanied by shifts in the absorptive maxima of the rod pigments towards shorter wavelengths, shifts which are also complete early in migration. The shifts have been described in

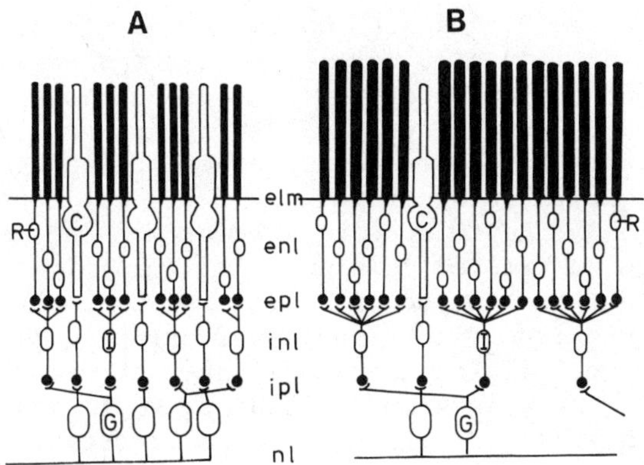

Figure 1. Diagram showing the major structural characteristics of the retina of the European eel at different life history stages. A. Sexually immature (I ≤ 6.5, mean oocyte diameter ≤ 0.10 mm[1]). B. Sexually maturing (I > 6.5, mean oocyte diameter > 0.10 mm[1]). C = cone, G = ganglion cell, I = inner nuclear layer cell, R = rod, elm = external limiting membrane, enl = external nuclear layer, epl = external plexiform layer, inl = inner nuclear layer, ipl = inner plexiform layer, nl = neural layer. Not to scale. ([1]Data from Pankhurst 1982a.)

the European eel by Carlise and Denton (1959) and Wald (1960) and in the American eel (*Anguilla rostrata*) by Beatty (1975). The expected effect of these alterations would be to increase the sensitivity of the eye. Changes in the dimensions of rods should increase the photon-capture success rate by increasing the receptor cross-sectional area and the light path-length through the visual pigment. Shifts in absorptive maxima of the pigments increase the absorptive efficiency in blue light, and changes in retinal convergence ratios enhance sensitivity by summing the input from an increased number of rods onto each bipolar cell. A consequence of this may be a reduction in acuity in the eyes of maturing eels due to the larger area on the retina occupied by receptors which function as a single unit. The effect of decreases in cone density has not been established, but if, as available information suggests is the case for *A. rostrata* (Gordon et al. 1978), cones are concerned with colour vision in *A. anguilla*, then a reduced ability to perceive colour might be expected in mature eels. The resulting visual system would appear to function best in dim monochromatic light. These photic conditions exist in the mesopelagic oceanic environment (Jerlov 1976), and mesopelagic fish commonly display retinal characteristics (Locket 1975) similar to those found in sexually maturing eels. The inference must be that migrating eels utilise the mesopelagic habitat.

Olfactory capabilities also appear to change with sexual maturation. In artificially matured eels a marked loss of mucous cells occurs from the olfactory epithelium and in some cases degeneration of the epithelium itself occurs (Pankhurst and Lythgoe 1983). The changes have not been observed in untreated eels and so may be an event which occurs after the fresh water phase of the migration. Equally they may be an artefact of the hormone treatment; however, in hormone treated eels in which significant increases in sexual development did not occur, there was no apparent degeneration of the olfactory epithelium. Sexually immature eels have a very acute sense of smell (Teichmann 1959), so that olfactory degeneration in sexually maturing eels seems to indicate a radical reorientation of the sensory hierarchy; olfaction is of importance during the fresh water period of the life cycle, while vision is well developed in the marine phase and olfaction may be of reduced importance.

INTEGUMENT

The integument of teleosts is intimately concerned with camouflage (Denton 1970), protection against physical abrasion (Van Oosten 1957; Burdak 1980; Mittal and Banerjee 1980) and maintenance of osmotic integrity (Van Oosten 1957; Olivereau and Lemoine 1971). The structural characteristics of the teleost integument are often specific to a particular habitat (Aleyev 1969). The response of migrating eels to the suggested changes in cryptic and protective requirements in the open ocean might therefore be expected to involve some changes in the integument. In the sexually immature eel the integument consists of a well-developed mucal epidermis overlying a thick dermis composed of orthogonally arranged sheets of collagen. Embedded in the dermis are small, discrete, hyaline scales, and at the inner and outer boundaries of the dermis layers of chromatophores occur. Associated with each of these is a layer of reflecting elements made up of the purines hypoxanthine and guanine (Jakubowski 1960; Leonard and Summers 1976; Pankhurst and Lythgoe 1982). The protective function is performed by the production of a copious mucus, and the camouflage function by avoidance (burrowing) and adoption of a range of olive-green to yellow tones appropriate to the spectral distribution of light in fresh water. These colours are the net result of absorption and reflection of incident light by each of the two chromatophore layers, compounded by additional scattering and absorption in the dermis and epidermis.

With sexual maturation a number of structural changes occurs in the skin. The thickness of the epidermis and the density of epidermal mucous cells decreases, while the dermal thickness increases. This is accompanied by a large increase in the area of the skin covered by, and the degree of overlap of, the scales (Pankhurst 1982c). These changes seem to make the skin more resistant to physical stress (author's unpublished data). Precisely why this is advantageous is not clear as during migration there is presumably reduced likelihood of abrasive contact with the substrate. The toughened skin may function as a defense against predators; however, in other teleosts, scales are not thought to confer any great protection against predation (Aleyev 1969; Burdak 1980). Energetic considerations may hold at least part of the answer. Epidermal mucous cells move up through the epidermis undergoing differentiation, at the end of which they are exfoliated (Mittal and Banerjee 1980). This involves loss of protein and polysaccharides and requires continual synthesis of new cells. Maintenance of an epidermal mucous sheath is likely to be energetically expensive and the adoption of a scale coat for the protective function may represent a significant energy economy during migration. Skin calcium measurements have shown that the increase in scale area does not appear to be associated with calcium storage (Pankhurst 1982c).

Changes also occur in colour-producing features of the integument of sexually maturing eels. Total concentrations of purine in the skin decrease, and the location of the reflecting elements changes. Skin colour changes to become dark dorsally and silver or white ventrally (Pankhurst and Lythgoe 1982). Other workers report similar changes in skin colour of sexually maturing eels (D'Ancona 1960; Sinha and Jones 1975; Tesch 1977). The colour changes increase the countershading of migrating eels. In the predominantly down-welling light of the open ocean the adoption of dark dorsal and light ventral surfaces is the best method available to the eel to offset the effects of directionally biased lighting and to become less visible. The oval body shape of the eel makes the "mirror" camouflage assumed by many pelagic fish (Denton 1970; Land 1972) difficult to achieve. Ultrastructural examination of the crystalline elements and angular reflectance measurements have confirmed that the colour changes found in migrating eels are concerned with counter-shading and not reflection (Pankhurst and Lythgoe 1982). Figure 2 is a summary of the changes occurring in the integument of sexually maturing eels.

MUSCLE

The energy source for migration and gonad development has usually been assumed to be lipid, which is reported to make up nearly 30% of total body weight at the beginning of migration (Bertin 1956; Meyer-Waarden 1965). There is good evidence that in the American eel this reserve is metabolised via an enlarged red (oxidative) muscle compartment, which is also attributed with functioning in prolonged swimming during migration (Hulbert and Moon 1978a,b,c). Initial reports on the red muscle of European eels suggested that the red muscle volume was smaller than in its American eel counterpart (Boddeke et al. 1959; Greer Walker and Pull 1975), although there are qualitative references to increased red muscle occurring in maturing

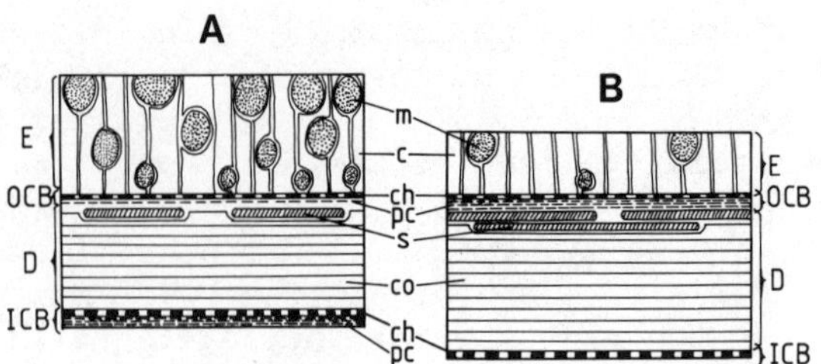

Figure 2. Diagram showing the principal structural characteristics of the flank integument of the European eel at different life history stages. A. Sexually immature (I ≤ 6.5, mean oocyte diameter ≤ 0.10 mm[1]). B. Sexually matring (I > 6.5, mean oocyge diameter > 0.10 mm[1]). D = dermis, E = epidermis, ICB = inner chromatophore bilayer, OCB = outer chromatophore bilayer, c = club cell, co = collagen, ch = chromatophore, m = mucous cell, pc = purine crystals, s = scale. Not to scale. ([1]Data from Pankhurst 1982a.)

European eels (Lewander et al. 1974). The apparently low volume of red muscle was seen as being at variance with the migratory habit of European eels (Hulbert and Moon 1978a). Examination of the body musculature over a range of maturity states shows that the red muscle volume of migrating European eels does, in fact, increase markedly (Pankhurst 1982b). The volume increase is caused by increases in fibre diameter and this in turn is due to the inclusion of extra- and intra-myofibrillar lipid. In addition the mitochondrial volume fraction of the red muscle increases, while that of white (anaerobic) muscle remains at a constant low value. The appearance of lipid droplets in the red muscle coincides with that of gonadal lipid. It appears that red muscle is involved in lipid mobilisation, and this is consistent with the liver-like function which has been ascribed to teleost red muscle (Braekkan 1956; Wittenberger and Diaciuc 1965). There is an intriguing suggestion that in sexually maturing American eels lipid inclusions partition the muscle to such an extent that the contractile function may be impaired (Hulbert and Moon 1978b). In this case white muscle would have to be capable of functioning at slow swimming speeds and for extended periods. White fibres are known to be active over a range of swimming speeds in other teleosts (Smit et al. 1971; Johnston et al. 1977), and Hulbert and Moon (1978c) have demonstrated that red muscle in American eels has the ability to oxidise lactate. The latter authors hypothesised that red muscle may finance sustained activity of white muscle through a lactate-pyruvate exchange system. A similar mechanism has been found in carp (*Cyprinus carpio*) (Wittenberger 1973; Wittenberger et al. 1975). Partitioning of red muscle myofibrils also occurs in maturing European eels, though what effect this has on contractile activity is not known. What is certain is that red muscle is active during maturation and that this activity involves utilisation of a lipid

TABLE 2. Structural differences in lateral body musculature, between sexually immature (fresh water) and sexually maturing (migratory) female *A. anguilla*. Numerical values represent the ranges of the measured parameters with means in parentheses. Data from Pankhurst (1982b).

Sexual Stage[1]	Muscle Volume (% total)		Mean Red Muscle Fibre Diameter (μm)	Volume Fractions			
	Red	White		Lipid (Red Muscle)	Myofibrils (Red Muscle)	Mitochondria (Red Muscle)	Mitochondria (White Muscle)
Immature	1.5–12.5	88–98.5	21–36	0–8.5	62–87	2.2–8.5	0.25–3.5
	(8.9)	(94.1)	(28.4)	(3.1)	(74.7)	(5.9)	(1.1)
Maturing	8–14	85–92	27–50	12.5–25	43–72	8.2–16.5	0.25–2.5
	(9.9)	(89)	(39.3)	(17.6)	(59.8)	(11.5)	(1.1)

[1]Immature: $I \leq 6.5$, mean oocyte diameter ≤ 0.10 mm; Maturing: $I > 6.5$, mean oocyte diameter > 0.10 mm (data from Pankhurst 1982a).

based energy store. Table 2 gives a synthesis of the structural changes found in the muscle of sexually maturing European eels.

FUTURE RESEARCH

The scope for future work is great. Research could be profitably extended in two directions. The first involves further development and refinement of the techniques for artificially maturing eels. Problems with the present methods include the long treatment period required to induce maturity and the possible generation of the artefacts discussed earlier. Both of these may be related to the use of heterologous hormones. A preferable treatment program would consist of administration of pituitary extracts or purified gonadotrophic hormone (GtH) from maturing eels. The eel pituitary is relatively large (Callamand 1943), and Fontaine (1969) reports that 1 g of dried hypophysial extract can be harvested from about a thousand eels. Work on other teleosts has shown that serum and pituitary GtH levels are often significantly elevated some time before full sexual maturity is reached (Hontela and Peter 1978; Peter and Crim 1979). In this respect pituitary extracts from downstream migrating eels may contain sufficient maturational hormones to be of use in artificial maturation. Another avenue has been opened by the recent discovery that pimozide (a dopamine antagonist) has a spectacular effect in potentiating the effects of mammalian GtH releasing factor in stimulating GtH secretion in goldfish (Chang and Peter in press). Experiments on the effectiveness of this technique in other species, including eels, are in progress (R.E. Peter personal communication).

The second line of enquiry involves further investigation of the maturational changes themselves. The preliminary work indicating that olfaction is of reduced importance should be extended. Pilot studies on sexually immature eels (unpublished data) have shown that conditioned responses to olfactory stimuli can be detected in the electrocardiograms (ECGs) using radiotelemetry. Such a method allows investigation of functional responses to complement morphological investigations. The teleost ECG response has been used to study a variety of sensory inputs (Randall 1966; Blaxter and Tytler 1972; Laming and Savage 1980) and is conceivably appropriate in eels for examination of any sensory modality where the stimulus can be quantified. Telemetric techniques are also likely to be of use in examining the functional roles of skeletal muscle types at various swimming speeds and maturity states. Electromyographic studies have been conducted on a number of teleost species using fish tethered to the electrodes in a flow system (Hudson 1973; Johnston et al. 1977; Johnston and Moon 1980). Practical difficulties which prevent similar experiments from being conducted on restrained eels could be overcome by the use of radiotelemetry to monitor muscle activity. Changes in skin structure and function require further study, particularly with respect to the function of the increased scale cover and its possible effect on locomotion. Degree of scale cover has been suggested to be of importance in locomotion in other teleosts (Aleyev 1969). In addition the attitude of the fish body during swimming is known to be of importance in lateral-line function (Roberts and Russell 1972), and the possibility that there is a change in swimming mode in mature eels to facilitate lateral-line receptivity warrants consideration.

The study of the changes occurring in artificially matured eels provides an insight into the maturational process in naturally maturing eels and also a focus for productive areas of research, so that optimal use can be made of "wild" marine-migrating eels should they become available.

ACKNOWLEDGEMENTS

Thanks are extended to R.E. Peter for a critical review of the manuscript. The author was supported by a Commonwealth Scholarship (University of Bristol) and a Killam Postdoctoral Fellowship (University of Alberta) during this study.

REFERENCES

Aleyev, Yu G. 1969. Function and gross morphology in fish. Keter Press, Jerusalem, Israel.

Beatty, D.D. 1975. Visual pigments of the American eel *Anguilla rostrata*. Vision Research 15:771-776.

Bertin, L. 1956. Eels - A biological study. Cleaver-Hume, London, England.

Bieniarz, K., and P. Epler. 1975. Investigations on accelerating sexual maturity of eel (*Anguilla anguilla*) males. International Council for the Exploration of the Sea C.M. 1975/M2, Anadromous and Catadromous Fish Committee.

Blaxter, J.H.S., and M.P. Jones. 1967. The development of the retina and retinomotor responses in the herring. Journal of the Marine Biological Association of the United Kingdom 47:677-694.

Blaxter, J.H.S., and M. Staines. 1970. Pure retina and retinomotor responses in larval teleosts. Journal of the Marine Biological Association of the United Kingdom 50:449-460.

Blaxter, J.H.S., and P. Tytler. 1972. Pressure discrimination in teleost fish. Symposia of the Society of Experimental Biology 26:417-443.

Boddeke, R., E.J. Slijper, and A. Van der Stelt. 1959. Histological characteristics of the body musculature of fishes in connection with their mode of life. Koninklijke Nederland Akadamie van Wetenschappen Series C 42:576-587.

Boëtius, I., and J. Boëtius. 1967. Studies in the European eel *Anguilla*

anguilla (L.). Experimental induction of the male sexual cycle and its relation to temperature and other factors. Meddelelser fra Danmarks Fiskeri-og Havundersøgelser 4:339-405.

Boëtius, I., and J. Boëtius. 1980. Experimental maturation of female silver eels, *Anguilla anguilla*. Estimates of fecundity and energy reserves for migration and spawning. Dana 1:1-28.

Boëtius, J., I. Boëtius, A.M. Hemmingsen, A.F. Brunn, and E. Moller-Christensen. 1962. Studies of ovarial growth induced by hormone injections in the European and American eel (*Anguilla anguilla* L. and *Anguilla rostrata* LeSueur). Meddelelser fra Danmarks Fiskeri-og Havundersøgelser 3:183-198.

Boehlert, G.W. 1979. Retinal development in post-larval through juvenile *Sebastes diploproa*: adaptions to a changing photic environment. Revue canadienne de biologie 38:265-280.

Braekkan, O.R. 1956. Function of red muscle in fish. Nature (London) 178:747-748.

Burdak, V.D. 1980. The protective function of scale cover of fishes. Journal of Ichthyology 19(4):101-107.

Callamand, O. 1943. L'Anguille européenne. Les basis physiologique de sa migration. Annales de l'Institut Océanographique 21: 361-440.

Carlisle, C.P., and E.J. Denton. 1959. On the metamorphosis of the visual pigments of *Anguilla anguilla* (L.). Journal of the Marine Biological Association of the United Kingdom 38:97-102.

Chang, J.P., and R.E. Peter. In press. Effects of pimozide and DesGly^{6} [D-Ala^{6}] luteinizing hormone-releasing hormone ethylamide on serum gonadotropin concentrations, germinal vesicle migration and ovulation in female goldfish, *Carassius auratus*. General and Comparative Endocrinology.

D'Ancona, U. 1960. The life cycle of the Atlantic eel. Symposia of the Zoological Society of London 1:61-77.

Denton, E.J. 1970. On the organisation of reflecting surfaces in some marine animals. Philosophical Transactions of the Royal Society B 258:285-313.

Fontaine, M. 1964. Sur la maturation des organes génitaux de l'Anguille femelle (*Anguilla anguilla* L.) et l'émission spontanee des oeufs en aquarium. Comptes Rendus des hebdomindaires Séances de l'Académie des Sciences, Paris 259: 2907-2910.

Fontaine, Y.A. 1969. La specificité zoologique des proteins hypophysaires capables de stimuler la thyroide. Acta Endocrinologica supplementum 136:1-154.

Gordon, J., R.M. Shapley, and E. Kaplan. 1978. The eel retina. Receptor classes and spectral mechanisms. Journal of General Physiology 71:123-138.

Greer Walker, M., and G. Pull. 1975. A survey of red and white muscle in marine fish. Journal of Fish Biology 7:295-300.

Hontela, A., and R.E. Peter. 1978. Daily cycles in serum gonadotropin levels in the goldfish: effects of photoperiod, temperature and sexual condition. Canadian Journal of Zoology 56:2430-2442.

Hudson, R.C.L. 1973. On the function of white muscles in teleosts at intermediate swimming speeds. Journal of Experimental Biology 58:509-522.

Hulbert, W.C., and T.W. Moon. 1978a. General characteristics and morphology of eel (*Anguilla rostrata*) red and white muscle. Comparative Biochemistry and Physiology A. Comparative Physiology A 61:377-382.

Hulbert, W.C., and T.W. Moon. 1978b. A histochemical, light and electron microscope examination of eel (*Anguilla rostrata*) red and white muscle. Journal of Fish Biology 13:527-533.

Hulbert, W.C., and T.W. Moon 1978c. The potential for lactate utilization by red and white muscle of the eel *Anguilla rostrata*. Canadian Journal of Zoology 56:128-135.

Jakubowski, M. 1960. The structure and vascularisation of the skin of the eel (*Anguilla anguilla* L.) and the viviparous blenny (*Zoarces viviparus* L.). Acta biologica Cracoviensa Series Zoologica 3:1-22.

Jerlov, N.G. 1976. Marine optics. Elsevier, London, England.

Johns, P.R., and S.S. Easter. 1977. Growth of the adult goldfish retina. II Increase in retinal cell number. Journal of Comparative Neurology 176:331-341.

Johnston, I.A., W. Davison, and G. Goldspink. 1977. Energy metabolism of carp swimming muscles. Journal of Comparative Physiology B. Metabolic and transport functions 114B:203-216.

Johnston, I.A., and T.W. Moon. 1980. Exercise training in skeletal muscle of brook trout (*Salvelinus fontinalis*). Journal of Experimental Biology 87:177-194.

Kokhnenko, S.V., V.A. Bezdenezhnykh, and S.L. Gorovaya. 1978. Maturation of the European eel when artificially reared. Journal of Ichthyology 17:878-883.

Laming, P.R., and G.E. Savage. 1980. Physiological changes observed in the goldfish (*Carassius auratus*) during behavioural arousal and fright. Behavioural and Neural Biology 29:255-275.

Land, M.F. 1972. The physics and biology of animal reflectors. Progress in biophysics and molecular biology 24:75-106.

Leonard, J.B., and R.G. Summers. 1976. The ultrastructure of the integument of the American eel *Anguilla rostrata*. Journal of Cell and Tissue Research 171:1-30.

Lewander, K., G. Dave, M.L. Johansson, A. Larsson, and V. Lidman. 1974. Metabolic and hematological studies on the yellow and silver phases of the European eel, *Anguilla anguilla* L. I. Carbohydrate, lipid, protein and inorganic ion metabolism. Comparative Biochemistry and Physiology B. Comparative Biochemistry 47B:571-581.

Locket, N.A. 1975. Some problems of deep-sea fish eyes. Pages 645-656 *in* A.M. Ali, editor. Vision in fishes. Plenum Press, New York, New York, USA.

Lythgoe, J.N. 1979. The ecology of vision. Clarendon Press, Oxford, England.

Meyer-Waarden, P.F. 1965. Die wundersame Lebensgesicht des Aales.

Pages 7-47 *in* J.A. Keune and H. Struck, editors. Der Aal. H.A. Keune Verlag, Hamburg, Federal Republic of Germany.

Mittal, A.K., and T.K. Banerjee. 1980. Keratinisation versus mucus secretion in fish epidermis. Pages 1-12 *in* R.I.C. Spearman and P.A. Riley, editors. The skin of vertebrates. Academic Press, London, England.

Nose, T. 1971. Spawning of eel in a small aquarium. Rivista Italiana di Piscicoltura e Ittiopatologica 6:25-27.

Olivereau, M., and A. Lemoine. 1971. Action de la prolactine chez l'anguille intacte et hypophysectomisée. VII. Effet sur la teneur en acid sialique (N-acétyl neuraminique) de la peau. Zeitschrift für vergleichende Physiologie 73:34-43.

Pankhurst, N.W. 1982a. The relation of visual changes to the onset of sexual maturation in the European eel *Anguilla anguilla* (L.). Journal of Fish Biology 21:127-140.

Pankhurst, N.W. 1982b. Changes in body musculature with sexual maturation in the European eel, *Anguilla anguilla* (L.). Journal of Fish Biology 21:417-428.

Pankhurst, N.W. 1982c. Changes in the skin-scale complex with sexual maturation in the European eel, *Anguilla anguilla* (L.). Journal of Fish Biology 21:549-561.

Pankhurst, N.W., and J.N. Lythgoe. 1982. Structure and colour of the integument of the European eel, *Anguilla anguilla* (L.). Journal of Fish Biology 21:279-296.

Pankhurst, N.W., and J.N. Lythgoe. 1983. Changes in vision and olfaction during sexual maturation in the European eel, *Anguilla anguilla* (L.). Journal of Fish Biology 23:229-240.

Peter, R.E., and L.W. Crim. 1979. Reproductive endocrinology of fishes: gonadal cycles and gonadotropin in teleosts. Annual Review of Physiology 41:323-335.

Randall, D.J. 1966. The nervous control of cardiac activity in the tench (*Tinca tinca*) and the goldfish (*Carassius auratus*). Physiological Zoology 34:185-192.

Roberts, B.L., and I.J. Russel. 1972. The activity of the lateral line efferent neurones in stationary and swimming dogfish. Journal of Experimental Biology 57:435-458.

Sandy, J.M., and J.H.S. Blaxter. 1980. A study of retinal development in larval herring and sole. Journal of the Marine Biological Association of the United Kingdom 60:59-71.

Sinha, V.R.P., and J.W. Jones. 1975. The European freshwater eel. Liverpool University Press, Liverpool, England.

Smit, H., J.M. Amelink-Koutstaal, J. Vijverberg, and J.V. van Vaupel-Klein. 1971. Oxygen consumption and efficiency of swimming goldfish. Comparative Biochemistry and Physiology A. Comparative Physiology 39A:1-28.

Stramke, D. 1972. Veränderugen am Auge des europäischen Aales (*Anguilla anguilla* L.) während der Gelb- und Blankaalphase. Archiv für Fischereiwissenschaft 23:101-117.

Teichman, H. 1959. Über die Leistung des Geruchssinnes beim Aal (*Anguilla anguilla* L.). Zeitschrift für vergleichende Physiologie 42:206-254.

Tesch, F.-W. 1977. The eel. Chapman and Hall, London, England.

Van Oosten, J. 1957. The skin and scales. Pages 207-244 *in* M.E. Brown, editor. The physiology of fishes, volume 1. Academic Press, New York, New York, USA.

Villani, P., and F. Lumare. 1975. Nota sull'accrescimento ovarico indotta in *Anguilla anguilla* L. Investigacion Pesquera 39: 187-197.

Wald, G. 1960. The distribution and evolution of visual systems. Pages 311-345 *in* M. Florkin and H.S. Mason, editors. Comparative biochemistry, volume 1. Academic Press, New York, New York, USA.

Wittenberger, C. 1973. Metabolic interaction between isolated white and red carp muscles. Revue roumaine de Biologie Serie de Zoologie 18:71-76.

Wittenberger, C., D. Corprean, and L. Morar. 1975. Studies of the carbohydrate metabolism of the lateral muscles in carp (influence of phloridzin, insulin and adrenaline). Journal of Comparative Physiology A. Sensory, Neural, and Behavioral Physiology 101A:161-172.

Wittenberger, C., and I.A. Diaciuc. 1965. Effort metabolism of lateral muscle in carp. Journal of the Fisheries Research Board of Canada 22:1397-1406.

FISH MIGRATIONS IN COASTAL AND ESTUARINE ENVIRONMENTS: A CALL FOR NEW APPROACHES TO THE STUDY OF AN OLD PROBLEM

William C. Leggett

Department of Biology
McGill University
1205 Avenue Docteur Penfield
Montreal, Quebec H3A 1B1 Canada

ABSTRACT

The proposition is advanced that studies of fish migration have been impeded by an uncritical acceptance of the hypothesis that a high degree of orientation by individual fish is essential to successful directed migration, by the general absence of quantitative models which yield rigorously-testable hypotheses, and by a concentration of research effort on the behavioral as opposed to the ecological aspects of migration. The need for, and potential advantages of, a change in these attitudes and approaches are discussed.

INTRODUCTION

This paper seeks to place the present status of studies on fish migration into perspective. My original mandate was to focus on coastal and estuarine fish migrations. While I have attempted to adhere to this request, I have taken some liberties. I have done so because I believe the progress of research on the migrations of fishes in coastal and estuarine environments and in other environments, while significant, is not commensurate with the money and effort being expended.

It is my thesis that the progress of research on fish migration has been impeded by:
1) an excessive fixation on the concept of precision of oriented movements by individual fish;
2) the paucity of generalized predictive theory, developed a priori,

which defines testable hypotheses, the time and space scales within which these hypotheses can be effectively evaluated, and the most practical means of doing so;
3) the general failure of fish ecologists to systematically evaluate those models that do exist; and
4) the paucity of collaborative studies involving physical and chemical oceanographers/limnologists and students of behaviour, a collaboration whose absence has consistently reduced the results of potentially useful studies to the level of speculation.
These shortcomings have seriously compromised the effectiveness of our use of the powerful new technologies that have become available for the study of fish migrations in recent years. The progress of our science has suffered as a consequence.

In making these comments I wish to stress that my own work has suffered from the same shortcomings. My conclusions are therefore drawn not from superior insight but from bitter experience and from the frequent frustration of attempting to draw meaningful generalities and make reliable predictions from my own experience and that of others. It is my hope that this critical assessment of some of our past successes and failures, and of our biases, will stimulate a more general questioning of these matters and a more open and creative approach to future research on fish migrations.

THE PROBLEM OF PRECISION

Early students of fish migration employed the time-honoured technique of mark-recapture, and the most celebrated examples were centered on salmon (reviewed in Foerster 1968; Harden Jones 1968). The homing ability of salmon was recognized early, and the concept of highly directed movement from distant tagging sites to the natal streams, generated in large part by truly-remarkable migrations by a comparatively few fish, was soon entrenched (Davidson 1937). This concept was reinforced by later studies involving open-ocean migrations that continued to report strongly homeward-oriented movement but which failed to properly assess the statistical probabilities of being captured in areas or directions inappropriate to home (e.g. Went 1958; Royce et al. 1968; Vreeland 1975). Our subconscious may have been further conditioned by the plotting of these movements by connecting mark and recapture sites with straight lines.

Inappropriate movements are not uncommon or unreported, but have been largely relegated to the grey literature and ignored (T. Quinn, personal communication; see also Leggett 1977). Our collective willingness to discount the importance of these "strays," as revealed in the absence of a single comparative study of their general frequency and pattern, reveals our bias. Quinn (1984--this volume) has made a beginning in this area and is to be commended for his recognition of its importance.

The concept of precision, as it relates to orientation and migration, has been further reinforced by the demonstration that fish can obtain directional information from the sun, from the earth's geomagnetic field, from induced electrical fields, from polarized light and from olfactory cues (reviewed in Leggett 1977). What we have consistently failed to do, and not for want of trying, is to demonstrate that these abilities are used to maintain constant headings over distances exceeding a few hundred meters to a few kilometers.

The concept of precision should have been shattered by the coming of the age of biotelemetry. For those of us who were involved in such studies from the beginning, and who were conditioned to the precision precept, the early results were an outstanding lesson in frustration. We refused to capitulate completely, however, and frequently managed to explain away the most confusing tracks, to downplay others, and to accentuate the positive (for examples see McCleave and Horrall 1970; McCleave and LaBar 1972; Madison et al. 1972; Leggett 1976; Tesch 1979). Nothing has changed. Some fish clearly can maintain constant headings for considerable distances. Comparatively few do.

I believe that by ignoring the exceptions to our preconceived notion of what should be we have failed to exploit an outstanding opportunity to advance our understanding of the migration process.

Saila (1961), Saila and Shappy (1963) and Patten (1964) warned us, showed us this fixation on precision was unnecessary and intellectually unhealthy, and showed us that a slight degree of directional bias was sufficient to predict the occurrence of much of what we saw in the way of long range movement and homing. We have responded by attempting to show that they were wrong in the particular and have once again failed to critically test the importance of the underlying message of their works. Their message has been too long ignored.

Evidence in support of the view that a limited degree of directional bias may be sufficient to explain many long and seemingly complex migrations continues to mount, however. I contend we should begin to take it seriously and to alter our investigative approaches accordingly. A few examples will illustrate my point.

A wide array of fish species, and of developmental stages ranging from larvae to adults, exploit tidal currents to achieve directed movement (Stasko 1975; de Veen 1978; Greer Walker et al. 1978; Weinstein et al. 1980; Staples 1980; Arnold 1981; Melville-Smith et al. 1981; McCleave and Kleckner 1982). This behaviour facilitates movement of adults from spawning to feeding areas and return, and of marine larvae into estuarine and coastal nursery areas. Many of these migrations involve transport over considerable distances (Staples 1980; De veen 1978; Arnold and Cook, 1984--this volume). Similar exploitation of estuarine currents and of Ekmann transport and ocean currents have recently been shown to be important determinants of migration patterns and of seasonal changes in centres of distribution (LaBar et al. 1978; McCleave 1978; Bailey 1981; Parrish et al. 1981; Fortier and Leggett 1982). No precise orientation by individual fish is involved, although the results of recent studies with plaice (*Pleuronectes platessa*) suggest the selection of appropriate currents may be quite precise (Arnold 1981; Harden Jones 1981). Tidal transport is a fact, and its behavioral basis is now beginning to be understood. However, the jury remains out on the extent to which this and related forms of transport are synchronized, directed or prolonged. These facts notwithstanding, tidal transport does facilitate orderly, timed migrations between feeding and reproductive areas. What is required to achieve these migrations is not highly-directed movement, but appropriate directional bias. Reversal of directional bias involves changes in behaviour which have recently been confirmed. The exact cause of such behaviour reversal is not known in each instance, but numerous studies have documented the role of environment and developmental changes in altering behavioral responses to specific stimuli.

A second example involves Dodson and Leggett's (1973) work with American shad (*Alosa sapidissima*). Forty-three unimpaired Connecticut River shad were tracked in Long Island Sound in the area of the Connecticut River. All but two moved in directions quite inappropriate to river entry even though subsequent recovery rates of these fish in the river did not differ significantly from those of fish tagged in the river itself. Clearly, precise orientation was not a part of the behavioral repertoire of these fish. Directional bias induced by responses to tidal currents, perhaps altered by the presence or absence of home water cues (Dodson and Leggett 1974; Dodson and Dohse 1984--this volume) did, however, lead to successful homing.

THE ABSENCE OF MODELS

The study of fish migrations has suffered from a paucity of quantitative models and, in particular, models developed a priori as guides to the development of testable hypotheses and to the measurement scales necessary to achieve meaningful results. Unlike several other areas of behaviour and ecology where theory has preceded, and frequently inspired, field study (two outstanding examples being theories of optimal foraging (reviewed in Pyke et al. 1977; Krebs 1978), and life-history strategy (reviewed in Stearns 1976, 1977)), those quantitative models of fish migration that do exist have, in the main, been developed a posteriori in an attempt to explain or understand the results of descriptive studies (Weihs 1973a,b,c, 1974, 1978; Trump and Leggett 1980). While this in no way diminishes their subsequent worth, a posteriori development frequently means that the initial study generates less precise data, and hence less understanding, than would be possible had the model come first and the study second.

An example drawn from personal experience illustrates this point. From 1968 until 1975 my colleagues and I were continuously engaged in studies of the migratory behaviour and migration energetics of American shad along the Atlantic coast, in Long Island Sound, and in the Connecticut, York and St. Johns Rivers (Dodson et al. 1972; Leggett and Whitney 1972; Dodson and Leggett 1973, 1974; Leggett 1976, 1977; Leggett and Trump 1978; Glebe and Leggett 1981a,b). These studies, while generally designed to evaluate existing theory, followed the conventional mode--sample and observe, analyse and interpret. Subsequently, in an attempt to more fully understand the behaviour we had observed, we developed an explicit quantitative model of the optimum behavioral response of migrating fish to currents (Trump and Leggett 1980). This model was in no way dependent on the results of our previous studies for its development, relying as it did on the well-documented metabolic relationship between swimming speed and energy expenditure in fish (reviewed in Webb 1975; Beamish 1978). There was not reason, short of lack of foresight, why it could not have been developed before I initiated these field studies. Had I modelled first and studied later, data central to the testing of several questions which remained unrecognized until our fieldwork had terminated would have been gathered; the quality of the data brought to bear on the questions we did address would have been substantially improved; and several unproductive avenues of research would have been avoided.

To be more specific by way of illustration, one of our interests in these studies was the energetic cost of migration. We chose the conventional approach and assessed these costs by documenting changes in fat and protein reserves during the nonfeeding, freshwater migration (Glebe and Leggett 1981a,b). Instructive as these studies were, especially as they related to the question of the importance of the cost of migrations in shaping life-history strategies (Leggett and Carscadden 1978), they provided only an integrated estimate of costs over the entire migration. These costs could not be effectively extrapolated to ocean migrations, nor could we evaluate the importance of different swimming behaviours in different sectors of the river, as revealed by ultrasonic telemetry studies, on the overall costs. Our modelling efforts (Trump and Leggett 1980) showed that greater attention to the determination of current velocities during both the marine and freshwater tracking studies would have allowed the extrapolation of these energetic studies to other areas and other species and greatly increased their precision and information content. Greater attention to detail in studies of currents would also have facilitated a more precise evaluation of the extent to which swimming speeds during freshwater and marine migrations represented optimum energetic responses to currents. Finally, predicted relationships between swimming speeds, swimming efficiency and body size (Trump and Leggett 1980) could have been readily evaluated with little extra effort.

These opportunities and others, which could have been exploited with relatively little increase in effort, were missed and may never be regained given the magnitude and diversity of the program in question. Had the modelling been completed first they would not have been missed, and the reports of the field studies would have contained significantly less speculation regarding the possible significance of various patterns of observed behaviour.

Further evidence of the positive effects of explicitly-stated models or hypotheses on the design of studies of migration is provided in the success of A.D. Hasler and his co-workers in their systematic, experimental evaluation of the olfactory basis of homing (reviewed in Leggett 1977).

Not only is the field lacking in a priori models, but as students of migration we have been remarkable in our apparent reticence to systematically evaluate the predictions of models provided for us. The almost total absence of systematic studies designed to evaluate the more general implications of the models of Saila (1961), Saila and Shappy (1963), Patten (1964) and Weihs (1973a,b,c, 1974, 1978) are cases in point.

Models can serve to guide us in the design of our studies, thereby increasing the probability that we will actually collect the data we require. Equally importantly they frequently identify the most effective and cost-efficient means of obtaining that data. This should always be an important element of experimental design, but in times of rapidly increasing costs and declining resources its importance is amplified. For example, the efficacy of slight directional bias on the part of individual fish in producing directed migrations of populations of fishes can probably be evaluated more quickly and less expensively by employing a properly designed conventional tagging study than by telemetry techniques. If directional bias is evident and adequate to explain the observed movement, the more sophisticated and expensive ultrasonic technology could be brought to bear on the question of how this bias is achieved. This approach has been effectively exploited by the Lowestoft group in their studies of selective tidal stream transport in plaice (reviewed in Harden Jones 1981).

Few students of migration, least of all I, consider themselves competent modellers. Our corporate reluctance to engage in such activities may therefore be understandable but remains unjustified.

The potential benefits in terms of insight, accuracy, and cost are so great that I do not believe we can afford to continue to underplay this approach. Moreover, because of the considerable observational data base that exists, many such models are capable of preliminary testing before fieldwork begins. The scope for the development of empirical models as guides to further work is also enormous. This approach has been effectively exploited by students of freshwater production processes in recent years (Dillon and Rigler 1974; Ryder et al. 1974; Oglesby 1977; Matuszek 1978; McCauley and Kalff 1981; Hanson and Leggett 1982).

In many instances it may be possible to find answers to important general questions of orientation and migration by concentrating our field studies more effectively on critical gaps in existing data identified by such empirical models. Given these potential benefits I strongly believe we should seek to develop these skills ourselves, to insist on their acquisition by our students, and to solicit the aid of those with requisite skills by revealing to them the challenges and opportunities that await. The contributions of Trump (Trump and Leggett 1980), Weihs (Weihs 1973a,b,c, 1974, 1978) and Cook (Arnold and Cook 1984-- this volume) show the advantages of such collaboration.

THE NEED TO INTEGRATE STUDIES OF ENVIRONMENT AND BEHAVIOUR

A rapidly growing body of literature is strong testament to the role of physical and biological features of the environment in shaping the rate, direction, timing, precision and annual variations in migratory behaviour of fishes and other aquatic organisms. I reviewed much of the literature in this area to 1976 (Leggett 1977) and will not repeat myself here. Significant additional understanding and insight has been achieved since that time, however. Its impact is to further emphasize the potential benefits of an even greater attention to physical and biological features of the environment in the design and execution of studies of fish migration.

The Role of Currents

One of the most dramatic developments in fish migration in recent years has been the repeated demonstration of the importance of currents as the agent of migration in fishes in coastal and estuarine areas. The underlying generality arising from these studies is that seasonal migrations are achieved less by directed lateral movements by individual fish than by temporal differences in the current regimes with which they associate. The selective tidal stream transport (Greer Walker et al. 1978) exhibited by plaice in the North Sea is perhaps the best documented example of this behaviour to date (reviewed in Arnold 1981). This behaviour is exhibited by both juvenile and adult plaice, by several other flatfish species (de Veen

1978) and by American eels (*Anguilla rostrata*) (McCleave and Kleckner 1982).

Selective use of currents to achieve directed movement has now been documented for a wide variety of aquatic organisms. Staples (1980) has shown the larvae of banana prawns (*Penaeus mergurensis*), which spawn off the Gulf of Carpentaria in northern Australia, employ both wind-driven and tidal currents to achieve their migrations to estuarine nursery areas. The hatching of larvae is coincident with a shift in seasonal wind patterns from offshore to onshore. Coastwise movement results from passive drift in wind-driven onshore currents. Movement into the estuary is achieved by passive drift on the flood tide and settling behaviour on the ebb. The cues used by banana prawn larvae to select the appropriate tide are unknown. Diel migrations are not responsible. One possible cue is salinity which is maximal during flood tide. Interestingly, this behaviour changes with size. Larger larvae, which are concentrated in the upper estuary feeding areas, remain in the deep residual flow and are retained in the upper estuary. A similar example of the selective exploitation of coastal currents to achieve landward migration is reported by Miller et al. (1984--this volume).

Weinstein et al. (1980) have shown post-larvae of Atlantic croaker (*Micropogonius undulatus*), spot (*Leiostomus xanthurus*) and flounder (*Paralichthys* spp.) employ tidal and deep, landward, residual currents in the Cape Fear River estuary to achieve landward penetration and retention in high productivity areas. Differences in the distribution of the three species in the estuary are related to species-specific reactions to tidal and nontidal currents. Their occurrence in the shallower surface waters during flood tides causes them to be carried laterally into marshes and tidal streams. Atlantic croakers achieve landward movement by remaining exclusively in the strong, deep, landward, residual current characteristic of this intensively-flushed estuary. Because they avoid the surface waters, croakers are not commonly transported to the shallow water areas ultimately frequented by the larvae of spot and flounder.

The response of larvae of *Gilchristella aestuarius* (family Clupeidae) to currents in the Sunday River estuary, South Africa parallels that of croakers (Melville-Smith et al. 1981). Larvae are found predominantly in deep, landward, residual current which transports them up the estuary. Change in the mean depth distribution of the larvae occurred only at slack tide when stability of the water mass is lowest, as would be expected of passively-drifting larvae (Fortier and Leggett 1982). There is evidence that the larvae also avoid ebb tide currents by settling out on the bottom.

We have recently shown (Fortier and Leggett 1983) that the direction and rate of migration of larvae of Atlantic herring (*Clupea harengus*) and capelin (*Mallotus villosus*) in the St. Lawrence Estuary and ultimate distribution of the larvae in the estuary and Gulf of St. Lawrence is determined by occupation of different water masses and currents. Graham (1972) reported similar findings regarding the retention of herring larvae in the Sheepscot Estuary, Maine.

Capelin, which are spawned in the upper St. Lawrence Estuary, are found almost exclusively in the upper 20 m of the water column. This association with the seaward surface currents causes them to be rapidly flushed from the estuary toward the Gulf. The path of this migration follows that of the dominant surface currents (Jacquaz et al. 1977). The larvae are initially transported passively (Fortier and Leggett 1982) but develop increased swimming ability as they grow, which allows them to maintain the centre of their distribution progressively closer to the surface (Fortier and Leggett 1983). Since seaward current velocities are greatest in the surface waters, a differential rate of migration based on size is achieved. The result is a separation of larvae by size and a faster rate of drift in larger larvae. All larvae are ultimately entrained in a large gyre in the western Gulf of St. Lawrence (Jacquaz et al. 1977), an area of high primary and secondary productivity relative to adjacent areas in the Gulf (Steven 1974).

Herring spawn demersally in the upper St. Lawrence Estuary near Isle Verte. The larvae of herring emerge into the deep residual current and are initially transported passively landward. Vertical migratory behaviour, which develops rapidly with increasing size, ultimately causes the larvae to cross the pycnocline during these vertical movements. As the duration of time spent in the surface seaward current increases, the rate of landward displacement is slowed, and the larvae ultimately become entrained in the upper estuary, a zone of intense upwelling and high production (Rainville and Lacroix 1976). The vertical migrations of herring, which modify the pattern of drift and ultimately define the retention area, are strongly linked to the diel migrations of their principal prey (Fortier and Leggett in press).

Adult Atlantic salmon (*Salmo salar*) also appear to achieve landward movement in estuaries by passively riding the flood tides and actively stemming the ebb tide currents (Stasko and Rommel 1977). The coastal movements of Atlantic salmon may also involve selective use of tidal currents. Atlantic salmon smolts, too, appear to achieve seaward displacement by drifting passively with the currents (LaBar et al. 1978; McCleave 1978).

The extent to which nontidal oceanic currents influence and facilitate fish migrations is less well known. However, given the documented importance of currents in coastal and tidal areas, it is difficult to believe that the energetic and other advantages enjoyed by coastal and estuarine fishes would not also accrue to open-ocean migrants and result in similar adaptations (see Trump and Leggett

[1980] for an amplification of this point).

Interannual differences in rates and direction of larval drift, induced by changes in current direction and speed, are known to have significant influences on larval survival and ultimately on recruitment to adult stocks of marine fishes. Ekmann transport is known to directly influence the rates of larval drift of Atlantic menhaden (*Brevoortia tyrannus*) and Pacific hake (*Merluccius productus*) on the east and west coasts of North America, respectively. Year-class strength in menhaden is positively correlated with the intensity of landward Ekmann transport during the period of larval drift. In years when transport rates are high, larvae are carried to their coastal nursery areas quickly and survival is enhanced (Nelson et al. 1979). Year-class strength in Pacific hake is negatively correlated with the intensity of offshore Ekmann transport. When transport rates are high, larvae are more rapidly advected offshore and away from the most productive nursery areas, thereby reducing survival (Bailey 1981). The importance of current patterns in the reproductive and migratory behaviour of fishes associated with the California Current is further considered in Parrish et al. (1981).

The demonstrated importance of currents in the migration of coastal and estuarine fishes clearly warrants a systematic examination of the generality of this role in other environments. This will require a much greater utilization of the tools and skills of physical oceanography/limnology than has heretofore been the case.

In this context the importance of the temporal and spatial scale of observations relating the physics of the environment with observed behaviour during migrations cannot be underestimated (Smith 1978). In our studies of the migrations of larval capelin and herring in the St. Lawrence Estuary (Fortier and Leggett 1983, in press) we employed hourly sampling at three depth intervals. This temporal resolution was necessary to distinguish between the effects of passive displacement, induced by the dynamic nature of current and vertical stability patterns in the area, and active vertical migration. Even at this scale not all components of this behaviour could be resolved.

Parrish et al. (1981) have recently shown daily changes in the magnitude and direction of Ekmann transport in the California Current system to be equally dramatic. A thorough understanding of the response of fish to currents and other physical variables will require that the physical data be gathered on a temporal and spatial scale relevant to the animal under study. This has been a major shortcoming of telemetry studies to date.

Temperature and Salinity

The temporal and spatial pattern of temperature, oxygen concentration and salinity is known to be highly variable. This variability is induced by circulation patterns, seasonal changes in atmospheric conditions, localized production/decomposition maxima, etc. I previously reviewed the relationship between such variation and seasonal movements of a variety of fish species (Leggett 1977). Evidence continues to build in support of the hypothesis that migrations are both caused and directed by such physical features of the environment. The seasonal movements of yearling amber fish (*Seriola quinqueradiata*) in the Japan Sea is now known to be influenced by seasonal shifts in the 16-17 C isotherm. Unusual shifts in distribution are also known to be consistent with atypically cold years (Watanabe 1978). The seasonal distribution of squid (*Loligo pealei*), which ranges over 800 km between winter and summer, also appears to be strongly linked to changes in the temperature of major ocean masses (Vovk 1978).

The role of ocean fronts, and associated gradients in temperature, salinity and production, as regulators of the distribution and movements of fishes has long been suspected (Leggett 1977). Magnusson et al. (1981) have recently demonstrated the effect of short term variations in the location of a Gulf Stream front off Cape Hatteras on the distribution of both fishes and macroinvertebrates in the area.

The role of frontal systems in regulating the migrations and distributions of albacore (*Thunnus alalunga*) has also been more clearly defined in recent years. Laurs and Lynn (1977; Lynn 1984--this volume) have demonstrated, by combined intensive oceanographic sampling and fishing effort (which should serve as a model for future studies of this type), that a major E-W transition zone in the North Pacific Ocean has a strong influence on the coastward migration. The transition zone is bounded by cool, low salinity sub-arctic water on the north and warm, high salinity, eastern North Pacific central water on the south. When this zone, characterized by sharp salinity and temperature gradients, is distinct, the shoreward migrations are slower and the distribution of fishes more concentrated. When the zone is broader, and hence less distinct, shoreward migration is faster and the distribution wider. Laurs and Lynn suggest these influences on migration rates and routes may be occasioned by temperature preferences and the effect of thermal gradients on thermoregulation, and perhaps to food availability linked to upwelling. W.H. Neill (personal communication) has recently provided additional support for this hypothesis by the successful modelling of albacore movements in the frontal zone based on temperature profiles and known temperature preferences of albacore.

The temporal and spatial distribution of specific temperature and salinity zones in the ocean is known to be variable on scales of months and years (see Lasker 1978). Clark et al. (1975) have used tree ring growth anomalies as indicators of changes in marine climate to demonstrate that annual water temperature anomalies in the northeast Pacific can explain up to 83% of the variance in the catch distribution of the albacore fishery on the Pacific coast of North America. Historical records of anomalous catches in the fishery also correspond well with tree growth anomalies which are induced by oceanic climate. Clear warm conditions during the spring and summer favour a more northerly and inshore distribution of tuna. The high heat storage in the ocean in such circumstances results in greater evaporation, cyclonic activity and rainfall in the fall and winter. This rainfall stimulates tree growth the following year and generates the observed correlation.

Knowledge of the degree of coupling between atmospheric and ocean environments has expanded greatly in recent years, and its implications for students of fish migration are becoming increasingly evident (Lasker 1978; Hartline 1980). The influence of such coupling on an annual basis has been discussed above. We have recently found that meteorological conditions can have a major and synchronous influence on temperature and salinity conditions over hundreds of miles of coastal Newfoundland at a scale of hours (Frank and Leggett 1982; Taggart and Leggett unpublished data). These variations are induced by wind-driven water-mass exchange. Such short term changes in temperature and salinity regimes may influence hour-to-hour behaviour of fishes and will obscure more general behaviour patterns if their influence is not considered in the interpretation of ultrasonic telemetry or other distributional data. L. Mysak (personal communication) has suggested that the effect of seiches on the thermal structure and small-scale current pattern of lakes may similarly confound attempts to interpret ultrasonic telemetry data on the movement of fishes in lakes if these effects are not monitored and incorporated into the analysis.

Food and Predators

The migrations of fishes have long been considered to be an adaptation to increase feeding opportunities and hence abundance (Nikolskii 1963; for review see Northcote 1978). Demonstration of a direct link between feeding activity and migrations has been elusive, however. My colleagues and I have recently documented two distinct coastal migrations which are strongly linked to the temporally variable availability of food. These examples derive from our studies of the ecology of capelin in eastern Newfoundland and involve the larvae of capelin, on the one hand, and the winter flounder (*Pseudopleuronectes americanus*), an important seasonal predator on capelin adults and eggs, on the other.

The initiation of larval drift in capelin is strongly tied to the occurrence of onshore winds (Frank and Leggett 1981), which, as previously described, induce coastal water-mass exchange (Frank and Leggett 1982). Surface waters, which dominate the nearshore during onshore winds, are characterized by having higher densities of zooplankton in the edible size range for larval capelin, $<250\mu m$ (Frank and Leggett 1982). The initiation of larval drift during onshore winds thus ensures the co-occurrence of capelin larvae with maximal concentrations of food in the appropriate size spectrum. A further advantage accrues from this behaviour. Predator densities are significantly reduced in the surface water mass, relative to the deeper upwelling waters which prevail during offshore winds (Frank and Leggett 1982). The initiation of larval drift during onshore winds thereby reduces predator impact. Preliminary enclosure experiments suggest the resulting decrease in predator-induced mortality may be up to 60%.

The occurrence of winter flounder in nearshore areas adjacent to capelin spawning sites is also highly variable. Scuba surveys conducted over two years (Frank and Leggett unpublished data) show the occurrence of flounder to be highly correlated to the presence of spawning capelin. This seasonal occurrence of flounder is size specific. The timing of arrival of larger flounder, which prey predominantly on adult capelin, in the nearshore area is most closely linked to the presence of mature adult capelin near the beaches prior to spawning. Peak numbers of smaller flounder, which feed predominantly on capelin eggs, occur slightly later following spawning. Both large and small flounder leave the nearshore area when adults and eggs of capelin become unavailable.

The coincidence of the onset of larval drift and the wind driven occurrence of food-rich, predator-poor surface waters in the nearshore area is not restricted to capelin. The abundance of larvae of demersal spawning species, other than capelin, in the nearshore waters of coastal Newfoundland is also strongly and positively linked to coastal water-mass replacement and co-occurs with the onset of larval drift in capelin (Frank and Leggett 1983). Three potential advantages accrue from this linking of the onset of larval drift to processes of water-mass exchange: 1) food levels and growth rates are enhanced; 2) the density of the associated predator field is reduced; and 3) predation rates on other species may be further reduced by association with the numerically dominant larvae of capelin which saturate the predator field. This co-occurrence of species of lesser abundance with a dominant species is widespread (Richards 1959; Pearcy and Richards 1962; Herman 1963; Chenoweth 1972; Faber 1976) and suggests the apparent synchrony in the timing of life stage events may be adaptive. This potential influence on fish distribution and migration has not previously been considered.

SUMMARY

Studies of fish migration have historically followed two main paths: 1) detailed, primarily laboratory-based studies of orientation mechanisms which have been strongly influenced in their design and execution by the largely untested assumption that a high degree of orientation is essential to successful directed migration and homing; and 2) predominantly descriptive field studies of the large- and small- scale movements of fishes in a variety of habitats. These field studies, while almost invariably purporting to test general theory have, in the main, failed to do so. In large part this failure stems from the failure of students of fish migration to develop, or encourage the development of, explicitly stated theories yielding a priori predictions which are both operational and testable. These problems have been compounded by the fact that the approach to the study of fish migrations has been, until quite recently, primarily ethological, whereas it is becoming increasingly apparent that many of the major factors regulating fish migrations are ecological.

It is not my intent to denigrate the work that has been conducted to date. It has yielded a rich mechanistic and observational data base and represents a valuable resource. The time has come, however, for a more synthetic and systematic approach to the problem. This will require a partial abandonment of existing ideologies, greater attention to the development of rigorous, testable models of fish migration, greater willingness to identify and evaluate the important predictions of these models, particularly those which have the potential of yielding generalizable results, and greater attention to the influence of fine- and large-scale variability in the physical and biological environment on the migratory behaviour of fishes. Several outstanding examples of the power of this approach are documented in this paper and in this volume.

REFERENCES

Arnold, G.P. 1981. Movements of fish in relation to water currents. Pages 55-79 *in* D.J. Aidley, editor. Animal migration. Cambridge University Press, Cambridge, England.

Arnold, G.P., and P.H. Cook. 1984. Fish migration by selective tidal stream transport: first results with a computer simulation model for the European continental shelf. Pages 227-261 *in* J.D. McCleave, G.P. Arnold, J.J. Dodson, and W.H. Neill, editors. Mechanisms of migration in fishes. Plenum Press, New York, New York, USA.

Bailey, K.M. 1981. Larval transport and recruitment of Pacific hake *Merluccius productus*. Marine Ecology Progress Series 6:1-9.

Beamish, F.W.H. 1978. Swimming capacity. Pages 101-187 *in* W.S. Hoar and D.J. Randall, editors. Fish physiology, volume 7. Academic Press, New York, New York, USA.

Chenoweth, S.S. 1972. Fish larvae of the estuaries and coast of central Maine. US National Marine Fisheries Service Fishery Bulletin 71:105-113.

Clark, N.E., T.J. Blasing, and H.C. Fritts. 1975. Influence of interannual climatic fluctuations on biological systems. Nature (London) 256:302-305.

Davidson, F.A. 1937. Migration and homing in Pacific salmon. Science (Washington DC) 86:1-4.

Dillon, P.J., and F.J. Rigler. 1974. The phosphorus-chlorophyll relationship in lakes. Limnology and Oceanography 19:767-773.

Dodson, J.J., and L.A. Dohse. 1984. A model of olfactory-mediated conditioning of directional bias in fish migrating in reversing tidal currents based on the homing migration of American shad (*Alosa sapidissima*). Pages 263-281 *in* J.D. McCleave, G.P. Arnold, J.J. Dodson and W.H. Neill, editors. Mechanisms of migration in fishes. Plenum Press, New York, New York, USA.

Dodson, J.J., W.C. Leggett, and R.A. Jones. 1972. The behaviour of adult American shad (*Alosa sapidissima*) during migration from salt to fresh water as observed by ultrasonic tracking techniques. Journal of the Fisheries Research Board of Canada 29:1445-1449.

Dodson, J.J., and W.C. Leggett. 1973. Behaviour of adult American shad (*Alosa sapidissima*) homing to the Connecticut River from Long Island Sound. Journal of the Fisheries Research Board of Canada 30:1847-1860.

Dodson, J.J., and W.C. Leggett. 1974. Role of olfaction and vision in the behaviour of American shad (*Alosa sapidissima*) homing to the Connecticut River from Long Island Sound. Journal of the Fisheries Research Board of Canada 31:1607-1619.

Faber, D.J. 1976. Hypo-neustonic fish larvae in the Northumberland Strait during summer 1962. Journal of the Fisheries Research Board of Canada 33:1167-1174.

Foerster, R.E. 1968. The sockeye salmon. Bulletin of the Fisheries Research Board of Canada 162.

Fortier, L., and W.C. Leggett. 1982. Fickian transport and the dispersal of fish larvae in estuaries. Canadian Journal of Fisheries and Aquatic Sciences 39:1150-1163.

Fortier, L., and W.C. Leggett. 1983. Vertical migrations and transport of larval fish in a partially mixed estuary. Canadian Journal of Fisheries and Aquatic Sciences 40:1543-1555.

Fortier, L., and W.C. Leggett. In press. Small scale covariability in the abundance of fish larvae and their prey. Canadian Journal of Fisheries and Aquatic Sciences.

Frank, K.T., and W.C. Leggett. 1981. Wind regulation of larval emergence in capelin (*Mallotus villosus*) and its population consequences. Canadian Journal of Fisheries and Aquatic Sciences 38:215-223.

Frank, K.T., and W.C. Leggett. 1982. Coastal water mass replacement: its effect on zooplankton dynamics and the predator-prey complex associated with larval capelin (*Mallotus villosus*). Canadian Journal of Fisheries and Aquatic Sciences 39:991-1103.

Frank, K.T., and W.C. Leggett. 1983. Multispecies larval fish associations: accident or adaptation? Canadian Journal of Fisheries and Aquatic Sciences 40:754-762.

Glebe, B.D., and W.C. Leggett. 1981a. Latitudinal differences in energy allocation and use during the freshwater migrations of American shad (*Alosa sapidissima*) and their life history consequences. Canadian Journal of Fisheries and Aquatic Sciences 38:806-820.

Glebe, B.D., and W.C. Leggett. 1981b. Temporal, intra-population differences in energy allocation and use by American shad (*Alosa sapidissima*) during the spawning migration. Canadian Journal of Fisheries and Aquatic Sciences 38:795-805.

Graham, J.J. 1972. Retention of larval herring within the Sheepscot estuary of Maine. US National Marine Fisheries Service Fishery Bulletin 70:299-305.

Greer Walker, M., F.R. Harden Jones, and G.P. Arnold. 1978. The movements of plaice (*Pleuronectes platessa*) tracked in the open sea. Journal du Conseil Conseil International pour l'Exploration de la Mer 38:58-86.

Hanson, J.M., and W.C. Leggett. 1982. Empirical prediction of fish biomass and yield. Canadian Journal of Fisheries and Aquatic Sciences 39:257-263.

Harden Jones, F.R. 1968. Fish migration. St. Martins, New York, New York, USA.

Harden Jones, F.R. 1981. Fish migration: strategy and factors. Pages 139-165 *in* D.J. Aidley, editor. Animal migration. Cambridge University Press, Cambridge, England.

Hartline, B.K. 1980. Coastal upwelling: physical factors feed fish. Science (Washington DC) 208:38-40.

Herman, S. 1963. Planktonic fish eggs and larvae of Narragansett Bay. Limnology and Oceaography 8:103-109.

Jacquaz, B., K.W. Able, and W.C. Leggett. 1977. Seasonal distribution, abundance, and growth of larval capelin (*Mallotus villosus*) in the St. Lawrence estuary and northern Gulf of St. Lawrence. Journal of the Fisheries Research Board of Canada 34:2015-2029.

Krebs, J.R. 1978. Optimal foraging: decision rules for predators. Pages 23-63 *in* J.R. Krebs and N.B. Davies, editors. Behavioral ecology. Sinauer Associates, Sunderland, Massachusetts, USA.

LaBar, G.W., J.D. McCleave, and S.M. Fried. 1978. Seaward migration of hatchery-reared Atlantic salmon (*Salmo salar*) smolts in the Penobscot River estuary, Maine: open-water movements. Journal du Conseil Conseil International pour l'Exploration de la Mer 38:257-269.

Lasker, R. 1978. Ocean variability and its biological effects - regional review - Northeast Pacific. Rapports et Procès-Verbaux des Réunions Conseil International pour l'Exploration de la Mer 173:168-181.

Laurs, R.M., and R.J. Lynn. 1977. Seasonal migration of albacore, *Thunnus alalunga*, into North American coastal waters, their distribution, relative abundance and association with Transition

Zone waters. US National Marine Fisheries Service Fishery Bulletin 75:795-822.

Leggett, W.C. 1976. The American shad (*Alosa sapidissima*) with special reference to its migration and population dynamics in the Connecticut River. Pages 169-225 *in* D. Merriman and L.M. Thorpe, editors. The Connecticut River ecological study: the impact of a nuclear power plant. American Fisheries Society Monograph Number 1.

Leggett, W.C. 1977. The ecology of fish migrations. Annual Review of Ecology and Systematics 8:285-308.

Leggett, W.C., and R.R. Whitney. 1972. Water temperatures and the migrations of American shad. US National Marine Fisheries Service Fishery Bulletin 70:659-670.

Leggett, W.C., and J.C. Carscadden. 1978. Latitudinal variation in reproductive characteristics of American shad (*Alosa sapidissima*) evidence for population specific life history strategies in fish. Journal of the Fisheries Research Board of Canada 35:1469-1478.

Leggett, W.C., and C.L. Trump. 1978. Energetics of migration in American shad. Pages 370-377 *in* K. Schmidt-Koenig and W.T. Keeton, editors. Animal migration, navigation, and homing. Springer-Verlag, New York, New York, USA.

Lynn, R. 1984. Aspects of physical oceanography as they relate to the migration of fish. Pages 471-486 *in* J.D. McCleave, G.P. Arnold, J.J. Dodson and W.H. Neill, editors. Mechanisms of migration in fishes. Plenum Press, New York, New York, USA.

Madison, D.M., R.M. Horrall, A.B. Stasko, and A.D. Hasler. 1972. Migratory movements of adult sockeye salmon (*Oncorhynchus nerka*) in coastal British Columbia as revealed by ultrasonic tracking. Journal of the Fisheries Research Board of Canada 29:1025-1033.

Magnusson, J.J., C.L. Harrington, D.J. Stewart, and G.N. Herbst. 1981. Responses of macrofauna to short-term dynamics of a Gulf Stream front on the continental shelf. Pages 441-448 *in* F.A. Richards, editor. Coastal upwelling. American Geophysical Union, Washington, District of Columbia, USA.

Matsuzek, J.E. 1978. Empirical predictions of fish yields of large North American lakes. Transactions of the American Fisheries Society 107:385-394.

McCauley, E., and J. Kalff. 1981. Empirical relationships between phytoplankton and zooplankton biomass in lakes. Canadian Journal of Fisheries and Aquatic Sciences 38:458-463.

McCleave, J.D. 1978. Rhythmic aspects of estuarine migration of hatchery-reared Atlantic salmon (*Salmo salar*) smolts. Journal of Fish Biology 12:559-570.

McCleave, J.D., and R.M. Horrall. 1970. Ultra-sonic tracking of homing cutthroat salmon, *Salmo clarki* in Yellowstone Lake. Journal of the Fisheries Research Board of Canada 27:715-730.

McCleave, J.D., and G.W. LaBar. 1972. Further ultrasonic tracking and tagging studies of homing cutthroat salmon (*Salmo clarki*) in Yellowstone Lake. Transactions of the American Fisheries Society 101:44-54.

McCleave, J.D., and R.C. Kleckner. 1982. Selective tidal stream transport in the estuarine migration of glass eels of the American eel (*Anguilla rostrata*). Journal du Conseil Conseil International pour l'Exploration de la Mer 40:262-271.

Melville-Smith, R., D. Baird, and P. Wooldridge. 1981. The utilization of tidal currents by the larvae of an estuarine fish. South African Journal of Zoology 16:10-13.

Miller, J.M., J.P. Reed, and L.J. Pietrafesa. 1984. Patterns, mechanisms and approaches to the study of migration of estuarine dependent fish larvae and juveniles. Pages 209-225 *in* J.D. McCleave, G.P. Arnold, J.J. Dodson and W.H. Neill, editors. Mechanisms of migration in fishes. Plenum Press, New York, New York, USA.

Nelson, W.R., M.C. Ingham, and W.E. Schaaf. 1979. Larval transport and year-class strength of Atlantic menhaden, *Brevoortia tyrannus* US National Marine Fisheries Service Fishery Bulletin 75:23-41.

Nikolskii, G.V. 1963. The ecology of fishes. Academic Press, New York, New York, USA.

Northcote, T.G. 1978. Migratory strategies and production in freshwater fishes. Pages 326-359 *in* S.D. Gerking, editor. Ecology of freshwater fish production. Blackwell, Oxford, England.

Oglesby, R.T. 1977. Relationships of fish yield to lake phytoplankton standing crop, production, and morphoedaphic factors. Journal of the Fisheries Research Board of Canada 34:2271-2279.

Parrish, R.H., C.S. Nelson, and A. Bakun. 1981. Transport mechanisms and reproductive success of fishes in the California Current. Biological Oceanography 1:175-203.

Patten, B.C. 1964. The rational decision process in salmon migration. Journal du Conseil Conseil Permanent International pour l'Exploration de la Mer 28:410-417.

Pearcy, W.G., and S.W. Richards. 1962. Distribution and ecology of fishes of the Mystic River estuary, Connecticut. Ecology 43:248-259.

Pyke, G.H., H.R. Pulliam, and E.L. Charnov. 1977. Optimal foraging: a selective review of theory and tests. Quarterly Review of Biology 52:137-154.

Quinn, T. 1984. Homing and straying in Pacific salmon. Pages 357-362 *in* J.D. McCleave, G.P. Arnold, J.J. Dodson and W.H. Neill, editors. Mechanisms of migration in fishes. Plenum Press, New York, New York, USA.

Rainville, L., and G. Lacroix. 1976. Étude comparative de la distribution verticale et de la composition des populations de zooplankton du Fjord du Saguenay et de l'estuaire maritime du Saint-Laurent. Groupe Interuniversitaire des Recherches Oceanographiques du Quebec, Rapporte Annales 1974-1975:49-51.

Richards, S.W. 1959. Oceanography of Long Island Sound. VI. Pelagic fish eggs and larvae of Long Island Sound. Bulletin of the Oceanographic Collection 17:95-124.

Royce, W.F., L.S. Smith, and A. Hartt. 1968. Models of ocean migrations of Pacific salmon and comments on guidance mechanisms. US National Marine Fisheries Service Fishery Bulletin 66:441-462.

Ryder, R.A., S.R. Kerr, K.H. Loftus, and H.A. Regier. 1974. The morphoedaphic index, a fish yield estimator -- review and evaluation. Journal of the Fisheries Research Board of Canada 31:663-688.

Saila, S.B. 1961. A study of winter flounder movements. Limnology and Oceanography 6:292-298.

Saila, S.B., and R.A. Shappy. 1963. Random movement and orientation in salmon migration. Journal du Conseil Conseil Permanent International pour l'Exploration de la Mer 28:153-166.

Smith, P.E. 1978. Biological effects of ocean variability: time and space scales of biological response. Rapports et Procès-Verbaux des Réunions Conseil International pour l'Exploration de la Mer 173:117-127.

Staples, D.J. 1980. Ecology of juvenile and adolescent banana prawns *Tenaeus merguiensis*, in a mangrove estuary and adjacent off-shore area of the Gulf of Carpenteria. I. Immigration and settlement of post larvae. Australian Journal of Freshwater Research 31:635-652.

Stasko, A.B. 1975. Progress of migrating Atlantic salmon (*Salmo salar*) along an estuary observed by ultrasonic tracking. Journal of Fish Biology 7:329-338.

Stasko, A.B., and S.A. Rommel Jr. 1977. Ultrasonic tracking of Atlantic salmon and eels. Rapports et Procès-Verbaux des Réunions Conseil International pour l'Exploration de la Mer 170:36-40.

Stearns, S.C. 1976. Life history tactics: a review of the ideas. Quarterly Review of Biology 51:3-47.

Stearns, S.C. 1977. The evolution of life history traits: a critique of the theory and a review of the data. Annual Review of Ecology and Systematics 8:145-171.

Steven, D.M. 1974. Primary and secondary production in the Gulf of St. Lawrence. McGill University Montreal Marine Sciences Centre Manuscript Report.

Tesch, F.W. 1979. Tracking of silver eels (*Anguilla anguilla* L.) in different shelf areas of the northeast Atlantic. Rapports et Procès-Verbaux des Réunions Conseil International pour l'Exploration de la Mer 174:104-114.

Trump, C.L., and W.C. Leggett. 1980. Optimum swimming speeds in fish: the problem of currents. Canadian Journal of Fisheries and Aquatic Sciences 37:1086-1092.

Veen, J.F. de. 1978. On selective tidal transport in the migration of North Sea plaice (*Pleuronectes platessa*) and other flatfish species. Netherlands Journal of Sea Research 12:115-147.

Vovk, A.N. 1978. Peculiarities of the seasonal distribution of the

North American squid *Loligo pealei* (Lesueur 1821). Malacological Review 11:130.

Vreeland, R.R., R.J. Wahle, and A.H. Arp. 1975. Homing behavior and contribution to Columbia River fisheries of marked coho salmon released at two locations. US National Marine Fisheries Service Fishery Bulletin 73:717-725.

Watanabe, K. 1978. Studies on the juvenile stage of the amber fish in the eastern Japan Sea. IV. The recovery of tagged yearlings in the Ryotsu Bay of Sado Island, with special reference to their migratory behavior. Bulletin of the Japan Sea Regional Fisheries Research Laboratory 29:89-102.

Webb, T.W. 1975. Hydrodynammics and energetics of fish propulsion. Bulletin of the Fisheries Research Board of Canada 190:158.

Weihs, D. 1973a. Hydromechanics of fish schooling. Nature (London) 241:290-291.

Weihs, D. 1973b. Optimal fish cruising speed. Nature (London) 245:48-50.

Weihs, D. 1973c. Mechanically efficient swimming techniques for fish with negative buoyancy. Journal of Marine Research 31:194-209.

Weihs, D. 1974. Energetic advantages of burst swimming of fish. Journal of Theoretical Biology 48:215-229.

Weihs, D. 1978. Tidal stream transport as an efficient method of migration. Journal du Conseil Conseil International pour l'Exploration de la Mer 38:92-99.

Weinstein, M.P., S.L. Weiss, R.G. Hodson, and L.R. Gerry. 1980. Retention of three taxa of post larval fishes in an intensively flushed estuary, Cape Fear River, North Carolina. US National Marine Fisheries Service Fishery Bulletin 78:419-436.

Went, A.E.J. 1958. Salmon movements around Ireland. VIII. From drift nets along the coast of County Donegal (1953-1957). Proceedings of the Royal Irish Academy Section B Biological, Geological and Chemical Science 59B:205-212.

THE ORIENTATION OF FISH AND THE VERTICAL STRATIFICATION AT FINE- AND MICRO-STRUCTURE SCALES

Håkan Westerberg
Department of Oceanography
University of Gothenburg
Box 4038
S-400 40 Gothenburg, Sweden

ABSTRACT

Some characteristic properties of vertical fine structure and microstructure in natural waters are surveyed. Ubiquitous thermal microstructure, by its anisotrophy due to shear in the fine-structure gradient layers, is a property that could be used by fishes to orient relative to the local current shear. The geostrophic shear causes a bias in the probability of breaking of internal waves, depending on the direction of wave travel relative to the shear. By this mechanism the direction of major ocean currents may be mirrored in local microstructure. The influence of the fine-structure field on the character of large-scale dispersion of an odorous substance leads to a hypothesized mechanism for olfactory orientation. Some results from ultrasonic tracking of Atlantic salmon and European eels are given to show that fish do perceive and react to vertical fine structure.

INTRODUCTION

Traditionally, fisheries hydrography deals with the behaviour of fishes in relation to such large-scale hydrographic features as permanent current gyres and areas of upwelling with length scales on the same order of magnitude as the ocean basin. The time scales involved are on the order of 1 year. With the advent of instruments for the continuous recording of salinity and temperature, it has become clear that the ocean is rich in small-scale structures with strong vertical variability all the way down to centimeter scales. These fine-structure and microstructure properties have been studied extensively during the last two decades, because they are important for the understanding of mixing processes in the ocean.

From the point of view of the individual fish, small-scale features constitute the aspect of the hydrography that is primarily accessible to the senses, but little attention has been given to fine structure in the study of fish behaviour. The notion of orientation to the "rheocline" put forward by Harden Jones (1968) is the only example of which I am aware.

The purpose of this article is to review some of the observations and interpretations of fine structure that have been made, to summarize the properties of fine structure that may be important in a fish's perception of its hydrographic environment and to argue the importance of fine structure for the orientation of fishes.

A crucial point with regard to orientation is the existence of directional cues that can be observed by the fish. The limitations in the ability of fish to detect current directions in midwater have been discussed by Harden Jones (1968) and Arnold (1974). I suggest that the current-shear field in the ocean can be detected by fishes, not through direct perception of the weak velocity gradients, but through perception of microstructure thermal fluctuations resulting from current shear. The fine-structure layers in the ocean are striped with fish-sized thermal filaments, pointing in the direction of the shear and with temperature differences well within the sensory limits of fishes.

There are several possible mechanisms by which fish might use the shear field in orientation. An essential point is that the vertical stratification in many cases mirrors the mean distribution of hydrographic parameters that exist at a much larger scale in the horizontal dimension. Being free to move in three dimensions, a fish is able to gather information in the vertical that can be used for horizontal orientation.

Some observations made with ultrasonic tracking are given to illustrate that fishes indeed perceive and seem to react to hydrographic fine structure, and a hypothesized mechanism for olfactory orientation is put forward and discussed for the case of salmon homing.

SMALL-SCALE PROCESSES IN THE OCEAN

The terms *fine structure* and *microstructure* are essentially descriptive and based on the vertical resolution of the sampling instrument used. Scales from the gross structure resolved by classical hydrographic cast (10–100 m) down to meter scales can be measured by modern sounding instruments lowered by wire, as CTD (conductivity-temperature-depth) equipment; features measured on the scale of 1-10 m comprise the fine structure. To study the microstructure, which is defined on scales less than 1 m, free-falling instruments normally are needed to avoid distortions by the heaving of the ship.

Continuous recordings of salinity and temperature profiles in all natural bodies of water reveal large perturbations of the gross vertical stratification. Typically, fine-structure gradients are at least an order of magnitude larger than the mean vertical gradient, amounting in many cases to "steps" in the profile, with almost homogeneous layers separated by sharp gradients. A seasonal thermocline typically will contain about 10 such steps, and local salinity-compensated temperature inversions occur frequently.

Most measurements of small-scale structure are those of temperature and salinity, which are dynamically significant variables in that they determine the density stratification. The distribution of other properties, such as dissolved gases and chemical substances, are concomitant to that of temperature, although the magnitude of the concentration change at a particular temperature step need not necessarily be related to the magnitude of the temperature change. The first descriptions of fine structure in the stratification of the deep ocean in fact were made using observations of oxygen and silica concentrations (Cooper 1967).

Vertical variability is the result of a large number of different processes. A distinction can be made between reversible fine structure, which is the temporary distortion of a smooth density gradient by periodic wave motions, and irreversible structures caused by mixing processes. According to the source of energy for the mixing, irreversible fine structure can be related to the following processes:

1. Surface mixing, where local temperature and salinity anomalies are caused by heating, precipitation and run-off to the surface layer. Mechanical stirring by the wind will cause well-mixed volumes that spread by gravity at their appropriate density levels.

2. Bottom-boundary mixing, where the stratified fluid of the interior is mixed vertically along the bottom topography and the mixed volumes separate and spread by gravity and advection into the interior.

3. Internal mixing from local shear-induced turbulence, or overturning of internal waves.

4. Double-diffusive mixing, where a stable density stratification can become convecting if the vertical gradient of either salt or temperature is destabilizing in its contribution to density.

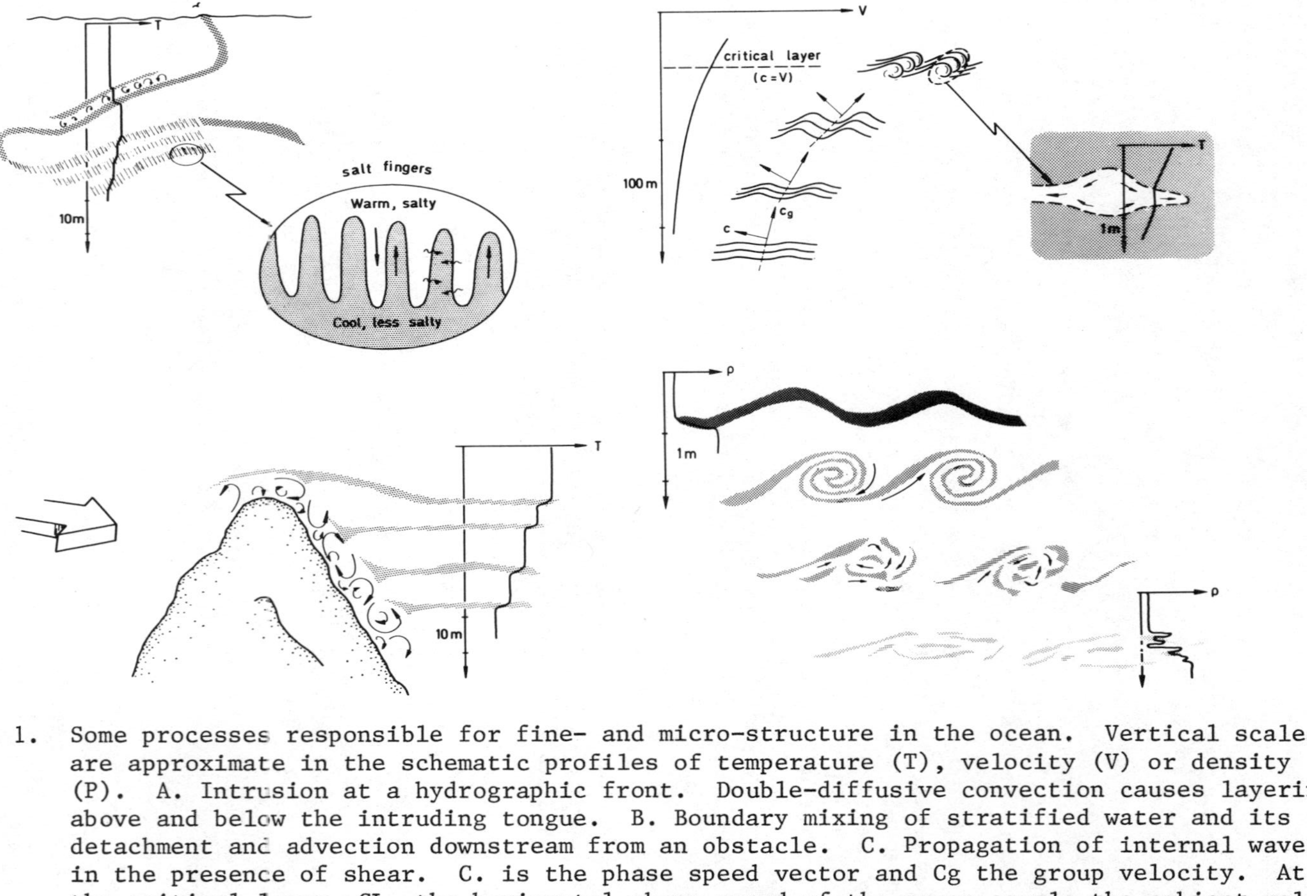

Figure 1. Some processes responsible for fine- and micro-structure in the ocean. Vertical scales are approximate in the schematic profiles of temperature (T), velocity (V) or density (P). A. Intrusion at a hydrographic front. Double-diffusive convection causes layering above and below the intruding tongue. B. Boundary mixing of stratified water and its detachment and advection downstream from an obstacle. C. Propagation of internal waves in the presence of shear. C. is the phase speed vector and Cg the group velocity. At the critical layer, CL, the horizontal phase speed of the waves equals the ambient velocity, and the waves are trapped and intensified, causing overturning and mixing. The insert shows the resulting mixed volume which flattens out in the ambient stratification, forming a blin. D. Stages in development of a Kelvin-Helmholtz instability.

Figure 1 illustrates some of the mechanisms involved.

The intent of this introduction to small-scale processes is to give a largely descriptive account, concentrating on properties that may be relevant to fish orientation. Several recent reviews cover the physics of the processes. Internal waves are reviewed by Garret and Munk (1979) and Phillips (1977) and fine and microstructure by Gregg and Briscoe (1979), Munk (1981) and Turner (1981).

Reversible Fine Structure

Wherever the ocean is stratified, internal waves are present--often in a most tangible way with typical amplitudes on the order of 10 m. In a two-layer system the internal waves behave much like ordinary surface gravity waves, propagating along the internal density discontinuity. The dominant wavelengths and periods are, however, much greater than for surface waves.

In a vertical gradient of density, which is more representative of natural waters, internal waves behave in a more exotic manner. If a parcel of fluid is moved vertically in a density gradient and then released, it will experience a buoyancy force, which acts to restore it to its original level in the stratification. Inertia will cause an overshoot and the parcel will oscillate vertically with a frequency called the Brunt-Väisälä frequency, N, which depends on the density ρ, the vertical density gradient $\frac{d\rho}{dz}$ and the gravitational acceleration g:

$$N = \left(\frac{g}{\rho}\ \frac{d\rho}{dz}\right)^{\frac{1}{2}} .$$

In a seasonal thermocline typical values of N correspond to a period of about 10 minutes; in the main thermocline of the ocean periods are around 1 hour.

If the displacement is not vertical, the restoring force is lower and the frequency of the oscillation is reduced. In the limit, when the displacement is nearly horizontal, the Coriolis force (due to the rotation of the earth) will act to restore the parcel. The frequency of oscillation is given by the Coriolis frequency, f, which depends on the earth's angular velocity Ω and the latitude ϕ:

$$f = 2\,\Omega \sin\phi .$$

This corresponds to periods of 12 hours at the poles, increasing towards the equator. At latitude 50° the period is approximately 16 hours.

The frequencies f and N give the permissible range of internal wave frequencies. There is no corresponding restriction on the range of wavelengths. Any single frequency allows all wavelengths, given that the ratio of the horizontal and vertical components of the wavelength is fixed. This independence between wavelength and period makes it difficult to describe the internal wave field in a simple way. Statistically, the observed wave energy increases gradually towards both lower frequencies and larger wavelengths. There is an accumulation of energy at the low frequencies close to f, and these frequencies dominate even at small vertical wavelengths. The orbital motions close to f are nearly horizontal, with the water traveling in circular paths. The ratio of horizontal to vertical wavelength is large, which means that the motion is coherent over large horizontal distances even if the vertical variation is rapid. More details about the generation and properties of these so-called inertial waves can be found in Hasselman (1970).

Energy can also be enhanced at tidal frequencies if the tidal frequencies are higher than the local f. Internal tides are caused by interaction of tidal currents and topography (Bell 1975), and the contribution to the internal wave field will be more pronounced in coastal areas.

Internal waves are always present where there is stratification in the sea. Far from the surface and bottom boundaries the temporary wave-straining of an otherwise relatively smooth profile seems to dominate the fine-structure variability. In shallow and coastal regions the waves modulate and distort a stratification which contains much fine structure of other origin.

Intrusions

At the border of major current systems there are fronts where temperature and salinity change rapidly in the horizontal plane. At a lesser scale frontal regions are created by the passage of storms which produce horizontal heterogeneities by wind mixing, heat exchange and evaporation-precipitation. When the storm has passed, water with lower density will try to spread on top of adjacent denser water and a series of two or more nearly homogeneous layers will form in the surface layer. At the major fronts horizontal mixing and instabilities will create similar intrusions that spread horizontally by gravity (cf Fig. 1A).

The first measurements and discussion of these processes were made by Stommel and Federov (1967). The monograph by Federov (1976) contains many examples on intrusive fine structure. More detailed measurements are described in Voorhis et al. (1976), Horne (1978) and Gregg (1980). In general the intrusion is identified as a layer of temperature inversion with an accompanying salinity increase to maintain static stability. The size of the intrusions can be quite large; water formed at the Antarctic shelf spreads in an approximately 200-m thick intrusion at a depth of 2000 m and can be traced in a 100 x 800 km area (Carmack and Killworth 1978). More commonly, the thickness is on the order of 10 m and the horizontal dimensions are around 10 km.

The definition of an intrusion as a layer of thermal inversion allows lesser scales of fine structure to be part of the entity. In general there will be a step structure above and below a nearly homogeneous core. These temperature steps and homogeneous laminae with vertical scales of 1 m or less can be secondary intrusions on a smaller scale, or they may be formed by double-diffusive processes.

The spreading of an intrusion by gravity is arrested by the rotation of the earth. A homogeneous volume of water collapsing in a continuously stratified environment will reach an equilibrium shape where the ratio of the horizontal to the vertical dimensions is on the order of N/f (Gill 1981). This implies aspect ratios of from 10 to 100 in the ocean's main and seasonal thermoclines. Viscous effects, acting on time scales larger than one day, will increase the aspect ratio by a factor of 10, and an imposed vertical shear can increase the horizontal flattening further still. A ratio of 1000 for the horizontal extent to the vertical thickness is commonly reported.

The lifetime of an intrusion is related to the vertical scale; the thinner ones disintegrate faster due to mixing. Estimates range from several months for a 20-50-m thick intrusion to approximately 1 week for an intrusion with the thickness 5-10 m. During this time advection by the mean currents can carry the intrusive patch distances on the order of 10 times the horizontal dimensions.

Bottom-Boundary Mixing

Intrusions similar to the ones discussed above can penetrate into the stratified interior at any depth, if they are caused by local mixing at the bottom (cf Fig. 1B). This process is discussed by Ivey and Corcos (1982). Armi (1978) and Armi and D'Asaro (1980) have made measurements on such intrusions in the deep ocean. Examples from nearshore are given by Caldwell et al. (1977).

The formation of mixed water can in this case be a continuous process which feeds water to an intrusion that is carried into the interior of the ocean by large-scale advection. This means that the constraint put by geostrophy on the aspect ratio of intermittently formed intrusions does not apply to this process, and the horizontal extent of bottom-formed intrusions can be quite large. Armi (1978) describes an approximately 20-m thick layer of Norwegian Sea water that was traced for 3000 km into the North Atlantic Ocean.

The lesser mixing activity at greater depths means that bottom-formed intrusions disintegrate more slowly than those formed at the surface. There are no measurements of the time scales that apply. Destruction of a 1-m thick intrusive layer would take approximately a month by purely molecular diffusion.

Internal Mixing

Like surface waves, internal waves also can break when their amplitude becomes sufficiently great. As the wave slope increases the orbital speed in the crest eventually exceeds the propagation speed of the wave, and a region with a density instability forms. In this region small-scale turbulence will develop rapidly, and the end result is a well-mixed volume that flattens slowly by gravity, giving a homogeneous layer, called a blin (or pancake), with sharp density steps above and below (cf Fig. 1C). The process was analysed by Orlanski and Bryan (1969). If the wave propagates in the presence of a steady vertical shear, the shear will add to the orbital motions so that breaking occurs at lower wave slopes.

In the presence of a sharp density step internal waves can cause mixing by shear instability at the interface. This is called Kelvin-Helmholtz instability; a description of the process is given by Thorpe (1973). The instability starts as short waves at the interface with a wavelength on the same order of magnitude as the interface thickness. The crests are perpendicular to the shear. These waves grow rapidly, extracting energy from the shear, and roll up into vortices or "billows." The billows will subsequently collapse and the end result is a slightly thicker interface that contains much microstructure due to incomplete mixing (cf Fig. 1D).

The interaction of internal waves with differing frequencies and propagation directions allows other modes of breaking and mixing (McEwan and Robinson 1975). In a situation with strong stratification and intermittent and weak mixing, as that caused by wave breaking, there are indications that the turbulence will enhance a discontinuous, step-like density profile rather than smooth out the vertical gradients, as would be the case if the turbulence dominated (Linden 1979; Turner 1981). Much of the connection among internal waves, turbulence and layer structures in the interior of the ocean, however, is not clearly understood.

Double Diffusion

In a large part of the ocean the basic stratification is such that both temperature and salinity increase upwards. The density decrease with increasing temperature is larger than the increase due to the salinity difference. Because salt diffuses approximately 100 times more slowly than heat, this stratification can be unstable. If a volume of seawater is moved downward, it will cool by giving off heat to its surroundings, but retain its salt and in this way become denser than the surrounding fluid and tend to sink further still. If the disturbance is upwards the parcel of fluid heats and will continue to rise. A number of alternatively sinking and rising narrow columns of fluid will form (Fig. 1A). These convection cells are called salt fingers. The process was first discussed by Stern (1960), and has been studied extensively, both empirically and theoretically. A summary is given in Turner (1981).

Typical vertical gradients of salt and temperature in the ocean produce salt fingers with diameters in the range of 1-10 cm. In an undisturbed state the fingers form an array of long, squarely packed cells. Shear tends to tilt the structure, and lamellae form parallel to the shear (Linden 1974). In the ocean salt fingering is a source of fine structure by producing almost homogeneous convecting layers of 1-10 m thickness, alternating with thinner gradient layers containing the fingers. In some regions this process creates a regular staircase structure in the temperature profile, and the layers can be of large horizontal extent. Under the Mediterranean Sea out-flow into the North Atlantic Ocean individual layers show continuity over as much as 50 km (Elliot et al. 1974). Recent measurements show that salt fingering may be a more widespread and important phenomenon in the ocean than was thought earlier (Gargett and Schmitt 1982). In regions with stratification stable in both temperature and salinity components conditions for fingering can develop locally at the interfaces of intrusions. This seems to be an important process in modifying intrusions at frontal regions (Schmitt and Georgi 1982).

Where both temperature and salt increase with depth, layers are apt to form owing to local convective instability. The reason is, as with salt fingering, that heat can diffuse more readily than salt. The theory of such layer formation was worked out by Veronis (1965).

Conditions favourable for this regime of double-diffusive convection are less widespread in the ocean than for the salt fingering. Layers that probably are generated by this process have been observed at the upper face of intrusions and in high-latitude regions. Characteristically, the homogeneous layers (one half to several meters thick) are separated by sharp gradient layers, typically a few centimeters thick. In undisturbed conditions the horizontal extent can be quite large; gradient sheets of approximately 10 cm thickness have been traced for at least 17 km horizontally under the Arctic Ocean ice (Neshyba et al. 1971).

THE HYDROGRAPHIC NEARFIELD OF A FISH

As can be seen from the preceding survey, small-scale vertical structure in the ocean can be generated by a large number of processes. In general several of the processes operate simultaneously at a given location and time, and the resulting state is

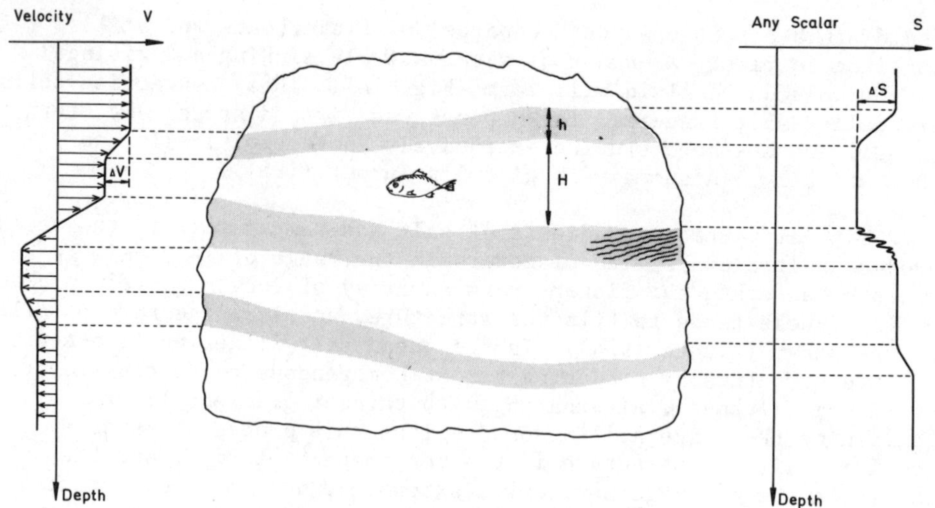

Figure 2. An idealized picture of the hydrographic nearfield of a fish.

complex and irregular but not completely random. An idealized model of the stratification surrounding a fish is given in Figure 2.

The picture is one of distinct layers--almost homogeneous layers alternating with layers containing strong vertical density gradients. This basic structure changes slowly in time, and for times on the order of several minutes to, say, 1 hour, it can be regarded as steady. An important feature is that the horizontal extent of the layering is much greater than the thickness; typically the aspect ratio is 100 to 1000.

The magnitude of the thickness of homogeneous layers, H, varies widely. Statistically, H follows a log-normal frequency distribution with a mean on the order of 1 m. The mean of H will be small, say 0.5 m, in a situation with strong density stratification in coastal waters and will tend to larger values in waters with a weaker mean stratification. The gradient layer thickness, h, behaves similarly, but the mean of h is less than that of H, at least in regions close to energy inputs, such as the surface and bottom boundaries. There is no clear relation between H and h in adjacent layers.

S in Figure 2 shows the distribution of some scalar quantity, such as temperature or the concentration of a solute. The frequency distribution of ΔS will also tend toward log-normality; a typical range of ΔS is 1/10 to 1/100 of the maximum range encountered in the water column. In the case of temperature, this means 1-0.1 C in the seasonal thermocline.

The vertical variation of the current, the shear, also is exemplified in Figure 2. It should be noted that a shear can be produced by rotation of the current direction, even if the magnitude of the current speed is constant with depth. There are few measurements of the fine structure of current shear. Those that have been made (e.g., Woods 1968; Simpson 1975; Van Leer and Rooth 1975) show that the shear is concentrated in the density-gradient layers. Typical values of the rate of change are 0.01 sec^{-1} (1 cm/sec/m) to 0.1 sec^{-1}.

If the shear across the density gradient layer is too large, the layer becomes unstable and turbulent mixing decreases the thickness of the layer. The criterion for shear instability is that the Richardson number, Ri, becomes less than 0.25. The Richardson number is defined as:

$$Ri = \frac{N^2}{(\frac{dv}{dz})^2} .$$

In difference form this becomes:

$$Ri = \frac{g\Delta\rho h}{\rho(\Delta V)^2} ,$$

which, for the step to be stable, means that:

$$\Delta V \leq 2(\frac{g\Delta\rho h}{\rho})^{\frac{1}{2}}.$$

This relation allows an estimate of the maximum probable velocity differences that are associated with an observed fine-structure profile.

The wiggles on the S-profile in the lower gradient layer (Fig. 2) are to indicate the presence of microstructure. Basically the flowfield in the model is laminar, or weakly turbulent, in the homogeneous layers, which seems to be true for a large part of the ocean interior. When turbulence occurs it is highly intermittent and localized to patches. In the upper ocean a major source of turbulence is local Kelvin-Helmholz instability (Woods and Wiley 1972; Gregg 1975), and this will generate centimeter-scale irregular structures in the gradient layers. The microstructure will exist on some fraction of the layer surface, typically 20-70%. In the deep ocean, and where double-diffusive processes are at work, the distribution of microstructure fluctuations is not coupled to the gradient layers (Gargett 1976).

DIRECTIONAL CUES

In an orientation process the fish has to observe some directional quantity in the environment. The hydrography can provide directional cues either through horizontal gradients of scalar quantities or directly through the current field. The possibility of extracting directional information from large-scale horizontal gradients seems remote. Traveling along a horizontal path a fish will encounter fine-structure gradient layers undulated by irregular internal waves, and the mean horizontal gradient will be obscured by step-like fluctuations that are much larger than the mean. As Arnold (1974) noted there are no known means by which current direction can be sensed without access to fixed points of reference. The current-induced geoelectric field at a point in the sea will be determined essentially by the difference between the local current vector and that of the vertically averaged, conductivity weighted, current (Sanford et al. 1978). This means that orientation with respect to geoelectric fields (Rommel and McCleave 1973) is one possible mode of orientation with respect to the current shear.

The current shear could be detected in several other ways. Apart from the geoelectric effects, Harden Jones (1968) discussed visual detection of the differential motions of particles above and below a fine-structure step, detection of the acceleration experienced upon crossing the step and tactile detection of differential water velocity across the body. There are observations of fine-structure layers associated with very strong shear, say 10 cm/sec across a 10 cm interface (Woods and Wiley 1972), and at those layers direct detection seems possible. The magnitude of the shear across a fine-structure layer is in most cases considerably weaker, however. Observations and estimates by the relation to the density gradient discussed above show values of a few millimeters per second across most density steps. The mechanical and visual stimuli provided by such low shears are below most observed rheotactic thresholds.

A possible detection mechanism could be provided by the effect of the shear on microstructure fluctuations of temperature in the gradient layers. Any horizontal variation in the layer will be stretched in the direction of the shear, and this effect is integrated in time, so that even a weak shear will have a noticeable effect in elongating microstructure elements.

The horizontal to vertical anisotropy of microstructure is well documented (Elliot and Oakey 1975), but anisotropy in the horizontal plane has not been investigated, as it has little relevance for the study of the turbulent vertical fluxes, which is the aim of most microstructure measurements. The processes that are likely to produce horizontal anisotropy are the decay of turbulent patches and salt fingering. Salt fingers tilt or reorganize into lamellae parallel to the shear. In this case the cross-stream scale is a few centimeters and the temperature fluctuations are on the same order of magnitude as the vertical temperature difference across the fingering layer.

A common cause of local turbulence is Kelvin-Helmholtz instability. In the early stages of development of a Kelvin-Helmholtz billow the density-gradient layer is wrapped up, entraining fluid from above and below, but little actual mixing occurs. Eventually the available energy for the instability is exhausted, and the motion in the billow is dissipated by viscosity. At this stage gravitational instability leads to an essentially three-dimensional collapse, causing horizontal temperature gradients that initially are of the same magnitude as those in the vertical dimension. This field of temperature eddies soon is distorted by the shear; at the same time they restratify and decay by molecular diffusion. The details of the restratification process are poorly understood. A lower bound on the readjustment time is given by the local N^{-1}. Because $R_i \sim 0.25$, we get $\Delta V \sim 2hN$, and in the time N^{-1} the stretching of the shear will cause a 1:3 ratio of cross-stream to downstream dimensions of a microstructure element. A theoretical model for such "fossil" turbulence given by Gibson (1982a,b) indicates that restratification is a slower process, as the effects of molecular diffusion oppose convective motion. He estimates the restratification time to be ten or a hundred times N^{-1}, which would give a quite strong horizontal anisotropy due to shear. The details of the horizontal gradients produced in this way are uncertain. The cross-stream scale will be on the same order of magnitude as the layer thickness, or smaller, and the temperature differences probably less than the total temperature difference across the layer by a factor of two to ten. An additional sharpening of the horizontal gradients could arise by double-diffusive convection superimposed on the vertical microstructure layering--the "triple-diffusion" discussed by Williams (1981).

The detection of such downstream filaments could be effected either by olfaction or by sensing the temperature field. If the fine-structure field contains an olfactant which changes its concentration across the gradient layer, gradients of the olfactant will be part of the horizontal microstructure, and orientation would be possible by klinotaxis or even tropotaxis. The lower molecular diffusion of solutes means that a microstructure of odour can be present even when the temperature field has been conducted away.

Direct detection of the thermal microstructure is also a possibility, given the great sensitivity of fish to temperature differences. Both Bull (1936) and Bardach and Björklund (1957) obtained conditioned responses in several species to temperature changes as small as 0.03-0.05 C. It should be noted that in neither of the experiments was a threshold sensitivity established; the quoted temperature differences were the smallest that could be produced and monitored in the apparatus. Nevertheless, if 0.03 C were the threshold level, fish could sense a large fraction of the microstructure fluctuations in the ocean.

The orientation of the shear is mirrored in the elongation of the microstructure elements, but it is unclear if and how the up- and down-stream directions could be discriminated. In the patches of active billow turbulence the signature of the billows would give a clear indication of the shear direction, and one possibility is that these patches serve as signposts, showing the way up or down the trail of microstructure filaments. The details of the three-dimensional organization of decaying fossil turbulence are largely unknown, and there may exist characteristics that make determination of the sign of the shear possible everywhere.

ORIENTATION TACTICS

With the assumption that the local shear, irrespective of its strength, is detectable, a number of hypotheses can be made about how fishes could use properties of the fine structure to orientate in midwater. Some will be outlined below to suggest that the hydrographic nearfield may contain information useful both in long-distance migration and in local orientation and food search. Gravity is a dominant force in the ocean and surfaces of equal density and other properties tend to be almost horizontal. This means that the vertical distribution of a property on a small scale often is similar to a horizontal distribution on a much larger scale. A single dive can in many cases take a fish through hydrographic regimes corresponding to a basin-wide horizontal distribution. A vertical search is in this way more efficient than a horizontal one, and this principle is invoked in some of the models.

Orientation with Respect to Major Currents

Apparently the oceanic migrations of fishes follow the major current gyres. This can be attributed partly to passive advection, but in several instances the observed ground speeds cannot be accounted for by drift alone (Arnold 1981). That oceanic motions at the scale of the large currents could be observed in centimeter-sized microstructure filaments seems intuitively unlikely, but there are couplings between the large-scale motions and the processes that produce the microstructure which could make such a scheme plausible.

The large-scale vertical shear due to the ocean currents gives an ambient shear field which is weak compared to the fluctuating fine-structure shear, but which has a profound effect on the propagation and breaking of internal waves. In general a packet of internal waves will travel at some angle to the horizontal and gradually change its depth. In the presence of a mean shear in the horizontal traveling direction, the packet will gradually change its inclination, and the vertical travel becomes arrested at a "critical layer" where the phase speed of the waves equals the local current speed (Booker and Bretherton 1967). As the wave approaches the critical layer the vertical wavelength decreases and the shear in the wave motion increases, with the result that the waves are likely to break (cf Fig. 1C). The critical layer is not at one particular depth in the ocean; for a particular case, the critical layer depth depends on the frequency and phase speed of the wave.

In the presence of a random wave field, with internal waves traveling in all horizontal directions, the shear refraction will single out those waves that travel in the direction of the shear and increase the probability that these waves will be absorbed at a critical layer and possibly break. A discussion of this selection mechanism can be found in Ruddick (1980). It should be noted that this process is much more efficient than the increased probability of breaking due simply to the addition of the mean shear to the wave shear.

The result for the fine- and micro-structure is that blini formed by breaking of internal waves preferentially will be the product of waves traveling in the direction of the mean shear and hence associated with a strong wave shear of the same direction as the much weaker large scale shear. In this way the statistical distribution of the direction of microstructure of layers can be strongly peaked in the direction of the mean shear, even if the fluctuating shear field due to internal waves is essentially isotropic.

Aggregation

The distribution of fishes in the sea is uneven with strong concentrations at particular places and hydrographic features; this unevenness forms the basis for the great fisheries. The way in which a particular bank or hydrographic front is found by fishes is unclear; often the area of aggregation is local and random search seems inefficient. Both topographic features, like islands and banks, and the fronts between water masses are sources of intrusions, and one possibility is that those could be used for orientation during the final approach when the fish is within the general area of the source of the intrusion. The correlation between the direction of the shear at the upper and lower faces of the intrusion and the direction to the origin of the intrusion is evidently strong close to the source. In this area a vertical search to find an intrusive layer and subsequent orientation along microstructure filaments should take the fish in the right direction. This would be an efficient strategy for homing to a bank if a fish is carried away by currents or strays in some other way.

For intrusions originating in a turbulent bottom layer, advection by the ambient current is the primary means of spreading. In this case the preferential orientation of microstructure parallel to the mean current shear could make orientation possible at essentially all distances where the intrusions can be detected. The situation is more complicated with instrusions originating in fronts. The initial spreading motion will decrease with time and the instantaneous shear at an old intrusion need not be related at all to the direction to the front.

The most likely method for discriminating intrusions from fine structure of other origin, as well as for discriminating between intrusions from different sources, is by olfaction. The case of orienting in a field of intrusions far from a front, where the correlation between shear and source directions is low, is similar to the problem of olfactory orientation to a river mouth in the coastal zone. An orientation mechanism that could be used in this case is discussed in the following section.

Olfactory Orientation

The spreading and diffusion of a passive substance in the sea is strongly affected by the fine structure. The results from diffusion experiments with point release of a tracer like rhodamin B gives an instructive picture of how an olfactory trail from a small source, such as a prey, may look. Tracer experiments are described by Kullenberg et al. (1973), Kullenberg (1974) and Ewart and Bendiner (1981). The main features of an instantaneous vertical section are shown in Figure 3. From a source in a homogeneous layer the tracer is spread by weak turbulence in the layer. When it reaches the top

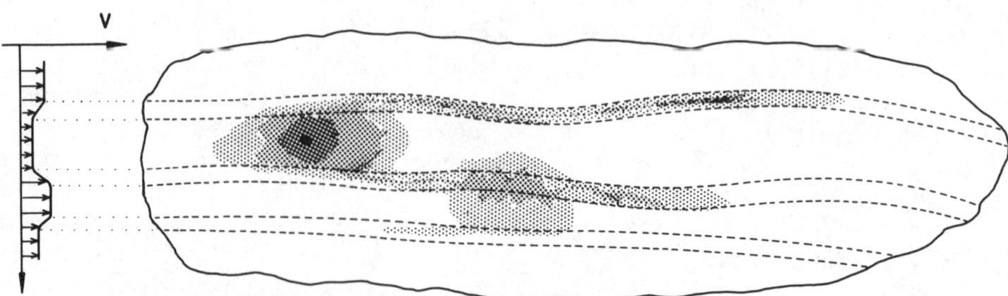

Figure 3. Schematic section through an odour cloud spreading from a source located in a homogeneous fine-structure layer. The horizontal scale is compressed.

and bottom gradient layers the spreading is by a combination of slow vertical diffusion and dispersion by the shear. Two thin veils of tracer will trail down-shear from the thicker patch. As indicated in the figure, the tracer can "leak" into an adjacent homogeneous layer if a patch of turbulence is present at the gradient layer, and a complicated situation can develop with several levels marked with dye.

The extent of the veils is much larger than that of the patch containing the bulk of the tracer. This implies that food search by olfaction in open water would be much more efficient if it is carried out by a combination of horizontal and vertical search than by horizontal motion alone. This may account for the diving behaviour of skipjack tuna (*Euthynnus pelamis*) observed by Strasburg (1961). The shear in the layer containing a veil of odour could then be used to guide the fish to the source.

The vertical distribution of odour from a large-scale, diffuse and continuous source, such as a river, is the product of the processes that generate fine structure, not a perturbation caused by a preexisting fine-structure field. In this case the homogeneous layers are filled with odour of differing concentrations, and the vertical distribution is concomitant with other variables such as temperature and salinity. I have proposed a scheme for olfactory orientation in this situation in the context of salmon homing (Westerberg 1982). The general argument is that during the coastal approach a salmon could orient to the river mouth by moving against the shear in fine-structure gradient layers. The layer is chosen by a vertical olfactory search through the water column, and the criterion should be such that the probability for the local shear to be in the direction of the river mouth is large. The searching dives are made through the part of the water column that contains any odour variation and are repeated with a spacing comparable to the horizontal coherence length of the fine structure. In this way the salmon can detect and switch to a new gradient layer, which locally is the more optimal according to the criterion used. This method would take the salmon along a path which is irregular and may contain detours but which in the mean points in the right direction.

Several possibilities exist for what could be an optimum odour concentration signature in the choice of gradient layer. The relative motion across the layer with the largest total concentration difference should have a good correlation to the direction of the source if the odour concentration decays with time. If the substance is conservative or decays slowly, a better correlation may be found at the layer where the odour concentration equals the sensory threshold. The argument for this is that the largest shear in the mean estuarine circulation, driven by the river discharge and mixing from above, is found at the interface to a compensatory inflow of ambient water below the spreading plume.

OBSERVATIONS OF BEHAVIOUR IN RELATION TO FINE STRUCTURE

During a study of the effects of discharges from power plants on migrating Atlantic salmon (*Salmo salar*) and European eels (*Anguilla anguilla*) we have made ultrasonic tracking experiments. Our objective was to learn how the migrating fish behave in relation to the undisturbed thermal stratification in coastal waters. The swimming depth of the fish was recorded continuously, using a depth-sensitive transmitter with high resolution and stability, and the thermal stratification along the track was measured with a free-falling thermistor sond with a sufficiently rapid response to resolve the thermal fine structure. The calibration of the transmitter was made by lowering it together with the thermistor sond, using the output of the pressure transducer of the sond as depth scale. In this way the relative error in comparing the observed swimming depth and the measured thermal structures could be kept to approximately 1% of the full-scale range of the transmitter (20 or 50 m for the experiments with salmon). A detailed description of the instrumentation and data processing, together with a more complete account of the experiments with salmon is given in Westerberg (1982).

Atlantic Salmon

A characteristic behaviour, which occurred in all salmon tracked with depth transmitters, was brief dives of 5-15 m amplitude. Normally, dives occurred approximately once or twice per hour. The dives were directed towards the thermocline and lasted for a few

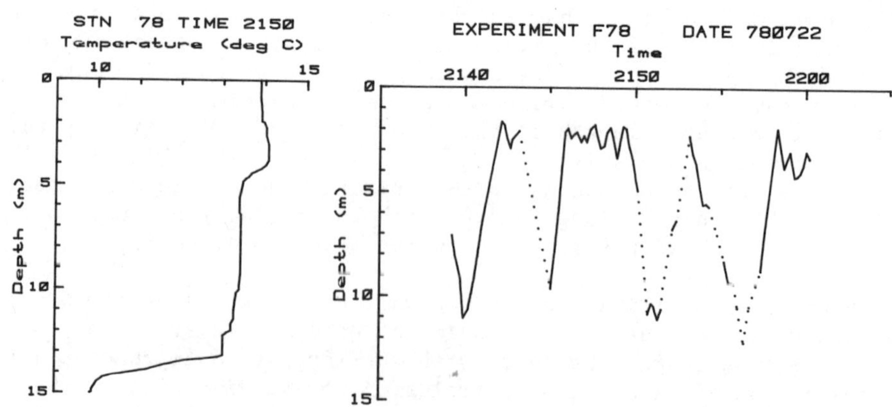

Figure 4. Time-depth diagram for a telemetered Atlantic salmon, swimming in water with the vertical thermal structure shown at left. The continuous line shows successive 10-sec averages of the recorded swimming depth; the dotted portions indicate periods of low signal-to-noise ratio and data drop-out. The time scale shows local time.

minutes. Occasionally there was a "burst" of dives as exemplified in Figure 4. When the swimming depth during the interval between dives was compared to simultaneous temperature recordings it was frequently associated with fine-structure layers. Figure 5 shows an example to illustrate that the motion of the salmon seems to be along the thermal field rather than at a constant depth.

In most cases the spacing of the temperature profiles was too great to trace individual fine-structure features between the stations. A second boat was used for the temperature measurements during one experiment, and, bolstered by the somewhat higher sampling rate of temperature in this case, an attempt was made to show the motion of the salmon in relation to the fine structure during a longer period (Fig. 6). The salmon moved steadily in open water where the depth was approximately 50 m. During the first half of the record the swimming depth was around that of a gradient between 14.5-15.0 C. Shortly after noon the salmon dived; the first part of the dive is missing in the recording, due to the sudden drop in signal strength during the dive. The signal was still audible, however, and the maximum depth reached by the fish was approximately 20 m. On the way up the salmon continued to the top of the thermocline and then switched to a new gradient region within the temperature interval 11-12 C. At the time of the dive and change of level in the temperature field there was also a slight change of swimming direction as can be seen in the track chart (Fig. 6).

Observations of this type form the basis for the hypothesis about olfactory orientation described above. Further support for the interpretation of the brief dives as olfactory exploration was given by an experiment with an anosmic salmon. Upon release this salmon made a long burst of dives through most of the water column, and then it moved along the bottom, making forays up through the thermocline at approximately two times the frequency observed in normal fish.

European Eels

Analysis of our tracks of silver (migratory) eels is still in progress, and the possible interpretations remain unclear. The behaviour differed much from that of salmon, both in the trajectories, that normally were more straight, and in the diving activity. Many eels showed long periods of almost uninterrupted dives, with sharp turns close to the surface and bottom, giving a triangle-wave appearance to the time-depth diagram. The horizontal motion was continuous and steady during this time with a speed of approximately five times the vertical.

The turns were not always at the surface or bottom. On many occasions turns occurred in midwater at the depth of a fine-structure gradient layer. Another variation was that the eel stopped the dive

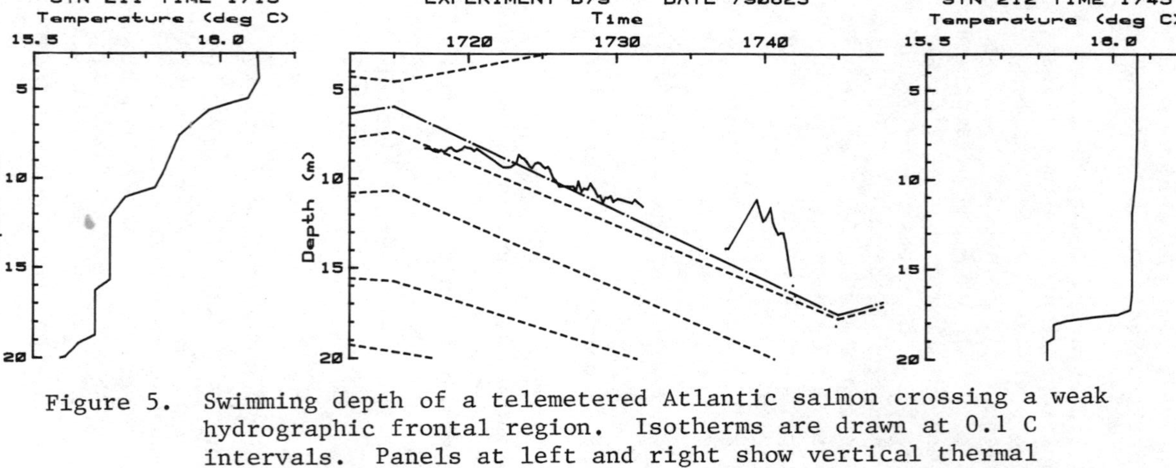

Figure 5. Swimming depth of a telemetered Atlantic salmon crossing a weak hydrographic frontal region. Isotherms are drawn at 0.1 C intervals. Panels at left and right show vertical thermal structure at two points along the track.

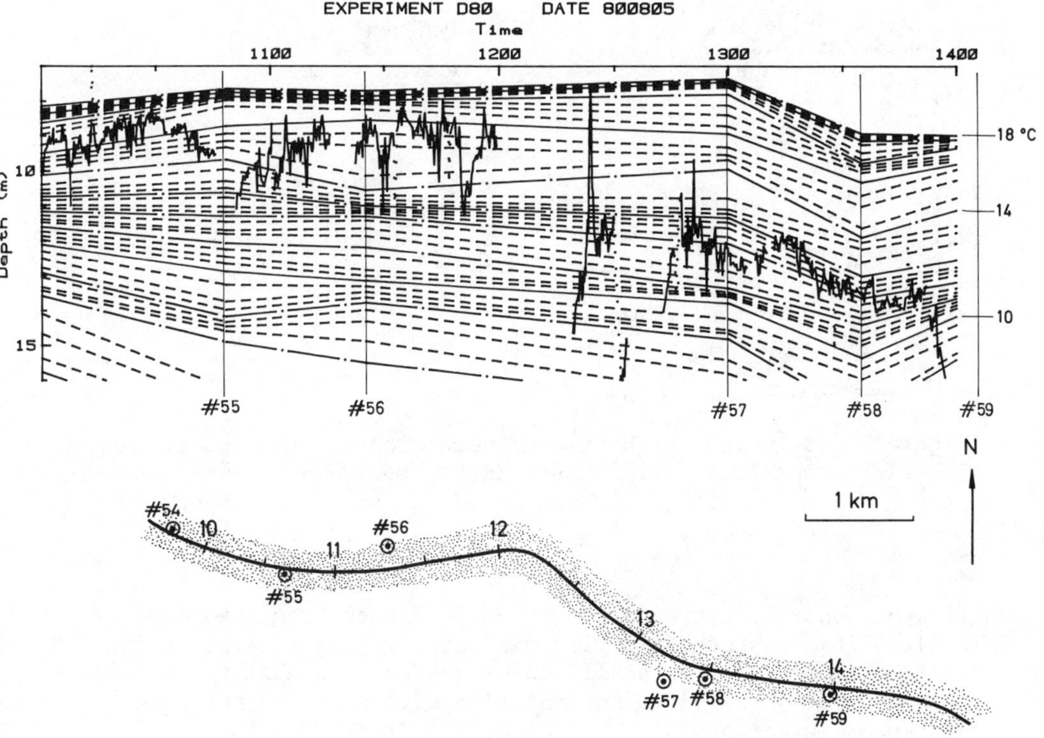

Figure 6. Time-depth diagram showing swimming depth and vertical thermal structure along the path of a telemetered Atlantic salmon and the plan view of the track for the same time interval. Observed temperature profiles are indicated with a vertical line and a station number in the temperature section; the corresponding positions are shown in the plan view. The shading along the track indicates the extent of uncertainty about the salmon's path.

temporarily at a gradient layer, followed it for a few minutes, and then continued up or down (Fig. 7).

DISCUSSION

The observations on salmon and eels suggest that these species indeed can perceive and react to fine-structure gradient layers. Whether this behaviour has any significance in connection with migration mechanisms is an open question. The purpose of this article has been to point out that the known properties of the fine- and micro-structures in the ocean are such that they could allow a fish to extract from their immediate surroundings directional

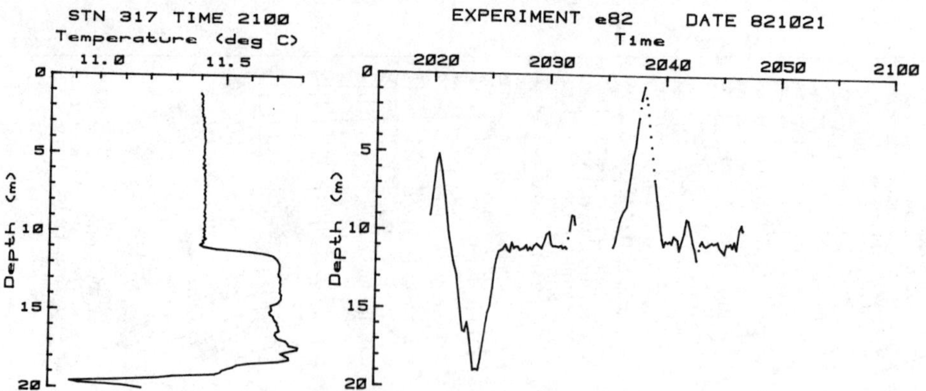

Figure 7. Swimming depth of a silver European eel, and accompanying vertical thermal structure. Bottom depth was approximately 25 m. Ground speed of the eel was about 40 cm/second.

information relevant to orientation on a much larger horizontal scale. This cascade of directional cues from the large hydrographic horizontal features to small-scale vertical variability makes it possible for purely local orientation mechanisms to guide long-distance migrations.

Demonstration of such mechanisms in fish behaviour remains to be done. It is technically a difficult task to adequately map the fine structure along the track of a fish. The undulations by the background of internal waves and the relatively short distances over which a particular fine-structure layer can be identified require a very dense network of vertical profiles if one is to compare swimming depth and fine structure. The ultimate goal, to observe the movements of the fish in relation to the fine structure of the current field, seems remote.

A way of avoiding the difficulties of pinpointing the true position of the fish in a map of the fine structure would be to let the fish carry the measuring equipment. An interesting experiment would be to telemeter the instantaneous vertical and horizontal temperature gradients across a fish.

REFERENCES

Armi, L. 1978. Some evidence for boundary mixing in the deep ocean. Journal of Geophysical Research 83:1971-1979.

Armi, L., and E. D'Asaro. 1980. Flow structures of the benthic ocean. Journal of Geophysical Research 85:469-484.

Arnold, G.P. 1974. Rheotropism in fishes. Biological Reviews of the Cambridge Philosophical Society 49:515-576.

Arnold, G.P. 1981. Movements of fish in relation to water currents. Pages 55-79 *in* D.J. Aidley, editor. Animal migration. Cambridge University Press, Cambridge, England.

Bardach, J.E., and R.G. Björklund. 1957. The temperature sensitivity of some American freshwater fishes. American Naturalist 91:233-251.

Bell, T.H. 1975. Topographically generated internal waves in the open ocean. Journal of Geophysical Research 80:320-327.

Booker, J.R., and F.P. Bretherton. 1967. The critical layer for internal gravity waves in shear flow. Journal of Fluid Mechanics 27:513-539.

Bull, H.O. 1936. Studies on conditioned responses in fishes. Part VII. Temperature perception in teleosts. Journal of the Marine Biological Association of the United Kingdom 21:1-27.

Caldwell, D.R., J.M. Brubaker, and V.T. Neal. 1977. Thermal microstructure on a lake slope. Limnology and Oceanography 23:372-374.

Carmack, E.C., and P.D. Killworth. 1978. Formation and interleaving of abyssal water masses off Wilkes Land, Antarctica. Deep-Sea Research 25:357-370.

Cooper, L.H.N. 1967. Stratification in the deep ocean. Science Progress 55:73-90.

Elliot, A.J., M.R. Howe, and R.I. Tait. 1974. The lateral coherence of a system of thermohaline layers in the deep ocean. Deep-Sea Research 21:95-107.

Elliot, J.A., and N.S. Oakey. 1975. Horizontal coherence of temperature microstructure. Journal of Physical Oceanography 5:506-515.

Ewart, T.E., and W.P. Bendiner. 1981. An observation of the horizontal and vertical diffusion of a passive tracer in the deep ocean. Journal of Geophysical Research 86:10974-10982.

Federov, K.N. 1976. The thermohaline finestructure of the ocean. Pergamon Press, Oxford, England.

Gargett, A.E. 1976. An investigation of the occurrence of oceanic turbulence with respect to finestructure. Journal of Physical Oceanography 6:139-156.

Gargett, A.E., and R.W. Schmitt. 1982. Observations of salt-fingers in the central waters of the Eastern North Pacific. Journal of Geophysical Research 87:8017-8029.

Garrett, C., and W.H. Munk. 1979. Internal waves in the ocean. Annual Review of Fluid Mechanics 11:339-369.

Gibson, C.H. 1982a. Alternative interpretations for microstructure patches in the thermocline. Journal of Physical Oceanography 12:374-383.

Gibson, C.H. 1982b. On the scaling of vertical temperature gradient spectra. Journal of Geophysical Research 87:8031-8038.

Gill, A.E. 1981. Homogeneous intrusions in a rotating stratified fluid. Journal of Fluid Mechanics 103:275-295.

Gregg, M.C. 1975. Microstructure and intrusions in the California Current. Journal of Physical Oceanography 5:253-278.

Gregg, M.C. 1980. The three-dimensional mapping of a small thermohaline intrusion. Journal of Physical Oceanography 10:1468-1492.

Gregg, M.C., and M.G. Briscoe. 1979. Internal waves, finestructure, microstructure and mixing in the ocean. Reviews of Geophysics and Space Physics 17:1524-1548.

Harden Jones, F.R. 1968. Fish migration. Arnold, London, England.

Hasselman, K. 1970. Wave-driven inertial oscillations. Geophysical Fluid Dynamics 1:463-502.

Horne, E.P.W. 1978. Interleaving at the subsurface front in the slope water off Nova Scotia. Journal of Geophysical Research 83:3659-3671.

Ivey, G.N., and G.M. Corcos. 1982. Boundary mixing in a stratified fluid. Journal of Fluid Mechanics 121:1-26.

Kullenberg, G. 1974. An experimental and theoretical investigation of the turbulent diffusion in the upper layer of the sea. København Universitet Institut for Fysisk Oceanografi Report 25:1-212.

Kullenberg, G., C.R. Murthy, and H. Westerberg. 1973. An experimental study of diffusion characteristics in the thermocline and hypolimnion regions of Lake Ontario. Pages 774-790 *in* Proceedings 16th Conference on Great Lakes Research International Association of Great Lakes Research. Ann Arbor, Michigan, USA.

Linden, P.F. 1974. Salt-fingers in a steady shear flow. Geophysical Fluid Dynamics 6:1-27.

Linden, P.F. 1979. Mixing in stratified fluids. Geophysical and Astrophysical Fluid Dynamics 13:3-23.

McEwan, A.D., and R.M. Robinson. 1975. Parametric instabilities of internal waves. Journal of Fluid Mechanics 67:667-687.

Munk, W. 1981. Internal waves and small-scale processes. Pages 264-291 *in* B.A. Warren and C. Wunsch, editors. Evolution of physical oceanography. Massachusetts Institute of Technology Press, Cambridge, Massachusetts, USA.

Neshyba, S., V.T. Neal, and W. Denner. 1971. Temperature and conductivity measurements under Ice Island T-3. Journal of Geophysical Research 76:8107-8120.

Orlanski, I., and K. Bryan. 1969. Formation of the thermocline step structure by large-amplitude internal gravity waves. Journal of Geophysical Research 74:6975-6993.

Phillips, O.M. 1977. The dynamics of the upper ocean. Second edition. Cambridge University Press, Cambridge, England.

Rommel, S.A., and J.D. McCleave. 1973. Prediction of oceanic electric fields in relation to fish migration. Journal du Conseil Conseil International pour l'Exploration de la Mer 35:27-31.

Ruddick, B. 1980. Critical layers and the Garrett-Munk spectrum. Journal of Marine Research 38:135-145.

Sanford, T.B., R.G. Drever, and J.H. Dunlap. 1978. A velocity profiler based on the principles of geomagnetic induction.

Deep-Sea Research 25:183-210.

Schmitt, R.W., and D.T. Georgi. 1982. Finestructure and microstructure in the North Atlantic Current. Journal of Marine Research 40: Supplement:659-705.

Simpson, J.H. 1975. Observations of small scale vertical shear in the ocean. Deep-Sea Research 22:619-627.

Stern, M.E. 1960. The "salt-fountain" and thermohaline convection. Tellus 12:172-175.

Stommel, H., and K.N. Federov. 1967. Small-scale structure in temperature and salinity near Timor and Mindanao. Tellus 19:306-325.

Strasburg, D.W. 1961. Diving behaviour of Hawaiian skip-jack tuna. Journal du Conseil Conseil International pour l'Exploration de la Mer 26:223-229.

Thorpe, S.A. 1973. Experiments on instability and turbulence in a stably stratified shear flow. Journal of Fluid Mechanics 61:731-751.

Turner, J.S. 1981. Small-scale mixing processes. Pages 236-262 *in* B.A. Warren and C. Wunsch, editors. Evolution of physical oceanography. Massachusetts Institute of Technology Press, Cambridge, Massachusetts, USA.

Van Leer, J.C., and C.G. Rooth. 1975. Shear observations in the deep thermocline. Deep-Sea Research 22:831-836.

Veronis, G. 1965. A finite amplitude instability in thermohaline convection. Journal of Marine Research 23:1-17.

Voorhis, A.D., D.C. Webb, and R.C. Millard. 1976. Current structure and mixing in the shelf/slope water front south of New England. Journal of Geophysical Research 81:3695-3708.

Westerberg, H. 1982. Ultrasonic tracking of Atlantic salmon: I Movements in coastal regions, II Swimming depth and temperature stratification. Institute of Freshwater Research Drottningholm Report 60:81-120.

Williams, A.J. 1981. The role of double diffusion in a Gulf Stream frontal intrusion. Journal of Geophysical Research 86:1917-1928.

Woods, J.D. 1968. Wave-induced shear instability in the summer thermocline. Journal of Fluid Mechanics 32:791-800.

Woods, J.D., and R.L. Wiley. 1972. Billow turbulence and ocean microstructure. Deep-Sea Research 19:87-121.

LARGE-SCALE INTERANNUAL FLUCTUATIONS IN OCEAN PARAMETERS AND THEIR INFLUENCE ON FISH POPULATIONS

L. A. Mysak

Departments of Mathematics and Oceanography
University of British Columbia
Vancouver, British Columbia V6T 1W5 Canada

ABSTRACT

Cross-spectral analysis of physical oceanographic variables (sea level, sea-surface temperature, and salinity) along coastal regions of the Northeast Pacific Ocean produced coherent signals with periods of 5-6 years (among other frequencies). Annual sockeye salmon (*Oncorhynchus nerka*) and Pacific herring (*Clupea harengus pallasi*) recruitments were coherent with the 5-6 year signal from British Columbia to southeastern Alaska. Research is proposed to determine if interannual temperature, salinity and current oscillations associated with baroclinic waves affect the structure of fish populations by altering migration speeds and productivity of the food supply for fish populations.

INTRODUCTION

From the cross-spectral analysis of 40-80 years of coastal sea level, sea-surface temperature and salinity data in the Northeast Pacific Ocean, Mysak et al. (1982) found coherent signals at frequencies of 0.09, 0.17-0.20, 0.33 and 0.43 cycles per year, corresponding to periods of 11, 5-6, 3, and 2.3 years. The strong 5-6 year oscillation, besides appearing in the sea level cross-spectrum, was manifested as a sea level-temperature oscillation near Vancouver Island and as a sea level-salinity oscillation farther north. This signal was found to propagate northward with a phase speed roughly comparable to that of a coastally-trapped baroclinic Kelvin wave (LeBlond and Mysak 1978). The 3-year oscillation, which appeared to be cut off at higher latitudes, could be due to a westward propagating baroclinic Rossby wave (LeBlond and Mysak 1978).

The source of these interannual oceanic fluctuations is not known for certain, but it is conceivable that they may be related to the "Southern Oscillation" in the atmosphere which is characterized by global-scale interannual oscillations in the air temperature and air pressure at sea level (Horel and Wallace 1981; Van Loon and Madden 1981).

The relationship between these signals, which were interpreted as baroclinic waves, and fish populations in the Northeast Pacific was also investigated by Mysak et al. (1982). Cross-spectral analyses of the annual sockeye salmon (*Oncorhynchus nerka*) catch and Pacific herring (*Clupea harengus pallasi*) recruitment with the physical variables were performed. Both the sockeye salmon catch and the Pacific herring recruitment were coherent with the 5-6 year signal from British Columbia up to southeastern Alaska. A more detailed study of the Fraser River sockeye salmon showed that both the average weight per fish and the total number of fish caught were coherent with the 5-6 year signal.

PROPOSED RESEARCH

It is speculated that the interannual temperature, salinity and current oscillations associated with baroclinic waves may affect fish populations and fish weight, especially in the Northeast Pacific Ocean. To test this hypothesis the following research work is proposed:

1. The actual mechanisms by which the interannual waves affect fish growth and, ultimately, fish catch must be identified. Possible mechanisms or questions which deserve further study are:
    a. Do wave-induced fluctuations of interannual time scales in temperature or salinity at a particular ocean depth significantly affect primary production and hence the food supply for fish?
    b. Associated with the longshore-travelling Kelvin waves are coastally aligned currents which flow back and forth over a wave period ($\sim$5 years). In the upper 100-200 m these longshore currents could be of order 10 cm/sec ($\simeq$9 km/day). When the juvenile salmon migrate northward along the shelf before heading out to the open ocean, they would encounter these currents. If the young fish happened to migrate northward on a wave cycle which had a northward current near the surface, then they would more readily get to the open ocean and hence would return as larger and healthier fish about two years later. Thus the wave motions in this case would certainly enhance fish growth and catch. Is there a way in which this mechanism could be demonstrated?

2. Further spectral analyses between oceanographic data and fish

catch data should be performed for other parts of the world. Meteorological variables (e.g. rainfall and air temperature) could also be used in such analyses.

3. Interannual variability of large-scale oceanic fronts (as represented by horizontal temperature gradients, for example) should be examined and correlated with the structure of fish population and other biological populations.

REFERENCES

Horel, J.D., and J.M. Wallace. 1981. Planetary-scale atmospheric phenomena associated with the southern oscillation. Monthly Weather Review 109:813-829.

LeBlond, P.H., and L.A. Mysak. 1978. Waves in the Ocean. Elsevier Scientific Publications, New York, New York, USA.

Mysak, L.A., W.W. Hsieh, and T.R. Parsons. 1982. On the relationship between interannual baroclinic waves and fish populations in the Northeast Pacific. Biological Oceanography 2:63-103.

Van Loon, H., and R.A. Madden. 1981. The southern oscillation. Part I: Global associations with pressure and temperature in northern winter. Monthly Weather Review 109:1150-1162.

PATTERNS, MECHANISMS AND APPROACHES TO THE STUDY OF MIGRATIONS OF ESTUARINE-DEPENDENT FISH LARVAE AND JUVENILES

John M. Miller[1,2], James P. Reed[1]
and Leonard J. Pietrafesa[2]

Departments of [1]Zoology and [2]Marine, Earth and Atmospheric Sciences
North Carolina State University
Raleigh, North Carolina 27650 USA

ABSTRACT

Five species (<10%) contribute over 85% of the annual commercial catch of finfish off North Carolina. All five spawn in winter near the Gulf Stream and migrate about 100 km to major inlets in the barrier islands and then up to another 100 km to juvenile estuarine nursery areas. The vertical distribution of pelagic larvae of menhaden (*Brevoortia tyrannus*) differs significantly from that of more benthic-oriented larvae such as spot (*Leiostomus xanthurus*), croaker (*Micropogonias undulatus*) and flounder (*Paralichthys lethostigma* and *P. dentatus*). Analysis of the shelf current regime suggests that differences in vertical distribution and season will subject larvae to markedly different currents, and therefore, different mechanisms for migration are required. Calculations of water movements using Ekman's original transport equations are probably wrong in relatively shallow shelf waters subject to cross-shelf winds and density currents. Major differences probably exist, therefore, between the pelagic larvae of the west coast (e.g. Pacific sardine, *Sardinops sagax*) and estuarine dependent species of the east coast--both with respect to the importance of microscale processes (e.g. food patchiness) and drift. Predator avoidance is likely to be a more important determinant of migration pattern than has been suggested to date. Problems of determining migration vectors and mechanisms from highly variable larval and juvenile fish distributions are discussed. A research strategy dealing with the above is outlined, and testable implications of hypotheses are presented.

INTRODUCTION

The primary objective of this paper is to develop insights into migration mechanisms that determine the relative abundance of commercial fish species off North Carolina, USA. In particular we want to see if patterns emerge which can be related to known physical processes in the respective environments of the various life history stages. Of special concern are those physical processes which may determine the success of movements between, and survival within, habitats utilized by larval and juvenile stages. The second objective is to identify gaps in either the physical or biological knowledge which, if filled, could lead to substantial improvement in understanding the mechanisms and consequences of these movements.

ECOLOGICAL PERSPECTIVE

We begin with the observation that as size increases, specific growth rate declines but mortality rate declines even faster, and with the assumption that behavior (including migration) has evolved as an optimization strategy. It follows that migration will be keyed to factors relating to survival in larvae and as fish grow will be progressively more keyed to factors relating to growth.

Now any testable implication of the above hypothesis must be considered in the perspective that the ability of young fish to behaviorally regulate their environment by swimming is limited. That is, larval distributions are likely to exhibit a higher stochastic component and be more directly a function of currents than adult distributions. Also, because larvae and juveniles are characterized by high mortality rates, the deductions of migratory patterns from changes in distribution (abundance) must include subtraction of the effects of mortality.

Relative to adults, available evidence suggests larvae and juvenile fish can tolerate greater environmental extremes (Holliday 1965). Their estuarine nursery areas are just such habitats, being characterized by more variable temperatures and salinities than either freshwater or marine habitats (Stickney 1959). It is, therefore, possible that estuaries serve as refuges from predation, provided the expense of tolerating the potentially stressful conditions is not excessive, and that larvae and juveniles with their lower vagility could avoid lethal conditions (Miller and Dunn 1980). These do not exclude the possibility of greater food resources; however, comparison in the fall of the sizes of juvenile spot, croaker, flounder and menhaden with other species which do not inhabit estuarine nurseries and, in fact, spawn later, suggests that there is not a particular growth advantage. Whether or not the food resources of other environments could support similarly dense populations of juveniles in estuaries is debatable.

Our general approach in this paper will be to examine selected data sets (some unpublished) which serve to suggest mechanisms and testable implications.

LIFE HISTORY

The Dominant Species

We examined the life histories of the 25 most important (by weight) commercial species delivered to North Carolina processors to see if a dominant life history pattern was reflected in their apparent relative success. This was necessary because no better estimates exist of the relative abundance of fishes such as would be provided by systematic sampling or CPUE (catch per unit effort). While, as Leggett (1977) pointed out, catch statistics often represent fishing patterns better than abundance of fishes, we felt that the ranks of at least the dominant species could be estimated accurately enough to support the arguments presented in this paper.

The annual commercial deliveries to North Carolina processors have been reported since 1950 (Chesnut and Davis 1975, and in the Annual Report of the North Carolina Division of Marine Fisheries [P.O. Box 769, Morehead City, NC, 28857, USA]). Because long-term averages (e.g. tens of years) and single-year catches probably reflect trends in effort and chance, respectively, we summed the annual statistics of the most recent 5 years available (1977-1981). These sums were ranked for our estimate of the relative abundance of the adult stocks.

Seven species of fishes comprise 92% of the total commercial catch. Five spawn in winter, are estuarine-dependent and migratory and make up 85% of the catch. The other two (*Cynoscion regalis* and *Alosa pseudoharengus*) spawn in spring but are also migratory. Thus the bulk of the commercial catch is made up of winter-spawning migratory species. None of the other commercially-important species spawns in winter, although some are estuarine-dependent or migratory. Thus, the dominant pattern, representing over 85% of the commercial fish catch, is that of five species: menhaden (*Brevoortia tyrannus*), croaker (*Micropogonias undulatus*), flounder (*Paralicthys dentatus* and *P. lethostigma*) and spot (*Leiostomus xanthurus*). This pattern (Fig. 1) is as follows: 1) after hatching near the Gulf Stream, larvae migrate onshore during December-March; 2) juveniles migrate through inlets and sounds to estuarine nursery areas in early spring; 3) subadults migrate out of juvenile nursery areas during fall; and 4) adults migrate offshore in fall or winter to spawn. A few fish may spawn at the end of one year, but most probably do not spawn until the second. The relative importance of subsequent spawnings is unknown, but it probably diminishes rapidly owing to heavy exploitation of older fish.

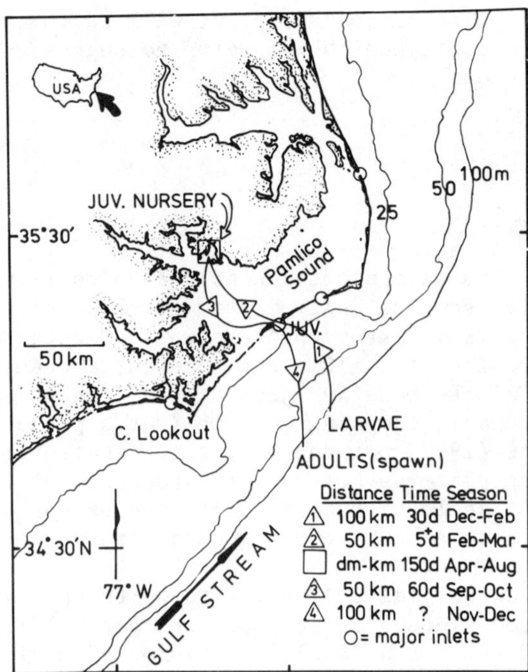

Figure 1. Migration pattern of five dominant fish species off North Carolina, USA: menhaden (*Brevoortia tyrannus*); spot (*Leiostomus xanthurus*); croaker (*Micropogonias undulatus*); and flounder (*Paralichthys dentatus* and *P. lethostigma*).

Why should this pattern be so successful relative to others? We hypothesize that the answer lies in two features of this pattern--winter instead of other spawning seasons and estuarine instead of offshore nursery areas--both of which are primarily related to survival, not growth.

The Advantages of Winter Spawning

There seem to be three potential advantages of winter spawning: 1) because water temperatures are low, survival times at low rations are longer; 2) predation pressure on larvae is at a minimum in winter; and 3) larval transport toward shore is favored by winter currents.

The first is certainly true, although food resources in winter may not be particularly depleted. Standing stocks of zooplankton on the continental shelf off North Carolina are about equal in winter and summer with mean concentrations of 0.12 ml/m^3 (November-February) and 0.1 ml/m^3 (April-October) (Clark et al. 1969).

The second is probably true but needs to be assessed further. In addition to generally lower activity of potential predators (including feeding) at low temperatures, many of them are summer residents of shelf waters and sounds, and they move either offshore or south with cold temperatures. It is not known, however, how the total predation pressure (abundance of predators X their feeding rate) varies with season.

The third is true, at least for larvae in shelf waters out to the edge of the Gulf Stream (about 100 km). We will concentrate on this potential advantage.

Nelson et al. (1977) examined 15 years of menhaden data and found a correlation between year-class strength and the calculated volume of shoreward (westward, off North Carolina) Ekman transport during January to March. Years when onshore transport was great were years which produced large numbers of recruits. Since menhaden spawn in winter near the Gulf Stream, Nelson et al. (1977) concluded that higher survival occurred in years with favorable shoreward transport conditions. They also suggested that the favorable conditions for onshore transport were winter conditions, i.e. larvae in other seasons would be less likely to be carried shoreward by currents.

While the above may be at least partly responsible for transport of pelagic larvae toward shore in winter, recent data show that the circulation patterns on the continental shelf are considerably more complex than this and further, that conventional Ekman-type calculations may be considerably in error.

Shoreward transport as calculated from an Ekman model for the east coast and used by Nelson et al. (1977) has at least two problems. First, the model cannot accommodate wind-driven processes which are perpendicular to shore, and it relies strongly on alongshore winds (NE-SW off North Carolina). Second, neither density-driven currents nor shallow-water friction is considered, both of which are important on the continental shelf off North Carolina.

CONTINENTAL SHELF CURRENTS

There are three principal physical-oceanographic seasons on the continental shelf of North Carolina. They are: *winter*, from mid-December to late February or early March; *spring/summer*, from March to August; and *fall*, from September to December. These three seasons feature dramatically different diabathic flow patterns (Figs. 2, 3).

During fall, winds are generally northeasterly to northwesterly. These winds tend to be aligned with the coastline of North Carolina north and south of Cape Hatteras. As a consequence of these winds,

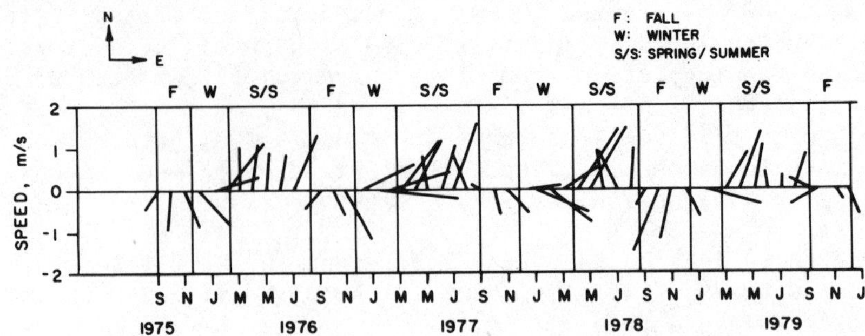

Figure 2. Monthly mean vectors of wind velocity for the North Carolina continental shelf during the period September, 1975 - January, 1980. Vectors are plotted in direction towards which wind is blowing.

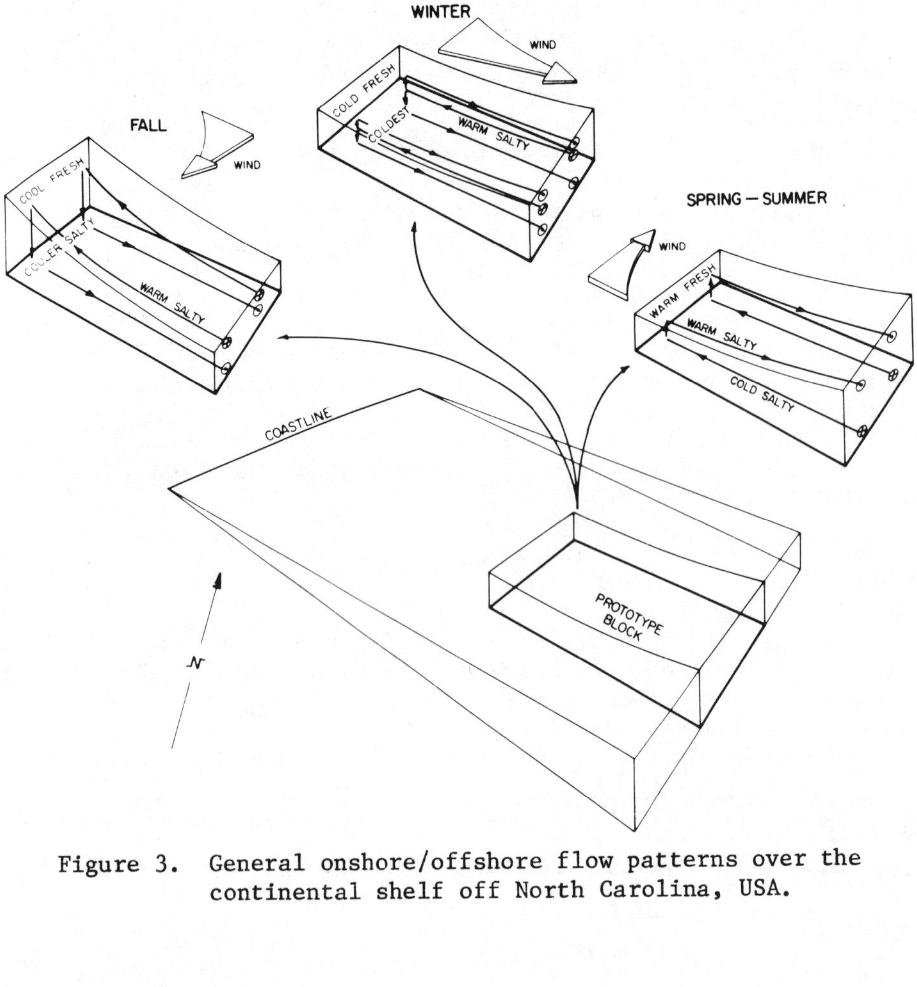

Figure 3. General onshore/offshore flow patterns over the continental shelf off North Carolina, USA.

surface waters are blown toward the coastline. Nearshore waters pile up against the coast, and a southerly (longshore) geostrophic flow results in the interior. This southerly interior flow then must be brought to zero at the bottom and, consequently, exacts a bottom stress which drives a bottom boundary layer flow offshore. Since these winds tend to have appreciable magnitude (Saunders 1977; Weber and Blanton 1980; Weisberg and Pietrafesa in press), the surface frictional layer, which moves shoreward and somewhat to the south, is of the order of 5-15 m thick. The near-bottom offshore-flowing layer is approximately 1-5 m thick. Surface flows are generally of the order of 10-20 cm/s with wind speeds of 0.6 to 1.8 m/s.

In winter, mean regional winds tend to be directed offshore. During this time of year cold fronts pass over the coastal ocean and extract an enormous amount of heat from shelf waters. This is, in fact, the major mechanism by which the coastal ocean is cooled. It presents the combined possibility of both wind and buoyancy flux or thermohaline-driven currents.

Superimposed upon, reinforcing, and possibly producing the correct conditions for the thermohaline-driven flow, is the offshore winter wind which drives surface currents offshore. Pietrafesa (1973) and Pietrafesa and Janowitz (1979) showed in separate sets of mathematical models that an offshore wind, in concert with a negative flux of buoyancy at the surface, will cause surface waters to move offshore in a layer 2-5 m thick at speeds of 5-15 cm/s. Because inner-shelf water is cooled more rapidly than offshore waters, it can sink, by gravity, and flow offshore in a thin, bottom boundary layer. As a kinematical consequence of conserving mass and volume an onshore interior flow will occur. This flow, although density-driven, helps to maintain the heat and salt balance of the shelf. It is likely that this interior layer occupies about 70% of the water column and moves at speeds of the order of 3-8 cm/s and brings warm, salty waters onshore.

Spring and summer merge as one in terms of wind influence on shelf circulation along the North Carolina coast. Winds rotate to become northeastward in March and persist towards the northwestward to northeastward directional quadrant until August (Fig. 2). This situation is one of conventional Ekman coastal upwelling and features an offshore flow in the surface layer and an onshore flow in a bottom boundary layer. The surface layer is 2-7 m thick, and the bottom layer is less than 1-3 m thick. Offshore flow speeds are about 10 cm/s while onshore flows are about 5 cm/s. Interior flow is northeastward and parallel to the coastline. This flow pattern (Fig. 2) persists until August when the transition to fall occurs, and the yearly cycle begins anew.

Thus, conditions favoring onshore transport of larvae are as follows: Fall - in the 5-15 m thick surface layer; Winter - in the intermediate layer which occupies 50-70% of the water at any depth on the shelf; and Spring and Summer - in the thin, 1-3 m thick, bottom layer. Larvae occupying other layers will not tend to be transported onshore. The most persistent of these three layers is the intermediate layer in winter because it is partly density-driven and, therefore, less responsive to wind-forcing. Thus, it is concluded that the best time for onshore larval transport (in the sense of predictability) is winter--and this fits the general pattern of the relative success of winter spawners. The next best time is fall if larvae remain in the surface layer. The worst time is spring and summer when larvae would have to be in the relatively thin (1-2 m) bottom layer to reach the shore.

Even though larvae in the surface layer in the fall would generally be transported onshore, it is important to realize that surface currents are responsive to the wind, which is variable in direction, especially during changing seasons.

ONSHORE MIGRATION

Spawning and Larval Drift

We will now examine the larval data to see if the above current patterns are consistent with larval success. Like menhaden and flounders, spot and croaker both spawn offshore in the vicinity of the edge of the Gulf Stream. Peak spawning of croaker occurs in October (R.M. Lewis and M.H. Judy, personal communication) and continues through February, while spot spawn primarily in December and January. As they grow the larvae of both species move toward inlets where they arrive about 60-90 days after hatching (Warlen 1981 and personal communication). The relative numbers of larvae, however, do not parallel the relative abundance of adults. In a study off Beaufort, North Carolina, spanning the spawning seasons of two years (1972-73) spot larvae were about twice as abundant as croaker larvae (R.M. Lewis and M.H. Judy, personal communication). Corresponding relative abundance of adults in the same years showed about equal abundance in the two years. In another study (Warlen 1981 and unpublished data) conducted off Beaufort from November 1979 through March 1980, spot larvae were three times as abundant as croaker larvae. During this same year spot adults were only about one third as abundant as croaker in the commercial catch.

The general pattern emerges that spot larvae are generally more abundant in relation to the adult stocks than croaker. This is, in fact, consistent with the testable implication of an hypothesis that larvae in shelf waters between December and February would be more successful than those present before. Recall that croakers spawn from October through February, with peaks in October and November, whereas spot spawning peaks in December and January. It is suggested that the earlier-spawned croaker larvae may contribute relatively less to an annual cohort than the later-spawned larvae.

Warlen's (1981 and unpublished) data showed that the ages of both spot and croaker larvae increase systematically toward shore (Fig. 4). The standard errors of these ages are generally ± a day or so and are means of five collections over a 5-month span (November-March), so the pattern is both remarkably persistent and precise. From their ages larvae seem to be transported shoreward at a rather uniform speed of about 4 cm/s (about 3.5 km/day). Of course the different-aged larvae could have reached their respective sampling points via parallel trajectories (say to the N) at higher speeds. The apparent speed of about 4 cm/s is well within the speed range of the intermediate layer moving onshore. Although Warlen's (1981 and unpublished) data were from oblique tows, the limited vertical distribution data available from discrete depth tows taken in March, 1975 (Kjelson et al. 1976) showed that spot and flounder larvae were caught in about equal abundance near the bottom, at mid-depth and near the surface at a station about 40 km S (30 m depth) of the Beaufort Inlet. Although certainly not definitive, if larvae are

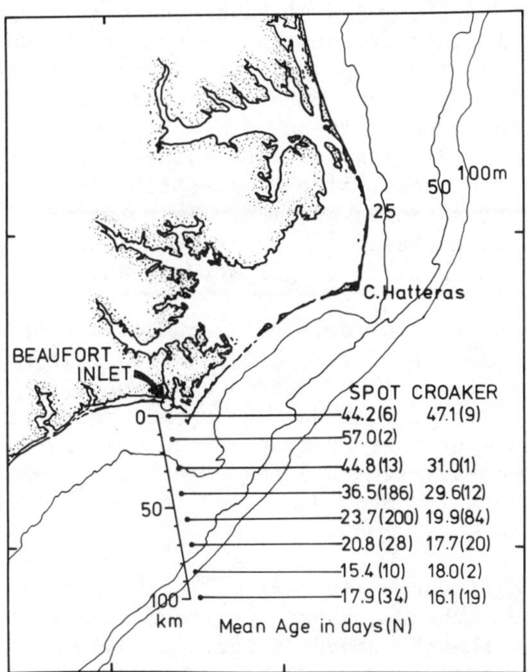

Figure 4. The age distribution of larval spot (*Leiostomus xanthurus*) and croaker (*Micropogonias undulatus*) off North Carolina, USA. Combined samples from five cruises, November-March (redrawn from Warlen 1981 and unpublished data).

nearly uniformly distributed with depth, the majority would be in the relatively thick intermediate layer moving onshore.

The other testable implication of the shelf current regime is the prediction that the ages of larvae, such as menhaden, which inhabit the surface waters (Kjelson et al. 1976) during onshore transport, should be more mixed than those which are deeper (spot, flounder and croaker). Berrien et al. (1978) summarized the results of ichthyoplankton collections taken on 14 transects from the shelf waters of Massachusetts to Cape Lookout, North Carolina. Ten samples were in the data from the three transects between Cape Hatteras and Cape Lookout, which contained both menhaden and spot larvae in sufficient numbers (21-378) for comparison. These transects (three-five stations each) were sampled in 1-2 days in December 1965 and February 1966. The range in length of menhaden in all 10 comparisons exceeded that of spot in the samples. This difference in range increased with mean size up to a factor of six. For example, the range of total lengths of menhanden was 7.1-29.8 mm in one sample (mean 13.77 mm), whereas the range of standard lengths of spot was 10.1-14.2 mm (mean 12.44 mm). Nelson (1975) examined the menhaden larval data from monthly cruises in Onslow Bay, North Carolina, from October 1972-April 1974. Stations (16-19) were sampled out to a depth of about 50 m (83 km) S of Beaufort Inlet. Also included were data from several other cruises in the same vicinity in those years. According to Nelson (1975, p. 38) "...menhaden larvae were thoroughly mixed, but still exhibited a general increase in size as onshore movement occurred." Examination of his figures shows menhaden mean size to be unevenly distributed with respect to distance from shore, with as many as four modal size-classes (18, 13, 8 and 5 mm total length) present at a single station. Thus, as predicted, the distribution of sizes of menhaden larvae is much less precise and persistent than the distribution of ages of spot and croaker larvae.

We would expect, however, that menhaden are relatively less successful than spot, at least, because the optimum time for onshore transport of surface-oriented larvae would seem to be in fall, not winter. Alternatively, menhaden may spend enough time in the intermediate layer in winter to be transported onshore at what would most likely be a slower average rate than spot and croaker. Unfortunately, no age data for menhaden exist to test this or alternative hypotheses.

Thus, the differences in vertical distribution between the more surface-oriented menhaden larvae and the deeper spot and croaker larvae, and also the abundance differences between spot and croaker, support the hypothesis advanced above. The mechanisms, other than an apparent difference in depth distribution, are unknown. Menhaden larvae may be passively drifting, but it is difficult to imagine that the observed age distribution of spot and croaker could be generated or maintained without some more active and oriented response. The simplest response for all species would seem to be a vertical temperature response. Larvae would remain in those waters moving onshore both in fall and winter if they were attracted to the warmest water available to them. During their onshore transport larvae would experience gradually decreasing temperatures (<1 C/10 km), the magnitude of which would probably preclude active selection of the warmest water in the horizontal plane and in any case larvae would need to swim against the onshore currents to move offshore to warmer waters. The vertical temperature gradient is considerably steeper (about 1 C/10 m) than the horizontal (except near shore). It would seem that larvae could detect and select temperatures in the vertical dimension considerably more easily than in the horizontal (see Westerberg 1984--this volume). Since croaker spawn from October to February, a testable implication of the above is that croaker larvae should be higher in the water column in fall than in winter.

Regardless of the mechanism, it is clear that additional resolution of both the currents and distribution of larvae (both horizontally and vertically in each case) is necessary. With a vertical array of current, temperature and salinity recorders in place along a transect perpendicular to shore, a rapid larval sampling response capability, and a capability of tracking ocean fronts with shipboard satellite communications, data could be collected before, during, and after the passage of an event along a specific pathway. This approach would obviate the usual extensive array of meters and sampling required to provide adequate coverage in time and space when the position of a front is unknown. Ideally, a parcel of water and its attendant larvae would be tagged. The difference between the movements of water and larvae would be the activity of the larvae. The reduction in scale (to a few hours and kilometers) would add significantly to the resolution of vectors, not just the resultant distributions, of migrations.

Migration Through Inlets

As spot and croaker larvae approach the inlets they apparently become more demersal (Kjelson et al. 1976). Near the inlets they metamorphose into juveniles. By remaining near the bottom, currents would eventually transport them through the inlets into the sounds. Brown shrimp (*Penaeus aztecus*) post-larvae make the same migration through inlets into the sounds, having been spawned in the nearby shelf waters, also in winter. We conclude that, if shrimp post-larvae can successfully enter the estuaries with much less vagility than juvenile fishes, a more complex mechanism is not necessary for fishes. Apparently some shrimp accomplish such directed migrations by moving up into the water column on flood tides then down to the bottom on ebb tides (Penn 1975). We postulate that juvenile spot, croakers and flounder accomplish a similar movement into North Carolina estuaries by remaining near the bottom. The mechanism in the case of menhaden may be more complex since they apparently remain in the upper portion of the water column even after metomorphosing later. Fore and Baxter (1972) suggested menhaden larvae may escape strong ebb currents by aggregating along the quieter edges of inlet channels.

Migration from Inlets to Juvenile Nursery Areas

Having traversed the inlets to Pamlico Sound (Fig. 1), how do the juvenile fish move some 50 km to nursery areas such as Rose Bay (= juvenile nursery on Fig. 1)? Apparently no more complex behavior is necessary than to remain near the bottom, at least in winter. Winter storms characterized by relatively strong northwesterly winds blow Pamlico Sound water against the barrier islands. This creates a hydrostatic imbalance which is restored by bottom currents in the opposite direction to the winds. In shallow sounds such bottom currents can develop velocities of the order of 10% of the wind speed, usually within 24 h. Thus, a 5-day storm with wind speeds of only about 1 m/s would transport juveniles the 50 km to the nursery area. Such storms occur about once every 2-3 weeks in winter. In seasons other than winter, however, prevailing winds would produce bottom currents which are decidedly less favorable to westerly transport to upstream nursery areas. It should be noted that other mechanisms for transport, such as riding upstream transport of salt water in response to river flows in drowned river estuaries and currents caused by lunar tides, are not major forcing functions in shallow sounds with narrow inlets and low lunar-tidal amplitudes (about 10 cm) such as Pamlico Sound.

While the above mechanism of riding wind-induced bottom currents seems sufficient to account for the transport of spot, croaker and flounder juveniles to their estuarine nursery areas, the transport of menhaden juveniles, which seem to remain near the surface in currents which generally parallel the wind would not be facilitated.

The Nursery Areas: Immigration and Emigration

The pattern of immigration into the juvenile nursery areas such as Rose Bay from inlets is similar for the dominant species. However, certain differences in their utilization patterns within these nurseries are apparent and are illustrated by spot and croaker. Both species begin arriving in February (Fig. 5) with peak immigration rates in March and early April. Peak numbers of spot are typically an order of magnitude greater than croaker. After May, numbers decline--early, due to mortality and later, due to emigration. Their distribution within the bay is different. Juvenile spot are more evenly distributed with depth, and their intra-bay movements are significantly correlated with changes in food abundance. Croaker juveniles, on the other hand, are always more abundant in the shallower regions and tributaries of the bay. Eighty-four and 55% of the juvenile croakers and spot, respectively,

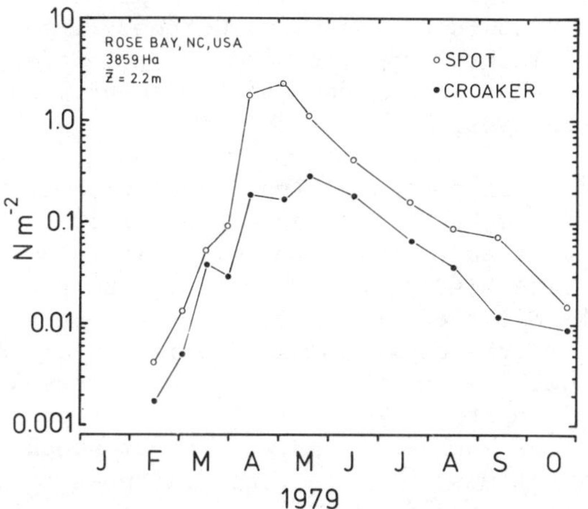

Figure 5. Pattern of entry of juvenile spot (*Leiostomus xanthurus*) and croaker (*Micropogonias undulatus*) into estuarine nursery areas, as exemplified by Rose Bay, North Carolina.

caught in 1979 were in the shallower half of Rose Bay (J.M. Miller and B.M. Currin unpublished data). In contrast to the ten-fold greater biomass of spot in May, by July the biomass of croaker (12.9 g/m^2) exceeded that of spot (10.3 g/m^2). This reversal was due primarily to the lower mortality rate of croaker. From mid-May to mid-July the instantaneous daily mortality rates of spot and croaker were -0.0313 and -0.0237, respectively. The growth rate of croaker was lower (0.195 g/d) than that of spot (0.0237 g/d). So the apparently more efficient feeding of spot was more than counter-balanced by their higher mortality.

Because spot utilized the deeper portions of the bay, where we suspect greater numbers of predators inhabit the waters which have more stable salinities, we hypothesize the greater mortality of spot was due to predation. In contrast, croaker juveniles, by remaining in the shallower areas, although less favorable for growth in terms of both food abundance and salinity and temperature variation, apparently avoided the greater predation pressure. Thus, during the juvenile nursery phase, croaker begin to catch up with spot. Recall croaker adult biomass generally exceeded that of spot by a factor of three, so croaker production may exceed that of spot after migrating from the primary nurseries (bays) to the more open areas of Pamlico Sound.

It is not known what triggers the emigration from the primary nurseries. As fish grow they become progressively less vulnerable to capture by small trawls, and it becomes increasingly difficult to separate mortality from emigration. Emigration may be related to ontogenetically increased sensitivity to salinity changes or decreasing food resources. By about October large numbers of small (70-100 mm total length) spot, croaker and flounders are observed near inlets and, thereafter, move offshore--apparently in response to declining temperatures.

We conclude this portion of our paper with the observation that most of the fish stocks of the eastern and gulf coasts of the USA are estuarine dependent and, therefore, migration plays an important part in their early life histories. This is in sharp contrast to most important stocks off the west coast which are pelagic at all stages. In these latter stocks environmental variation on the micro- or meso-scale may well be determinants of year-class success (Lasker 1981). Finding a microscale patch of food may pose a similar problem for early larvae of both east and west coast stocks, but for larvae and juveniles which must migrate long distances to estuarine nurseries, it would seem that microscale processes may be considerably less important as determinants of year-class strength than success of migration. At the very least, studies of important marine stocks on the east coast must include the additional dimension of migration.

DIRECTION OF FUTURE RESEARCH

In accordance with the purpose of this institute we suggest that the following research directions should lead to a greater understanding of larval and juvenile fish movements.

1. Because success in early life is probably more a function of survival than growth, migration should also be viewed as a strategy to avoid predators or lethal environmental conditions rather than simple energetic optimization.

1a. A common currency for predator avoidance and energetic optimization needs to be developed, as well as further investigation of possible cues involved in predator avoidance.

1b. The distribution of predator pressure needs to be determined in both time and space.

2. Migratory behavior of larvae and juveniles cannot be studied with conventional tags. Since movements may have to be deduced from changes in distribution, such research must be conducted in concert with estimates of mortality, which is high.

2a. Age determination of larvage by using otoliths is a powerful tool for estimating both vectors of migration and short-term mortality.

3. Because the vagility of larvae and juveniles is limited, movements--if keyed to environmental gradients--may be imprecise and, to a greater extent than adults, a function of currents. The passive distributional responses of larvae and juveniles to these currents must be subtracted to determine any active component.

3a. A more precise resolution of depth distribution of larvae and juveniles is necessary to determine the current regime to which these stages are exposed.

3b. Studies of short-term movements in response to ocean fronts may be economical if these fronts can be adequately monitored physically and followed with the aid of shipboard satellite imagery, which can greatly reduce the data coverage required.

4. It seems likely that the greater tolerance of larvae and juveniles of environmental extremes is a physiological alternative to behavioral regulation. Ontogenetic changes in costs of tolerance need to be estimated relative to the costs of movements.

ACKNOWLEDGEMENTS

This work was supported, in part, by the Office of Sea Grant, NOAA, US Department of Commerce under Grant No. NA81AA-D-00026, the US Department of Energy under Contract No. DOE-A809-76-EY00902, and the University of North Carolina Water Resources Research Institute. The authors are especially grateful to Dr. Stanley M. Warlen, National Marine Fisheries Service, Beaufort, North Carolina, USA, for providing us with some of his unpublished data on spot larval ages.

REFERENCES

Berrien, P.L., M.P. Fahay, A.W. Kendall, Jr., and W.A. Smith. 1978. Ichthyoplankton from the R/V *Dolphin* Survey of Continental Shelf waters between Martha's Vineyard, Massachusetts and Cape Lookout, North Carolina, 1965-66. US National Oceanic and Atmospheric Administration Northeast Fisheries Center Sandy Hook Laboratory, Technical Series Report 15:1-152.

Chesnut, A.F., and H.C. Davis. 1975. Synopsis of marine fisheries. University of North Carolina Sea Grant Publication UNC-SG-75-12:1-425. Raleigh, North Carolina, USA.

Clark, J., W.A. Smith, A.W. Kendall, Jr., and M.P. Fahay. 1969. Studies of estuarine dependence of Atlantic coastal fishes. Data Report 1: Northern Section, Cape Cod to Cape Lookout. R.V. *Dolphin* cruises 1965-66. Zooplankton volumes, midwater trawl collections, temperatures and salinities. US Fish and Wildlife Service Technical Papers 28:1-132.

Fore, P.L., and K.N. Baxter. 1972. Diel fluctuations in the catch of larval gulf menhaden, *Brevoortia tyrannus*, at Galveston Entrance, Texas. Transactions of the American Fisheries Society 101:729-732.

Holliday, F.G.T. 1965. Osmoregulation in marine teleost eggs and larvae. California Cooperative Oceanic Fisheries Investigations Reports 10:89-95.

Kjelson, M.A., G.N. Johnson, R.L. Garner, and J.P. Johnson. 1976. The horizontal-vertical distribution and sample variability of ichthyoplankton populations within the nearshore and offshore ecosystems of Onslow Bay. Pages 287-341 *in* Atlantic Estuarine Fisheries Center Annual Report to ERDA (Energy Research and Development Administration), US National Marine Fisheries Service, Beaufort, North Carolina, USA.

Lasker, R. 1981. The role of a stable ocean in larval fish survival and subsequent recruitment. Pages 8-87 *in* R. Lasker, editor. Marine fish larvae. University of Washington Press, Seattle, Washington, USA.

Leggett, W.C. 1977. The ecology of fish migrations. Annual Review of Ecology and Systematics 8:285-308.

Miller, J.M., and M.L. Dunn. 1980. Feeding strategies and patterns of movement in juvenile estuarine fishes. Pages 437-438 *in* V.S. Kennedy, editor. Estuarine perspectives. Academic Press, New York, New York, USA.

Nelson, W.R. 1975. Larval transport and year-class strength of Atlantic menhaden. Doctoral dissertation. University of Southern Mississippi, Hattiesburg, Mississippi, USA.

Nelson, W.R., M.C. Ingham, and W.E. Schaaf. 1977. Larval transport and year-class strength of Atlantic menhaden, *Brevoortia tyrannus*. US National Marine Fisheries Service Fishery Bulletin 75:23-41.

Penn, J.W. 1975. The influence of tidal cycles on the distributional pathway of *Penaeus latisulcatus* Kishinouye in Shark Bay, Australia. Australian Journal of Freshwater Research 26:93-102.

Pietrafesa, L.J. 1973. Steady baroclinic circulation on a continental shelf. Doctoral dissertation. University of Washington, Seattle, Washington, USA.

Pietrafesa, L.J., and G.S. Janowitz. 1979. On the effects of buoyancy flux on continental shelf of North America. Journal of Physical Oceanography 9:911-918.

Saunders, P.M. 1977. Wind stress on the ocean over the eastern continental shelf of North America. Journal of Physical Oceanography 7:555-566.

Stickney, A.P. 1959. Ecology of the Sheepscot River estuary. US Fish and Wildlife Service Special Scientific Report Fisheries 309:1-21.

Warlen, S.M. 1981. Age and growth of larvae and spawning time of Atlantic croaker in North Carolina. Proceedings of the Annual Conference, Southeastern Association of Fish and Wildlife Agencies 34:204-214.

Weber, A.H., and J.O. Blanton. 1980. Monthly mean wind fields for the

South Atlantic Bight. Journal of Physical Oceanography 10:1256-1263.

Weisberg, R.H., and L.J. Pietrafesa. In press. Kinematics and correlation of the surface wind field in the South Atlantic Bight. Journal of Geophysical Research.

Westerberg, H. 1984. Vertical small-structure of natural waters: a source of orienting cues for migratory fish? Pages 179-203 *in* J.D. McCleave, G.P. Arnold, J.J. Dodson and W.H. Neill, editors. Mechanisms of migration in fishes. Plenum Press, New York, New York, USA.

FISH MIGRATION BY SELECTIVE TIDAL STREAM TRANSPORT: FIRST RESULTS WITH A COMPUTER SIMULATION MODEL FOR THE EUROPEAN CONTINENTAL SHELF

G. P. Arnold and P. H. Cook[1]

Ministry of Agriculture, Fisheries and Food
Fisheries Laboratory
Lowestoft, Suffolk NR33 OHT England

ABSTRACT

The demonstration of selective tidal stream transport in species as diverse as plaice, sole, cod, dogfish and eels suggests that the phenomenon may be of widespread significance in the migration of fishes on the European continental shelf. A computer simulation model has been written which interpolates the speed and direction of the tidal stream from the data given for the tidal diamonds on the appropriate British Admiralty Charts. The model allows for the cyclical change of current speed between neap and spring tides and includes three behavioural options for the fish: 1) semidiurnal vertical migration initiated at either slack water with a variable period of transport in midwater; 2) diurnal vertical migration with a variable ratio of time spent in midwater to time spent on the bottom; and 3) continuous transport (i.e. drift) in midwater. The aims of the model are to generate simulated tracks which: 1) taken individually, may lead to predictions about the behaviour of individual fish, which can be tested by further tracking experiments; and 2) taken collectively, may lead to predictions about the resulting distributions of whole populations, which can be tested by conventional tagging experiments. The first simulations have been

---

1 Natural Environment Research Council postgraduate research student based at Fisheries Laboratory, Lowestoft. Present address Institute of Oceanographic Sciences, Wormley, England.

concerned with the prespawning and postspawning migrations of plaice in the Southern Bight and English Channel and with the spawning migrations of silver European eels crossing the European shelf on their return to the Sargasso Sea. Some simulations have also been made of the tracks of plaice exhibiting diurnal vertical migrations on various feeding, spawning and staging grounds in the southern North Sea and English Channel.

Details are given of a series of semidiurnal tracks running from the southern North Sea through the Strait of Dover into the English Channel, and it is shown that the four centres of plaice spawning off Flamborough Head, in the Southern Bight and in the English Channel lie along a common streampath. A second series of semidiurnal tracks is used to predict the distributions of spent plaice migrating out of the Southern Bight into the central North Sea. The distributions are plotted on a lunar time base, and it is shown that fish contained within the area of peak spawning will use the western migration route. The diurnal tracks show that a fish adopting the diurnal mode of vertical migration will remain within a relatively confined locality even in areas of fast and directional tidal streams.

The simulated silver eel tracks show that a fish leaving the River Severn, England, should be able to reach the edge of the continental shelf to the southwest of the British Isles in about 40 days by selective tidal stream transport. A fish from the River Humber, England, should similarly be able to reach the edge of the shelf by the northern North Sea in about 100 days. But there is a problem for eels using tidal stream transport from the River Elbe, West Germany. They would appear to become permanently trapped in the system of east-west tidal streampaths in the central North Sea and might fail to reach the edge of the continental shelf in sufficient time to complete the subsequent oceanic phase of the migration. An analogy is drawn between this apparent "tidal gyre" in the North Sea and the areas of the central North Pacific from which Pacific salmon fail to return to North America. An eel from Texel, The Netherlands, relying on tidal stream transport would similarly reverse in the English Channel in the area of dividing tidal streampaths to the west of the Cherbourg peninsula, France, and return to the North Sea. Reversal is also relevant to the migrations of the plaice and appears to be limited to tracks passing through the Strait of Dover within 20 km of Cap Gris Nez, France.

Future work will be concerned with extending the simulations to the distributions of other plaice stocks on the European shelf--and to other species such as cod and haddock--and to predictions aimed at furthering our understanding of the mechanism of selective tidal stream transport itself. The phenomenon of reversal, for example, suggests the possibility of conducting a sea-going experiment to determine whether the underlying mechanism is, as we suspect, based on a succession of temporal cues or instead involves an absolute sense of direction on the part of the fish.

INTRODUCTION

Selective tidal stream transport has been demonstrated in five species of European fishes and appears therefore to be a significant mechanism of migration on the continental shelf. The objectives of this paper are threefold: 1) to summarize our knowledge of selective tidal stream transport; 2) to outline briefly the development of a computer simulation model; and 3) to describe the results of the first simulations.

SELECTIVE TIDAL STREAM TRANSPORT

Although previously known in penaeid shrimps, tidal stream transport was first described in fish by Creutzberg (1961), who observed elvers of the European eel, *Anguilla anguilla*, migrating inshore on the flood tide during the final stages of the long migration from the Sargasso Sea. Subsequent work at the Fisheries Laboratory, Lowestoft, has shown that the phenomenon occurs in plaice (*Pleuronectes platessa*), sole (*Solea solea*), Atlantic cod (*Gadus morhua*), dogfish (*Scyliorhinus canicula*) and adult European eels. Movements of individual fish have been described from tracks in the open sea using transponding acoustic tags (Mitson and Storeton-West 1971) and sector scanning sonar (Voglis and Cook 1966); movements of natural populations have been inferred from midwater trawling experiments (Harden Jones et al. 1979).

The Essential Features

The behaviour, which we have called *selective tidal stream transport* (Greer Walker et al. 1978), is a consistent pattern of semidiurnal[1] vertical migration, in which the vertical movements of the fish from the bottom up into midwater are linked to the tidal streams. The fish leaves the bottom at, or shortly after, slack water and is carried downstream for the duration of the transporting tide. It returns to the bottom at the next slack water and remains there for the duration of the opposing tide, making no significant movement in the direction of overall transport. Descriptions of the tracks of individual plaice have been given by Greer Walker et al. (1978), Harden Jones (1977, 1980), Harden Jones et al. (1978), Arnold (1981) and Harden Jones and Arnold (1982). Here the essential

[1] The terms *diurnal* and *semidiurnal* are used here as in physical oceanography (Defant 1958, page 48) to describe tides and other cyclical phenomena with periods of about 24 h and 12 h, respectively.

features of tidal stream transport are illustrated with the track of a silver eel released in the North Sea off East Anglia (J.D. McCleave, M. Greer Walker, and G.P. Arnold, unpublished data). This eel was taken, with 51 other silver eels, on the night of 17 September 1979 in a downstream trap on the River Ure, a tributary of the River Humber, in Yorkshire. It was kept in the laboratory at Lowestoft until it was taken to sea on R.V. CLIONE on 26 October. The 97-cm long eel was released from a cage on the sea bed at 1515 h, 28 October at a position 22.8 km east by north of Orfordness, after 4.5 h adaptation at a depth of 35 m. It was tracked for 34.5 h and showed selective tidal stream transport moving to the north on three

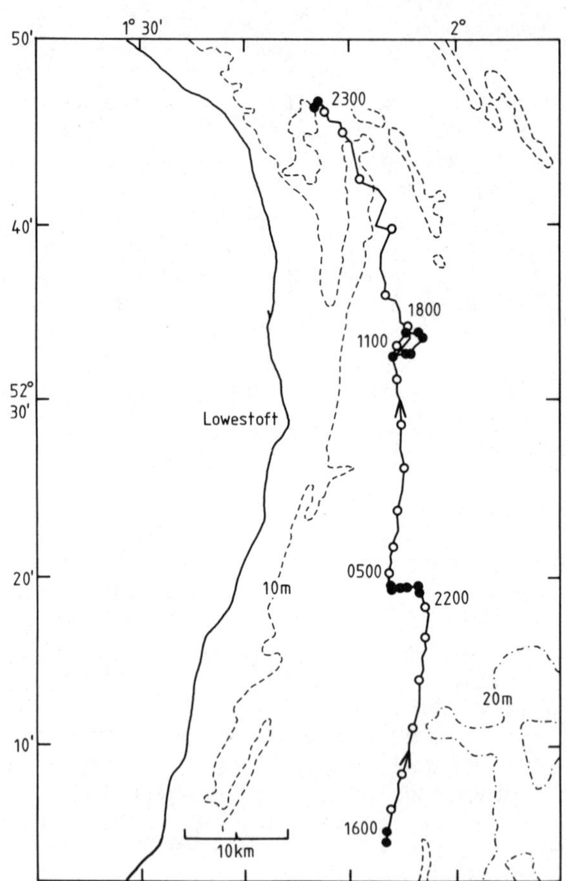

Figure 1. Track chart of a silver European eel released at 1515 h, 28 October 1979. The position of the fish is plotted at hourly intervals with open circles indicating the north-going tide and filled circles the south-going tide. The 10-m and 20-m isobaths are indicated (J.D. McCleave, M. Greer Walker and G.P. Arnold, unpublished data).

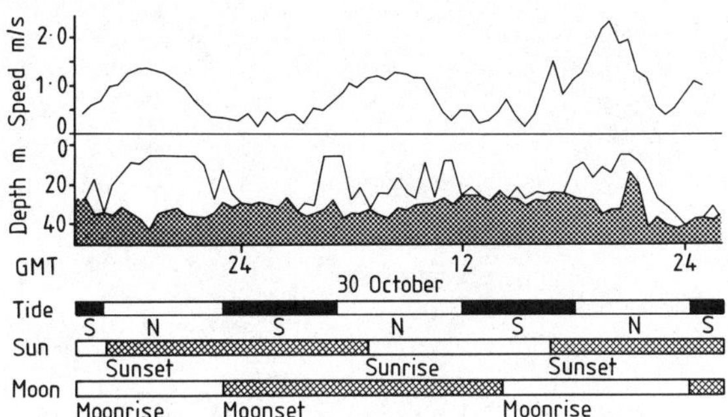

Figure 2. The ground speed and depth of the eel in relation to the tide and other environmental factors (J.D. McCleave, M. Greer Walker and G.P. Arnold, unpublished data).

successive north-going tides (Fig. 1) and going to the bottom on the two intervening south-going tides (Fig. 2).

The Life History and Migrations of the Plaice

The plaice, which is found on the sandy parts of the continental shelf of Europe, is a typical demersal flatfish, whose life is greatly influenced by both tidal and residual currents. It spawns in winter producing pelagic eggs, which drift with the residual current. Egg and larval development occur during the planktonic phase, after which the larvae metamorphose, become asymmetric and drop to the bottom. They then move inshore onto intertidal nursery grounds. There are several stocks of plaice on the European shelf and for each stock there are discrete spawning, feeding and nursery grounds. In the northern North Sea there are spawnings off the Scottish east coast and in the Moray Firth. In the southern North Sea there are three stocks, which spawn in the Southern and German Bights and off Flamborough Head (Fig. 3). There is also an intermediate spawning in the central North Sea and two spawnings in the English Channel. The centres of these spawnings--whose positions appear to be stable over several decades--are indicated in Figure 4. In the summer the Southern Bight, German Bight and Flamborough spawners occupy different feeding areas with only limited overlap in the central North Sea (de Veen 1962).

Conventional tagging experiments (reviewed by Harden Jones 1968) have shown that in November and December maturing plaice migrate down the western side of the Southern Bight from the feeding grounds to the south of the Dogger Bank to the spawning area centered on the Hinder Ground (Fig. 3). The migration, which is 200-300 km in extent, probably also extends into the eastern English Channel: some 60% of the fish spawning in the eastern Channel are thought to be of North Sea origin (Houghton and Harding 1976). Peak egg production

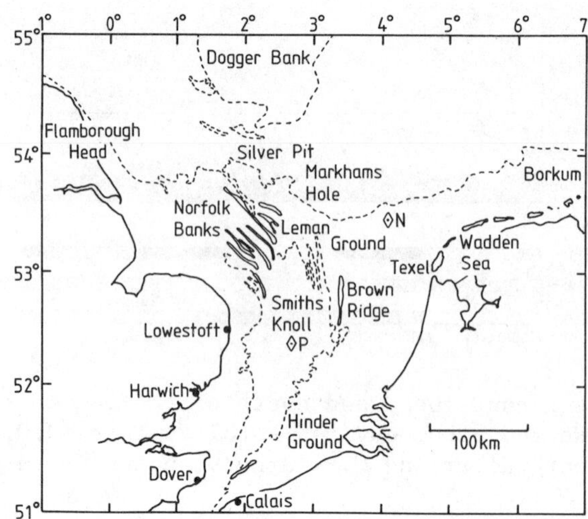

Figure 3. Place names in the southern North Sea of interest in connection with plaice migrations. The positions of tidal stations N and P (Table 3) are indicated by diamonds. The 30-m isobath is shown.

occurs in January in the Southern Bight, and spent fish return to the north in January, February and March, again along the western route. Meanwhile the pelagic eggs and larvae drift to the northeast at a rate of about 3 km day^{-1} with the residual current entering the North Sea through the Strait of Dover. Depending on temperature the larvae take about 5-8 weeks to reach metamorphosis. The point at which they then drop to the bottom is governed by the speed and direction of the residual current. Most of the larvae take to the bottom along the Dutch coast, frequently some distance offshore but in some years close to Texel (Harding et al. 1978, their Fig. 107), where under certain conditions there may be a residual inshore current at the sea bed (Cushing 1972). Many of these young plaice enter the Wadden Sea--an extensive area of tidal mudflats interspersed with deep creeks--through the passages between the Dutch Frisian Islands. Other important nursery grounds are the estuaries of the rivers Maas and Scheldt and the sandy beaches of the open coast.

After spending the first summer in depths of 2-3 m the young plaice move to deeper water in the autumn, returning to shallower water again the following spring or early summer. There is an overall movement to deeper water in the second and third years of life. Some of the larger plaice, mostly males, reach first maturity in their third year and join the mature fish of earlier year-classes on the spawning ground towards the end of the season. These first-time spawners migrate down the eastern side of the Southern Bight between Texel and the Brown Ridge (Fig. 3). After spawning they

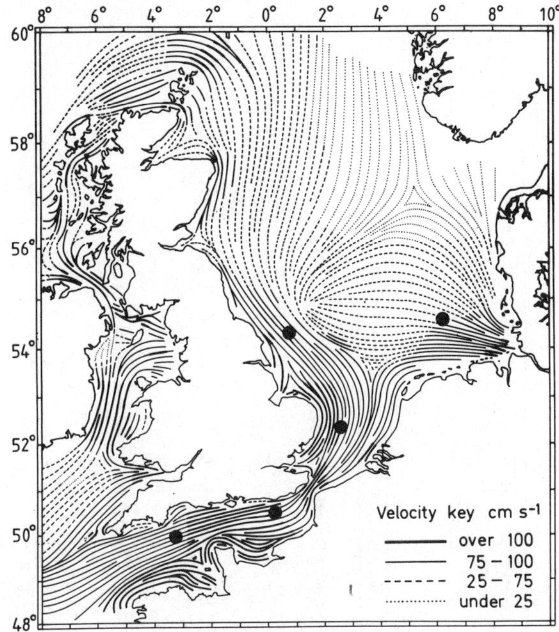

Figure 4. A tidal streampath chart for the North Sea and adjacent areas (Harden Jones et al. 1978). The centres of the Flamborough, German Bight, Southern Bight and eastern and western English Channel plaice spawnings are indicated by the filled circles.

return to the northerly feeding grounds with the repeat spawners. Most of the remaining immature fish spawn for the first time in the following year and recruitment to the adult stock is then almost complete. Plaice spawn annually until they die but with present levels of exploitation few individuals are likely to spawn more than ten times.

The Role of Tidal Stream Transport in the Migrations of Southern Bight Plaice

Tidal stream transport has been shown to be an important feature throughout the life history of the plaice in the Southern Bight of the North Sea. Midwater trawling experiments, using a paired-haul technique, have shown that in November and December ripening plaice migrating into the Southern Bight by the western route use tidal stream transport from at least as far north as the Leman Ground (Table 1a). Paired tows made in February show that spent female plaice return to the north by the same transport system at least as far as Smiths Knoll (Harden Jones et al. 1979). Similar fishing experiments with towed or anchored nets have shown that juvenile

Table 1. Catches of plaice by night for paired tows with an Engel midwater trawl on consecutive north-going and south-going tides in the Southern Bight of the North Sea.

Months	Years	Number of paired tows	Number of plaice caught on each tide		Catch rate on each tide		Paired t-test		
			N	S	N	S	t	df	P
(a) Along the line of the western migration route (Harden Jones et al. 1979)									
Nov, Dec	1974–5	18	35	149	1.9	8.3	4.42	17	0.001
Feb	1976–7	8	84	11	10.5	1.4	4.36	7	0.01
(b) On the spawning grounds (unpublished data for running males and maturing, ripe and spent females)									
Jan	1982	6	90	116	15.0	19.3	1.08	5	NS
Jan	1983	7	30	43	4.3	6.1	1.04	6	NS

plaice enter their nursery grounds on the Dutch coast by the same mechanism. This has been demonstrated for newly metamorphosed stage V larvae entering the Wadden Sea (Creutzberg et al. 1978) and the Eastern Scheldt (Rijnsdorp and van Stralen 1982) and for 1-group fish reentering the Wadden Sea after spending the first winter offshore in deeper water (de Veen 1978). Tidal stream transport thus appears to be a mechanism enabling the fish to make rapid directional movements between feeding, spawning and nursery grounds, probably without the need for any navigational ability. Significant savings of energy--which in the case of the prespawning migration of Southern Bight plaice may be expected to have a selective advantage (Harden Jones 1980)--may also accrue from the use of this mechanism (Weihs 1978).

At other times in the life history rapid directional movements are inappropriate, and the semidiurnal pattern of vertical migration linked to the tides is replaced by a diurnal pattern of vertical migration related to day and night. This has been demonstrated by tracking individual fish on their summer feeding grounds (Arnold 1981) and by fishing experiments on both feeding and spawning grounds.

Three plaice tracked in the vicinity of Markhams Hole (Fig. 3) for up to 82 h (G.P. Arnold, M. Greer Walker, and F.R. Harden Jones, unpublished data) remained on the bottom throughout the day but made a vertical excursion into midwater each night. These excursions began at, or shortly after, sunset and the fish swam freely in midwater for 1-3 h. Apart from an initial period after release in the case of one individual, movement over the ground was restricted to periods when the fish were in midwater. As with plaice using tidal stream transport off the East Anglian coast, the direction of transport was that of the tidal stream when the fish was in midwater. However, the scale and speed of movement was substantially less (by a factor of 4 to 5), partly as a result of the slower tidal streams--the speeds are approximately half those encountered off East Anglia--but primarily because of the diurnal periodicity of the shorter midwater excursions.

Midwater trawling experiments in the same area have confirmed that a small proportion of the plaice population on the summer feeding grounds is also in midwater by night (Table 2a). The majority of these fish have empty stomachs, and the results are consistent with the hypothesis that this nightly movement represents a change of local feeding ground within a larger feeding area. Similar fishing experiments on the spawning grounds in the vicinity of 52°15'N 2°40'E[1] have demonstrated a comparable diurnal pattern of

---

[1] This position was chosen in preference to one on the Hinder Ground to avoid the navigational difficulties of fishing in shipping routes.

Table 2. Midwater-trawl catches of plaice by day and night: (a) on the summer feeding grounds in May 1981 and July 1982 (sunrise 0330 h, sunset 2000 h, GMT); and (b) on the spawning grounds in January 1982 and 1983 (sunrise 0755 h, sunset 1600 h, GMT).

Period	Number of hauls	Number of plaice	Number of plaice per haul
(a) Sunrise to sunset	27	2	0.1
Sunset to midnight	13	42	3.2
Midnight to sunrise	6	17	2.8
(b) Sunrise to sunset	20	32	1.6
Sunset to midnight	19	247	13.0
Midnight to sunrise	13	169	13.0

vertical migration with spawning plaice up in midwater by night (Table 2b) and with apparently no tidal effect (Table 1b).

The extensive use by the plaice throughout its life history of tidal streams for transport raises a number of behavioural problems. For local movements within a feeding area direction is probably unimportant and a diurnal vertical migration, occurring independently of the tide, will probably achieve adequate dispersion. But direction--and thus choice of tide--must be of paramount importance when plaice enter or leave a nursery ground or when they migrate between feeding and spawning grounds. The fish must then be able at the outset to "lock-on" to the appropriate tide; some behavioural mechanism must maintain the semidiurnal pattern of vertical migration, so that the fish repeatedly joins the appropriate tide; and the fish must leave the transport system on arrival at the appropriate destination. But whatever the solutions to these problems it is clear that a fish joining the transport system must switch from a diurnal to a semidiurnal pattern of vertical migration, and it seems that this switching ability must be an essential feature of the plaice's life history.

A COMPUTER SIMULATION MODEL

Although direct evidence of the use of selective tidal stream transport by natural populations is so far limited to the plaice which spawn in the Southern Bight, it seems inherently likely that the mechanism will occur in most of the other stocks of plaice on the European shelf. To demonstrate it in even one or two more of these would, however, require considerable research vessel effort, and at this juncture an analytical approach is preferable to a purely descriptive one.

The original tracking work with individual plaice led to the formulation of simple hypotheses for the migratory behaviour of the population as a whole, and these were readily substantiated for the Southern Bight spawners by specific and highly effective fishing experiments. The predictive technique thus has much to commend it, particularly if it can be extended to other stocks--and even other species--for which tracking data are not yet available.

On the European shelf the tidal currents are almost entirely deterministic, behaving in a regular and predictable way (Gould 1978) and conforming to a fairly simple pattern, such that a realistic simulation of tidal stream transport appears to be feasible. We have therefore written a computer model which interpolates the speed and direction of the tidal stream in space and time from the data given on the appropriate British Admiralty Charts.

The aims of the model are to generate simulated tracks which: 1) taken individually, may lead to predictions about the behaviour of individual fish, which can be tested by further tracking experiments; and 2) taken collectively, may lead to predictions about the resulting distribution of whole populations, which can be tested by conventional tagging experiments.

The Tidal Streampaths

The tidal streams around the British Isles are generally fast and directional as shown in the tidal streampath chart (Fig. 4) constructed by Harden Jones et al. (1978), using the data given in Chart 3 of the Rostock Atlas (Anonymous 1968). The streampaths represent the lines of maximal tidal flow and indicate routes along which fish might be transported if the fish were to adopt selective tidal stream transport.

The streampaths along the east coast of England clearly link the plaice that spawn on the east coast of Scotland with those that spawn off Flamborough and, indeed, with those that spawn in the Southern Bight and English Channel. By contrast, plaice that spawn in the German Bight would appear to be contained within a separate system of streampaths, where the expected movement would be largely east and west.

The Data Base

British Admiralty Charts present tidal stream data for selected tidal stations, whose positions are indicated by diamonds. There may be 20 or more diamonds on a chart covering a large area such as the

Table 3. Speeds (1 knot = 0.51 m s^{-1}) and directions of predicted tidal streams at Admiralty tidal stations N and P in the Southern Bight of the North Sea (Fig. 3). The data are taken from British Admiralty Chart 2182A (International Chart 1403) and the tidal streams are referred to the time of high water at Dover. Speeds are given for both spring (Sp) and neap (Np) tides.

		N	53°27.2'N 4 05.1 E		P	52°24.0'N 2 41.8 E	
			Rate (knots)			Rate (knots)	
Hours		Dir	Sp	Np	Dir	Sp	Np
Before HW	6	043	0.9	0.5	000	0.6	0.4
	5	071	0.8	0.4	215	0.4	0.3
	4	102	0.7	0.4	195	1.2	0.8
	3	136	0.6	0.3	194	1.8	1.2
	2	170	0.5	0.3	192	2.0	1.3
	1	200	0.5	0.3	190	1.7	1.1
HW		219	0.6	0.3	181	0.9	0.6
After HW	1	243	0.7	0.4	047	0.4	0.3
	2	261	0.6	0.4	020	1.5	1.0
	3	292	0.5	0.3	017	1.8	1.2
	4	346	0.6	0.4	009	1.7	1.1
	5	008	0.9	0.5	006	1.5	1.0
	6	032	0.9	0.5	003	0.9	0.6

southern North Sea. A table at the margin of the chart lists for each diamond the predicted speed and direction of the tidal stream by hourly intervals from 6 h before to 6 h after the time of high water at an appropriate standard port. Speeds in knots (1 knot = 0.51 m s^{-1}) are listed separately for mean neap and mean spring tides. The tidal stream data for diamonds N and P (Fig. 3) taken from British Admiralty Chart 2182A (North Sea, southern sheet 1981) are shown in Table 3. Tidal streams on this chart are referred to the time of high water at Dover.

Our aim is to compile a data file comprising all the available tidal diamond data for the European shelf, and a substantial part of this work is already done. Data have been compiled for the North Sea, English Channel, Bristol Channel, Celtic Sea and the northern part of the Hebridean shelf, which leaves only the Irish Sea and Hebrides to be completed. The distribution of tidal diamonds in the southern North Sea and eastern English Channel is indicated in Figures 5 and 6.

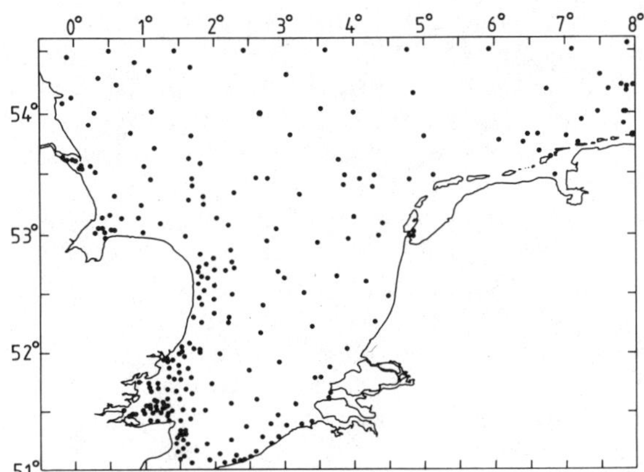

Figure 5. Positions of tidal diamonds in the southern North Sea taken from the appropriate British Admiralty Charts.

The tidal stream data taken from the Admiralty Charts are corrected, where necessary, to high water Dover, using the time delays for the standard ports obtained from the Bidston Observatory of the Institute of Oceanographic Sciences. In the English Channel, for example, high water occurs at Cherbourg and Le Havre 3 h 5 min and 1 h, respectively, before high water Dover. But in the Southern Bight the times of local high water reflect the anticlockwise rotation of the tidal wave around an amphidromic point at 52°30'N 3°03'E. Thus local high water occurs at Lowestoft 1 h 33 min *before* high water Dover and at the Hook of Holland 2 h 54 min *after* high water Dover. The flood tide is therefore the south-going tide on the western side of the Southern Bight and the north-going tide on the eastern side.

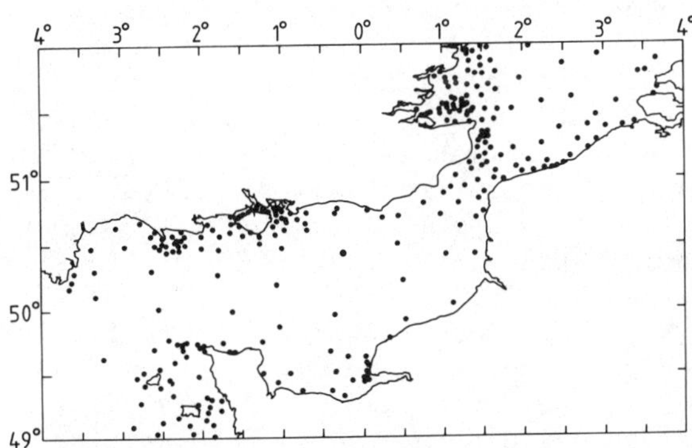

Figure 6. Positions of tidal diamonds in the eastern English Channel taken from the appropriate British Admiralty Charts.

The Interpolation of Current Speed and Direction

The distribution of the tidal diamonds is irregular, reflecting the interests of mariners rather than those of migrating fish. It is essential, therefore, to interpolate the speed and direction of the tidal stream at positions intermediate between the diamonds but not necessarily equidistant from them. It is also necessary to allow for the cyclical change in speed of the tidal stream between neap and spring tides.

The speed and direction of the tidal stream is determined in the model by a piecewise linear interpolation on a triangular finite element array. The computer program first searches for the nearest five diamonds and then selects the smallest triangle of diamonds containing the point of interest. Speed and direction are interpolated separately and direction is split into rectangular coordinates (x and y components). Speeds at neap and spring tides are calculated together. The variation of speed through the spring-neap cycle is accounted for subsequently by assuming a sinusoidal curve with a period of 14 days and an amplitude determined by the interpolated speeds at neaps and springs.

At the start of a simulated fish track the program calculates a full set of tidal diamond data for the chosen "release" point. It repeats this full-scale interpolation for each position at which the fish returns to the bottom and these data are used to identify the slackwater at which the fish should leave the bottom again. When the fish is in midwater the calculations are restricted to the interpolation of a series of tidal stream vectors for successive half-hourly periods.

The Behavioural Options

The model incorporates a number of options, which allow for three different patterns of vertical migration: 1) semidiurnal vertical migration initiated at either low-water or high-water slack tide with a variable period of transport in midwater; 2) diurnal vertical migration with a variable ratio of time spent in midwater to time spent on the bottom; and 3) continuous transport (i.e. drift) in midwater. Whenever the fish is in midwater its velocity over the ground is assumed to equal the tidal velocity.

In the semidiurnal option the period of transport can be specified as extending from one local slack tide to the next or as the number of hours spent in midwater. There is also a variant of this option designed to model the movements of fish in and out of the nursery grounds on the Dutch coast. In this variant successive tidal stream vectors are calculated at 0.25-h intervals.

Tidal stream vectors are predicted for the surface layers, but no attempt has been made to allow for the decrease in water speed with depth, firstly because acoustically tagged plaice and other fish have frequently been observed to swim in the upper half of the water column, and secondly because of the lack of a proper description of the speed profile in the open sea.[1] It is assumed that when moving the fish is always outside the surface and bottom boundary layers.

The starting point of each simulated track is specified by latitude and longitude in decimal degrees. The duration of the track is specified in days and hours. The model calculates the position of the fish at each half-hourly interval, together with the distance and speed from the previous position. It lists the latitude and longitude of each point at which the fish returns to the bottom, together with the distance covered (m), the mean ground speed ($m\ s^{-1}$) and the elapsed time (h). Total track length is calculated in kilometres. The starting point is also specified in relation to the spring-neap cycle using a simple code, in which 0 represents the spring tide, 7 the neap tide and the numbers 1-6 the intervening days.

RESULTS OF THE FIRST SIMULATIONS

The first simulations have been concerned with the prespawning and postspawning migrations of plaice in the Southern Bight and eastern English Channel and with the spawning migrations of silver eels crossing the European shelf on their return to the Sargasso Sea. Some simulations have also been made of the tracks of plaice exhibiting diurnal vertical migrations on various feeding, spawning and staging grounds in the southern North Sea.

Prespawning Migrations of Southern Bight Plaice

A series of 15 simulated tracks was started at latitude 53°55'N, just to the south of Flamborough Head, in the summer feeding area of plaice that spawn in the Southern Bight. The origins were spaced at intervals of 0.1° between longitudes 0.5°E and 2.5°E. Six tracks (A-F) from this series are shown in Figure 7, with origins at 0.7°, 1.0°, 1.2°, 2.0°, 2.2° and 2.3°E. Two other simulated tracks are shown starting at arbitrary points on the eastern side of the

---

[1] The error is likely to be small and can be estimated on the assumption that the speed profile in the water column approximates to the parabola $V_h = a\ h^b$ described by van Veen (1938) for the Strait of Dover, where V is the speed at a height of h metres above the bottom and a is the speed at 1 m above the bottom. The exponent b has a value of approximately 0.2 so that the speed at mid-depth in the water column is approximately 0.9 of the surface speed.

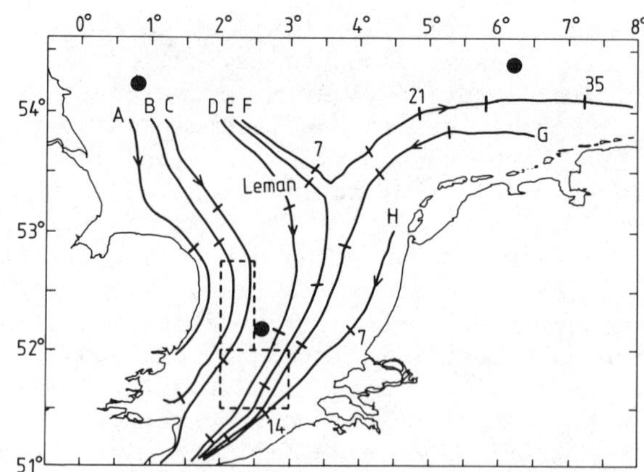

Figure 7. Simulated semidiurnal tracks (Table 4) relevant to the prespawning migration of plaice in the Southern Bight of the North Sea. Each track was commenced at slack water (LWS - tracks A-F; HWS - tracks G and H) on spring tides with a specified time of 6 h in midwater. Each track is based on plots of individual positions calculated at the end of each transporting tide. The direction of movement is indicated by an arrow and progression by cross ticks plotted at intervals of 7 days. The filled circles indicate the centres of the Flamborough, German Bight and Southern Bight plaice spawnings. The rectangles delimit the area of midwater trawling along the line of the western migration route in relation to the Hinder spawning area (Harden Jones et al. 1979, their Fig. 1).

Southern Bight at positions 53°48'N 6°30'E (track G) and 53°00'N 4°30'E (track H). All eight tracks were simulated using the semidiurnal behavioural option with a specified period of 6 h in midwater, and all were begun at peak spring tides. Track durations were initially determined by trial and error and subsequently specified in appropriate multiples of 7 days.

Tracks G and H, which were started at local high water slack (HWS) tide, run to the south and converge through the Strait of Dover together with tracks C, D and E, which were commenced at local low water slack (LWS) tide. Track D passes through both the Leman Ground and the area of the highest concentration of newly spawned (stage 1a) plaice eggs on the Southern Bight spawning ground. Tracks C, E and G pass through the larger area delimited by the next lower contour of stage 1a eggs (Fig. 10). Tracks B and C both pass through the area off the East Anglian coast in which substantial numbers of prespawning plaice were caught in midwater trawls on the south-going tide (Harden Jones et al. 1979), but whereas track C passes through the Strait of Dover, track B enters the Thames estuary. Track A similarly swings into the coast off the estuaries of the Rivers Stour and Orwell, which enter the sea at Harwich. Tracks originating to the west of track A on latitude 53°55'N enter the Wash and are not shown in Figure 7. Track F, after initially running almost parallel to track E in a south-easterly direction, turns abruptly to the northeast and progresses offshore around the Frisian Islands into the German Bight, where it ends at a position 53°53.8'N 8°41.3'E after covering 467 km in 42 days (Table 4).

The simulations of tracks C, D, E, G and H were all continued through the Strait of Dover and into the English Channel (C, D and H shown in Fig. 8). Track C, after passing close to Dover and Dungeness, progresses westward along the south coast of England to Start Point. After covering 953 km in 56 days it ends at 49°37.5'N 6°2.5'W southeast of the Scilly Isles. Tracks D and H pass close to Cap Gris Nez and traverse the eastern Channel to the Cherbourg peninsula. Once west of Cap de la Hague (the western tip of the Cherbourg peninsula), however, both tracks show an abrupt reversal of direction and return to the east. In each case the return leg follows closely the line of the outward portion of the track. Track D ends in the eastern Channel at 50°11.1'N 0°27.1'W after covering 1098 km in 56 days. Track H covers 1088 km in the same time and returns through the Strait of Dover to end in the Southern Bight at 51°27.5'N 2°53.6'E. Tracks E and G (not shown in Fig. 8) also reverse in the eastern Channel (at 49°56.5'N 2°18.1'W and 49°53.6'N

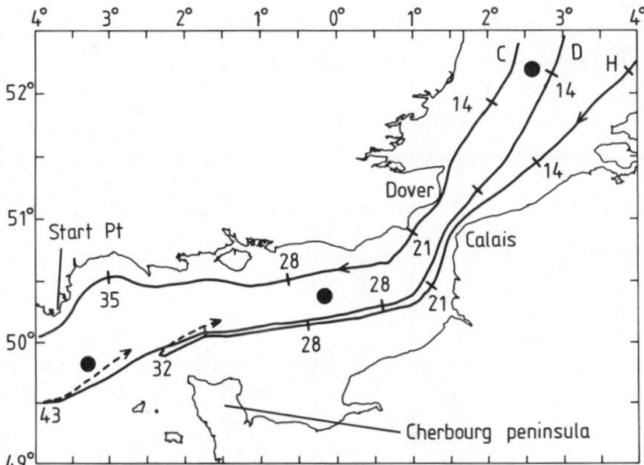

Figure 8. The continuation of simulated semidiurnal tracks C, D and H (Table 4) from the Southern Bight into the English Channel. The direction of movement is indicated by arrows and progression by cross ticks plotted at intervals of 7 days. Tracks D and H reverse after periods of 43 and 32 days, respectively. The filled circles indicate the centres of plaice spawning in the Southern Bight and eastern and western Channel.

Table 4. Distances covered by simulated semidiurnal plaice tracks in the Southern Bight and English Channel (Figs. 7, 8 and 9). Reversing tracks are identified by an asterisk, and distances which include parts of the track after reversal are underlined. Tracks D, E, G and H reverse at 43, 40, 52 and 32 days after covering distances of 833, 739, 835 and 644 km, respectively. Tracks A, B and F reach the coast within shorter periods.

Track	Distances (km) covered in each consecutive 7 day period									Overall		
	0-7	7-14	14-21	21-28	28-35	35-42	42-49	49-56	56-63	Distance (km)	Duration (days)	Speed (km day^{-1})
A	143	-	-	-	-	-	-	-	-	257	12	21.7
B	132	164	-	-	-	-	-	-	-	316	15	21.1
C	97	156	140	131	171	97	64	97	-	953	56	17.1
D*	106	120	125	144	167	155	$\underline{102}$	$\underline{179}$	-	1098	56	19.8
E*	91	103	110	161	118	$\underline{193}$	$\underline{189}$	$\underline{131}$	-	1097	56	19.7
F	82	69	58	64	91	$\underline{103}$	-	-	-	467	42	11.3
G*	80	78	71	107	116	144	164	$\underline{150}$	$\underline{154}$	1065	63	17.0
H*	100	119	154	126	$\underline{159}$	$\underline{160}$	$\underline{117}$	$\underline{153}$	-	1088	56	19.4
I	104	118	141	141	$\underline{160}$	$\underline{147}$	$\underline{87}$	-	-	899	49	18.4
J	112	118	144	134	183	141	91	-	-	922	49	18.8

2°19.3'W) having covered 739 and 835 km in the first 40 and 52 days, respectively. Track G ends in the Channel (at 50°14.5'N 0°45.2'E), while track E returns through the Strait of Dover to a position 51°7.2'N 1°57.7'E in the Southern Bight.

Reversal appears to be limited to those tracks which pass through the Dover Strait on the French side of the English Channel at a distance of less than 20 km from Cap Gris Nez, and it occurs in the area of dividing tidal streampaths (Fig. 4) immediately to the west of the Cherbourg peninsula. Most simulated tracks to date have reversed in the vicinity of a point some 30 km northwest of Cap de la Hague and to the north of the island of Alderney, but two have gone further south and west and reversed at positions 49°38.2'N 3°21.1'W and 49°15.8'N 3°32.5'W. These two positions lie to the west of the Channel Islands but are still within the general area of the dividing streampaths (Fig. 4).

In the southern North Sea the tidal speeds generally decrease from west to east, and this is reflected in the distances covered during the first 14 days of tracks A-E (Table 4), which decrease progressively with distances offshore. The distance covered during the first 21 days of track G, too, are correspondingly less than those of track E, although those closer to the Dutch coast (track H) are a little higher. Indeed, a fish following track G needs 31 days to reach the centre of the Hinder spawning ground box shown in Figure 7, compared with 21 days for a fish following track E, 16.5 days for a fish following track D and 14 days for a fish following track C. The four distances are 395, 304, 282 and 253 km, respectively, and the corresponding speeds are 12.6, 14.4, 17.1 and 18.2 km day^{-1}. Fish following the western migration route into the Southern Bight can thus travel significantly faster than those using the eastern route.

With the exception of some of the reversing sections of tracks D and E, the highest transport speeds occur off the East Anglian coast and in the middle section of the English Channel. Track C, for example, covers 156 km between days 7 and 14 (mean speed 23 km day^{-1}) off the East Anglian coast and 171 km between days 28 and 35 (mean speed 25 km day^{-1}) off the south coast. Corresponding speeds for individual tides during this track range from 0.3-0.7 m s^{-1} (days 7-14) and 0.2-1.2 m s^{-1} (days 28-35).

The timing of the start of the tracks in relation to the neap-spring cycle appears to have little effect on either the route or the transit time. Tracks I and J (Fig. 9) were started in the area of the Flamborough spawning ground at spring tides and neap tides, respectively. The two tracks follow almost identical courses, covering total distances of 899 and 922 km in 49 days and ending at positions 49°38'N 4°39'W and 49°32.8'N 4°53.7'W. The transit times between the Flamborough and Southern Bight and between the Southern

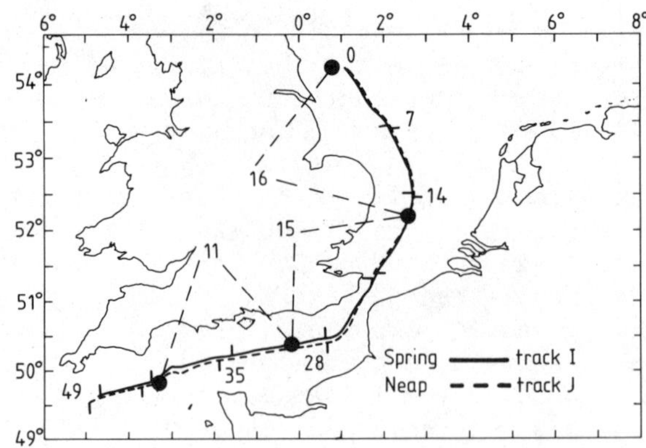

Figure 9. The tracks of simulations I and J (Table 4) which were commenced at spring and neap tides, respectively, using the semidiurnal option with a specified period of 6 h in midwater. The tracks were begun at the same position (54°13'N 1°5.6'E) in the area of the Flamborough spawning and pass through the three centres (filled circles) of plaice spawning in the Southern Bight and English Channel. Progression is indicated by cross ticks plotted at intervals of 7 days, and the transit times between the spawning centres are also shown.

Bight and eastern English Channel spawning centres are 16 and 15 days, respectively. That between the eastern and western Channel spawning centres is 11 days. A fish migrating south by continuous selective tidal stream transport from a position just to the south of the centre of the Flamborough spawning will thus reach the centre of the Southern Bight spawning in 14 days and the centre of the eastern Channel spawning in 28 days. Making an allowance for the errors in determining the centres of the several egg patches, we are tempted to suggest that the duration of migration may be under the control of a lunar- or semi-lunar rhythm. On this hypothesis one would predict that an individual fish would join the transport system at one spring tide, remain with the transport system for 14-28 days, and leave it on another spring tide. If there were any truth in this suggestion, one might expect the distances along the tidal streampaths between feeding and spawning grounds to be equivalent to transit times of the order of 14, 28 or even 42 days. But there are inherent difficulties with this hypothesis in that fish starting from adjacent positions (e.g. tracks D and E) may take rather different times to reach the same destination.

Postspawning Migrations of Southern Bight Plaice

The second series of simulations is related to the northward migration of spent plaice out of the Southern Bight into the central North Sea. As a first approximation it is assumed that the distribution of spawning fish on the ground is represented by the distribution of newly spawned eggs in the water column, and a number of tracks has therefore been run, originating from the contours enclosing the highest concentrations of stage la eggs. These are newly spawned eggs with a stage duration of 2 days at a typical winter temperature of 6.5 C.

The distribution of plaice eggs in the Southern Bight is well known, having been sampled many times between 1911 and the present (Simpson 1959; Harding et al. 1978). A synoptic chart of the distribution of recently spawned plaice eggs at the time of peak spawning on each of the various spawning grounds around the British Isles is given in Sheet 3.041 of Lee and Ramster (1981). For the purposes of this paper we have taken the two highest contours of stage la plaice eggs in the Southern Bight in January 1971, which was a typical year. These contours enclose, respectively, stations with egg concentrations in excess of three and nine eggs beneath a square metre of the sea surface (Fig. 10, for day 0 contours). Peak egg concentration in the centre of the patch was 15 eggs m^{-2}.

Two sets of simulations have been made with the origins of the individual tracks distributed at regular intervals around the two egg contours. The first set consists of 10 tracks based on the higher of the two contours (9 eggs m^{-2}) and the second of 24 tracks based on the lower contour (3 eggs m^{-2}). The results are shown in Figure 10, not as individual tracks, but as areas predicting the distributions of spent fish at successive intervals of 7 days. The areas are delimited by straight lines joining the positions reached by each of the constituent tracks at the appropriate time interval. Each track was begun at local HWS and is based on the semidiurnal behavioural option with a specified period of 6 h in midwater.

The results show that all fish contained within the higher of the two contours--and thus within the area of peak spawning--will leave the Southern Bight by the western route. The majority of fish (85% by area) within the lower contour will also use this route. The minority (15% by area) will migrate up the eastern side of the Southern Bight and move around the Dutch coast into the German Bight. The predicted distributions again reflect the higher tidal speeds on the western side of the southern North Sea, and it is clear that in some instances the destination of a fish migrating solely by tidal stream transport will depend quite critically upon its point of origin.

Diurnal Vertical Migration and Local Movements of Plaice

The third series of simulations concerns the diurnal vertical movements of plaice on various feeding, spawning and staging grounds

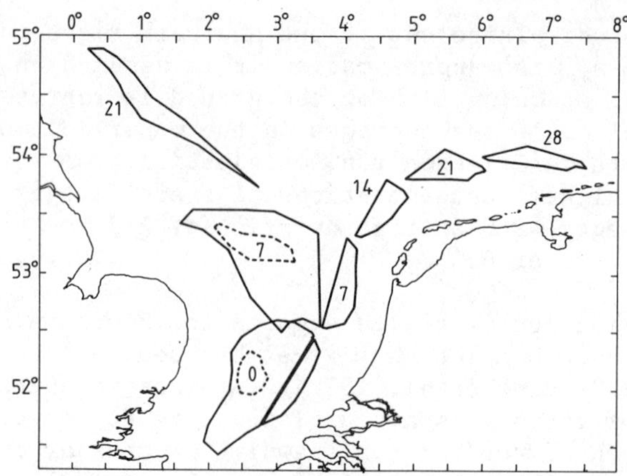

Figure 10. Predicted distributions for spent plaice migrating northwards from the Southern Bight spawning ground by selective tidal stream transport. The inner (dashed line) and outer contours are based on two sets of semidiurnal tracks with a specified time of 6 h in midwater. The origins of the tracks were spaced at regular intervals around the two contours for day 0, which enclose the highest concentrations of freshly spawned (stage 1a) plaice eggs contoured at 3 and 9 eggs m^{-2}, respectively.

in the southern North Sea and English Channel. In each case the track was commenced at spring tides and was run for 28 days with a set period of 2 h in midwater out of each 24 h. Two tracks (K and L; Fig. 11) were begun at positions 49°49'N 3°17'W and 50°22'N 0°10'W corresponding to the two centres of spawning in the English Channel (Fig. 4), while track M was started at 51°45'N 2°30'E in the centre of the Hinder spawning box (Fig. 7). Track O was started at the position of tidal diamond N on Admiralty Chart 2182A (Fig. 3, Table 3) and track P at an arbitrary point (53°48'N 6°E) on the eastern migration route between the Southern and German Bights. Track N was commenced at a position 54°06'N 1°30'E midway between the centre of the Flamborough spawning (Fig. 4) and the Silver Pit (Fig. 3).

The overall distances covered by the diurnal tracks (Table 5) are of the order of 50-100 km, but because the paths are contorted, the total area encompassed by each track is small (Fig. 11) as is the next distance covered during 28 days (Table 5).

The path followed by each track reflects the shape of the local tidal ellipse as can be seen by comparing tracks M and O (Fig. 11) with the data in Table 3. The distances covered during the individual midwater excursions similarly reflect the speed of the local tidal stream. They also vary considerably, because they depend

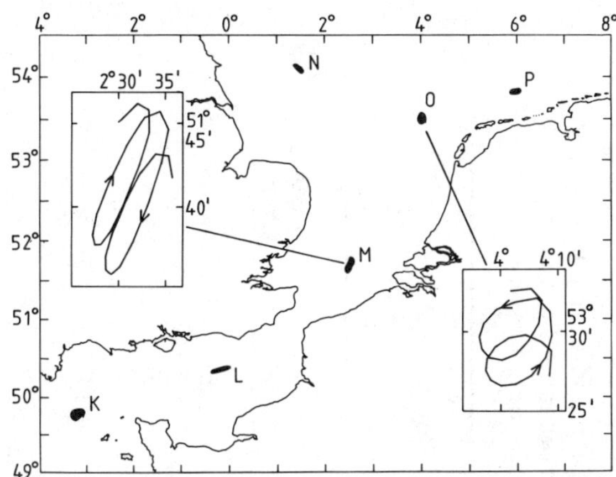

Figure 11. Areas covered by simulated diurnal plaice tracks (Table 5) on various feeding, spawning and staging grounds in the Southern Bight and English Channel. Each track was commenced at spring tides and run for 28 days with a set period of 2 h in midwater out of each 24 h. The inset figures show the details of tracks M and O, each of which reflects the shape of the local tidal stream ellipse.

as well on the timing of the excursion in relation to the state of the tide. The greatest distance covered during a single 2 h excursion is 7.3 km (track L) and the least 0.03 km (track P) but most (76%) are between 1 and 4 km in extent.

It can thus readily be seen that, even in areas of fast and directional tidal streams, a fish adopting the diurnal mode of vertical migration will remain within a relatively confined area. This is consistent with the hypothesis that the nightly movements in midwater represent a mechanism for the local redistribution of a small and hungry proportion of the population on the summer feeding grounds. The scale of the movement is such, too, that a plaice remaining in midwater for a few hours each night would appear to have little difficulty in remaining within the limits of any of the spawning areas in the southern North Sea and English Channel. It seems reasonable, therefore, to extend the local redistribution hypothesis to the spawning grounds, where the change of position each night would represent not a search for food--plaice do not feed during the spawning season--but for other plaice that are ready to spawn. Spawning occurs by pairing, with the male and female fish swimming up into midwater together and remaining in close contact with each other (Forster 1953; J.D. Riley personal communication). The males are generally thought to be more active and to remain on the spawning grounds for much longer than the females, so it may be reasonable to postulate that the males do the searching. If this

Table 5. Distances covered by simulated diurnal plaice tracks in the southern North Sea and English Channel (Fig. 11). The net distance is the direct distance between the starting and finishing point of each track.

Track	Distances (km) covered in each consecutive 7 day period				Overall		Net	
	0-7	7-14	14-21	21-28	Distance (km)	Speed (km day^{-1})	Distance (km)	Direction (°T)
K	22	21	23	21	87	3.1	18	124
L	28	29	28	26	111	4.0	6	97
M	22	19	22	19	82	2.9	9	135
N	16	15	16	14	60	2.1	4	192
O	13	14	13	14	55	2.0	11	335
P	13	13	12	12	50	1.8	5	32

were so it might explain the predominance of males that we have observed on occasion in midwater trawl catches on the southern Bight spawning grounds.

Migrations of Silver Eels Crossing the European Shelf

The final series of simulations concerns the spawning migrations of silver eels crossing the European shelf on their supposed return to the Sargasso Sea. The intention is to predict the routes that might be followed from each of the major eel rivers and to estimate the time that it would take to reach the edge of the continental shelf, as defined by the 200 m contour. Five such simulations have been run starting from the mouths of the rivers Seine, Severn, Humber, Weser and Elbe. A sixth track was started off Texel to simulate the movements of eels leaving the IJsselmeer by the freshwater sluices in the Afsluitdijk that encloses the old Zuider Zee. A seventh track was started off the East Anglian coast at the starting point of the eel track illustrated in Figure 1.

All seven tracks were commenced at HWS on the assumption that migrating silver eels would be expected to leave the estuary on the ebb tide. Each track was begun at spring tides and was based on the semidiurnal behavioural option with a set period of 6 h in midwater. Details are given in Table 6.

If the eel of Figure 1 had continued to migrate to the north by selective tidal stream transport for 106 days, it would have arrived at a position some 50 km to the east of the Shetland Islands (track S, Fig. 12). Track U suggests that an eel leaving the River Humber could similarly arrive at a point to the southeast of the Shetlands in 85 days. Tidal diamonds are rather sparse in the extreme North Sea, such that these tracks are not readily extended to the 200 m contour. But track X suggests that another 20 days would be sufficient to take them to the shelf edge. An eel from the River Severn (track R) would similarly appear to be able to reach the edge of the shelf by a direct route to the southwest in about 40 days. An eel from the River Weser (track V), too, would seem set to reach the edge of the shelf by the northern route, but taking a substantially longer time to do so than one starting from the Humber on approximately the same latitude (Table 6).

In contrast there would appear to be problems for eels from the Seine, Texel and Elbe. Track Q (Fig. 12) ends on the French coast after only 14 days, while track T reverses in mid-Channel after 36 days. Even more strikingly, track W turns abruptly back on itself and appears to become permanently trapped in the system of east-west tidal streampaths in the central North Sea (Fig. 4). It reverses at 55°6'N 1°30.6'E after 66 days and again in almost exactly the same position (55°5.6'N 1°30.8'E) after 188 days (Table 6). (Only the first 177 days of the track are shown in Fig. 12.) It is possible

Table 6. Distances covered by simulated silver eel tracks (Fig. 12) in successive 28 day periods. Reversing tracks are identified by an asterisk, and distances which include parts of the track after reversal are underlined.

Track	Distances (km) covered in each consecutive 28 day period						Overall		
	0–28	28–56	56–84	84–112	112–140	140–168	Distance (km)	Duration (days)	Speed (km day^{-1})
R	434	-	-	-	-	-	492	37	13.2
S	449	205	167	-	-	-	942	106	8.9
T*	507	573	-	-	-	-	1150	60	19.2
U	331	210	178	-	-	-	728	85	8.6
V	325	246	182	160	-	-	995	125	8.0
W*	297	184	170	172	254	181	1454	200	7.3
X	268	-	-	-	-	-	268	28	9.7

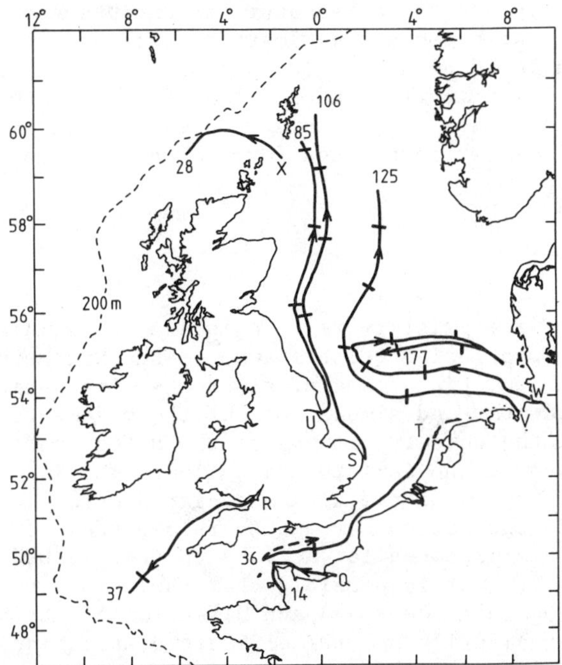

Figure 12. Simulated semidiurnal tracks (Table 6) for silver eels leaving the mouths of the Rivers Seine (track Q), Severn (track R), Humber (track U), Weser (track V) and Elbe (track W). Tracks S and T were commenced off the East Anglian coast at the point of release of the eel of Figure 1 and off Texel, respectively. Track X was commenced at 59°21'N 1°30'W just to the east of the Orkney-Shetland Channel. Each track was commenced at HWS with a specified time of 6 h in midwater. The direction of movement is indicated by arrows and progression by cross ticks plotted at intervals of 28 days.

that, if extended, track T might succeed in leaving the North Sea by the northern route. It seems equally possible that eels migrating from the Elbe might[1] be substantially delayed by a tidal gyre in the central North Sea. Indeed an eel caught in such a system might fail to reach the edge of the continental shelf in sufficient time to

---

[1] There is no direct evidence to show that silver eels do migrate by selective tidal stream transport, and we know of only one published record describing the capture of a silver eel in midwater on the European Shelf. This was a 24.8 cm male *Anguilla anguilla* which was caught by R.V. TRIDENS in the Bay of Biscay in November 1979 (de Groot 1981 and personal communication).

allow it to complete the subsequent oceanic phase of its migration. If silver eels were to become "lost" in this way the "tidal sink" in the North Sea would be directly analogous to those areas in the central North Pacific from which salmon fail to return to North America (Harden Jones 1984a--this volume, his Figs. 11-14) and indeed to the "swell gyres" postulated for the North Atlantic by Cook (1984--this volume).

FUTURE WORK

The immediate priority is to complete the simulations relevant to the plaice population which spawns in the Southern Bight. This will involve simulations of diurnal tracks with periods of 3, 4, 5 and 6 h in midwater and simulations of the movements of juvenile fish in and out of the nursery grounds on the Dutch coast, using the special variant of the semidiurnal option. The intention here is to determine the extent of the sink associated with each of the entrances into the Wadden Sea and to relate these areas to those in which newly metamorphosed larvae take to the sea bed at the end of the larval drift. It is possible that there is a predictable relationship between the speed and direction of the prevailing wind during the larval drift and the ultimate destination of the recently metamorphosed larvae within the nursery ground. Such predictions might well be relevant to our understanding of the factors governing variations in year class strength and thus to practical problems of fishery management. Thereafter we intend to extend the simulations on the one hand to the distributions of other plaice stocks on the European shelf--and indeed to other species such as cod and haddock--and on the other to predictions aimed at furthering our understanding of the mechanism of selective tidal stream transport itself.

Predictions About the Behaviour of Individual Fish

Several features of the simulated tracks of individual prespawning plaice described above are relevant to the behavioural and physiological mechanisms underlying tidal stream transport. Fish migrating to the south in the autumn on the eastern side of the Southern Bight must commence their migration at HWS, whereas those using the western route must commence their migration at LWS. In the one case transport occurs on the ebb tide and in the other on the flood tide. Furthermore fish using the western route into the Southern Bight must switch tides if they are to pass successfully through the Dover Strait because the south-going tide on the western side of the Southern Bight is the flood tide and the west-going tide in the English Channel is the ebb tide. It seems unlikely therefore that hydrostatic pressure per se is the cue triggering the onset of the semidiurnal vertical migration, although it seems quite possible that a fish would be capable of detecting the difference in water level between high and low tides.[1]

The reversing tracks, too, are of particular interest because they offer the possibility of conducting a sea-going experiment to determine whether the underlying mechanism of tidal stream transport is, as we suspect, based on a succession of temporal cues or instead involves an absolute sense of direction on the part of the fish. Preliminary experimental work in a tidal stream tank shows that individual plaice will allow themselves to be carried several times in succession around an enclosed annulus and on the basis of this observation it seems likely that plaice moving to the west on the southern side of the English Channel should reverse as the simulations predict. But a tracking experiment to prove the point at sea is a highly desirable adjunct to the laboratory work and indeed essential to show that fish moving to the south in the Southern Bight by tidal stream transport can pass through the Strait of Dover with no interruption in the pattern of semidiurnal vertical migration. In the short term, however, navigational difficulties might preclude such an experiment in the Strait itself with existing short-range tracking equipment.

Further simulations may of course reveal other more convenient areas in which track reversal can be predicted, and there are a number of other situations, too, which warrant further investigation. Of particular interest are the Bay of the Seine, another area of dividing streampaths, and the Bristol Channel and German Bight, in each of which the tidal streampaths converge to a dead end. Conventional tagging experiments have shown that plaice released off North Wales are not infrequently recaptured within a few months in the Bristol Channel and the western English Channel[2] (Macer 1972). Similarly, plaice tagged off Texel (Rauck 1977) prior to spawning in the Southern Bight are recovered after spawning first in the German Bight and then along the Danish coast. Both situations could involve a switch of transporting tide but unlike the situation in the Dover

---

1 Over much of the southern North Sea the mean spring range of the tide is 2 m (British Admiralty Chart 5059, Southern North Sea: co-tidal and co-range chart, 1971), which represents a change of 7% in hydrostatic pressure between high and low tides in a depth of 30 m. In contrast the variation in pressure produced at the sea bed at the same depth by a wave with height of 2 m and a period of 8 s is only ± 0.9% of the ambient hydrostatic pressure. This is a rather severe wave for the southern North Sea. The pressure variation produced by a more typical wave with a height of 1 m and a period of 5 s is a mere ± 0.03% of ambient.

2 The number of fish tagged in the English Channel and moving in the opposite direction into the Celtic and Irish Seas is very small (Houghton and Harding 1976).

Strait would presumably involve a temporary cessation of movement. In the German Bight, for example, a change would appear to be necessary from a semidiurnal vertical migration triggered at LWS to one triggered at HWS. But some local redistribution would also appear to be necessary and an interim period of diurnal vertical migration might perhaps be implicated.

Other simulations are needed to investigate the effect of the duration of the midwater excursion on the semidiurnal tracks and the possibility that the fish may be capable of detecting the local slackwater at the end of the transporting tide, while the fish is still up in midwater. This might possibly occur in response to some physical stimulus such as a reduction in turbulence or a cessation of the noise generated by the tide flowing over the sea bed.

Two other factors need to be taken into account. One is the possibility that some plaice may move on the south-going tide but only at night. Such a pattern of behaviour has recently been observed on the Southern Bight spawning grounds (M. Greer Walker, unpublished data) and is consistent with the diurnal variation in midwater trawl catches[1] observed along the line of the western migration route (Harden Jones et al. 1979). If fish were to migrate in this way on alternate tides, the effect would be to approximately double the calculated transit times. A more sophisticated version of the model might be useful for such predictions, particularly when it comes to extending the simulations to fish like herring (*Clupea harengus*), which appear to have a pronounced diurnal pattern of vertical movement during their migration, coming up into midwater at night. A variant of the model which took into account the time of sunrise and sunset in relation to the time of slackwater would enable us to extend the simple model postulated by Harden Jones (1968, his Figs. 73-75) for the migrations of Downs herring in the Southern Bight. This stock is of particular interest because its spawning grounds, which extend through the Strait of Dover into the Bay of the Seine, appear to be located along a common tidal stream path and its migration circuit--as deduced from tagging experiments and the movements and catches of the fishing fleets (Harden Jones 1968, his Fig. 40)--appears to fit closely the pattern of the tidal streampaths along the western side of the North Sea (Harden Jones 1980).

The second factor is the velocity of the fish itself through the water. Vector analysis has shown that plaice can maintain a surprisingly consistent course in midwater for periods of up to 2 h after leaving the bottom (Greer Walker et al. 1978), and this capacity, which has been investigated further using a telemetering

---

[1] The alternative explanation is that the lower catches of plaice in midwater by day are accounted for by net avoidance.

compass tag (Harden Jones 1981, Harden Jones and Arnold 1982), is the subject of another paper in this volume (Harden Jones 1984b). Cod and yellow (nonmigrating) eels, too, can maintain a consistent course through the water, which significantly alters the track of the fish over the ground. Acoustically tagged cod tracked off the coast of East Anglia (M. Greer Walker, G.P. Arnold, and L.S. Emerson, unpublished data) swim offshore in an easterly direction and these movements in conjunction with those of the tide produce a sinusoidal track over the ground. Acoustically tagged yellow eels (J.D. McCleave, M. Greer Walker, and G.P. Arnold, unpublished data) move inshore in a comparable fashion. The sinusoidal offshore cod tracks have been successfully simulated using the existing model, but cod tracked further off the coast have been observed to swim along the bottom against the adverse tide in between periods of midwater transport with the favourable north-going tide. No attempt has yet been made to take account of this variant of the behaviour in the simulations.

Predictions About the Resulting Distributions of Populations

Few stocks of fish on the European shelf can, as yet, be distinguished genetically. The ability to describe their distributions in hydrodynamic terms could therefore be of considerable practical significance and predictions of the type described in Figure 10 are thus of some interest.

It is our intention to extend these simulations first to the plaice spawnings in the eastern and western English Channel and subsequently to the Flamborough spawners. We envisage a series of overlapping distributions extending through the Channel and Southern Bight and up the east coast of England and Scotland as far as the Moray Firth. Thereafter we plan to carry out comparable simulations for the other major plaice spawning grounds in the North Sea--the German Bight and Intermediate spawnings--and the rather smaller spawnings off the Scottish east coast and in the Irish Sea and Bristol Channel.

Conventional tagging experiments suggest that in the North Sea, at least, the majority of plaice tagged on their spawning grounds segregate in subsequent years to the same grounds at spawning time (de Veen 1962; Harden Jones 1968, his Fig. 53). But at other times of the year there is some degree of overlap in the distributions when the various spawning groups are dispersed on the feeding grounds. The degree of overlap will obviously depend on the duration of the post-spawning migration, which will thus in turn influence the degree of mixing between the various spawning groups. At present the factors which control the duration of the migration and cause the fish to revert from the semidiurnal to the diurnal pattern of vertical migration are unknown.

We propose to examine the problem of mixing with a series of predicted distributions based on semidiurnal tracks and plotted on a lunar time base and a series of diurnal tracks commenced in the areas of overlapping distributions. Small scale local redistribution within these areas might be sufficient to account for the degree of mixing of 1-2% required to produce genetic similarity between the various spawning groups. Mixing may also occur on the spawning grounds in some instances and such a situation could arise in the Southern Bight as shown in Figure 10. A systematic series of diurnal tracks started on either side of the line dividing the outer egg contour (3 eggs m^{-2}) for day 0 should allow us to calculate the probability of mixing of plaice from the German Bight with those from the Leman Ground and other feeding areas on the western side of the North Sea. An alternative source of mixing could arise from those few fish which appear to have undergone a 180° phase shift and to have moved along the tidal streampaths in totally the opposite direction to the bulk of the population. Examples of such anomalies are provided by recaptures in the English Channel of individual plaice tagged in the North Sea. One such individual tagged at 51°35'N 2°10'E on 10 January 1975 was recaptured 111 days later at 50°15'N 4°10'W to the west of Start Point (de Clerck 1977 and personal communication). Another tagged as a juvenile on the Filey Bay nursery ground immediately to the north of Flamborough Head was recaptured in Rye Bay in the eastern English Channel (S.J. Lockwood, personal communication).

Improvements and Extensions to the Model

Work is currently in hand to compile a data file for the tidal streams in the various entrances to the Wadden Sea between the Frisian Islands, as given in the several atlases published by the Netherlands Hydrographer (e.g. Anonymous 1978). Thereafter there should be scope to improve the data base with the production by West Germany and Belgium, respectively, of improved tidal stream atlases for the German (Anonymous 1983) and Southern Bights. There is a rather greater need to improve the tidal stream data in certain other areas, especially in the northern North Sea, and in the first instance we may be able to incorporate some data from the Rostock Atlas (Anonymous 1968). In the longer term it might be possible to use tidal stream predictions from finite difference tidal models. Two such models, a two-dimensional one with a grid of 0.1° latitude by 0.2° longitude and a three-dimensional one with a grid of 0.3° latitude by 0.5° longitude, are described by Davies (1979, 1981) for the northern North Sea and the European Shelf, respectively. A similar numerical model exists for the North Frisian Wadden Sea (Duwe and Hewer 1982).

Three-dimensional tidal stream data would enable us to make allowance for the height of the fish in the water column, and extra data in appropriate places would obviate problems with coastlines.

There are at present no boundaries in the model, and in certain places tracks can cross the coastline because of the distribution of the tidal diamonds and the nature of the interpolation of the tidal streams. Track S exemplifies this problem turning abruptly west from the last position indicated in Figure 12 and cutting across mainland Shetland rather than passing to the north of Unst, the most northerly island of the Shetland group.

REFERENCES

Anonymous. 1968. Atlas der Gezeitenströme für die Nordsee, den Kanal und die Irische See. Seehydrographischer Dienst der Deutschen Demokratischen Republik, Second edition. Rostock, Democratic Republic of Germany.

Anonymous. 1978. Stroomatlas Waddenzee, westelijk deel. Dienst der Hydrografie van de Koninklijke Marine. 's-Gravenhage, The Netherlands.

Anonymous. 1983. Atlas der Gezeitenströme in der Deutschen Bucht. Deutsches Hydrographisches Institut. Hamburg, Federal Republic of Germany.

Arnold, G.P. 1981. Movements of fish in relation to water currents. Pages 55-79 *in* D.J. Aidley, editor. Animal migration. Society for Experimental Biology. Seminar series 13. Cambridge University Press, Cambridge, England.

Clerck, R. de. 1977. The migration of plaice on the spawning grounds 'Noord-Hinder.' International Council for the Exploration of the Sea. C.M. 1977/F:40. Demersal Fish (Northern) Committee (mimeo).

Cook, P.H. 1984. Directional information from surface swell: some possibilities. Pages 79-101 *in* J.D. McCleave, G.P. Arnold, J.J. Dodson and W.H. Neill, editors. Mechanisms of migration in fishes. Plenum Press, New York, New York, USA.

Creutzberg, F. 1961. On the orientation of migrating elvers (*Anguilla vulgaris* Turt.) in a tidal area. Netherlands Journal of Sea Research 1:257-338.

Creutzberg, F., A. Th. G.W. Eltink, and G.J. van Noort. 1978. The migration of plaice larvae *Pleuronectes platessa* into the western Wadden Sea. Pages 243-251 *in* D.S. McLusky and A.J. Berry, editors. Physiology and behaviour of marine organisms. Pergamon Press, Oxford, England.

Cushing, D.H. 1972. The production cycle and the numbers of marine fish. Pages 213-232 *in* R.W. Edwards and D.J. Garrod, editors. Conservation and productivity of natural waters. Symposia of the Zoological Society of London 29. Academic Press, London, England.

Davies, A.M. 1979. Role of 2D and 3D models in JONSDAP'76. Pages 1085-1103 *in* Proceedings of the 16th Coastal Engineering Conference, Hamburg, 1978, volume 1. American Society of Civil Engineers, New York, New York, USA.

Davies, A.M. 1981. Three dimensional hydrodynamic numerical models. Part 1, A homogeneous ocean-shelf model. Part 2, A stratified model of the northern North Sea. Pages 370-426 *in* R. Saetre and M. Mork, editors. The Norwegian coastal current, volume 2. University of Bergen, Bergen, Norway.

Defant, A. 1958. Ebb and flow. The University of Michigan Press, Ann Arbor, Michigan, USA.

Duwe, K.C., and R.R. Hewer. 1982. Ein semi-implizites Gezeitenmodell für Wattgebiete. Deutsche Hydrographische Zeitschrift 35:223-238.

Forster, G.R. 1953. The spawning behaviour of plaice. Journal of the Marine Biological Association of the United Kingdom 32:319.

Gould, W.J. 1978. Currents on continental margins and beyond. Philosophical Transactions of the Royal Society, London. A. 290:87-98.

Greer Walker, M., F.R. Harden Jones, and G.P. Arnold. 1978. The movements of plaice (*Pleuronectes platessa* L.) tracked in the open sea. Journal du Conseil Conseil International pour l'Exploration de la Mer 38:58-86.

Groot, S.J. de. 1981. Dutch observations on rare fish in 1979. Annales Biologiques. Conseil International pour l'Exploration de la Mer 36 (1979):196.

Harden Jones, F.R. 1968. Fish migration. Arnold, London, England.

Harden Jones, F.R. 1977. Performance and behaviour on migration. Pages 145-170 *in* J.H. Steele, editor. Fisheries mathematics. Academic Press, London, England.

Harden Jones, F.R. 1980. The migration of plaice (*Pleuronectes platessa*) in relation to the environment. Pages 383-399 *in* J.E. Bardach, J.J. Magnuson, R.C. May and J.M. Reinhart, editors. Fish behavior and its use in the capture and culture of fishes. ICLARM, Manila, Philippines.

Harden Jones, F.R. 1981. Fish migration: strategy and tactics. Pages 139-165 *in* D.J. Aidley, editor. Animal migration. Society for Experimental Biology. Seminar series 13. Cambridge University Press, Cambridge, England.

Harden Jones, F.R. 1984a. A view from the ocean. Pages 1-26 *in* J.D. McCleave, G.P. Arnold, J.J. Dodson and W.H. Neill, editors. Mechanisms of migration in fishes. Plenum Press, New York, New York, USA.

Harden Jones, F.R. 1984b. Could fish use inertial clues when on migration? Pages 67-78 *in* J.D. McCleave, G.P. Arnold, J.J. Dodson and W.H. Neill, editors. Mechanisms of migration in fishes. Plenum Press, New York, New York, USA.

Harden Jones, F.R., M. Greer Walker, and G.P. Arnold. 1978. Tactics of fish movement in relation to migration strategy and water circulation. Pages 185-207 *in* Henry Charnock and Sir George Deacon, editors. Advances in oceanography. Plenum Press, New York, New York, USA.

Harden Jones, F.R., G.P. Arnold, M. Greer Walker, and P. Scholes. 1979. Selective tidal stream transport and the migration of plaice (*Pleuronectes platessa* L.) in the southern North Sea. Journal du Conseil Conseil International pour l'Exploration de la Mer 38:331-337.

Harden Jones, F.R., and G.P. Arnold. 1982. Acoustic telemetry and the marine fisheries. Pages 75-93 *in* C.L. Cheeseman and R.B. Mitson, editors. Telemetric studies of vertebrates. Symposia of the Zoological Society of London 49. Academic Press, London, England.

Harding, D., J.H. Nichols, and D.S. Tungate. 1978. The spawning of plaice (*Pleuronectes platessa* L.) in the southern North Sea and English Channel. Rapports et Procès-verbaux des Réunions. Conseil International pour l'Exploration de la Mer 172:102-113.

Houghton, R.G., and D. Harding. 1976. The plaice of the English Channel: spawning and migration. Journal du Conseil Conseil International pour l'Exploration de la Mer 36:229-239.

Lee, A.J., and J.W. Ramster. 1981. Atlas of the seas around the British Isles. Ministry of Agriculture, Fisheries and Food, Lowestoft, England.

Macer, C.T. 1972. The movements of tagged adult plaice in the Irish Sea. Fishery Investigations, London. Series 2, 27(6):1-41.

Mitson, R.B., and T.J. Storeton-West. 1971. A transponding acoustic fish tag. The Radio and Electronic Engineer 41:483-489.

Rauck, G. 1977. Two German plaice tagging experiments (1970) in the North Sea. Archiv für Fischereiwissenschaft 28:57-64.

Rijnsdorp, A.D., and M. van Stralen. 1982. Selective tidal migration of plaice larvae (*Pleuronectes platessa* L.) in the Easterscheldt and the western Waddensea. International Council for the Exploration of the Sea. C.M. 1982/G:31 Demersal Fish Committee (mimeo).

Simpson, A.C. 1959. The spawning of the plaice (*Pleuronectes platessa*) in the North Sea. Fishery Investigations, London. Series 2, 22(7):1-111.

Veen, J. van. 1938. Water movements in the Straits of Dover. Journal du Conseil Conseil International pour l'Exploration de la Mer 13:7-36.

Veen, J.F. de. 1962. On the subpopulations of plaice in the southern North Sea. International Council for the Exploration of the Sea. C.M. 1962 Near Northern Seas Committee 94 (mimeo).

Veen, J.F. de. 1978. On selective tidal transport in the migration of North Sea plaice (*Pleuronectes platessa*) and other flatfish species. Netherlands Journal of Sea Research 12:115-147.

Voglis, G.M., and J.C. Cook. 1966. Underwater applications of an advanced acoustic scanning equipment. Ultrasonics 4:1-9.

Weihs, D. 1978. Tidal stream transport as an efficient method for migration. Journal du Conseil Conseil International pour l'Exploration de la Mer 38:92-99.

A MODEL OF OLFACTORY-MEDIATED CONDITIONING OF DIRECTIONAL BIAS IN FISH MIGRATING IN REVERSING TIDAL CURRENTS BASED ON THE HOMING MIGRATION OF AMERICAN SHAD (*ALOSA SAPIDISSIMA*)

Julian J. Dodson and Lothar A. Dohse

Départément de Biologie
Université Laval
Québec, Québec G1K 7P4 Canada

ABSTRACT

A computer simulation of anadromous fish behavior is presented to illustrate the hypothesis that an olfactory-regulated rheotaxis can account for unidirectional displacement in reversing tidal currents in the absence of odorous concentration gradients. It is hypothesized that fish can learn tide-specific rheotactic behavior that increases migration rate, if fish behave in such a way that leads to the synchrony of tidal and natal-river olfactory cues at the detectable edge of the diffuse olfactory field. Such synchrony may lead to stimulus substitution whereby cues associated with current detection acquire the capacity to elicit behavioral responses similar to those produced by olfactory stimuli. It is demonstrated that the behavioral modifications leading to synchrony could determine the mechanism by which migrating American shad approach the Connecticut River.

INTRODUCTION

Olfactory clues have long been considered as the basis of a mechanism of migratory orientation leading to home stream recognition and upstream migration in a variety of anadromous fish species (Leggett 1977; Hasler et al. 1978; Kleerekoper 1982). To guide migrants to their home, the olfactory clue must provide a concentration gradient steep enough to elicit a klino- or tropo-taxis (Hara 1970). However, such steep gradients would only be expected at the boundaries of different water masses or close to the source of the odorous substance. Except for the latter case, there is no directional component to the olfactory clue at all. Migrants utilizing olfactory cues may be able to move into areas containing olfactory cues indicative of the home stream, but they will not obtain any further information as to the direction in which to swim.

This problem has been largely accounted for by research demonstrating that olfactory clues are not necessarily orienting factors in themselves. Rather, their presence or absence acts to reinforce or alter the basic rheotactic response, while reference to water currents provides the necessary directional clues (Mathewson and Hodgson 1972; Kleerekoper 1982). For example, Emanuel and Dodson (1979) demonstrated that in a circular stream tank male rainbow trout (*Salmo gairdneri*) moved upstream in a plume of ovarian fluid emanating from a source channel. Ovarian fluid was obtained from spawning female rainbow trout. In an optomotor tank male trout exhibited an increase in the frequency of upstream orientation and, more significantly, in the frequency of upstream movement in the presence of diffuse, nondirectional ovarian fluid at concentrations equivalent to those existing in the plume of the stream tank. Thus, the response of trout observed in the stream tank was not solely due to the presence of an odor gradient. Rather, the mechanism of orientation revealed in the optomotor tank was related to the regulation of rheotaxis, whereby the ovarian fluid acted as a stimulus regulating the orientational and kinetic components of the response to the simulated water current.

Although this mechanism of orientation is appealing to the student of fish migration and homing to account for upstream migration in unidirectional currents in the absence of odor gradients, its application to account for unidirectional displacement in reversing tidal currents is more problematic. Fish experiencing a reversal in current direction every six hours or so and using olfactory clues to attain their goal could profit from such an olfactory-rheotactic mechanism, if the presence of the odor was uniquely associated with only one part of the tidal cycle. For example, Creutzberg (1961, 1963) described a mechanism of upstream migration for elvers of the European eel (*Anguilla anguilla*) based on a strong positive rheotaxis in the presence of freshwater and the absence of this response when in saltwater. During the flooding tide elvers rise from the bottom to be carried upstream, whereas during the ebbing tide they either swim against the current or drop to the bottom. This tide-specific behavior is related to the presence during ebb tides of an odorous substance originating in freshwater and the absence of the odor during flood tides. Miles (1968) observed a similar rheotactic response to freshwater by elvers of the American eel (*A. rostrata*). However, the proposed mechanism of upstream migration is a naive interpretation of experimental observations. In nature, if an elver settles to the bottom of an estuary during an ebbing tide due to the presence of an odor associated with freshwater, the elver should remain on the bottom during the first part of the ensuing flood tide as recently flushed river water is pushed back up through the estuary. Once seawater moves back over it, the elver rises to be carried upstream by the flooding tide. If the elver is more or less passive, it will remain a captive of the water mass and will be flushed from the estuary on the next ebbing tide without ever encountering the freshwater odorous cue necessary to elicit the positive rheotaxis. The fundamental problem with the proposed mechanism of upstream migration is that the odor/no-odor transition is not necessarily synchronized with the tidal transition.

The unidirectional displacement of many fish species migrating under the influence of tidal currents necessarily involves differential tide-specific behavior. In the case of American shad (*Alosa sapidissima*) homing to the Connecticut River from Long Island Sound (Dodson and Leggett 1973), sonic-tagged fish exhibited a well-defined behavior pattern highly correlated to tidal conditions. Shad tended to orient into the current at all times, but the precision and directivity of the countercurrent orientation differed depending on the direction of the tidal current. In the case of shad exhibiting a net westerly displacement in Long Island Sound (Fig. 1), the precision of countercurrent orientation was greater and the magnitude of angular changes in bearing less during ebb tide than flood tide. Shad also demonstrated a significant positive correlation between swimming speed and tidal current speed. However, shad swam west at a speed significantly greater than the ebb tide current speed but during flood tide swam at a speed approximately equal to tidal speed (Fig. 1). The combination of the orientational and kinetic components of this rheotactic behavior resulted in active westerly displacement during the ebb and a tendency to drift to the west during the flood. This rheotactic behavior caused most shad to move west beyond the Connecticut River. The fact that tagged shad were subsequently recaptured in the Connecticut River indicates that this westerly bias was reversed at some point in western Long Island Sound.

The sonic tracking of olfactory-occluded shad (Dodson and Leggett 1974) demonstrated that this westerly bias was significantly impaired when fish were unable to detect odors. However, the strong reversing tidal currents of Long Island Sound dilute and disperse Connecticut River water throughout eastern Long Island Sound (Fig. 1) such that an olfactory clue should surely be detectable on both tides. Although tide-specific behavior cannot therefore be related to a tide-specific olfactory cue, the loss of the directional bias following occlusion of the nares nevertheless implicates an olfactory-rheotactic mechanism.

The purpose of this paper is to illustrate the hypothesis that an olfactory-regulated rheotaxis can account for unidirectional displacement in reversing tidal currents. It is hypothesized that

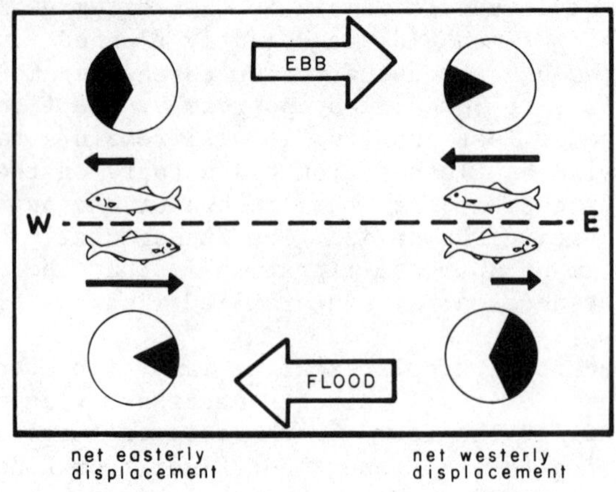

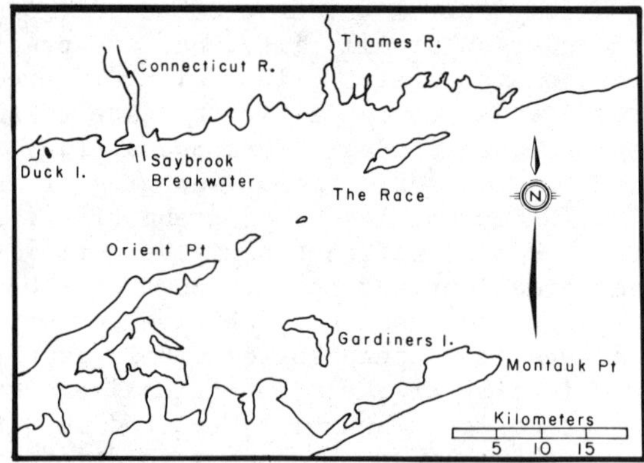

Figure 1. (Upper panel) Schematic representation of the rheotactic behavior of American shad homing to the Connecticut River from Long Island Sound. Large black sectors of circles indicate low precision of countercurrent orientation and large angular changes in bearing. Small black sectors of circles indicate high precision of countercurrent orientation and small angular changes in bearing. Long black arrows indicate fish swim faster than tidal current speed, short black arrows indicate fish swim at same speed as current. The combination of large sectors and short arrows results in displacement in the direction of the tide. The combination of small sectors and long arrows results in displacement in the opposite direction to the tide. Right side of panel presents the behavior pattern of the majority of shad tracked in Long Island Sound resulting in a net westerly displacement. Left side of panel presents behavior of shad moving east after overshooting the Connecticut River. (Lower panel) Location map of eastern Long Island Sound.

olfactory-mediated behavior can be transposed, by learning, to tide-related behavior at the detectable edge of an odor field, if the fish behaves in such a way that leads to the synchrony of tidal and olfactory events. It is further hypothesized that the behavioral modifications leading to such synchrony define the subsequent tide-related behavior leading to net unidirectional displacement towards the source of the odor field. These hypotheses are examined with a simple computer simulation of fish behavior.

THE HYPOTHESIS

It is fundamental that adult fish migrating through an area of reversing tidal currents and exhibiting rheotactic behavior may increase the rate of displacement towards a spawning site only if the behavior differs between the two tides. The hypothesis underlying the present simulation is that fish can learn tide-specific rheotactic behavior that increases the migration rate, if they behave in such a way that leads to the synchrony of appropriate olfactory stimuli and tide-related stimuli. Such synchrony is a criterion for associative learning, whereby a response regulated by a change in olfactory stimuli (Event 1) can be transferred to a tide-related cue (Event 2). This is the principal of stimulus substitution, whereby exposure to an E1-E2 association endows E2 with the properties of E1 (Dickinson 1980). Thus, cues associated with current detection may acquire the capacity to elicit behavioral responses similar to those produced by olfactory stimuli.

It has been argued that the pattern of fish movement is a process of optimization of environmental conditions leading to the selection of optimal physiological or neurological states (Balchen 1976; Leggett 1977). It is assumed for the present simulation that the detection of an odor imprinted during early life signals an optimal neurological state for the reproductive migrant, which will subsequently behave in such a manner as to maintain this neurological state. But it is unlikely that the fish will encounter an olfactory field with simple concentration gradients of sufficient steepness to elicit a klino- or tropo-taxis. The olfactory cue will probably be dispersed and fragmented by turbulent diffusion resulting in a pattern of numerous meandering filaments (Okubo 1980). The odor concentration will thus fluctuate at random and only on the average tend to increase toward the source. There is no evidence to support the idea that fish possess the necessary computational abilities to integrate concentration information over time and space so as to detect this average trend.

It is assumed that the nondirectional olfactory field represents a preferred zone in which reproductive migrants will attempt to stay. Animals usually congregate in such zones when the rate of unidirectional displacement declines as a result of decreasing swimming speed or increasing turning angle (Fraenkel and Gunn 1961). Thus it is proposed that once a migrant detects the appropriate odor its rate of unidirectional displacement will be reduced so as to maintain it within the olfactory field. As the model described below is one-dimensional, decreased displacement along the tidal axis could be due to reduced swimming speed, increased turning angle, or a combination of both (cf Rholf and Davenport 1969).

THE MODEL

The following symbols are used in describing the model (all units are in centimeters and seconds):

- $V_w(t)$ = tidal current velocity with respect to the ground at time t
- t = time (t = 0 at low slack water)
- k = constant equal to 3600
- x = the 'dummy' variable used in the integral equations 2, 7 and 9.
- p = frequency of oscillation of current speed ($p = 2\pi / 12.4k$)
- A = maximum speed of tidal current
- W = speed of residual current
- $D_o(t)$ = position of the olfactory front at time t ($D_o(t) = 0$ at the mean position of the front)
- $V_f(t)$ = fish swimming velocity with respect to the ground at time t
- i = an index to indicate whether or not the simulated fish is in the olfactory field (i = 1 when fish is in field, i = 2 when fish is out of field)
- a_i = proportionality constant used to describe the swimming speed of fish by model A
- b_i = fish swimming speed with respect to the ground used in model B
- c_i = fish swimming speed with respect to water current used in model C
- $D_f(t)$ = position of the fish at time t ($D_f(t) = 0$ when fish located at the mean position of the olfactory front)
- L = a constant defining the starting position of the fish relative to the olfactory front
- n = number of discrete time units per tidal cycle
- τ = length of one time unit, $\tau = 12.4k/n$
- Δt = time the simulated fish spends out of the olfactory field during each tidal cycle after the front has been crossed for the first time.

The scenario of the simulation involves Connecticut River shad migrating in open continental waters off Montauk Point (Fig. 1), moving north along the supposed oceanic migration route of the species. Individual fish swim against reversing tidal currents which exhibit sinusoidal velocities of semidiurnal periodicity superimposed on a constant residual current which flows in the direction of the ebbing tide. The model is one-dimensional and thus there are only two directions: upstream (+, direction of flood) and downstream (-, direction of ebb). The current speed at any time t is thus:

$$V_w(t) = A \sin(pt) - W. \tag{1}$$

If a drift bottle were placed in the water, its displacement would be given by:

$$\int_0^t V_w(x)\, dx = (A/p)[1-\cos(pt)] - Wt. \tag{2}$$

Prior to detecting olfactory cues indicative of the natal river, the simulated fish swims faster than, but always against, the tidal current. Thus, its net displacement is in the opposite direction to the residual current (Fig. 2). At some point in time the fish arrives at the detectable edge of the olfactory field indicative of the Connecticut River. The position of the olfactory front fluctuates with the tides in a sinusoidal fashion but exhibits no residual displacement, thus behaving as a simplified saline front. The location of the olfactory front is thus:

$$D_o(t) = (A/p)[1-\cos(pt)]. \tag{3}$$

The displacement of the olfactory front is defined relative to its mean position ($D_o(t) = 0$). Displacement of the front upstream from the mean position is positive, whereas downstream displacement from the mean position is negative.

Fish behavior is governed by two assumptions: the fish always swims against the current, and the fish decreases its swimming speed when the olfactory field is entered. Three models are used to calculate fish swimming speed as a function of current speed. Model A assumes that fish swimming speed is proportional to current speed such that:

$$V_f(t) = -a_i\, V_w(t). \tag{4}$$

Model B assumes that fish swimming speed is constant relative to the ground such that:

$$V_f(t) = -b_i, \text{ if } V_w(t) > 0 \quad \text{(flood tide)}, \tag{5a}$$

$$V_f(t) = b_i, \text{ if } V_w(t) < 0 \quad \text{(ebb tide)}. \tag{5b}$$

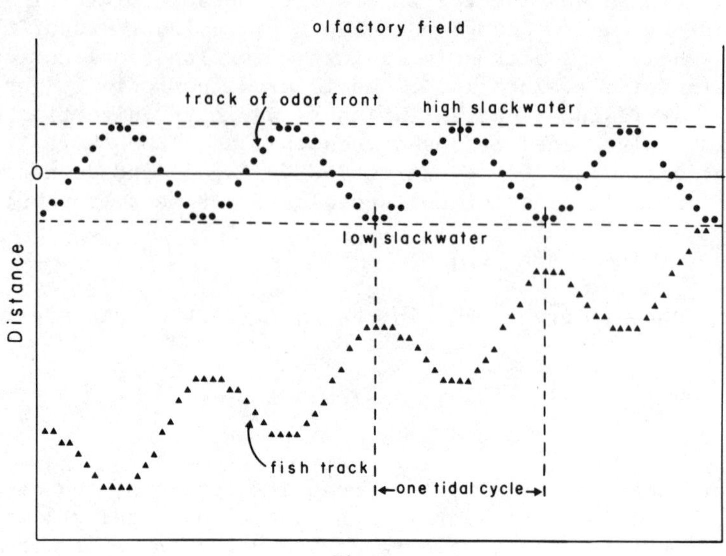

Figure 2. Fictitious simulation printout to describe major features of model. Before encountering the odor front, fish swim faster than the tidal currents resulting in a net displacement in the opposite direction to the residual current. The mean position of the odor front, which shows no residual displacement, is indicated by the horizontal solid line; the extent of the interface zone, the area delimited by the amplitude of displacement of the olfactory front, is indicated by the horizontal dashed lines.

As the basic behavior pattern involved in the simulation is rheotaxis, it is unreasonable to assume that fish maintain a constant swimming speed in the absence of a water current. Thus, fish swimming speed is equated to zero during slack water such that:

$$V_f(t) = 0, \text{ if } V_w(t) = 0. \tag{5c}$$

Model C assumes that fish swimming speed is constant relative to the water current such that:

$$V_f(t) = V_w(t) - c_i, \text{ if } V_w(t) > 0 \quad \text{(flood tide)}, \tag{6a}$$

$$V_f(t) = V_w(t) + c_i, \text{ if } V_w(t) < 0 \quad \text{(ebb tide)}. \tag{6b}$$

As in Model B, it is unreasonable to assume that fish maintain a constant swimming speed in the absence of a water current. Thus, fish swimming speed is equated to zero during slack water such that:

$$V_f(t) = 0, \text{ if } V_w(t) = 0. \tag{6c}$$

It is now possible to derive a formula for the fish's displacement and subsequent location relative to the olfactory front:

$$D_f(t) = \int_0^t V_f(x)\, dx + L. \tag{7}$$

As it is not realistic to assume that fish instantaneously recognize the transition from no-odor to odor, or vice versa, time is divided into n units per tidal cycle. During each time unit the behavioral parameters remain unchanged. Thus, the fish's location can be calculated with a modified version of equation 7. Let τ be the length of one time unit such that:

$$t_n = n\tau + t_0, \tag{8}$$

where t_0 is the beginning of the simulation and t_n is the time after n discrete time units. Using Model A as an example, the fish's location relative to the olfactory front is thus:

$$D_f(t) = D_f(t_n) - a_i \left[\int_{t_n}^{t} V_w(x)\, dx \right], \tag{9}$$

where $t_n < t < t_{n+1}$ and $i = 2$ if $D_f(t_n) < D_o(t_n)$; otherwise $i = 1$.

THE SIMULATION

The model described above is a simulation model, and unlike an analytical model, exact solutions to all equations are not provided. Rather, the above set of equations is put on a computer and the consequences of different numerical values for the parameters are explored. The following values of the parameters describing the currents were used for all simulations. They are representative of the real situation in eastern Long Island and Block Island Sound (Riley 1959):

maximum speed of tidal current (A) = 97 cm/sec
mean speed of tide = 64 cm/sec
speed of residual current (W) = 10 cm/sec.

The value τ can be considered as the resolution of the model. For all runs n = 30, such that τ is approximately 25 minutes; the positions of the fish and the olfactory front are thus plotted every 25 minutes. This value of τ was adopted as it approximates the resolution of the sonic-tracking program (fish positioned every 30 minutes) used to describe the migratory behavior of shad on which the simulation is based (Dodson and Leggett 1973). A variety of values for a_i, b_i, c_i (i = 1,2) were used, based on the range of swimming speeds of shad observed in Long Island Sound. The mean swimming speed, relative to the water, of shad tracked in Long Island Sound in 1970 was 75.3 cm/sec (SD=42.7) and in 1971 was 71.9 cm/sec (SD=36.6) (Dodson and Leggett 1973). Simulated fish tracks were started from a position 7.0 km downstream of the olfactory front at t = 0 (low slack water) and the position of the fish plotted every 25 minutes over 30 tidal cycles. For all three models of swimming behavior fish first entered the olfactory field during the ebbing tide (Fig. 3). Fish achieved an equilibrium position, remaining

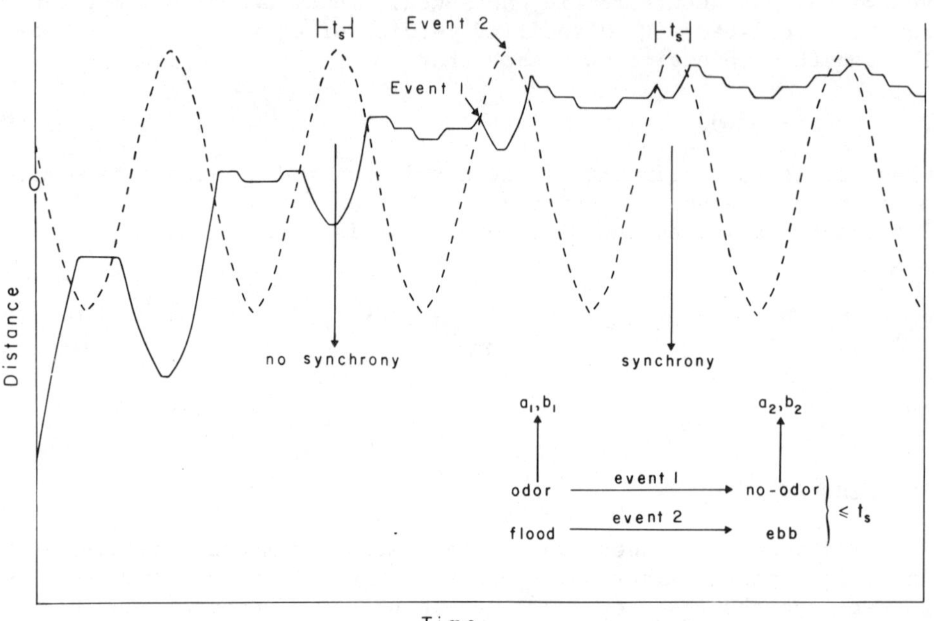

Figure 3. Track of simulated fish (solid line) attaining equilibrium position in interface zone. Track of odor front represented by dashed line. Positions of fish and front are plotted 30 times per tidal cycle. Synchrony is achieved when the fish experiences odor/no-odor transition coincident with flood/ebb transition. After a series of such pairings, stimulus substitution may occur whereby cues associated with the tidal transition acquire the capacity to elicit behavioral responses (a_2, b_2) similar to those produced by the loss of the odor. Ebb/flood transition is always associated with presence of the odor and thus olfactory-mediated behavior (a_1, b_1).

indefinitely within the interface zone (zone delimited by the amplitude of displacement of the olfactory front; Fig. 2), if they drifted with the current or held ground when in the olfactory field but gained ground against the current when out of the field. This produced a pattern of alternating periods in and out of the olfactory field, drifting out during the flooding tide and swimming back in during the ebbing tide. The position of the fish within the interface zone is reflected by Δt, the time the fish spends out of the olfactory field during each tidal cycle after the front is crossed for the first time. The smaller the value of Δt, the closer the fish is to the limit of the tidal incursion into the olfactory field.

Model C of swimming behavior requires unrealistically high swimming speeds at all times for fish to remain within the interface zone. Although fish swimming speed is equated to zero during slack water (equation 6c), model C nevertheless requires that fish maintain high swimming speeds during periods of weak currents before and after slack water. As a result, the position of the fish fluctuates erratically around periods of slack water. These observations caused us to reject model C as a viable simulation.

The fundamental criterion for fish to learn to associate the olfactory and tidal events is the synchrony of the two events. Referring to Figure 3, consider Event 1 to be the odor/no-odor transition and Event 2 to be the flood/ebb transition. The response, a_2 or b_2, to Event 1 will be associated with Event 2 only if Event 2 occurs more frequently in the presence of Event 1 than in its absence. Therefore, a fish may attain an equilibrium position within the interface zone, but if it does not experience the odor/no-odor transition at the same time as the flood/ebb transition stimulus substitution cannot occur. However, it is unreasonable to imagine that such environmental events occur instantaneously in the fish's perceptual field. Fish probably detect the tidal transition over a period of time when currents are weak and of variable direction. Therefore, a parameter t_s was defined as that period of time during which fish could detect the tidal transition. The criterion of synchrony for stimulus substitution was considered to be satisfied if the odor/no-odor transition occurred within t_s. Thus, t_s may be considered as the largest value of Δt during which the two events must occur for stimulus substitution to occur. The actual value of t_s, given the speculative nature of the model, is somewhat arbitrary. As the resolution of the model is in increments of 25 minutes, t_s is equated to approximately 2 hours corresponding to one positioning during slack water and two positionings before and after mid-slack water.

Table 1. Swimming speed parameter values used in model A of swimming behavior. a_1 = proportionality constant used to describe the swimming speed of fish relative to the ground when in the olfactory field; a_2 = proportionality constant used to describe the swimming speed of fish relative to the ground when out of the olfactory field; ss = equivalent mean swimming speed (cm/sec) of fish relative to the water for each value of a_1 (ss_1) and a_2 (ss_2). Once stimulus substitution occurs, mean swimming speed of simulated fish will be the mean of ss_1 and ss_2.

a_1	ss_1	a_2	ss_2
-0.56	28.0	0.34	85.0
-0.28	46.0	0.68	107.0
0.00	64.0	1.02	129.0
		1.36	150.0

RESULTS

Model A

Fish will reach a stable equilibrium position within the interface zone if $a_1 \leq 0.0 < a_2$. The simulations of fish behavior using three values of a_1 and four values of a_2 (Table 1, Fig. 4) reveal that Δt initially declines rapidly and then levels out at a value indicative of the equilibrium position. The smaller the value of Δt, the closer the fish is to the synchrony of olfactory and tidal events and the value t_s. If the fish drifts slightly in the olfactory field (a_1 = -0.28, Fig. 4) but makes significant headway outside the field (a_2 = 1.02, 1.36, Fig. 4), the equilibrium position falls within the t_s value (Δt = 2 h). If the fish holds ground in the olfactory field (a_1 = 0.00, Fig. 4) but makes headway outside the field (a_2 = all values, Fig. 4), the equilibrium position falls well within the t_s value ($\Delta t \leq 1$ h). The time taken to achieve synchrony ranges from 7-14 tidal cycles.

Model B

Fish will reach a stable equilibrium position within the interface zone if $b_1 \leq 0.0 < b_2$. The simulations of fish behavior using three values of b_1 and four values of b_2 (Table 2, Fig. 5) reveal that fish achieve synchrony more frequently than in the case of model A with a similar range of parameter values, taking 8-15 tidal cycles. If the fish drifts slightly (b_1 = -9.17) or holds ground (b_1 = 0.00) in the olfactory field but makes headway outside the field (b_2 = all values, Fig. 5), the equilibrium position falls well within the t_s value ($\Delta t \leq$ 1 h) in most cases.

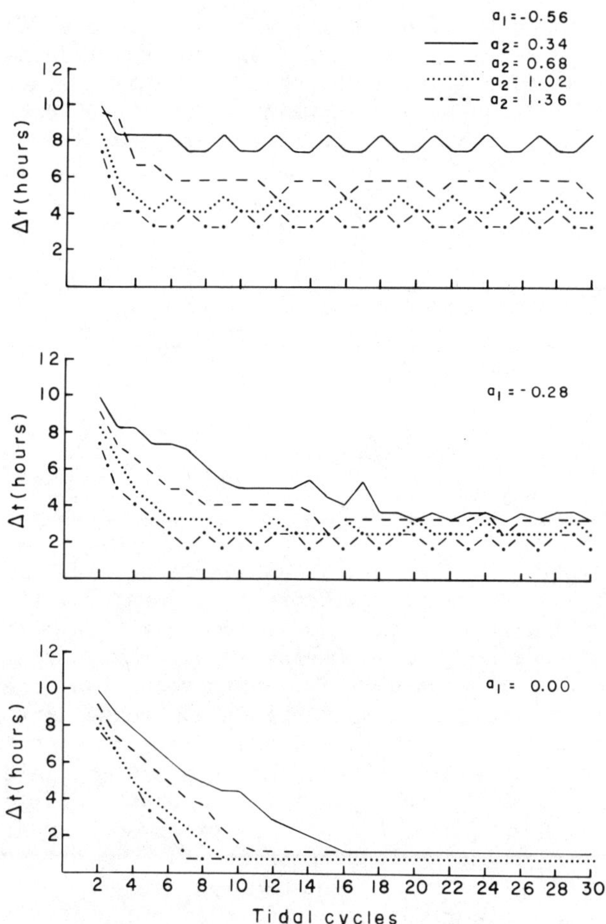

Figure 4. Results of simulations using model A of swimming behavior. a_1 = proportionality constant used to describe the swimming speed of fish when in the olfactory field; a_2 = proportionality constant used to describe the swimming speed of fish when out of the olfactory field; Δt = time (in hours) fish spends out of field after having crossed odor front for first time.

DISCUSSION

It is concluded that the simulated fish achieve synchrony between the two environmental events thus fulfilling the basic criterion for stimulus substitution. Thus the flood/ebb transition may acquire the capacity to elicit the behavioral response initially elicited by the odor/no-odor transition. If stimulus substitution occurs, we would expect the fish to continue to swim faster than the ebbing current despite re-entry into the olfactory field. The

Table 2. Swimming speed parameter values used in model B of swimming behavior. b_1 = swimming speed of fish (cm/sec) relative to the ground when in the olfactory field; b_2 = swimming speed of fish (cm/sec) relative to the ground when out of the olfactory field; ss = equivalent mean swimming speed (cm/sec) of fish relative to the water for each value of b_1 (ss_1) and b_2 (ss_2). Once stimulus substitution occurs, mean swimming speed of simulated fish will be the mean of ss_1 and ss_2.

b_1	ss_1	b_2	ss_2
-18.61	45.0	20.84	85.0
-9.17	55.0	41.67	105.0
0.00	64.0	62.51	126.0
		83.34	147.0

ebb/flood transition, however, is always associated with the presence of the odor once the olfactory front is crossed for the first time, and no such stimulus substitution would occur. Therefore, on the ensuing flood tide swimming behavior should revert to the initial olfactory-mediated homeward drift. Therefore, the behavioral modifications responsible for synchrony and subsequent stimulus substitution produce a directional bias that not only insures that the migrant remains within the olfactory field, but also determines the mechanism by which the migrant increases its migration rate towards the odor source in the absence of an odor gradient. There is no need to hypothesize goal orientation.

Evidence in support of this hypothesis is provided by the tracking of anosmic shad in Long Island Sound (Dodson and Leggett 1974). Following occlusion of the olfactory organ the swimming speed of these fish during the ebbing tide continued to exceed current speed as with sensory intact shad. This supports the suggestion that stimulus substitution had occurred during the ebbing tide as loss of the odor did not significantly alter migratory behavior. However, during the flooding tide anosmic shad exhibited swimming speeds in excess of current speed rather than drifting as in the case of sensory intact shad. Thus, the loss of directional bias observed in anosmic shad was principally due to behavioral changes during the flooding tide, that part of the tidal cycle associated with olfactory-mediated behavior in the present hypothesis.

The model of fish behavior presented here contains several simplifying assumptions that may be considered as influencing the validity of the conclusions. The simulation of tidal currents and the displacement of an olfactory front in open waters is an oversimplification of a complex hydraulic system. The restriction of

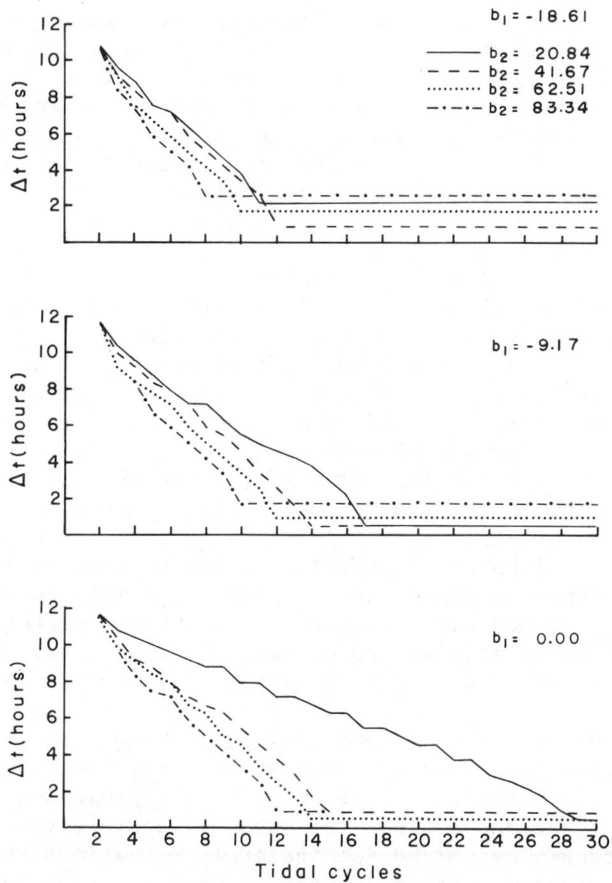

Figure 5. Results of simulations using model B of swimming behavior. b_1 = swimming speed of fish (cm/sec) with respect to the ground when in the olfactory field; b_2 = swimming speed of fish (cm/sec) with respect to ground when out of the olfactory field; Δt as in Figure 4.

fish movement to one dimension is also an oversimplification as evidence exists to indicate that shad undertake vertical migrations (Neves and Depres 1979). However, the feasibility of the hypothesis is demonstrated by the consistency of the simulation results generated by model A and B over a range of parameter values based on empirical observations. Model C provided unsatisfactory results, but it is felt that the assumption of maintaining a constant high speed and heading along the tidal axis during periods of weak and variable currents is biologically unfounded in the case of fish using rheotaxis as the basis of their migratory mechanism. Further confidence in the hypothesis is provided by the observation that model B provided the most consistent results in terms of achieving synchrony, and model B conforms to the optimal swimming strategy of fish migrating in detectable oscillating currents proposed by Trump and Leggett (1980) based solely on energetic considerations. It is concluded that a more exhaustive model of current dynamics and fish behavior may change the estimates of Δt and the time taken to achieve synchrony, but the general conclusion concerning the potential role of stimulus substitution will be upheld as long as the fish makes headway when out of the odor field but holds ground or drifts slightly when in the odor field.

The hypothesis and model presented here serves to identify three areas of study that merit greater research efforts in the coming years. The first and most obvious is the question of the role of associative learning in fish migration. For example, although synchrony between olfactory and tidal events fulfills the basic criterion for stimulus substitution, we can at present only speculate on the number of such pairings fish must experience before stimulus substitution occurs. Numerous psychological, physiological and abiotic factors influence such a learning process (cf Dickinson 1980). Extensive laboratory experiments in the area of animal learning (cf Bitterman 1984--this volume) using migratory species and appropriate environmental stimuli are necessary to test the validity of the present hypothesis and to explore the behavioral consequences of such a mechanism of migratory orientation.

Although water currents have long been thought to provide unambiguous directional clues to migrating fish, the sensory system(s) involved is still a matter of much speculation. Optical or tactile stimuli are responsible for many of the reported cases of rheotaxis, but they cannot be implicated in situations where the fish have no fixed reference points. Other mechanisms have been proposed and studied (Arnold 1981), yet no direct evidence exists to demonstrate how fish out of sight of the bottom can detect the current.

Related to this problem is the question of how fish may discriminate between ebb and flood tides. For example, an emergent property of the present simulation is that fish must be capable of discriminating between high and low slack water if they are to achieve stimulus substitution only during the ebbing tide. A similar discriminative capacity is an essential element of the mechanism of selective tidal stream transport used by plaice (*Pleuronectes platessa*) to insure that fish move in a direction appropriate to their migration (Arnold 1981; Arnold and Cook 1984--this volume). How fish exhibiting tide-specific migration behavior can discriminate between high and low slack water remains largely an unanswered question.

A third area of study identified by the present hypothesis is the observation of fish behavior in and around detectable physical discontinuities. Although it is not at present feasible to directly identify olfactory fronts in estuaries or open continental shelf waters, it is possible to identify temperature and salinity discontinuities through which migratory fish can be followed using sonic tracking. The documentation of behavioral changes and the resultant synchrony of different environmental stimuli experienced by fish in these zones and the behavior of fish once these zones are left could provide direct evidence of the role of stimulus substitution in migratory orientation. For example, shad entering the Connecticut River exhibit extensive meandering in the region of the saltwater/freshwater interface (Dodson et al. 1972). This movement resulted from a combination of passive drift and relatively inactive swimming that retained shad near the leading edge of salt water for periods of 24-53 hours. Dodson et al. (1972) concluded that the meandering was due to the obvious physiological stress associated with adaptation to freshwater. However, it is important to note that once shad leave the interface zone and enter freshwater, upstream migration is continuous through conditions of zero and reverse flow and no obvious changes in behavior related to tidal conditions are observed. Thus, behavior after the saltwater/freshwater transition is different from that observed in Long Island Sound suggesting that the meandering phase may also have contributed to the development of the unique migratory behavior observed in freshwater. It is proposed that the migratory behavior of fish should be observed in areas of physical discontinuities, which do not have major physiological consequences for the organism (pycnoclines, for example), but which could provide excellent situations in which to test the validity of stimulus substitution as an important factor in conditioning directional bias.

REFERENCES

Arnold, G.P. 1981. Movements of fish in relation to water currents. Pages 55-79 *in* D.J. Aidley, editor. Animal migration. Cambridge University Press, Cambridge, England.

Arnold. G.P., and P.H. Cook. 1984. Fish migration by selective tidal stream transport: first results with a computer simulation model for the European continental shelf. Pages 227-261 *in* J.D. McCleave, G.P. Arnold, J.J. Dodson and W.H. Neill, editors. Mechanisms of migration in fishes. Plenum Press, New York, New York, USA.

Balchen, J.G. 1976. Modelling of the biological state of fishes. Foundation for Scientific and Industrial Research (SINTEF), Trondheim, Norway. Technical note number 62.

Bitterman, M.E. 1984. Migration and learning in fishes. Pages 397-420 *in* J.D. McCleave, G.P. Arnold, J.J. Dodson and W.H. Neill, editors. Mechanisms of migration in fishes. Plenum Press, New York, New York, USA.

Creutzberg, F. 1961. On the orientation of migrating elvers (*Anguilla vulgaris* Turt.) in a tidal area. Netherlands Journal of Sea Research 1:257-338.

Creutzberg, F. 1963. The role of tidal streams in the navigation of migrating elvers (*Anguilla vulgaris* Turt.). Ergebnisse der Biologie 26:118-127.

Dickinson, A. 1980. Contemporary animal learning theory. Cambridge University Press, Cambridge, England.

Dodson, J.J., and W.C. Leggett. 1973. Behavior of adult American shad (*Alosa sapidissima*) homing to the Connecticut River from Long Island Sound. Journal of the Fisheries Research Board of Canada 30:1847-1860.

Dodson, J.J., and W.C. Leggett. 1974. Role of olfaction and vision in the behavior of American shad (*Alosa sapidissima*) homing to the Connecticut River from Long Island Sound. Journal of the Fisheries Research Board of Canada 31:1607-1619.

Dodson, J.J., W.C. Leggett, and R.A. Jones. 1972. The behavior of adult American shad (*Alosa sapidissima*) during migration from salt to freshwater as observed by ultrasonic tracking techniques. Journal of the Fisheries Research Board of Canada 29:1445-1449.

Emanuel, M.E., and J.J. Dodson. 1979. Modification of the rheotropic behavior of male rainbow trout (*Salmo gairdneri*) by ovarian fluid. Journal of the Fisheries Research Board of Canada 36:63-68.

Fraenkel, G.S., and D.L. Gunn. 1961. The orientation of animals. Kineses, taxes and compass reactions. Dover, New York, New York, USA.

Hara, T.J. 1970. An electrophysiological basis for olfactory discrimination in homing salmon: a review. Journal of the Fisheries Research Board of Canada 27:565-586.

Hasler, A.D., A.T. Scholz, and R.M. Horrall. 1978. Olfactory imprinting and homing in salmon. American Scientist 66: 347-355.

Kleerekoper, H. 1982. The role of olfaction in the orientation of fishes. Pages 199-225 *in* T.J. Hara, editor. Chemoreception in fishes. Elsevier Scientific Publication Company, Amsterdam, The Netherlands.

Leggett, W.C. 1977. The ecology of fish migrations. Annual Review of Ecology and Systematics 8:285-308.

Mathewson, R.F., and E.S. Hodgson. 1972. Klinotaxis and rheotaxis in orientation of sharks toward chemical stimuli. Comparative Biochemistry and Physiology A. Sensory, Neural, and Behavioral Physiology 42A:79-84.

Miles, S.G. 1968. Rheotaxis of elvers of the American eel (*Anguilla rostrata*) in the laboratory to water from different streams in Nova Scotia. Journal of the Fisheries Research Board of Canada 25:1591-1602.

Neves, R.J., and L. Depres. 1979. The oceanic migration of American shad, *Alosa sapidissima*, along the Atlantic coast. US National Marine Fisheries Service Fishery Bulletin 77:199-212.

Okubo, A. 1980. Diffusion and ecological problems: mathematical models. Springer-Verlag, Berlin, Federal Republic of Germany.
Riley, G.A. 1959. Oceanography of Long Island Sound. Bulletin of the Bingham Oceanographic Collection, Yale University 17:9-30.
Rohlf, F.J., and D. Davenport. 1969. Simulation of simple models of animal behavior with a digital computer. Journal of Theoretical Biology 23:400-424.
Trump, C.L., and W.C. Leggett. 1980. Optimum swimming speeds in fish: the problem of currents. Canadian Journal of Fisheries and Aquatic Sciences 37:1086-1092.

DOES THE ROUTE OF SEAWARD MIGRATION OF FRASER RIVER SOCKEYE SALMON (*Oncorhynchus nerka*) SMOLTS DETERMINE THE ROUTE OF RETURN MIGRATION OF THE ADULTS?

C. Groot, L. Margolis, and R. Bailey

Department of Fisheries and Oceans
Fisheries Research Branch
Pacific Biological Station
Nanaimo, British Columbia V9R 5K6 Canada

ABSTRACT

Sockeye salmon (*Oncorhynchus nerka*) can choose a northern or a southern route around Vancouver Island during their homing migration to the Fraser River. The percentage of fish taking the northern route has varied from 10-70% during the last 29 years. This has created problems in the management of sockeye salmon in southern British Columbia, Canada.

The purpose of this study is to test the hypothesis that adult sockeye return to the Strait of Georgia and the Fraser River by retracing the same route taken by the juveniles during seaward migration. Proportions of sockeye smolts from the different lake systems leaving the Strait of Georgia by the northern and southern exits will be used to predict the ratio of adult sockeye of the same brood year returning two years later by each of these routes. The plan includes determining the stock origin of the juveniles leaving via the northern and southern exits of the Strait of Georgia by differences in parasite fauna.

Results so far have indicated that (i) appreciable differences exist in the parasite fauna of sockeye smolts from different lake systems, (ii) sockeye smolts use both the southern and northern routes to migrate out of the Strait of Georgia, and (iii) these movements essentially occur northward along the mainland coast and southward along the east and west shores of the Gulf Islands.

INTRODUCTION

Sockeye salmon of the Fraser River system in British Columbia, Canada, normally have a four year life cycle. During their life these fish experience a number of major habitat changes (Groot 1982), migrating between breeding grounds in fresh water and feeding areas in lakes and the ocean (Fig. 1). Sockeye typically spawn in streams adjacent to lakes and the juveniles spend one year, sometimes two years, in the lake before migrating to salt water as smolts.

After two years in the Pacific Ocean adult Fraser River sockeye, varying in numbers from 2 - 19 million, return to their home stream to spawn (Fig. 1). To reach the Fraser River they can choose two routes around Vancouver Island: the northern passage through Queen Charlotte and Johnstone Straits and the southern approach through the Strait of Juan de Fuca (Fig. 2). Until about 5 years ago 80% (range 65-90%) of the fish migrated to the Fraser River by the southern route. In three of the five years since 1978 the percentage using the Johnstone Strait route has increased dramatically to highs of 70% in 1980 and 69% in 1981 (Fig. 2). This variability in approach

FRASER RIVER SOCKEYE SALMON

Age	Habitat change	Life cycle stage	Timing
1st Year	1. egg to gravel	–alevin	winter
	2. gravel to river	–fry	spring
	3. river to lake	–fry-fingerling	summer
2nd Year	4. lake to river to estuary	–smolt	spring
	5. estuary to winter grounds	–immature	sum/fall
3rd Year	6. winter grounds to summer grounds	–immature	spring
	7. summer grounds to winter grounds	–immature	fall
4th Year	8. winter grounds to estuary of home river	–maturing adult	spring/summer
	9. estuary to river to lake	–maturing adult	summer
	10. lake to river spawning grounds	–maturing adult	late summer/fall

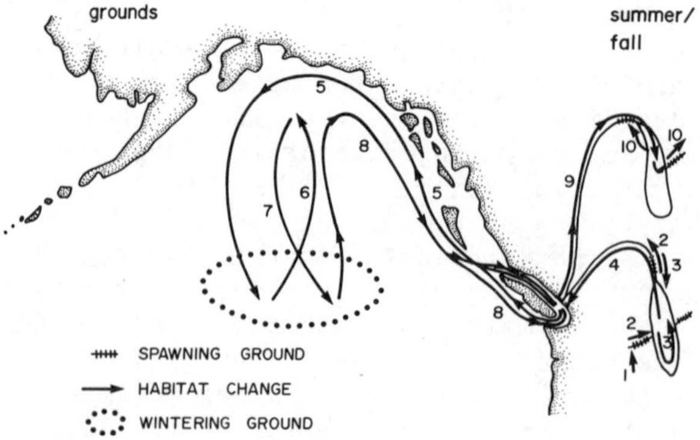

SPAWNING GROUND
HABITAT CHANGE
WINTERING GROUND

Figure 1. Habitat changes experienced by sockeye salmon of the Fraser River system during their life cycle.

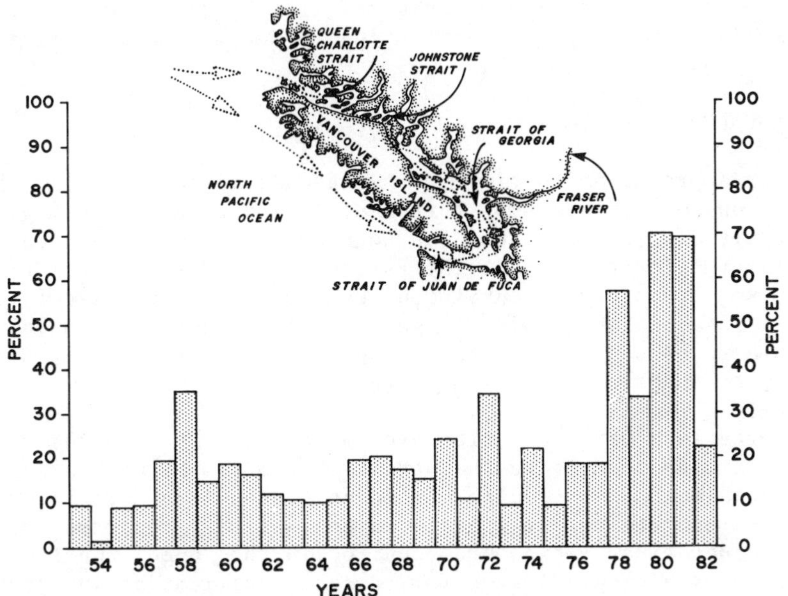

Figure 2. Map: The migratory routes of adult sockeye salmon on their way to the Fraser River and other southern British Columbia and Washington State spawning grounds.
Graph: The percentage of sockeye salmon approaching the Fraser River through the northern passage from 1953 to 1982.

routes to the Fraser River creates problems in the management of the fisheries on these sockeye stocks.

Attempts have been made to relate the changes in migratory behavior of Fraser River sockeye to offshore oceanographic indices (Royal and Tully 1961; Favorite 1961) and to use the relationship to predict the proportion of fish taking the northern route (IPSFC 1976-1981; Wickett 1977). However, these relationships have not held consistently.

THE HYPOTHESIS

A study has been initiated in which we propose to test the hypothesis that the proportions of adult Fraser River sockeye salmon moving through Johnstone Strait and the Strait of Juan de Fuca are predetermined by the routes followed by the seaward migrating juveniles out of the Strait of Georgia. The underlying assumption is that the juveniles during the outward migration record in time and space sequences of characteristics of the migratory route, which are then replayed in reverse order during the adult homeward migration.

THE APPROACH

The approach in testing the hypothesis is to determine the proportions of sockeye smolts from different stocks that move out of the Strait of Georgia by the southern or northern routes, and to use this information to predict the ratios of adult sockeye of the same stocks and brood year returning by each of these routes two years later. In addition we plan to gain a better understanding of the routes used by sockeye smolts in migrating from the Fraser River estuary to Johnstone and Juan de Fuca Straits.

Three sets of information are needed to calculate the ratio of sockeye smolts that leave the Strait of Georgia by each exit. Firstly, a method is required to identify the stock origin of the juveniles caught in salt water. Margolis (1982) reported that specific parasites, which are acquired in the "home" rearing lake, can serve as biological tags for stock identification in sockeye salmon. For purposes of this study the parasite tags need only survive in or on the fish for about 1-2 months after the fish have entered salt water. Additional information on stock origin of the smolts can be derived from scale pattern analysis (Henry 1961).

Secondly, estimates of numbers of smolts leaving each major nursery lake draining into the Strait of Georgia is required. More than 90% of the sockeye salmon migrating through the Strait of Georgia, either as juveniles or adults, are of Fraser River origin. Some sockeye originate from a number of small lake systems in southern British Columbia and Washington State (Fig. 3). For some of the Fraser and other lakes counts of smolt output are available from respective fisheries agencies responsible for management of sockeye salmon (Canada Department of Fisheries and Oceans, International Pacific Salmon Fisheries Commission, Washington State Department of Fisheries). For those lakes for which such information is absent smolt output values will have to be estimated from adult escapements.

Thirdly, samples of smolts must be obtained in and at both ends of the Strait of Georgia during their out-migration. These samples will be collected by using a purse seine and a two-boat trawl. The seine is 216 m long and 25 m deep, with a 6 mm woven mesh bunt, and the trawl measures 17.7 m long, with a mouth opening of 6.1 m wide and 3 m deep. Fishing efforts during 1982 were alternated between northern and southern parts of the Strait of Georgia and were concentrated in near shore areas, bays, and passes, to increase the chances of catching smolts.

Figure 3. Sockeye salmon nursery lakes of the Fraser River system and other systems draining into Georgia Strait and adjacent waters.

The proportions of smolts leaving the Strait of Georgia by the northern and southern routes in a simple case with smolts coming from two lakes (A and B) can be obtained by solving the following four equations:

$$N = A_n / B_n \qquad A = A_n + A_s$$

$$S = A_s/B_s \qquad B = B_n + B_s,$$

where

A = smolt output of Lake A;
B = smolt output of Lake B;
N = ratio of smolts from Lakes A and B in the north;
S = ratio of smolts from Lakes A and B in the south;
A_n, A_s = number of smolts from Lake A leaving the Strait of Georgia by the northern and southern routes, respectively; and
B_n, B_s = number of smolts from Lake B leaving the Strait of Georgia by the northern and southern routes, respectively.

The analysis in this simple case can be extended to the more complex real situation in which many more than two stocks are involved.

PRELIMINARY RESULTS

Only some preliminary information from the first year of the project is available at present. To date over 200 smolts from five systems have been examined for parasites (Table 1). Four of these systems are part of the Fraser River drainage; the fifth, the Nimpkish system, drains into Johnstone Strait, where its stocks mix with those of the Fraser, both as juveniles and as returning adults. The results of this analysis showed that the juvenile Nimpkish fish are distinguishable from those Fraser stocks examined, and also that the four Fraser stocks are, at least in part, separable from each other. Samples of smolts for the remaining Fraser River and other stocks must be examined before definitive conclusions can be drawn on the usefulness of parasites for distinguishing among all the stocks in question.

A total of 94 purse seine and 18 two-boat trawl sets were made between May 11 and June 25, 1982. About 1,000 smolts were caught in both southern and northern parts of the Strait of Georgia. Smolts were primarily captured near shore and in bays and inlets. In the northern part of the Strait of Georgia the smolts mainly occurred along the coast of the mainland and in the south around the islands (Gulf Islands) along the southeastern coast of Vancouver Island. These results suggest that sockeye smolts primarily move close to shore and in the northern part of the Strait of Georgia hug the eastern shore along the mainland and in the southern part of the Strait they swim along the east and west shores of the outer Gulf Islands on their migration towards the Strait of Juan de Fuca.

The smolt migration into the Strait of Georgia in 1982 was assumed to consist of about 50-60% Chilko Lake smolts, because the escapement to the spawning areas of this lake system received about

Table 1. Prevalence of certain parasites in sockeye smolts of four Fraser River Lakes and the Nimpkish River.

Stock	Number of smolts sampled	Percent frequency of occurrence of parasites								
		A	B	C	D	E	F	G	H	I
Nimpkish River	50	92	20	100	4	84	4	0	4	10
Fraser River										
Seton Creek	50	0	0	0	66	52	0	0	0	0
Chilko Lake	50	0	0	0	0	78	0	10	6	0
Francois Lake	47	0	0	0	57	15	0	6	6	30
Fraser Lake	20	0	0	0	5	80	0	55	55	45

A = *Myxobolus neurobius* (Myxosporea) - brain
B = *Myxidium* sp. (Myxosporea) - kidney and urinary bladder
C = *Philonema oncorhynchi* (Nematoda) - wall of swimbladder and mesenteries
D = *Philonema agubernaculum* (Nematoda) - body cavity
E = *Diphyllobothrium* sp. plerocercoids (Cestoda) - encysted on stomach wall
F = *Diphyllobothrium* sp. plerocercoids (Cestoda) - musculature
G = *Diplostomulum* sp. metacercaria (Trematoda) - eye
H = *Tetracotyle* sp. metacercaria (Trematoda) - viscera, pericardium
I = *Salmincola californiensis* (Crustacea) - external surfaces of body

499,000 sockeye compared with about 848,000 for the whole Fraser River system in brood year 1980 (IPSFC 1981). In addition three other lake systems, Harrison, Lillooet, and Fraser Lakes, may have contributed significantly (about 26%) to the Fraser River smolt output, judging from their respective escapements in 1980 of 57,900, 90,900 and 72,100 sockeye.

Calculation of the percentages and proportions of smolts in northern and southern parts of the Strait of Georgia during their migration to the Pacific Ocean must await final results of the parasite tag analysis.

CONCLUDING REMARKS

This is the first documented information that sockeye smolts leave the Strait of Georgia via the northern passage. Most other investigators (Chapman 1970; Barraclough and Phillips 1978; Healey 1978, 1980) have concentrated their fishing during May and June more in the southern parts of the Strait. Combining our results with these earlier efforts reveals the following migratory patterns.

Firstly, sockeye smolts enter the Strait of Georgia during late April and early May at an average length of 8.5 cm (Barraclough and Phillips 1978). By the end of June and early July most smolts have left the Strait of Georgia for the Pacific Ocean. Their average length then is between 11 and 12 cm (Healey 1978). The timing of peak catches in the Fraser River plume suggests that the smolts take 20 to 30 days to move through the Strait of Georgia (Healey 1980).

Secondly, the smolts leave the Fraser River estuary in two main directions. One movement is westward across the Strait of Georgia toward the Gulf Islands and southward along east and west sides of the outer island chain to the Strait of Juan de Fuca (Fig. 4). A second movement occurs northward along the coast of the mainland towards the passes leading to Johnstone Strait (Fig. 4). This

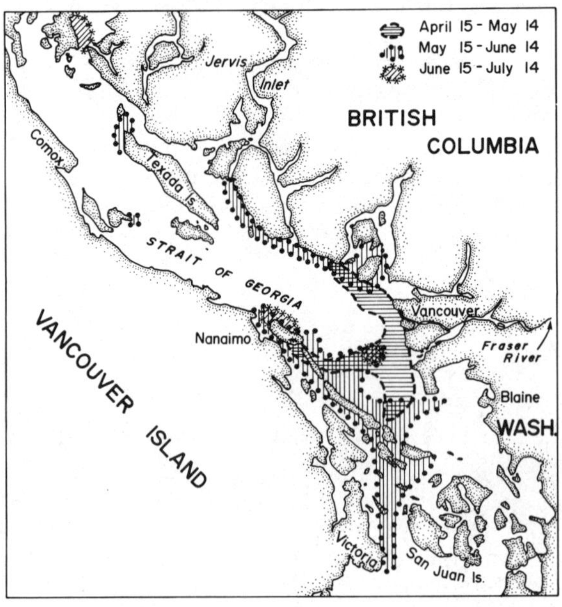

Figure 4. The distribution of seine and two-boat trawl catches with more than 5 smolts per set between April 15 - May 14, May 15 - June 15, and June 15 - July 14 as complied from a number of sources (see text for details).

strongly suggests that sockeye smolts use both the southern and northern routes to leave the Strait of Georgia.

ACKNOWLEDGMENTS

We gratefully acknowledge the help of Graeme Ellis, Robert Ball and Scott Macey with the collection of sockeye smolt samples in the Strait of Georgia. We thank staff of the International Pacific Fisheries Commission in New Westminster, British Columbia, Canada, for their support and for supplying smolt specimens of Fraser River Lakes for parasite tag analysis. Smolt samples from Nimpkish Lake were obtained through the courtesy of Terry Gjernes, Pacific Biological Station, Nanaimo. We thank Lauri Mackie and Sherry Greenham for preparing the figures.

REFERENCES

Barraclough, W.E., and A. Phillips. 1978. Distribution and abundance of juvenile salmon in the southern Strait of Georgia during the period April to July from 1966 to 1969. Fisheries and Marine Service Technical Report 826, Department of Fisheries and the Environment, Pacific Biological Station, Nanaimo, British Columbia, Canada.

Chapman, A. 1970. Pink salmon marine investigation, 1970. International Pacific Salmon Fisheries Commission, New Westminster, British Columbia, Canada, Internal Report.

Favorite, F. 1961. Surface temperatures and salinity off the Washington and British Columbia coasts, August 1958 and 1959. Journal of the Fisheries Research Board of Canada 18:311-319.

Groot, C. 1982. Modifications on a theme. A perspective on migratory behavior of Pacific salmon. Pages 1-21 *in* E.L. Brannon and E.O. Salo, editors. Proceedings of the Salmon and Trout Migratory Behavior Symposium, University of Washington College of Fisheries, Seattle, Washington, USA.

Hartt, A.C. 1980. Juvenile salmonids in the oceanic ecosystem. The critical first summer. Pages 25-57 *in* W.J. McNeil and D.C. Himsworth, editors. Salmonid Ecosystems of the North Pacific. Oregon State University Press, Corvallis, Oregon, USA.

Healey, M.C. 1980. The ecology of juvenile salmon in Georgia Strait, British Columbia. Pages 203-229 *in* W.J. McNeil and D.C. Himsworth, editors. Salmonid Ecosystems of the North Pacific. Oregon State University Press, Corvallis, Oregon, USA.

Healey, M.C. 1978. The distribution, abundance, and feeding habits of juvenile Pacific salmon in Georgia Strait, British Columbia. Fisheries and Marine Service Technical Report 788, Department of Fisheries and the Environment, Pacific Biological Station, Nanaimo, British Columbia, Canada.

Henry, K.A. 1961. Racial identification of Fraser River sockeye salmon by means of scales and its application to salmon management. International Pacific Salmon Fisheries Commission Bulletin 12, New Westminster, British Columbia, Canada.

International Pacific Salmon Fisheries Commission (IPSFC). 1976-1981. Annual Reports for 1975-1980.

Margolis, L. 1982. Parasitology of Pacific salmon - an overview. Pages 135-226 *in* E. Meerovitch, editor. A Festschrift dedicated to the fiftieth anniversary of the Institute of Parasitology of McGill University, 1932-1982. McGill University, Montreal, Quebec, Canada.

Royal, L.A., and J.P. Tully. 1961. Relationship of variable oceanographic factors to migration and survival of Fraser River salmon. California Cooperative Oceanic Fisheries Investigations Reports 8, Department of Fish and Game, California, USA.

Wickett, W.P. 1977. Relationship of coastal oceanographic factors to the migration of Fraser River sockeye salmon (*Oncorhynchus nerka* W.). International Commission for Exploration of the Sea, CM 1977/M:26.

MIGRATION IN CORAL REEF FISHES: ECOLOGICAL SIGNIFICANCE AND ORIENTATION MECHANISMS

John C. Ogden

West Indies Laboratory
Fairleigh Dickinson University
Teague Bay, Christiansted, St. Croix
U.S. Virgin Islands 00820 USA

Thomas P. Quinn

Department of Fisheries and Oceans
Fisheries Research Branch
Pacific Biological Station
Nanaimo, British Columbia V9R 5K6 Canada

ABSTRACT

Coral reef fishes are in general rather sedentary and territoriality and home range behavior patterns are highly developed. However, many species migrate, often relatively long distances compared to body size and often with spectacular precision. Migrations in reef fishes may be associated with: 1) life history--movements of planktonic larval stages to reefs or movements of juveniles from nursery areas to reefs; 2) seasons--precisely timed spawning aggregations drawing fishes to particular locations from wide areas of the reef; and 3) diel patterns--movements to and from feeding or resting areas associated with dawn and dusk. Of these migratory patterns the latter are the best known in coral reef fishes.

Certain coral reef fishes seem ideal as a model system for studies of orientation mechanisms in fishes. Grunts (Haemulidae), for example, have relatively short, highly regular migrations that occur in all seasons. The fish are numerous, small, and able to withstand tagging and handling. Preliminary work on orientation mechanisms of grunt migrations has indicated the potential use of local landmarks, a distance sense and a sun compass. The persistence of migration routes in grunts and the preliminary demonstration of socially facilitated learning of routes by new recruits sets up the possibility of studies of the development of a "map" in grunts combining a variety of cues. The goal of studies in such a model system will be to develop testable hypotheses that may be applied to the orientation mechanisms guiding fish migrations in general.

INTRODUCTION

Coral reef fishes form one of the most diverse community associations of animals in nature. A nearshore fish fauna in excess of 2,000 species is known from the Philippines, the generally acknowledged center of diversity of coral reef fishes. The Caribbean region, at a higher latitude, geographically smaller, and climatically more variable, has more than 500 species. Within-habitat diversity on coral reefs is high and correlates with measurable parameters of reef structure (Gladfelter et al. 1980) but not necessarily with regional diversity (Sale 1980). While the origin and maintenance of this high diversity is controversial (e.g. Sale 1977; Ogden and Ebersole 1981), coral reef fishes have evolved a high degree of specialization in morphology and behavior most often demonstrated in studies of resource partitioning, particularly space and food.

Coral reef fishes are appealing objects of study and have been subjected to increasing attention in the last 20 years, particularly with the development of scuba diving as a field research tool. They are easily observed at any hour of the day or night and may be relatively easily manipulated. There are numerous recent reviews on coral reef fish ecology, including population biology (Ehrlich 1975), community structure (Helfman 1978), reproductive biology (Johannes 1978) and aspects of social behavior and coexistence (Sale 1977, 1980).

The purpose of this paper is to review what is known of migration in coral reef fishes in relationship to the current status of knowledge of orientation mechanisms in fishes in general. It is proposed that coral reef fishes provide an opportunity to develop a model system for testing hypotheses which will have broad application to the study of fish orientation mechanisms.

CORAL REEF FISH MIGRATIONS

Coral reef fishes are in general rather sedentary with territoriality and home range behavior patterns highly developed in many species. The reef may be viewed as a complex mosaic of spaces in which many species of fishes have rather precisely defined daily activity patterns. However, many species migrate, often relatively long distances (particularly in relation to body size) and often with spectacular persistence and precision. Migrations in reef fishes may be associated with: 1) life history--movements of planktonic larval stages to reefs or movements of juveniles from nursery areas to reefs; 2) seasons--precisely timed spawning aggregations drawing fishes to particular locations on the coral reefs from wide areas; and 3) diel patterns--movements to and from feeding or resting areas associated principally with dawn and dusk. Of these migratory patterns the latter are the best known in coral reef fishes. The evolution of these migratory patterns in coral reef fishes is probably associated with two principal evolutionary pressures: 1) the intensity of predation in tropical waters; and 2) the importance of partitioning time, space and food in a complex fauna.

Life History Migrations

The larval and juvenile stages of most coral reef fishes occupy areas completely different from the adults. A great majority of coral reef fishes exhibit a planktonic larval stage which may last from days to more than a year. While the larval life of most coral reef fishes is poorly known, some species have been studied. Damselfishes (Pomacentridae), for example, lay demersal eggs within a nest in the territory. Upon hatching the larval damselfishes enter the plankton for as brief a period as several days before settling in an area removed from that of their parents (Sale 1980). In contrast, the balloonfish (*Diodon holocanthus*) is known to recruit to nearshore areas in huge schools from the open ocean (J.C. Ogden unpublished data). Based on otolith ring counts (E.B. Brothers personal communication) the age of several individual *Diodon* has been determined to be in excess of 450 days at the time of recruitment into coastal waters. Between these extremes, a more usual pattern of larval life in the plankton might be several weeks, and reef fishes can likely adjust the duration of larval life to a certain extent in response to local conditions (Sale 1980).

The "nursery" phenomenon is common in coral reef fishes. For example, stands of mangroves and seagrass beds selectively attract larvae of certain coral reef fishes, such as snappers (Lutjanidae) and grunts (Haemulidae). These fishes move to coral reefs later in life, usually in a pre-adult stage. The recruitment pattern in grunts has been followed by Brothers and McFarland (1981), McFarland (1979) and W.N. McFarland, E.B. Brothers, J.C. Ogden and M.J. Shulman (unpublished data) in the Virgin Islands. After a larval life of about 14 days (determined from counts of otolith daily rings) post-larval grunts settle in seagrass beds and gradually associate in schools. These schools slowly move to the base of the coral reef over a period of several months arriving at the reef in a juvenile stage at which time they join the pre-adult schools commonly found there in persistent sites.

Seasonal Migrations

Seasonal patterns of migration are widespread in fishes. In the temperate zone these movements are usually associated with changes in climate, water temperature, food supply or similar seasonal variation in resources. In the tropics seasonal migrations are most commonly associated with reproductive cycles. Smith (1972) noted spawning aggregations of Nassau grouper (*Epinephelus striatus*) in the Bahamas. This yearly phenomenon, well known to local fishermen, is commonly observed in groupers throughout the Caribbean. One particular reef area draws fish from an entire region and spawning activity proceeds over a period of several weeks. Smith (1972) suggested that the fish migrate to the spawning sites by following depth contours on the coral reefs and that these traditional spawning sites are maintained as younger fish have a tendency to follow older more experienced fish. Some areas of the reef are particularly advantageous for spawning. These probably coincide with favorable patterns of current and tide which assist in the dispersal of the newly released eggs. Johannes (1978) reviewed seasonal spawning movements of fishes in the Pacific and found a striking coincidence of these movements with seasonal variations in oceanic current patterns. This coincidence of behavior with long term oceanographic phenomena may assure the entrainment of larvae within the coastal zone rather than their random dispersal into the deep ocean. P.S. Lobel (unpublished data) has continued this work in a detailed examination of spawning patterns of coral reef fishes and seasonal current patterns in the Hawaiian Islands.

Diel Migrations

It is well established that communities of coral reef fishes undergo major diel changes, especially around dawn and dusk. Diurnal and nocturnal feeders exchange roles, migrating between refuge areas and feeding areas. Studies have detailed the temporal changes in community structure during these transition periods, emphasizing the sequence of appearance of various species and the predator-prey interactions (Hobson 1972, 1973; Starck and Davis 1966; Domm and Domm 1973). The time of appearance of particular species is highly predictable, based on light levels (Hobson and Chess 1973; Gladfelter 1979; McFarland et al. 1979). Not only species but size classes and maturity states within species have readily distinguishable migration patterns (Fishelson et al. 1971; Ogden and Ehrlich 1977).

Other less spectacular migrations are associated with daily spawning patterns. The wrasse *Thalassoma bifasciatum*, for example, spawns every day (Warner et al. 1975). Terminal-phase males break off feeding activities and move directly over considerable distances to a favored spawning site where spawning activities are initiated at particular times each day (Warner and Hoffman 1980). Such spawning sites have been known to persist for long periods of time. Colin and

Clavijo (1978) showed that a daily spawning site originally described by Randall and Randall (1963) for *Sparisoma rubripinne* persisted for over 17 years.

Thus large numbers of nocturnal, diurnal and crepuscular species of coral reef fishes migrate daily. These migrations are spatially and temporally predictable and are an important part of the reef ecosystem. In contrast to the attention given to the timing and ecological role of the migrations, the guidance mechanisms underlying them have been scarcely considered. The daily migrations of parrot-fish (Scaridae) between diurnal feeding areas and nocturnal resting areas (Ogden and Buckman 1973; Dubin and Baker 1982; Clavijo, unpublished data) are apparently guided in part by a time-compensating sun compass (Winn et al. 1964). With the exception of this study virtually nothing is known about the mechanisms guiding reef fish migrations. Waterman and Forward (1970) demonstrated a polarized light response in tropical halfbeaks (Hemirhamphidae), but it is not clear how these fish use polarization patterns.

MECHANISMS OF ORIENTATION AND NAVIGATION IN FISHES

In contrast to coral reef fishes the mechanisms of migration in other fishes and birds have been studied in some detail. These studies have generally fallen into two categories: observations of the movements of released individuals and analysis of the spontaneous or conditioned directional movements of animals in experimental areas.

A consideration of the vast bird literature (see reviews by Emlen 1975; Schmidt-Koenig 1979; Wallraff 1984--this volume) reveals two factors which may have been critical to success in this discipline. First, experimental research has included many species but has focused on relatively few. The indigo bunting (*Passerina cyanea*), European robin (*Erithacus rubicula*), and especially the homing pigeon (*Columba livia*) have been favored. In-depth understanding of orientation in a few species can then be extrapolated to species with similar migratory patterns. Second, the wide range of techniques used in laboratory and field research prevents undue emphasis on the results of any single technique or experiment and allows more complete understanding of the total process of orientation. This total process must include such aspects as the initial selection of direction, maintenance of direction, compensation for natural or artificial displacement, and termination of migration.

Though bird studies are more numerous than those with fishes, a considerable body of literature on fish orientation exists. Most numerous and relevant to this discussion are tracking studies and arena experiments. Hasler et al. (1958) were among the first to study compass orientation of fish in the field, using bobbers attached to the backs of white bass, *Morone chrysops*. Recently, tracking studies have tended to involve ultrasonic telemetry (reviewed by Stasko and Pincock 1977). In some cases multichannel transmitters relay information on swimming speed, depth, and ambient light and temperature (Standora and Nelson 1977) or compass heading (Harden Jones 1981), but usually the transmitters only indicate the position of the fish. Sonic-tracking studies tend to provide information on a small number of fish over a brief period. Transmitter size limits distance and duration of signal, but the studies are often limited by the endurance of the trackers and the distribution of the subjects. A serious drawback in most tracking studies is the lack of detailed, accurate information on the immediate environment of the fish. Studies of fish movements have seldom employed direct observation because the fish are generally too swift and the water too turbid to permit useful visual tracking.

The phenomenon of homing and home-site fidelity has been used in some studies designed to distinguish the cues used by fishes in orientation. Green and Fisher (1977) and Goff and Green (1978) looked at homing in a benthic stichaeid using normal, blind and anosmic fish. Homing appears to involve a complex set of olfactory and visual cues.

Arena tests of fish orientation have provided a fruitful way to overcome some of the problems of field work. Naturally- or artificially-conditioned sun-compass orientation has been indicated in mosquitofish (*Gambusia affinis*) by Goodyear (1973), in the starhead topminnow (*Fundulus notti*) by Goodyear (1970), in largemouth bass (*Micropterus salmoides*) by Loyacano et al. (1977) and bluegill (*Lepomis macrochirus*) by Goodyear and Bennett (1979).

In addition to the sun-compass studies various species have been shown to orient to light polarization patterns. Spontaneous orientataion to polarized light has been shown in hemirhamphids (Waterman and Forward 1970) and goldfish (*Carassius auratus*) (Kleerekoper et al. 1973). Conditioned orientation has been shown in a cichlid, *Pseudotropheus macrophthalmus* (Davitz and McKaye 1978) and in sockeye salmon (*Oncorhynchus nerka*) (Dill 1971).

Most of the above species have limited migratory tendencies, but Atlantic anguillid eels and Pacific salmon (*Oncorhynchus* spp.) have been the subjects of tracking and arena orientation studies. Adult silver European eels (ocean phase) (*Anguilla anguilla*) have been tracked with ultrasonic equipment (Tesch 1978; Westerberg 1979), but no consensus has been reached regarding guidance mechanisms. Compass orientation in the absence of solar cues has been indicated in adult American eels (*Anguilla rostrata*) (Miles 1968) and European eels (Tesch 1974), but attempts to establish electric or magnetic sensitivity in adult and juvenile eels have met with mixed success (see discussion in McCleave and Power 1978).

Perhaps the greatest amount of fish orientation research has been conducted on Pacific salmon. Tracking studies of adults in coastal areas generally revealed some passive individuals and some which were actively swimming homeward, usually along tidal axes (Stasko et al. 1973; Groot et al. 1975; Stasko et al. 1976). Besides tracking individual fish, salmon biologists have set up sonar stations to monitor the movements of yearling sockeye salmon leaving a large lake (Groot 1972). While all of these studies have provided useful information, they have the limitations of remote-sensing work described earlier. To date there appears to be no published report of adult salmon tracking on the high seas. The ocean distribution of salmon has been roughly mapped out by research fishing and mark-recapture work (Neave 1964; Royce et al. 1968), but the details of movement patterns are still unknown.

The tagging and tracking work in open water with adult salmon has not been complemented with arena orientation studies (except the work of Churmasov and Stepanov 1977, and Stepanov et al. 1979). However, the open-water orientation of juvenile salmon has been studied in some detail. The migration of sockeye salmon smolts to the outlet of Babine Lake, British Columbia, has been shown to be a directed movement by genetically distinct populations within the lake, cued by the sun's position, light-polarization patterns, and magnetic fields (Groot 1965; Simpson 1979; Quinn and Brannon 1982).

Young sockeye salmon not only show compass orientation when leaving their lake but also when entering it. Brannon (1972) showed that fry in experimental six-armed arenas would orient in the appropriate direction for their lake movements. These findings have been confirmed, and it is now clear that fry orient using a combination of celestial cues and the magnetic field in directions representing inherited adaptations to the geography of particular lake-river systems (Quinn 1980; Brannon et al. 1981).

Although the arena experiments have revealed a great deal about the sensory basis of juvenile salmon orientation, their interpretation has been hindered by the lack of detailed information on the pattern of salmon movements in the lakes and estuaries. While some progress has been made in this regard (in Lake Washington, for example, Woodey 1972), most of the sockeye salmon lakes are remote and largely unstudied. A second drawback in studying salmon is the unavailability of fish during much of the year. In a typical fry or smolt run most of the fish enter or leave their lake over a 2-3 week period. This greatly restricts the number of experiments which can be carried out, because it is difficult to catch the fish when they are in the lake itself, and in any case they would not be in the proper state of migratory motivation.

Briefly the current understanding of compass orientation by fishes is as follows. Several non-migratory species show a shoreline-related (Y-axis), predator-conditioned, sun-compass orientation. Two species of hemirhamphids, a cichlid, goldfish and sockeye salmon have been shown to orient using light-polarization patterns. However, the function of this response is not clear. A stingray, *Urolophus halleri*, has been conditioned to use magnetic fields for orientation (Kalmijn 1978), but this ability has not been related to migrations in this species. Compass orientation has been shown in eels, but the importance of electric, magnetic and other cues is not clear at this time. Compass orientation in juvenile Pacific salmon (especially sockeye) is now well established, and some progress has been made towards understanding the sensory inputs and genetic basis of the behavior. However, the short field season and lack of information on the behavior of eels and salmon at sea are drawbacks to understanding the role of compass orientation in ocean migrations. Thus the species whose migrations have been most intensively studied have, in many ways, been difficult ones, selected more for the spectacular nature of their migrations and their economic importance than for their convenience or appropriateness as models of fish migration.

CORAL REEF FISHES AS A MODEL SYSTEM FOR THE STUDY OF MECHANISMS OF ORIENTATION

Fish migration research has been unduly confined to tagging studies of large fish and controlled studies of small ones. The former reveal too little about guidance mechanisms, and the latter (too often conducted on non-migrants) can only establish the potential use of a particular cue, not its actual role in guiding the fish through the complex environment.

In order to understand the role of compass orientation and active navigation in the migrations of fish a battery of field and arena tests must be conducted on the same species. An ideal subject would be of modest size, display spatially and temporally regular migrations, be available for study during much of the year, have a well-known distribution pattern, and live in an area permitting displacement studies and direct observations. The feeding migration patterns of juvenile French grunts and white grunts (*Haemulon flavolineatum* and *H. plumieri*) appear to meet these criteria. These fishes are small (<15 cm total length), display relatively short but very regular migrations to and from feeding areas, are available almost all year, and can be readily studied in the field as well as in the laboratory. They can be easily collected on patch reefs, whose promixity to each other facilitates displacement experiments. The clear water makes it possible to observe the behavior of fish released after displacement or other experimental manipulations.

Diel Migrations of Grunts

Ogden and Ehrlich (1977) documented the existence of persistent diel migration routes in French and white grunts leading from patch reefs into surrounding seagrass beds in the Caribbean. The behavior on these routes was very regular and correlated with light level (McFarland et al. 1979). The routes followed were linear for 50 to 100 m, and then migrating groups fragmented in a dendritic pattern until single individuals remained at a particular location, feeding through the night on benthic invertebrates. In the morning individuals and then small groups apparently aggregated, returning with precision along the initial route at dawn (Ogden and Zieman 1977; Ogden and Ehrlich 1977). Mark-release-recapture experiments showed that grunts did not reaggregate randomly on reefs following nightly feeding forays but appeared to recognize a home reef and further a home schooling-site on a particular reef. Translocation experiments showed homing over distances ranging from 0.1-2.8 km from the home reef (Ogden and Ehrlich 1977).

Helfman et al. (1982) have shown that the development and persistence of diel migration routes in grunts is dependent upon social facilitation. Particular resting sites on coral reefs have persistent groups of juvenile and subadult grunts. These groups are joined continually by recruiting juveniles which have settled elsewhere and move gradually into the sub-adult schooling sites. Young fish have been shown in a series of observations and experiments to readily follow older, more experienced fish and thus the persistence of the routes appears to be assured.

McFarland et al. (1979) felt that avoidance of predators provides the most obvious adaptive significance for these precisely staged and timed migrations. Predation surrounding coral reefs is intense at dawn and dusk, and predators gather at grunt migration routes before their departure. The migration takes place in a narrow range of light levels, and changes in body coloration and photo-mechanical movements in the retina affords the grunts a maximum visual advantage over their predators. Once clear of the immediate vicinity of the reefs, the predation rate drops, and by the time individual fish reach their feeding areas, night has interrupted any advantage of continued schooling.

A second and less thoroughly examined consequence of migration patterns is the partitioning of feeding areas by grunts. Given the presumed necessity for solitary feeding in grunts, the dendritic migration route pattern with the extraordinary stereotypy of behavior of the fish assures that individual fish will arrive in an isolated part of the seagrass bed to feed. Furthermore, Ogden and Zieman (1977) and M. Robblee (unpublished data) have shown that the migration routes from a series of patch reefs several hundred meters apart surrounded by seagrass beds tended to be non-overlapping.

Thus, a large foraging area may be effectively partitioned by large numbers of several species of feeding fishes.

Preliminary Experiments

J.C. Ogden (unpublished data) conducted a series of preliminary experiments on orientation mechanisms in juvenile grunts (5-10 cm long). Fish were trapped in a seagrass bed near a patch reef on their dawn migration routes at varying distances from their home reefs and were subsequently released and followed visually. Three kinds of releases took place at varying times following capture: 1) at several locations along the route; 2) within approximately 100 m of the route in various compass directions; and 3) in totally unfamiliar territory in a seagrass bed over 5 km from the home reef.

Fish released on the route immediately after capture at dawn continue their migration rapidly and directly back to their home reef regardless of the distance they were from the home reef when captured. If the migrating fish are delayed until approximately noon, they return to their home reef but more slowly and with some meandering. If the sun angle serves as a reference point for compass orientation, then the delay appears to result in a substitution of knowledge of local landmarks as the major cue to homing, although the apparent lack of landmarks in a seagrass bed is striking.

Fish released at a site displaced up to 100 m from the route just after dawn capture move in the compass direction they were taking when captured and then meander, in most cases eventually arriving at the home reef. Fish released near noon tend to meander and generally do not reach their home reef. These observations indicate that fish appear to recognize their routes. On several occasions fish would approach the route from the side and immediately turn in the appropriate direction as they crossed it. The routes are through thick waving seagrass and are not perceptible to a human observer.

Fish trapped at dawn, displaced rapidly to an unfamiliar seagrass bed 5 km to the west, and immediately released, follow the compass course appropriate to their normal route over a distance approximately equal to the distance at which they were trapped from the patch reef and then meander. Fish released later in the day meandered, either returning to the release site or moving slowly away before being lost to sight.

FUTURE WORK

These preliminary experiments demonstrate some of the possibilities of observation and manipulation that may be used in the study of orientation in coral reef fishes. While the scale of these movements is small when compared to the long distance movements of other coastal and open ocean species, the generality of, for instance, a sun compass in fish may allow studies in such a model system to be extended to other species.

The routes of grunts, featureless to our eyes, are clearly perceived by them. Some form of socially facilitated learning is implied in the persistence of routes over long periods of time. New recruits readily follow older school members and learn the parameters of what likely amounts to a "map"--combining a compass sense with visual and possibly olfactory landmarks. The ontogeny of development of this map could conceivably be studied using observation and manipulation of grunts of different ages. The results of releases far from the home reef suggest that individual fish have a "time-distance" sense. This is implied also in the repeated observation of particular tagged fish on particular sections of seagrass bed following evening migrations. The destination of migrating individuals within the route system is apparently not random. Finally, homing over long distances following displacement occurs in grunts as in many other species. The results of studies on such a model system using a coral reef fish will not only lead to an understanding of the mechanisms orienting reef fish migrations, but also to the generation of testable hypotheses of the orientation mechanisms guiding fish migrations in general.

ACKNOWLEDGEMENTS

P. Ehrlich, J. Llibre, and I. Clavijo contributed greatly to preliminary studies of orientation in haemulids. J. Ogden has profited from a long association at West Indies Laboratory with numerous students of coral reef fishes, particularly E. Birmingham, E. Brothers, R. Clarke, I. Clavijo, J. Ebersole, J. Fallows, E. Gladfelter, W. Gladfelter, G. Helfman, P. Lobel, W. McFarland, M. Robblee, M. Shulman, and N. Wolf. Portions of the work were supported by the National Science Foundation (grant number OCE-7601304) and the National Oceanographic and Atmospheric Administration, Caribbean Undersea Research Program. This is Contribution Number 95 of the West Indies Laboratory, Fairleigh Dickinson University.

REFERENCES

Brannon, E.L. 1972. Mechanisms controlling migration of sockeye salmon fry. International Pacific Salmon Fisheries Commission Bulletin 21.

Brannon, E.L., T.P. Quinn, G.L. Lucchetti, and B.D. Ross. 1981. Compass orientation of sockeye salmon fry from a complex river system. Canadian Journal of Zoology 59:1548-1553.

Brothers, E.B., and W.N. McFarland. 1981. Correlations between otolith microstructure, growth, and life history transitions in newly recruited French grunts (*Haemulon flavolineatum* (Desmarest), Haemulidae). Rapports et Procès-Verbaux des Réunions Conseil Internationl pour l'Exploration de la Mer 178:369-374.

Colin, P.L., and I.E. Clavijo. 1978. Mass spawning by the spotted goatfish, *Pseudupeneus maculatus* (Bloch) (Pisces: Mullidae). Bulletin of Marine Science 28:780-782.

Churmasov, A.V., and A.S. Stepanov. 1977. Sun orientation and guideposts of the humpback salmon. Soviet Journal of Marine Biology 2(5):55-63. Translated 1978. Plenum Publishing Corporation, New York, New York, USA.

Davitz, M.A., and K.R. McKaye. 1978. Discrimination between vertically and horizontally polarized light by the cichlid fish *Pseudotropheus macrophthalmus*. Copeia 1978:333-334.

Dill, P.A. 1971. Perception of polarized light by yearling sockeye salmon (*Oncorhynchus nerka*). Journal of the Fisheries Research Board of Canada 28:1319-1322.

Domm, S.B., and A.J. Domm. 1973. The sequence of appearance at dawn and disappearance at dusk of some coral reef fishes. Pacific Science 27:128-135.

Dubin, R.E., and J.D. Baker. 1982. Two types of cover-seeking behavior at sunset by the princess parrotfish *Scarus taeniopterus* at Barbados, West Indies. Bulletin of Marine Science 32:572-583.

Ehrlich, P.R. 1975. The population biology of coral reef fishes. Annual Review of Ecology and Systematics 6:211-247.

Emlen, S.T. 1975. Migration: orientation and navigation. Pages 129-219 *in* D.S. Farner and J.R. King, editors. Avian biology, volume 5. Academic Press, New York, New York, USA.

Fishelson, L., D. Popper, and N. Gunderman. 1971. Diurnal cyclic behavior of *Pempheris oualensis* Cuv. & Val. (Pempheridae, Teleostei). Journal of Natural History 5:503-506.

Gladfelter, W.B. 1979. Twilight migrations and foraging activities of the Copper Sweeper *Pempheris schombergki* (Teleostei: Pempheridae). Marine Biology 50:109-119.

Gladfelter, W.B., J.C. Ogden, and E.H. Gladfelter. 1980. Similarity and diversity among coral reef fish communities: A comparison between tropical Western Atlantic (Virgin Islands) and tropical Central Pacific (Marshall Islands) patch reefs. Ecology 61:1156-1168.

Goff, G.P., and J.M. Green. 1978. Field studies of the sensory basis of homing and orientation to the home site in *Ulvaria subbifurcata* (Pisces: Stichaeidae). Canadian Journal of Zoology 56:2220-2224.

Goodyear, C.P. 1970. Terrestrial and aquatic orientation in the starhead topminnow, *Fundulus notti*. Science (Washington DC) 168:603-605.

Goodyear, C.P. 1973. Learned orientation in the predator avoidance behavior of mosquitofish, *Gambusia affinis*. Behaviour 45:191-224.

Goodyear, C.P., and D.H. Bennett. 1979. Sun-compass orientation of immature bluegill. Transactions of the American Fisheries Society 108:555-559.

Green, J.M., and R. Fisher. 1977. A field study of homing and orientation to the home site in *Ulvaria subbifurcata* (Pisces: Stichaeidae). Canadian Journal of Zoology 55:1551-1556.

Groot, C. 1965. On the orientation of young sockeye salmon (*Oncorhynchus nerka*) during their seaward migration out of lakes. Behaviour Supplement 14:1-198.

Groot, C. 1972. Migration of yearling sockeye salmon (*Oncorhynchus nerka*) as determined by time-lapse photography of sonar observations. Journal of the Fisheries Research Board of Canada 29:1431-1444.

Groot, C., K. Simpson, I. Todd, P.D. Murray, and G.A. Buxton. 1975. Movements of sockeye salmon (*Oncorhynchus nerka*) in the Skeena River estuary as revealed by ultrasonic tracking. Journal of the Fisheries Research Board of Canada 32:233-242.

Harden Jones, F.R. 1981. Fish migration: strategy and tactics. Pages 136-165 *in* D.J. Aidley, editor. Animal migration. Cambridge University Press, Cambridge, England.

Hasler, A.D., R.M. Horrall, W.J. Wisby, and W. Braemer. 1958. Sun-orientation and homing in fishes. Limnology and Oceanography 3:353-361.

Helfman, G.S. 1978. Patterns of community structure in fishes: summary and overview. Environmental Biology of Fishes 3:129-148.

Helfman, G.S., J.L. Meyer, and W.N. McFarland. 1982. The twilight migration patterns in grunts (Pisces: Haemulidae). Animal Behaviour 30:317-326.

Hobson, E.S. 1972. Activity of Hawaiian reef fishes during the evening and morning transitions between daylight and darkness. U S National Marine Fisheries Service Fishery Bulletin 70:715-740.

Hobson, E.S. 1973. Diel feeding migrations of tropical reef fishes. Helgolander Wissenschaftliche Meeresuntersuchungen 24:361-370.

Hobson, E.S., and J.R. Chess. 1973. Feeding oriented movements of the atherinid fish *Pranesus pinguis* at Majuro Atoll, Marshall Islands. U S National Marine Fisheries Service Fishery Bulletin 71:777-786.

Johannes, R.E. 1978. Reproductive strategies of coastal marine fishes in the tropics. Environmental Biology of Fishes 3:65-84.

Kalmijn, A. 1978. Experimental evidence of geomagnetic orientation in elasmobranch fishes. Pages 347-353 *in* K. Schmidt-Koenig and W.T. Keeton, editors. Animal migration, navigation and homing. Springer-Verlag, Berlin, Federal Republic of Germany.

Kleerekoper, H., J.H. Matis, A.M. Timms, and P. Gensler. 1973. Locomotor response of the goldfish to polarized light and its e-vector. Journal of Comparative Physiology 86:27-36.

Loyacano, H.A., J.A. Chappell, and S.A. Gauthreaux. 1977. Sun-compass orientation of juvenile largemouth bass, *Micropterus salmoides*. Transactions of the American Fisheries Society 106:77-79.

McCleave, J.D., and J.H. Power. 1978. Influence of weak electric and magnetic fields on turning behavior in elvers of the American eel *Anguilla rostrata*. Marine Biology 46:29-34.

McFarland, W.N. 1979. Observations on recruitment in haemulid fishes. Proceedings of the Gulf and Caribbean Fisheries Institute 32 (1979):132-138.

McFarland, W.N., J.C. Ogden, and J.N. Lythgoe. 1979. The influence of light on twilight migrations of grunts. Environmental Biology of Fishes 4:9-22.

Miles, S.G. 1968. Laboratory experiments on the orientation of the adult American eel, *Anguilla rostrata*. Journal of the Fisheries Research Board of Canada 25:2143-2155.

Neave, F. 1964. Ocean migrations of Pacific salmon. Journal of the Fisheries Research Board of Canada 21:1227-1244.

Ogden, J.C., and N.S. Buckman. 1973. Movements, foraging groups, and diurnal migrations of the striped parrotfish *Scarus croicensis* Bloch (Scaridae). Ecology 54:589-596.

Ogden, J.C., and P.R. Ehrlich. 1977. The behavior of heterotypic resting schools of juvenile grunts (Pomadasyidae). Marine Biology 42:273-280.

Ogden, J.C., and J.P. Ebersole. 1981. Scale and community structure of coral reef fishes: A long-term study of a large artificial reef. Marine Ecology Progress Series 4:97-103.

Ogden, J.C., and J.C. Zieman. 1977. Ecological aspects of coral reef-seagrass bed contacts in the Caribbean. Proceedings: Third International Coral Reef Symposium, Biology I:377-382.

Quinn, T.P. 1980. Evidence for celestial and magnetic compass orientation in lake migrating sockeye salmon fry. Journal of Comparative Physiology A. Sensory, Neural and Behavioral Physiology 137A:243-248.

Quinn, T.P., and E.L. Brannon. 1982. The use of celestial and magnetic cues by orienting sockeye salmon smolts. Journal of Comparative Physiology A. Sensory, Neural and Behavioral Physiology 147A:547-552.

Randall, J.E., and H.E. Randall. 1963. The spawning and early development of the Atlantic parrotfish *Sparisoma rubripinne*, with notes on other scarid and labrid fishes. Zoologica 48:49-60.

Royce, W.F., L.S. Smith, and A.C. Hartt. 1968. Models of oceanic migrations of Pacific salmon and comments on guidance mechanisms. U S National Marine Fisheries Service Fishery Bulletin 66:441-462.

Sale, P.F. 1977. Maintenance of high diversity in coral reef fish communities. American Naturalist 111:337-359.

Sale, P.F. 1980. The ecology of fishes on coral reefs. Oceanography and Marine Biology Annual Review 18:367-421.

Schmidt-Koenig, K. 1979. Avian orientation and navigation. Academic Press, London, England.

Simpson, K.S. 1979. Orientation differences between populations of juvenile sockeye salmon. Canada Fisheries and Marine Service Technical Report 717.

Smith, C.L. 1972. A spawning aggregation of Nassau grouper *Epinephelus striatus* (Bloch). Transactions of the American Fisheries Society 101:257-261.

Standora, E.A., and D.R. Nelson. 1977. A telemetric study of the behavior of free-swimming Pacific angel sharks, *Squatina californica*. Bulletin of the Southern California Academy of Sciences 76:193-201.

Starck, W.A., and W.P. Davis. 1966. Night habits of fishes of Alligator Reef, Florida. Ichthyologica: The Aquarium Journal 38:313-356.

Stasko, A.B., R.M. Horrall, and A.D. Hasler. 1976. Coastal movements of adult Fraser River sockeye salmon (*Oncorhynchus nerka*) observed by ultrasonic tracking. Transactions of the American Fisheries Society 105:64-71.

Stasko, A.B., R.M. Horrall, A.D. Hasler, and D. Stasko. 1973. Coastal movements of mature Fraser River pink salmon (*Oncorhynchus gorbuscha*) as revealed by ultrasonic tracking. Journal of the Fisheries Research Board of Canada 30:1309-1316.

Stasko, A.B., and D.G. Pincock. 1977. Review of underwater biotelemetry, with emphasis on ultrasonic techniques. Journal of the Fisheries Research Board of Canada 34:1261-1285.

Stepanov, A.S., A.V. Churmasov, and S.A. Cherkashin. 1979. Migration direction finding by pink salmon according to the sun. Soviet Journal of Marine Biology 5(2):92-99. Translated 1980. Plenum Publishing Corporation, New York, New York, USA.

Tesch, F.W. 1974. Influence of geomagnetism and salinity on the directional choice of eels. Helgoländer Wissenschaftliche Meeresuntersuchungen 26:383-395.

Tesch, F.W. 1978. Telemetric observations on the spawning migration of the eel (*Anguilla anguilla* L.) in different shelf areas of the Northeast Atlantic. Rapports et Proces-Verbaux des Reunions Conseil International pour l'Exploration de la Mer 174:104-114.

Wallraff, H. 1984. Migration and navigation in birds: A present-state survey with some digressions to related fish behaviour. Pages 000-000 *in* J.D. McCleave, G.P. Arnold, J.J. Dodson and W.H. Neill, editors. Mechanisms of migration in fishes. Plenum Press, New York, New York, USA.

Warner, R.R., and S.G. Hoffman. 1980. Population density and the economics of territorial defense in a coral reef fish. Ecology 61:772-780.

Warner, R.R., D.R. Robertson, and E.G. Leigh. 1975. Sex change and sexual selection. Science (Washington DC) 170:633-638.

Waterman, T.H., and R.B. Forward. 1970. Field evidence for polarized light sensitivity in the fish *Zenarchopterus*. Nature (London) 228:85-87.

Westerberg, H. 1979. Counter-current orientation in the migration of the European eel. Rapports et Procès-Verbaux des Réunions Conseil International pour l'Exploration de la Mer 174:134-143.

Winn, H.E., M. Salmon, and N. Roberts. 1964. Sun-compass orientation by parrotfishes. Zeitschrift für Tierpsychologie 21:798-812.

Woodey, J.C. 1972. Distribution, feeding, and growth of juvenile sockeye salmon in Lake Washington. Doctoral dissertation, University of Washington, Seattle, Washington, USA.

MIGRATION OF THE "NORTHERN" ATLANTIC COD AND THE MECHANISMS INVOLVED

W. H. Lear

Department of Fisheries and Oceans
Fisheries Research Branch
St. John's, Newfoundland A1C 5X1 Canada

J. M. Green

Marine Sciences Research Laboratory
Memorial University of Newfoundland
St. John's, Newfoundland A1B 3X5 Canada

ABSTRACT

Northern (Labrador-eastern Newfoundland shelf) Atlantic cod (*Gadus morhua*) spawn in March-June at depths in excess of 250 m and at bottom temperatures near 3 C. The eggs develop as they drift in the Labrador Current to shallow nursery grounds 600-1,000 km to the south. Spent adults also move southward and westward toward the coast of Labrador and Newfoundland, perhaps homing to the same areas where they grew up. The shoreward migration, through a layer of colder (-1.5 to 0 C) water, may be guided by migrating capelin (*Mallotus villosus*) on which the cod feed. Adults are joined by fish maturing for the first time in a fall migration back to the spawning grounds.

Better understanding of these extensive migrations requires research on the factors that trigger migratory activity and guide the migrations. Particularly appropriate are studies of endocrine involvement in the cod's migratory responses, the role of capelin in the shoreward migration of cod, and movements of tagged and telemetered cod in relation to thermal fronts and stratification in the water column.

INTRODUCTION

The Atlantic cod (*Gadus morhua*) of the Northwest Atlantic Fisheries Organization (NAFO) Divisions 2J, 3K, and 3L (Fig. 1), commonly referred to as "northern cod," are managed as a stock complex. The management unit extends from north of Cape Harrison, Labrador, to Cape St. Mary's, Newfoundland. This stock complex is composed of several major stocks and smaller stocks as evidenced from tagging experiments (Lear 1982; Templeman 1974, 1979) and other studies (Templeman 1962a,b, 1966).

Cod off Labrador (Subarea 2) spawn mainly in March-April in deep water on the continental slopes with some spawning occurring in May (Templeman 1981). In Division 3K (northeastern Newfoundland) spawning occurs on the outer continental slopes, mainly in March-April with possibly some additional spawning in May or June in the colder deep channels close to the coast. Further south in 3L, on the northern and northwestern slope of the Grand Bank, cod spawn mainly in April-June. Spawning may be delayed in some years when unfavorably low temperatures extend deeper than usual. This is known to have occurred in 1961 (Templeman 1962a) and 1971 (Dias 1972). Spawning of cod in this stock complex generally occurs at depths in excess of 250 m and at bottom temperatures of about 3 C.

Based upon developmental periods from fertilization to hatching reported by Apstein (1909), Templeman (1981) estimated that cod eggs spawned off northern Labrador (Div. 2G) in March-April would take 50-60 days to hatch at surface temperatures of -1.5 C to 0 C and those spawned on the slopes of Hamilton Bank (Div. 2J) in March-April about 40 days at temperatures of -1 to 1 C. At an average speed of about 16 km/day for the Labrador Current (Iselin [1930], Smith [1931], Killerich [1939], Buzdalin and Elizarov [1962], Templeman [1966], Serebryakov [1967] and Postolaky [1975]) eggs spawned off northern Labrador would drift 800-960 km southward to the area off southern Labrador and those from the slopes of Hamilton Bank would drift about 640 km to the northern Grand Bank and the Avalon Peninsula (Templeman 1981). These estimates are consistent with the appearance of cod larvae in May off mid-Labrador and in April and May off southern Labrador (Serebryakov 1967).

During the southward drift over the northeast Newfoundland shelf the cod larvae are transported by the shoreward sweep of the Labrador Current resulting from the Coriolis force. Larvae, brought into bays and inlets of eastern Newfoundland, are found as juveniles during September-October at modal lengths of 6-9 cm with some specimens as small as 3-5 cm long (Fleming 1963). On the basis of average vertebral numbers it has been demonstrated by Lear and Wells (1982) that these larval cod are most likely the progeny of cod which spawned much farther north.

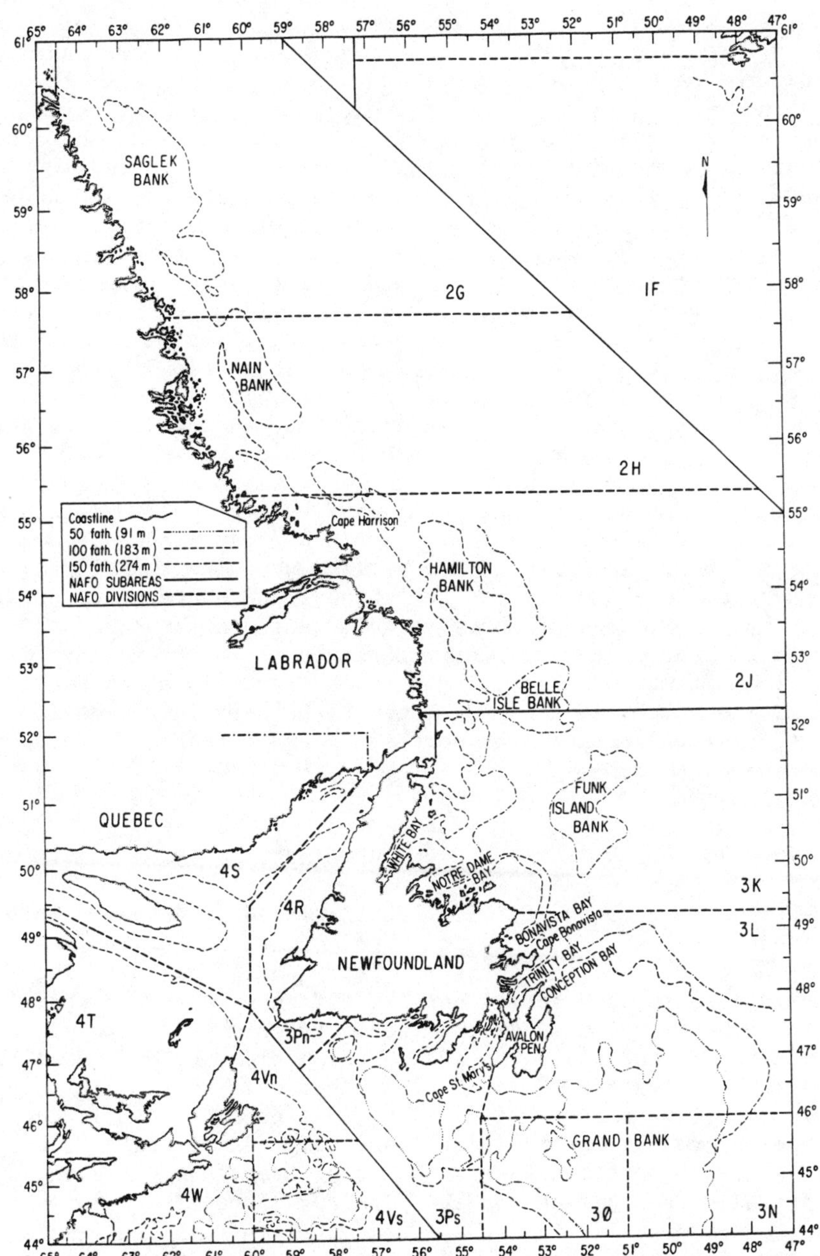

Figure 1. Area map showing NAFO Divisions and place names mentioned in the text.

After spawning is completed, schools of spent adult cod begin to disperse, probably during April and May, and move towards the coast. However, the main coastward migration occurs in June when the cod of the Labrador-East Newfoundland stock (Divs. 2J, 3K, 3L) move southward and westward toward the Labrador and Newfoundland coasts. This coincides with the spawning migration of capelin (*Mallotus villosus*) to shallow waters (Templeman 1979). It has been postulated that these adult cod are homing to the general areas where they grew up as immature fish (Templeman 1979). Adult cod remain in the inshore area until about October when offshore migration again begins; they congregate during November-December on the outer slopes of the continental shelf in the warmer, deeper waters remaining there to spawn during the winter and spring. Joining also in the migration at this time are cod maturing for the first time. It has been postulated that these cod are migrating back to the spawning grounds from which they were produced several years previously.

Given the nature of these long migrations of cod extending over several hundred kilometres, there remain many questions regarding factors which trigger migratory activity and guide the migrations. A successful inshore cod fishery in Labrador and eastern Newfoundland depends upon knowledge of those factors which cause annual variations in the migration patterns. Our understanding of the factors will enable the optimum benefit to be derived from this stock complex. The following sections describe some of the gaps in our basic understanding of migration of northern cod and possible approaches to resolution of them.

ENDOCRINE REGULATION OF MATURATION AND MIGRATION

Woodhead and Woodhead (1965) demonstrated a seasonal cycle of thyroid activity in adult and immature cod of the Barents Sea. In adults the thyroid became active at the start of the spawning migration and remained active until the fish reached the spawning grounds. The duration of thyroid activity coincided with migration. In immature cod the gland was active throughout the period of their overwintering migration. It is highly probable that increased thyroid activity is indicative of internal changes stimulating the migration of cod to the northern spawning grounds. Factors which govern activities of the thyroid and other regulatory organs, such as the adrenal cortex tissue and the pituitary, are less clearly understood, especially the role of external stimuli such as light intensity, seasonal variation of the photoperiod, rate of change of light intensity and temperature in cueing maturation and migration.

Ideally cod should be studied under varying environmental conditions to determine the effects of external stimuli on thyroid and related endocrine activity that may be involved in sexual maturation and orientation of cod to currents. Additionally cod endocrine tissue (thyroid, pituitary, and interrenal gland) should be studied to determine the seasonal cycle of activity of these organs. Only when we know more about internal controlling mechanisms of migration might we begin to predict the scope of the migration from year to year. This has practical application in that cod fail to appear along the coast in some years, and it should be determined if these failures are induced by external barriers to migration, such as temperature, or by decreased physiological motivation to migrate, induced by internal states during the feeding period.

VARIABILITY OF THE INSHORE MIGRATION ROUTE

Present knowledge indicates that spent cod from several overwintering areas migrate to specific geographic areas, and for some stocks the summer distributions cover a wide area, while for others it is somewhat more restricted (Lear 1982). At present we do not understand the role of offshore and inshore oceanographic structures as barriers to inshore migrations, whether the relative abundance of capelin inshore is important in "attracting" cod, or how the availability of juvenile capelin and other offshore resources might influence the inshore feeding migration of cod. Two strategies which the cod may adopt on their inshore migration are:

1. Cod swim shorewards in the upper surface layer (0-50 m) where during May off Trinity Bay the temperatures are generally 0-5 C. In this case the cod would have to leave an environment of 2-4 C if they migrated vertically immediately after spawning or 0-4 C if they dispersed westwards and southwards from the spawning area prior to migrating vertically through the cold layer (50-500 m) at temperatures of 0 to -1.5 C.

2. Cod swim in over the banks under the cold intermediate layer until they reach a depth of about 150-200 m, near the coast, where they encounter the cold-water barrier (0 to -1.5 C). If capelin were in the water column immediately above them, they might ascend into cold water in pursuit and follow the capelin migration into the inshore area in the upper layers.

These alternative hypotheses could be examined by using sonic tags in conjunction with a series of acoustic surveys during the migration period. This would be a challenging exercise but possibly a rewarding one. It would provide information on the vertical location of cod (on the bottom, midwater, near the surface) if done in conjunction with a few bottom or midwater net sets at selected depths based on sounder surveys to positively identify cod schools. Conventional tagging of the cod from the net catches could demonstrate whether the fish are from schools of shoreward-migrating cod. Collection of hydrographic data throughout the water column and specifically at the school depth would provide information on temperature at migration depths. Similarly, identification of migrating capelin could clarify the place and time at which cod encounter capelin. Are the cod led inshore by their pursuit of capelin, or do the two species just encounter each other along the way?

MIGRATION FROM NURSERY AREAS TO SPAWNING GROUNDS

Based on their high vertebral counts, the juvenile cod on the northeast coast of Newfoundland apparently are progeny of the Division 2J, 3K and 3L cod stock. It cannot be established from which component group they came (Funk Island Bank, Hamilton Bank, Belle Isle Bank, etc.). The eggs and larvae are postulated to have drifted southward with southerly currents and then shoreward by Coriolis deflection of this current westward onto the east coast of Newfoundland. These juvenile cod are currently being tagged inshore in the nursery areas to test the hypothesis that young cod, spawned offshore, using bays as nursery areas, return to spawn in areas from which they originated. These experiments will not tell us precisely that a cod larva hatched on Hamilton Bank and reared in Trinity Bay will upon maturity migrate back to Hamilton Bank, but in a general way the recapture data will enable us to make fairly obvious inferences.

REFERENCES

Apstein, C. 1909. Die Bestimmung des alters pelagisch lebender Fischeier. Mitteilungen des Deutschen Seefischerei - Vereins. 25:364-373.

Buzdalin, Y.I., and A.A. Elizarov. 1962. Hydrological conditions in the Newfoundland banks and Labrador areas in 1960. Pages 152-168 *in* Soviet Fisheries Investigations in the Northwest Atlantic, VNIRO-PINRO Moscow, USSR. (Translation for National Science Foundation, Washington, DC, by Israeli Program for Scientific Translation 1963.)

Dias, M.L. 1972. Portuguese research report, 1971. International Commission for the Northwest Atlantic Fisheries Redbook 1972 Part II: 71-78.

Fleming, A.M. 1963. Baby cod survey, 1962. Fisheries Research Board of Canada, Biological Station, St. John's, Newfoundland Report for 1962-63. Appendix number 5.

Iselin, C. 1930. A report on the coastal waters of Labrador, based on exploration by the Chance during the summer of 1926. Proceedings of American Academy of Arts and Science 66:1-37.

Killerich, A.B. 1939. A theoretical treatment of the hydrographical observation material. Godthaab Expedition, 1928. Meddelelser om Grønland 78(5):1-149.

Lear, W.H. 1982. Discrimination of the cod stock complex in Division

2J+3KL, based on tagging. Northwest Atlantic Fisheries Organization SCR Document 82/IX/89, Series Number N598.

Lear, W.H., and R. Wells. 1982. Vertebral averages of juvenile cod (*Gadus morhua*) from eastern Newfoundland and Labrador as indicators of stock origin. Northwest Atlantic Fisheries Organization SCR Document 82/IX/75, Series number N581.

Postolaky, A.I. 1975. Distribution and abundance of cod eggs in the South Labrador and Newfoundland areas in 1974. International Commission for the Northwest Atlantic Fisheries Research, Research Document Number 101, Series number 3593.

Serebryakov, V.P. 1967. Cod reproduction in the Northwest Atlantic. Trudy Polar Research Institute of Marine Fisheries and Oceanography (PINRO) 20:205-242. (Fisheries Research Board of Canada Translation Series Number 1133, 1968.)

Smith, E.H. 1931. Arctic ice, with special reference to its distribution in the North Atlantic ocean. The Marian Expedition to Davis Strait and Baffin Bay, 1928. Scientific results. Bulletin of the United States Coast Guard, No. 19(3):1-221.

Templeman, W. 1962a. Canadian research, 1961. A. Subareas 2 and 3. International Commission for the Northwest Atlantic Fisheries Redbook Part II:3-20.

Templeman, W. 1962b. Divisions of cod stocks in the Northwest Atlantic. International Commission for the Northwest Atlantic Fisheries Redbook Part III:79-129.

Templeman, W. 1966. Marine resources of Newfoundland. Bulletin of the Fisheries Research Board of Canada number 154.

Templeman, W. 1974. Migrations and intermingling of Atlantic cod (*Gadus morhua*) stocks of the Newfoundland area. Journal of the Fisheries Research Board of Canada 31:1073-1092.

Templeman, W. 1979. Migrations and intermingling of stocks of Atlantic cod, *Gadus morhua* of the Newfoundland and adjacent areas, 1947-71 and their use for delineating cod stocks. Journal of the Northwest Atlantic Fishery Science 2:21-45.

Templeman, W. 1981. Vertebral numbers in Atlantic cod, *Gadus morhua*, of the Newfoundland and adjacent areas, 1947-71, and their use for delineating cod stocks. Journal of Northwest Atlantic Fishery Science 2:21-45.

Woodhead, A.D., and P.M.J. Woodhead. 1965. Seasonal changes in the physiology of the Barents Sea cod, *Gadus morhua* L., in relation to its environment. I. Endocrine changes particularly affecting migration and maturation. International Commission for the Northwest Atlantic Fisheries Special Publication 6:691-715.

MECHANISMS OF FISH MIGRATION IN RIVERS

T.G. Northcote

Institute of Animal Resource Ecology
University of British Columbia
Vancouver, British Columbia V6T 1W5 Canada

ABSTRACT

The unidirectional flow of rivers provides a strong orientational cue and the confined channel provides fixed reference points for migrating fishes. Within a dendritic river system there may be great spatial and temporal variation in physical and chemical characteristics of the flow which though perhaps confounding might provide a means for branch recognition by migrants.

In rivers migration must ultimately involve both a downstream and an upstream component. The former is generally, but not always, a feature of early life-history stages and vice versa. Migration of a given life-history stage, however, may involve both downstream and upstream components. Downstream migration may be passive or active, but upstream migration must be active. Downstream migration of juveniles and adults is mainly nocturnal, but sometimes diurnal, especially in turbid rivers. Characteristically, maturing adults of riverine and anadromous species move upstream to spawn, but some species or populations may move downstream to spawn. Upstream migration may be nocturnal or diurnal.

Timing of migration in rivers depends on the physiological state of the fish, as influenced for example by thyroid and corticosteroid hormones, and external triggering factors. An environmental stimulus, such as water current, temperature or light, may alter fish orientation and act as a "director" of migration, or the stimulus may trigger movement or alter the intensity of movement and act as a "regulator" of migration. Celestial and magnetic cues seem to be involved in some migrations in river-lake systems, but use of landmarks has not been well documented. Orientation to water current is undoubtedly the most important "director" of upstream migration. Together with the co-role played by detection of unique stream odors and odors from conspecifics, it forms the most well-studied, and perhaps most important, guidance system for upstream migration, especially of anadromous salmonids. The precise nature of the odors, of the imprinting process (to both natural and synthetic substances), and of the orientation mechanism itself remain to be learned.

Population-specific, genetically determined factors may influence the responses to an environmental stimulus, e.g. the response to water current in fry of lake inlet and outlet spawning populations. The control mechanisms for riverine migrations are likely organized in sequences or hierarchies appropriate to the ontongeny and evolutionary history of a population. The adaptive significance of migration is enigmatic, but probably lies in the fact that optimal habitats for different functions (survival, growth, reproduction) are spatially, seasonally and onotogenetically separated.

INTRODUCTION

During the last decade there have been great advances in our understanding of animal migratory processes in general and of fish migration in particular. Nevertheless few researchers have focused specifically on mechanisms of fish migration in rivers or (perhaps wisely) have speculated on its adaptive significance.

Migration in this review is defined as those movements which result in an alternation between two or more separate habitats (i.e., a movement away from one habitat followed eventually by a return to it again) occurring with a regular periodicity (sometimes annual but certainly within the lifespan of an individual) and involving a large fraction of the population. Rivers include not only the largely surface-flowing inland waters but also the connecting lakes, marshes and other integral parts of watersheds from their uppermost elevations to sea level. Because many of the common fish migrants in rivers are anadromous or catadromous the riverine portion of their movements also will be covered.

RIVERS AS MEDIA FOR MIGRATION

The essential feature of a river with respect to fish migration is that it is a system of largely unidirectionally flowing water. Of course there are short term and small scale up-slope movements of water in backeddies or near the bottom because riverine flow is rarely if ever laminar (Hynes 1972). Because of the pervading directionality of current, fish movement of any appreciable length in a river must be either downstream or upstream. The flow can provide a strong orientational cue to migratory animals.

A second major attribute of rivers with respect to fish migration is that they usually flow in rather confined channels of relatively small vertical and horizontal dimensions. Again there are exceptions in areas of low gradient or near mouths of large rivers, but maximum depths of rivers are rarely more than a few metres and widths are rarely more than a few hundred metres. Therefore river-migrating fish are usually close to fixed reference points at the river edge or bottom, in contrast to those migrating in open oceans or continental shelves.

Another important feature of rivers of significance to migratory components of their fish communities is that they are highly branched systems which often may drain areas with considerable differences in physical and chemical characateristics. Therefore, upstream migrants can spread into a large number of tributaries often with very contrasting conditions whereas downstream migrants eventually reach one common point--the river mouth. Rivers also collect and transport downstream information on upstream features in their watersheds. Sediment, thermal and chemical plumes from tributaries may be carried surprising distances downstream before becoming thoroughly mixed with the main river discharge. Such plumes may serve as "trails" to guide upstream migrants into river branches.

Rivers exhibit greater temporal variation in conditions than do coastal or oceanic waters. Large year to year variations occur in discharge and other physical features. Rivers draining into the northeastern Pacific Ocean commonly exhibit seasonal differences of a thousandfold between maximum and minimum flow. In addition there often are marked day to day as well as diel changes in discharge, turbidity, temperature, water chemistry and other characteristics which probably make it difficult for river migrants to time movements to reliably repeated temporal characteristics.

Rivers provide a relatively simple means for fish movement by passive drift in one direction, downstream. In fact many rivers which contain rapids or waterfalls permit only downstream movement in some sections. Other rivers with large, diffuse marshes or with dams may restrict downstream as well as upstream movement at certain locations along their length.

MODES OF MIGRATION

Fish migration in rivers at some stage in the life history of the migrant must involve both downstream and upstream movement to complete the cycle. The migratory sequence depends on (1) the life-history stage being considered, (2) the location of the migrant and (3) the type of migration to be undertaken, i.e., feeding, refuge-

"seeking" or reproductive. Nevertheless it is convenient to group the patterns of migration into two major categories with respect to direction (downstream or upstream) and two subcategories with respect to life history stages (early--eggs, larvae, fry, juveniles, and late --subadults, adults, spawners, postspawners).

Downstream Migration

Early life-history stages. In some riverine fishes the downstream phase of migration begins with the fertilized egg as passive drift in the water current. Such must be the case in river-spawning populations of goldeye (*Hiodon alosoides*), a species with nonadhesive, semibuoyant eggs (McPhail and Lindsey 1970). Fertilized eggs of the shad *Hilsa ilisha* drift several hundred kilometers down Indian rivers to tidal reaches where the hatchlings thrive and grow (Ganapati 1973). Semipelagic eggs of the South American characin *Prochilodus scrofa* are swept down the River Mogi Guassu to temporary ox-bow rearing areas (Bayley 1973). The African cyprinid *Labeo victorianus* deposits semibuoyant, unattached eggs among grasses at the edge of tributaries to Lake Victoria, and these may be swept into midstream and transported downstream to the lake (Whitehead 1959).

Many fishes which spawn in rivers have adhesive eggs (e.g., all osmerids, many cyprinids and catostomids and some coregonids), so drift is delayed until hatching when the larvae are carried downstream to rearing habitats (river backwaters, lakes, lower reaches and estuaries or the sea). For those species that deposit and cover the eggs in nests on the river bottom, downstream movement is further delayed at least until the emergence of late alevins or early fry. Recently hatched larvae of the Japanese ayu (*Plecoglossus altivelis*) are carried down rivers to the estuary as minute larvae (Okada 1955) as are the eulachon (*Thaleichthys pacificus*) in the lower Fraser River of British Columbia.

Within the salmonids we find a broad spectrum of downstream migratory patterns (Northcote 1969a; Godin 1982). In some species such as pink and chum salmon (*Oncorhynchus gorbuscha*, *O. keta*) the fry migrate downstream almost immediately after emergence, and in others downstream movement may be delayed for several weeks or months after emergence as in chinook salmon (*O. tshawytscha*). In still others, while there may be an initial short phase of downstream movement in streams or rivers to reach the rearing habitat (mainstem rivers or lakes), the major downstream migration occurs one or more years after fry emergence and involves complex physiological changes associated with smolt transformation (Hoar 1976). Coho and sockeye salmon (*O. kisutch*, *O. nerka*), Atlantic salmon (*Salmo salar*) as well as anadromous forms of brown and rainbow trout (*S. trutta*, *S. gairdneri*) illustrate such migratory patterns. However, there is considerable stock plasticity, so young of some populations of sockeye salmon migrate down to the sea as fry, as do many young of some coho salmon populations (Skeesick 1970; Hartman et al. 1982). In some stocks of chinook salmon a large fraction of the young may rear for one or more years in upper mainstem rivers (Tutty and Yole 1978) or even lakes before starting seaward migration. Furthermore, in a single species such as rainbow trout various stocks illustrate the whole spectrum of downstream migration shown by the five species of eastern Pacific salmon (Northcote 1969a). Downstream movement of larvae or fry from upriver spawning areas also is a common migratory pattern in tropical rivers of South America (Bonetto et al. 1969; Bayley 1973), Africa (Whitehead 1959; Reynolds 1971; Welcomme 1969, 1975), and India (Rajyalakshmi 1973).

The early life-history stages of river-spawning fishes show similar diel patterns of downstream movement. Such movement, whether it occurs early at the alevin and recently emerged fry stage, or at later juvenile stages (parr, smolts), is usually nocturnal. Downstream movement of lamprey ammocoetes occurs mainly at night (Potter 1980), as does that of young rainbow trout (Northcote 1962, 1969a,b), cutthroat trout (*Salmo clarki*) (Shapley 1961), brown trout (Elliott 1966; Cuinat and Heland 1979; Heland 1980a,b), and Atlantic salmon (White 1939; Osterdahl 1969; Thorpe and Morgan 1978; Bakshtanskiy et al. 1980; Ruggles 1980) as well as pink salmon (McDonald 1960; Vernon 1966; Shershnev and Zhulkov 1979; Godin 1980), chum and sockeye salmon (McDonald 1960) and coho salmon (McDonald 1960; Chapman 1962; Ruggles 1966; Hartman et al. 1982). Similar nocturnal patterns also have been reported for juvenile cyprinids, catostomids, gobiids, clupeids and sturgeon (Lindsey and Northcote 1963; Geen et al. 1966; Pavlov et al. 1977). Exceptions to the usual nocturnal periodicity of young downstream migrants are found in highly turbid rivers. Pink salmon fry in the lower Fraser River move downstream during the day as well as at night (Vernon 1966), and over half of the daily catches of chum fry migrants in the Fraser can be taken during the morning (Todd 1966). Similarly in the highly turbid main channel of the Amur River over 90% of the young chum salmon move downstream during the day (Roslyj 1975). Although at times the downstream movement of Atlantic salmon and brown trout smolts in the River Piddle is nocturnal, apparently the majority descend diurnally during conditions of bright illumination (Solomon 1978a,b). Juvenile pike (*Esox lucius*) moving down a coastal stream in Sweden are said to follow a "day active rhythm" (Johnson and Müller 1978).

Most young migrants move downstream in mid- to near-surface waters away from the river banks (McDonald 1960; Todd 1966; Vernon 1966; Pavlov et al. 1977) where tactile association with fixed reference points on the bottom would be difficult.

Orientation of young fish moving down rivers is often said to be "random" as they drift passively in the water current (Scott and Crossman 1973; Davies and Thompson 1976; Solomon and Templeton 1976; Bakshtanskiy et al. 1980). This may be the case for buoyant larvae of the goldeye (McPhail and Lindsey 1970) and perhaps for the weakly swimming young larvae of many osmerids. However, there have been few attempts to directly observe such movement, no doubt because of its frequent nocturnal periodicity. For downstream movement of rainbow trout fry observed in streams by infra-red viewing, the majority were heading downstream and many were at or near the surface, swimming at a speed greater than the surrounding water as evidenced by wakes from the dorsal fin (Northcote 1962). Solomon (1978b) suggested that downriver progression of Atlantic salmon smolts in an English chalkstream was an active process not a passive displacement, but Thorpe et al. (1981) found evidence to the contrary in a Scottish river-reservoir system.

Subadults and adults. In strictly riverine fish populations, once the juveniles have established territories in feeding habitat, downstream movement of nonspawners seems to be minimal, although further "recruitment" movements, mainly downstream, may occur later (Solomon 1982). Other exceptions are found in environments where seasonal changes are so unfavourable that the fish must temporarily move to refuge habitats. Subadult and adult fish, particularly in arctic rivers, make summer feeding excursions downstream to the sea (Eddy and Lankester 1978).

Apart from eels (*Anguilla* spp.) there are other species such as inanga (*Galaxias maculatus*) (McDowall 1978), the threadfin chilka (*Eleutheronema tetradactyium*) (Kowtal 1972) and *Centropomus poeyi* (Chavez 1981) which at maturity move down rivers to spawn at mouths, estuaries or in some cases the sea (not inanga).

Surviving postspawners in streams typically move back downstream, as in the inconnu (*Stenodus leucichthys*) populations studied by Alt (1977), in the brown trout populations studied by Solomon and Templeton (1976) and by Huet and Timmermans (1979), in the Volga River sturgeon (Slivka and Dovgopol 1979), and in several species of freshwater fishes spawning in streams entering the oligohaline northern Bothnian Sea (Müller 1982).

Upstream Migration

Early life-history stages. Extensive upstream movement of young salmonids during stream-dwelling phases of their life history does not seem to be common (Bjornn and Mallet 1964; Bjornn 1971; Hartman et al. 1982) except for those populations emerging in lake outlet streams (Godin 1982). Upstream movement of juvenile brown trout following emergence is usually limited to small numbers and short distances (Solomon 1982). Nevertheless such movement does occur in coho salmon, even in recently emerged fry. Ruggles (1966) showed that coho upstream migrants took on residency to a greater degree than the fry which moved downstream upon emergence. Autumnal upstream movement of coho juveniles into tributary streams has been recorded by Skeesick (1970) and by Bustard and Narver (1975). In coastal British Columbia juvenile coho rear in small streams which do not support adult spawning runs so that the young must have moved upstream into these habitats. Stuart (1957) reported an example of yearling and older brown trout making definite spring upstream migrations from a lake into nursery streams. At Voss, western Norway, Atlantic salmon parr move up into small tributaries not used for spawning during spring and rear there over summer and autumn, but not winter when they return to Lake Vangsvatnet (B. Jonsson, personal communication). In fish whose adults move downstream at spawning time (e.g. the inanga) the young make a return upstream movement.

Subadults and adults. Characteristically in riverine fish populations maturing adults move upstream to spawn one or more years after the downstream movement of juveniles. Examples are provided by brook lamprey (*Lampetra planeri*) (Malmquist 1980), sturgeon (Slivka and Dovgopol 1979), inconnu (Alt 1977), brown trout (Solomon and Templeton 1976; Huet and Timmermans 1979) and the Mexican tetra (*Astyanax mexicanus*) (Edwards 1977). Frequently the return upstream movement of adults is considered to "compensate" for the earlier downstream displacement of young stages.

In contrast to the generally held view that salmonid spawners move steadily upstream at a fairly constant rate, Atlantic salmon adults tracked by radio telemetry (Power and McCleave 1980) showed much more complex and irregular patterns of movement with frequent reversals of direction combined with intermittent stops. Similar results for steelhead (*Salmo gairdneri*) have been reported by Spence (1980). However, internal radio tags themselves may cause delays and eventually downstream movement (e.g. in adult chinook salmon, Gray and Haynes 1979).

Just as seasonal patterns of upstream movement defy generalization, so too does its diel timing (Banks 1969). Some salmonids such as Atlantic salmon (Hayes 1953) and kokanee (*Oncorhynchus nerka*) (Lorz and Northcote 1965) start upstream progression at dusk and move mainly at night, whereas sockeye readily move up the clear Somass River during the day (Ellis 1962). Adult chum salmon migrate up the Chitose River, Japan principally in daylight (Mayama and Takahashi 1977), but discharge, turbidity and other factors affected the pattern (Mayama 1978).

Adult upstream movement in rivers may be associated with homing to feeding sites as in the arctic char (*Salvelinus alpinus*) (McBride 1980) and inconnu (Alt 1977) of Alaskan rivers or in brown trout to inlet tributaries of some Norwegian lakes (Jonsson 1981). Adults of anadromous arctic char (Eddy and Lankester 1978) in the Canadian polar region as well as cisco (*Coregonus artedii*) and lake whitefish (*C. clupeaformis*) of eastern James Bay (Morin et al. 1981) provide examples of upriver movements to over-wintering sites.

MECHANISMS OF MIGRATION

Physiological Preparation and Readiness

Control of the timing of migration depends on interaction between internal physiological state of the fish and external triggering factors in the environment (Hoar 1958, 1976; Fontaine 1975; Woodhead 1975; Meier and Fivizzani 1980). There is continuing evidence that thyroid hormones (Godin et al. 1974) and the corticosteroids are important not only in the preparation for migration but also in its orientation. Furthermore thyroid activity also may in part be timed by the lunar cycle (Grau 1982). Elevated levels of sodium and potassium adenosine triphosphate activity in gills of juvenile salmonids are associated with parr-smolt transformation and migratory tendency (Zaugg and McLain 1972; Giles and Vanstone 1976; Lorz and McPherson 1976; Ewing et al. 1979). However, recent work has shed some doubt on this relationship (Hart et al. 1981) indicating that the elevated activity may relate more to adaptation for seawater entry than to seaward migration (Ewing et al. 1980).

Environmental Directors and Regulators

Information from the environment received by the sensory system of a fish can act in two ways to affect its migratory behaviour. Firstly, an environmental stimulus such as water current, light or temperature may alter fish orientation and thereby act as a "director" of movement. Secondly, such stimuli may increase or decrease the intensity of movement, regardless of orientation, and so serve as a "regulator" of migration. Furthermore, these stimuli may operate in both ways and frequently in consort to bring about a complex control of migration.

Water current and discharge. The flow of water in rivers must in itself have an all-pervading influence on the orientation of fish (Arnold 1974) and on the intensity of their movement. The initiation and peaking of downstream movement of juvenile salmonids often has been associated with freshets (Bull 1931; Allen 1944; Svardson 1966; Churikov 1975; Pemberton 1976; Bagliniere 1979; Raymond 1979; Ewing et al. 1980). To what extent such correlations are the result of alterations in current response of the fish (negative rheotaxis), disturbance of the mechanisms used for maintenance of position (e.g., effects of increased water level on home "stations," increased turbidity) or "flushing-out" of fish in the appropriate physiological state is usually unknown. On the other hand, sharp reductions in discharge, perhaps by causing an alarm reaction, also may bring about increased movement of fish (Finnigan 1978; Johnson and Müller 1978;

Santos 1979). Furthermore, the directional response to water velocity of young may depend on their exact stage of development (Ottaway and Clarke 1981).

Upstream movement of fish, particularly of spawners, also has had a long history of proponents arguing for control mediated by freshets or associated conditions. For example, Whitehead (1959) noted that "flood factors" were indispensable to upstream spawning migrations in African rivers, and Ganapati (1973) associated the reduced migrations of *Hilsa* up Indian rivers with the lack of high discharges to the sea caused by dams or irrigation diversions. Upstream spawning migrations of trout are commonly thought to be induced by increases in river flow (Stuart 1957; Libovarsky 1976; Solomon and Templeton 1976), but Hellawell (1976) found no evidence for the concept of a discharge "gateway" (certain limits of flow within which most of the upstream movement occurs). High discharge may stimulate the upstream spawning migration of river lampreys (*Lampetra fluviatilis*) (Asplund and Sodergren 1974). The start of upstream movement of juvenile galaxids (whitebait) is said to be triggered by the first major flood of the spring (McDowall and Eldon 1980).

Water temperature. In addition to its effect in regulating intensity of fish movement when acting together with discharge changes, temperature often has been put forward as (1) the major or even sole orientating cue directing downstream migration as for example in that of silver-phase European eels (*Anguilla anguilla*) (Westin 1977; Westin and Nyman 1977), (2) a threshold stimulus for the initiation of downstream movement (Fried et al. 1978; Claridge and Gardner 1978; Raymond 1979), or (3) a day to day regulator of periodicity (Fried et al. 1978).

Commonly workers relating downstream fish movement in rivers to environmental conditions suggest that water-temperature increases act in combination with rises in discharge to stimulate egress (Bagliniere 1976; Nihouarn 1976; Solomon 1978b; Sopuck 1978; Raymond 1979; Potter 1980; Arawomo 1981), although downstream movement occurs sometimes during declining discharge but rising water temperature (Stauffer 1972; Alexander and MacCrimmon 1974). In other cases water temperature may set the broad seasonal timing of downstream movement, whereas freshets regulate its day to day fluctuation (Hartman et al. 1982). Coincidence of increases in discharge and water temperature may stimulate upstream movement in some rivers (Smith and Saunders 1958; Hartman et al. 1962; Campbell 1977; Robbins and MacCrimmon 1977; Barton 1980), but there are exceptions (Jellyman 1979; Malmquist 1980). Interactions between water temperature and photoperiod also may be involved in the mechanism of riverine fish migration (Northcote 1962; Dodson and Young 1977).

Light and photoperiod. Downstream movement of fish during

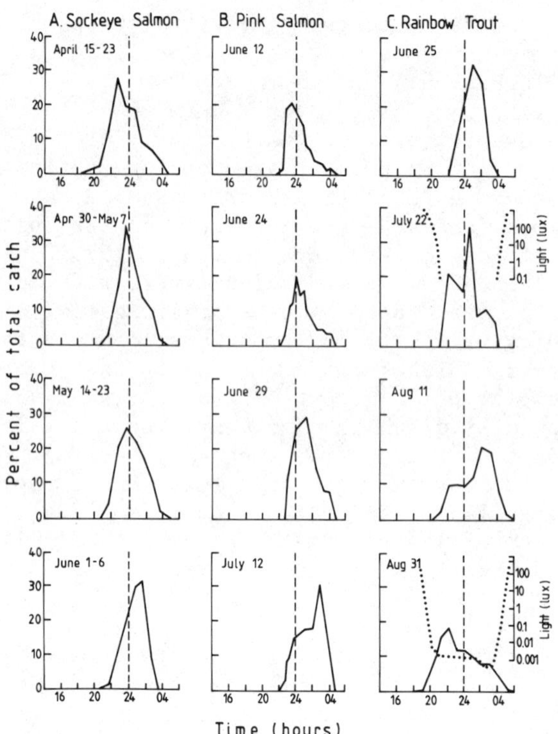

Figure 1. Seasonal change in the diel periodicity of downstream migration in young salmonids. A. Sockeye salmon fry in Williams Creek, British Columbia from McDonald (1960). B. Pink salmon fry in the Pritornaya River, Sakhalin from Shershnev and Zhulkov (1979). C. Rainbow trout fry in Loon inlet creek, British Columbia from Northcote (1962).

darkness usually involves loss of visual reference points and movement towards the water surface (Hoar 1976) as well as greater activity levels during night than day (Northcote 1962; Byrne 1971). Seasonal changes in the start and cessation of downstream movement clearly are controlled by shortening or lengthening of the night (Fig. 1). Similar changes in the timing of maximum descent are also evident, occurring progressively later each night as the length of darkness shortens (Fig. 1A,B) and occurring later each night as the length of darkness increases (Fig. 1C). Advanced photoperiod schedules also have been shown to shift seasonal timing of downstream movement of steelhead smolts ahead by at least a month (Zaugg 1981).

Size and age differences in downstream movement occur as in the case of the Volga River clupeids where recently emerged young show no clear diel pattern, whereas a definite nocturnal periodicity is

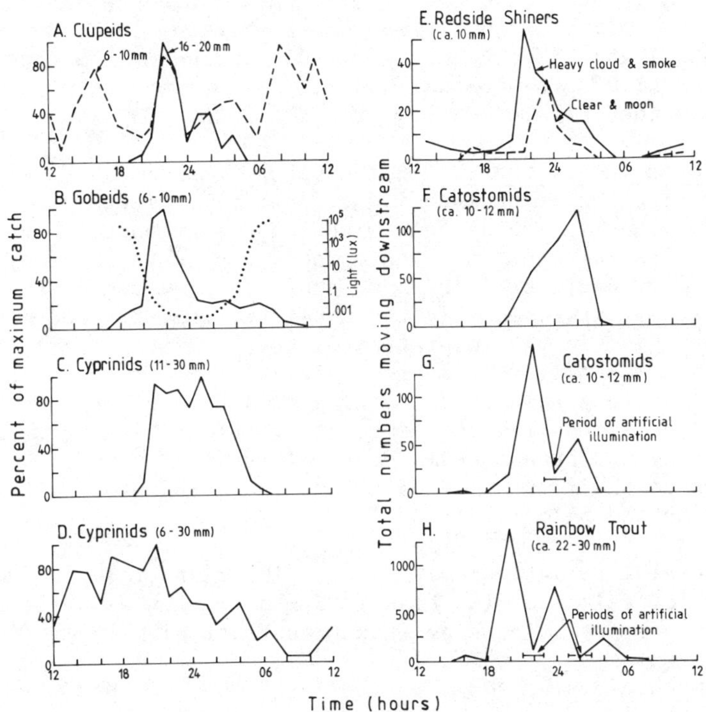

Figure 2. Diel periodicity of downstream movement of young fish in relation to size (A), water turbidity (B–D), and nocturnal illumination (E–H). Data for A–D adapted from Pavlov et al. (1977), for E from Lindsey and Northcote (1963), for F–G from Geen et al. (1966) and for H from Northcote (1962).

evident in slightly larger (older) individuals (Fig. 2A). Gobeids and cyprinids show a definite nocturnal downstream movement in the Volga River (Fig. 2B,C), but cyprinids in the highly turbid Kuban River do not (Fig. 2D).

Nocturnal light conditions regulate the intensity of downstream movement (Fig. 2E) with sharp reduction on clear moonlit nights. The majority of coho fry move downstream during low moonlight periods (Mason 1976). Changes in intensity of downstream movement of rainbow trout fry within a single night have been related to nocturnal illumination patterns (Fig. 1C), but more convincing evidence has been provided by field experiments using artificial illumination over the stream (Fig. 2F–H).

Wind driven currents. Strong onshore winds facilitate entrance of maturing adult fish from the sea into rivers (Banks 1969).

Similar influences have been demonstrated with the entry of lacustrine salmonids into tributaries (Mottley 1938; Lorz and Northcote 1965). Thorpe et al. (1981) indicated that out-migration of smolts from Loch Ness was greater during periods when winds increased the surface outflow from the loch.

Biotic interactions. In the view of Solomon and Templeton (1976), after an initial period of downstream movement to nursery areas by young salmonids following their emergence, there is a second phase of "recruitment" movement probably brought about by changing demands for food, territory and cover. Slaney and Northcote (1974) demonstrated experimentally the important role of food supply and territory size in the emigration of young rainbow trout from stream channels. Mason (1976) reported a similar response by under-yearling coho salmon to supplemental feeding in a natural stream. Hatchery-reared chum salmon fry released into river sections with an abundant food supply and stable environment may delay downstream movement in contrast to their usual seaward movement immediately after liberation (Kaeriyama and Sato 1979).

Social interactions may have an important role in downstream movement of salmonids (Hartman 1965; Hartman et al. 1982) as well as predator-prey relationships (Bakshtansky et al. 1980).

Visual Mechanisms

Instream cues. Eggs and buoyant larvae drifting down river have little opportunity to obtain and retain visual cues from topographic features of their environment which might be useful then or later in guiding their movements. However, in many rivers older young which move downstream after emergence do so mainly at night so that during daily periods of position maintenance they must gain visual as well as tactile association with particular stream reaches. Juveniles remaining in riverine rearing habitats for weeks to years before moving downstream would be expected to form strong visual association with temporary home areas. Whether or not such association is retained as part of the imprinting process and used later in recognition of particular stream reaches during return movement of adult stages is not known. Galaktionov (1978) considered that topographic features of the river bottom were used in the orientational behaviour of the European eel and that there was preferential movement along deepest sections of the river channel. Werner (1979) studied the return of mature white suckers (*Catostomus commersoni*) to inlet spawning streams and found no evidence that local landmark recognition was a major migratory cue. Nevertheless, Ellis (1962, 1966) observing the upstream movement of Pacific salmon spawners in the Somass River found that there were species-specific speeds and patterns of ascent through different types of water (pools, rapids). The migrants followed migration "routes" along the the deepest channel between regular holding pools so passage typically was made on one side, and not the other, of bedrock structures or boulders on the river bottom. On the other hand, radio-telemetry studies suggest that upriver progression of adults may be highly variable or erratic with long periods of pool holding or reversals in movement (Haynes et al. 1978; Power and McCleave 1980; Spence 1980; Haynes and Gray 1981).

Apparently migratory fish do not make use of river-bank features (trees, canyons, hills) as "landmarks" in recognition of particular reaches. However, Johnson and Groot (1963) from field observations and experiments suggest that lake shoreline or horizon outline might be used as secondary guiding cues for sockeye smolts migrating out of the Babine Lake system. Furthermore, in his study on seaward migration of sockeye smolts from complex lake-river systems Groot (1965) found evidence of lankmark orientation and noted that primary celestial and noncelestial (probably = magnetic compass) orientation mechanisms operate in relation to visual characteristics of the immediate environment, especially during periods of cloud cover.

Celestial cues. In moving down or up rivers fish often must migrate through complex lake systems, through lower reaches of large rivers where unidirectional water currents cannot be relied upon as primary orientational cues or through systems where there are reversals in direction of flow. Furthermore, in such areas the confinement and ready access to fixed reference points provided by the river banks and bottom are often lacking. Then visual cues provided by the sun, the plane of polarized light, the moon or other celestial information may be useful migratory cues.

Although sun orientation had been demonstrated earlier as a possible mechanism for homing in lake fishes (Hasler et al. 1958), the importance of celestial reference cues for passage of young fish out of complex lake-river systems first became clearly established by the comprehensive research of Groot (1965). He demonstrated the existence of three compass-orientation systems which sockeye salmon smolts used during their seaward migration out of Babine Lake and down the Babine River: 1) a sun compass system used under daytime cloud-cover conditions ranging from 0 to 70%; 2) a polarized light system useful during twilight periods when most of the smolt migration occurred and when neither sun, moon or stars would be available; and 3) an unknown (type-X) compass system utilized under completely enclosed conditions simulating high cloud-cover conditions at dusk.

Subsequent work on sockeye salmon smolts from the outlet of Babine Lake by Quinn and Brannon (1982) using different testing facilities has confirmed Groot's (1965) studies showing that visual orientation, with celestial cues such as the sun's position and light polarization patterns, was of primary importance during the day. Similar results have been obtained for young sockeye salmon from two other systems (Quinn 1980), and pink salmon also may use changes in the sun azimuth to adjust the direction of migration (Stepanov et al. 1979). Experiments have been made with sockeye salmon fry from the Weaver Creek-Harrison Lake population in British Columbia, a complex system where fry after emergence first must move downstream in one direction and then uptream in a different direction to reach their rearing lake (Brannon et al. 1981). Fry from Weaver Creek tested directly upon emergence showed a very weak directional orientation but later a generally appropriate directionality that was greatly enhanced by the presence of the moon, suggesting the importance of yet another celestial source of visual cues.

Magnetic-compass Mechanisms

The notion that magnetic fields might be perceived by vertebrates and utilized in migratory orientation is by no means recent, even for fishes. What has been lacking is clear demonstration that the fields are perceived, that they can be used in orientation, and that a physiological mechanism exists for their reception. Recent research by Quinn (1980), Quinn et al. (1981) and Quinn and Brannon (1982) provided evidence for the first two of these requirements. Young sockeye salmon apparently can perceive and utilize the earth's magnetic field to orient when no visual celestial cues are available (on heavily overcast days, at night especially in the absence of the moon, under ice and snow cover). Here then is a likely explanation for the type-X compass orientation demonstrated earlier by Groot (1965). Together with the celestial, visual mechanisms outlined previously, magnetic compass orientation would provide a reliable means for guidance of fish migrating through the large lakes and lower reaches of many river drainages where current patterns may be confused, tactile boundaries remote or turbidity high.

Olfactory Mechanisms

The possibility that fish use their sense of smell to detect and return to a parental spawning river was suggested over 100 years ago (Buckland 1880) and tested experimentally well over 50 years ago (Craigie 1926). Nevertheless, it was not until the early 1950's that an "olfactory hypothesis" for salmon homing was clearly enunciated (Hasler and Wisby 1951). It consisted of three basic requirements: 1) that each stream has a distinctive and persistent odor detectable by juvenile fish; 2) that juvenile fish have and can retain the ability to discriminate between the unique odors of different streams; and 3) that later as returning adults the fish detect and follow this distinctive home-stream odor to their juvenile home. The retention component of the second requirement, usually termed olfactory "imprinting" (after Lorenz's usage) refers to rapid, irreversible learning of an olfactory stimulus during a relatively short, sensitive period of juvenile development which subsequently affects behavior.

In over 30 years of research on olfactory homing in fishes, A.D. Hasler with his many students and associates amassed a large, generally convincing body of evidence for the hypothesis and its three tenets (for reviews see particularly Hara 1975; Hasler et al. 1978; Hasler and Scholz 1978; Johnsen and Hasler 1980). This work sparked great interest and additional research to probe and modify or amplify the hypothesis.

Nature and detection of stream odors. Early laboratory work by Hasler's group using conditioning experiments had shown that fish could detect olfactorily differences between stream waters, that these differences were persistent seasonally and probably were organic in nature. Following the electrophysiological approach to olfactory perception in fish by Hara et al. 1965, many studies have confirmed the specificity of fish response to home stream water at the neural level (Hara 1975). However, there are difficulties in interpretation of EEG responses and their relation to overall behavioral response of the fish (Oshima et al. 1969a,b; Cooper and Hasler 1973; Hara 1975; Barnett 1977). For example, nonhome waters also may elicit EEG responses, albeit lower ones than home waters.

Furthermore, the sources of the specific stream odor are not clearly identified. Hasler's group suggested that soils and plants in the stream watershed were involved and demonstrated that fish could distinguish between washes from different species of plants at very low dilution. Hain (1975) showed that washes from certain aquatic plants and leaf detritus were contributors to the upstream-swimming response of juvenile American eels (*Anguilla rostrata*). Following on the experimental demonstration by Brannon (1972) that young sockeye salmon respond behaviorally to odorant qualities of lake waters, Bodznick (1978a,b,c) has shown that there may be innate, as well as recently conditioned, preferences for lake water when compared to nonlake water, that this discrimination lies at the cellular level in the olfactory bulb, and that at least one component of the odorant is calcium ion, reliably detectable at 5×10^{-6} M concentrations, well below that in natural fresh waters. Suzuki (1978) also found that the calcium ion concentration affected olfactory reception at the cellular level in the lamprey *Entosphenus japonicus*.

That the fish themselves might contribute to the specific odor of streams was suggested by work in the early 1930's (Wrede 1932; White 1934), but was not taken seriously until the popularization of pheromones and chemical communication (see reviews by Barnett 1977; Solomon 1977; Liley 1982). Although there were suggestions by the late 1960's of olfactory cues from conspecifics (Miles 1968; Oshima et al. 1969a,b), it took the note by Nordeng (1971) to precipitate the spate of pheromone-related work on home stream odor over the last decade (e.g. Solomon 1973; Døving et al. 1973, 1974; Hoglund et al. 1975; Nordeng 1977; Selset and Døving 1980; Døving et al. 1980; Teeter 1980; Stabell 1982; Hara et al. 1984--this volume). From this we can note that fish from disparate taxonomic groups respond rheotactically to pheromones from individuals of their own species and even of their own population or perhaps smaller grouping, that the response can be detected by EEG techniques at the neural and cellular level of the olfactory bulb, that the skin mucus and some excretory products are important sources of pheromone, that certain amino acids and bile acids released by fish are detected by specific olfactory receptor cells, and that the bile acids have a number of characteristics which would make likely their involvement in the mechanism for homing in complex, branched river systems and also to specific territories. Nevertheless, as with the EEG studies considered earlier, there are interpretive problems, and homing can be induced to streams which do not contain a population of the species involved (Johnsen and Hasler 1980; Brannon 1982).

Imprinting and retention of stream odors. We now have good evidence that homing is not to parental, but rather to juvenile, rearing sites (Cooper et al. 1976). The explanatory mechanism proposed is that the young are imprinted to specific stream odors and that they retain this information for later use as returning adults. Imprinting by definition requires that the learning process occurs rapidly and at an early stage. But how early and how rapidly? Despite the pleas of Hara (1970, 1975), pointing out the dearth of knowledge in this area, there is still much to be done. Donaldson and Allen (1957) suggested that imprinting occurred in coho salmon juveniles during the first three months of their second year in freshwater. According to Jensen and Duncan (1971) coho salmon fingerlings were imprinted within 36 to 48 hours to a tributary over 240 km downstream from their hatching site. Commonly it is during the smolt stage that the critical time for imprinting is thought to occur (Scholz et al. 1976) and that shortly thereafter reproductive homing becomes far less precise (see citations in Scholz et al. 1978a,b, but also Nordeng 1971). In experiments with artificially imprinted coho salmon fingerlings a 5-week imprinting period extended from three weeks before to two weeks after the first signs of smolting behavior (Cooper et al. 1976). Nevertheless successful imprinting was obtained in one group with only a 2-day exposure. Imprinting was also reported for rainbow trout fingerlings using a 4-week exposure around the beginning of smolting (Cooper and Scholz 1976) and for 18-month old brown trout using a 34-day exposure period (Scholz et al. 1978b). Based on smolt transplantation experiments with anadromous Dolly Varden (*Salvelinus malma*), Armstrong (1974) suggested that imprinting for successful homing may occur within 1 day during the end of the smolt migration.

While the artificial imprinting experiments have provided information on the imprinting process itself, their major contribution to the olfactory homing hypothesis lies in the demonstration that fish can be imprinted to chemicals such as morpholine which do not occur naturally. However, there have been criticisms of some of the morpholine-imprinting work (Hara 1974; Hara and MacDonald 1975), rebuttals (Cooper and Hasler 1976), evaluations (Leggett 1977), further criticisms (Hara and Brown 1979) and comments (Cooper 1982; Hara and Brown 1982), mostly centering around the high concentrations of morpholine needed to elicit an EEG response in the olfactory bulb (see also Hara et al. 1984--this volume).

Oshima et al. (1969b) proposed that imprinting may occur sequentially as each specific stream odor was passed during the downstream movement of the juveniles, as suggested earlier by Harden Jones (1968). Leggett (1977) noted that this hypothesis was supported by studies indicating a reduced homing ability in juveniles that have been transported over part of their downstream journey. Two recent studies on this point gave opposite results. Ebel (1980) found that downstream transportation of chinook salmon and steelhead smolts did not significantly reduce their homing ability, whereas Power and McCleave (1980) showed for Atlantic salmon that headwater-released smolts homed better than those released near the river mouth. Clearly more work is needed on the sequential imprinting hypothesis as Barnett (1977) suggested.

Adult utilization of olfactory information. There is general agreement among most workers and reviewers of olfactory homing that we now have adequate evidence to support the third tenet of the hypothesis, namely that returning adults have retained, and utilize, information from earlier olfactory imprinting to discriminate among the various streams they encounter on their upstream passage to their juvenile home stream (Barnett 1977; Hara 1977; Leggett 1977; Hasler et al. 1978; Able 1980). The evidence comes from tagging and recapture of anosmic and control adults of both anadromous and freshwater populations of trout, salmon, char and suckers, from EEG studies on adults (but see Bodznick 1975), and from telemetric tracking of adults, which in some experiments included control and artificially-imprinted groups whose locomotory paths could be recorded at the mouths of streams with and without input of the imprinted chemical.

Hasler (1966) suggested that to avoid physiological fatigue of the olfactory system, upstream migrants tracking an odor trail might move in and out of odor plumes during their river ascent. Olfactory adaptation also might be avoided because of the turbulence created in the water as fish swing their heads from side to side (Woodhead 1975). Direct observation of adult coho salmon on reproductive migrations in tributaries to Lake Michigan supports this idea (Johnsen 1978). Not only did the fish confine their movements to whatever side of the stream was being scented with simulated homestream odor, but the fish zig-zagged in and out of the edge of the odor plume as they moved upstream. Johnsen's (1982, 1984--this volume) model for upstream migration and homestream selection proposes that: 1) in the presence of imprinted odor the fish show positive rheotaxis and move upstream; 2) if they encounter a boundary between water with and without odor, they commence zig-zagging along this interface, so being guided up and into odor-bearing plumes; and 3) if the imprinting scent is lost, the fish become negatively rheotactic until it is encountered again.

Other Sensory Mechanisms

Fish may receive information on their chemical environment from innervation associated with the mouth, pharynx, gill cavity, fins or barbels (cranial nerves 7, 9 and 10) forming the sense of taste and from free endings of the spinal nerves on the external body surface forming a common chemical sense (Hara 1971; Fontaine 1975). These senses may be useful in local movement and final stages of home recognition, but this possibility does not seem to have been seriously investigated (see Hara et al. 1984--this volume).

Fontaine (1975) noted that orientation might be effected by auditory cues from breaking waves. In turbulent reaches of rivers there is a profusion of noise from rock movement, from water passage through rapids and falls and from spawning fish, but whether or not this cacophony is useful in guiding the movement of fish seems unknown.

ENVIRONMENTAL AND GENETIC CONTROL

Recognition of an interplay between environmental and genetic factors in fish migration gradually emerged in the late 1950's and 1960's in part stimulated by attempts to unravel the intriguing patterns of migration shown by salmonids utilizing rivers for spawning and lakes for feeding (Northcote 1958, 1962, 1969a; Lindsey et al. 1959; Brannon 1967, 1972; Raleigh 1967). The overly simplistic control of lakeward movement of young rainbow trout from inlet and outlet spawning streams, thought to be mediated mainly by a temperature-photoperiod interaction (Northcote 1962), has now been shown to also involve genetic differences in response to water currents by stocks utilizing such streams for reproduction (Kelso and Northcote 1981; Kelso et al. 1981). Raleigh and Chapman (1971) and Brannon (1972) showed clear genetic differences in current response of young salmonids, and also indicated ways by which environmental factors, such as temperature, combined with developmental changes accompanying age of the young fish may interact with innate controls during lakeward movement. Genetic influence recently has been demonstrated in the growth and smolting of Atlantic salmon (Thorpe and Morgan 1978; Bailey et al. 1980; Thorpe et al. 1982), although here too there are environmental interactions.

Migratory behavior or the lack of it (residency) in populations of fish living in rivers below or above impassable waterfalls may be under a genetic control (Northcote et al. 1970; Jonsson 1982), and recent studies indicated the possibility of isozyme differences in lactate dehydrogenase affecting directional responses to water current of the young (Northcote and Kelso 1981). The progeny of above- and below-waterfall stocks of trout reared under the same environmental conditions showed differences in response to water current, growth rate and timing of maturity (Northcote 1981). Chum salmon adults return at maturity to the Tokachi River, Japan, in two distinct periods, and fish from each period have different isocitrate dehydrogenase frequencies (Okazaki 1978). Locally adapted genetic differences can have marked effects on the propensity for homing in pink salmon populations (Bams 1976).

HIERARCHIES AND SEQUENCES

In the control of migration of fish in rivers there is a hierarchical and sequential arrangement of several mechanisms, which may have arisen evolutionarily as a result of ontogenic changes in sensory capability and the "need" for different controls in the various habitats to be exploited. But there may be "dangers" in reliance on a single mechanism, particularly where migration routes extend over a broad range of habitats or through areas subject to unpredictable and variable environmental conditions. There must be advantages for redundancy and alternatives in the mechanisms which fish can use within a specific habitat and for shifts in the suite available at different ages, seasons or locations.

For two-stage longitudinal migrations between reproductive and feeding habitats (Fig. 3A,B) along a simple river system, only a short sequence of mechanisms may be used. The sequence could involve passive drift of early stages (fertilized eggs, larvae or recently emerged fry) from the reproductive habitat to the downstream feeding habitat with minimal involvement of environmental directors in orientation. Nevertheless, environmental regulation of migratory intensity may operate with visual mechanisms to control diel or daily periodicity of migration. After cessation of the downstream migration and a period of growth and maturation in the feeding habitat, upstream migration would commence. This movement could be timed by appropriate physiological preparation (hormonal control with environmental regulation), possibly with olfactory or other guidance mechanisms, if precise reproductive homing had developed, and if young stages had enough time to imprint on the reproductive habitat. A likely example is the migratory cycle of riverine goldeye populations.

A second level of complexity in riverine migration might involve a three- or four-stage migration, perhaps between two different

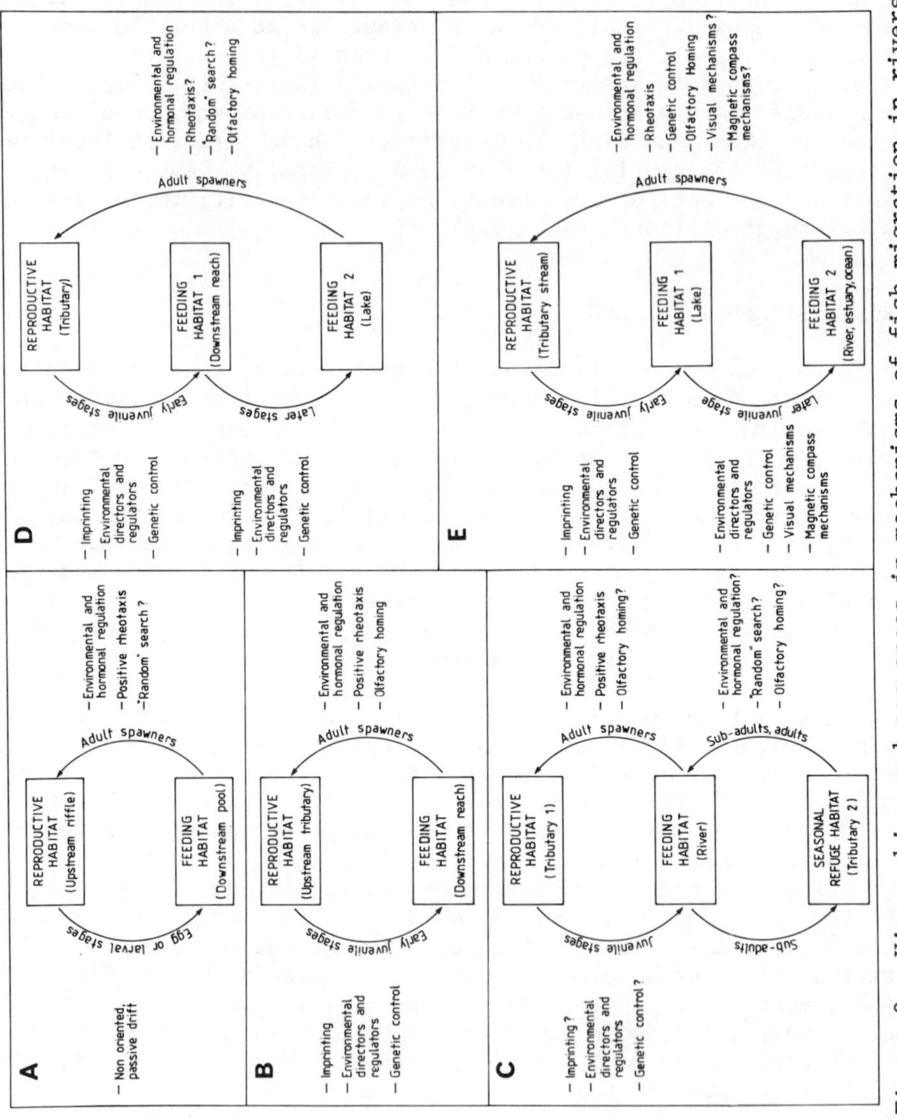

Figure 3. Hierarchies and sequences in mechanisms of fish migration in rivers.

feeding habitats or between a single feeding habitat and a seasonal refuge habitat as well as to and from the reproductive habitat (Fig. 3C,D). Most of the same sequence of mechanisms as above, combined with visual and olfactory mechanisms, could be used by the fish to locate the seasonal and reproductive habitats. Movements in northern Alaskan streams of arctic grayling (*Thymallus arcticus*) and arctic char (Craig and Poulin 1975; Craig 1978) possibly follow this sequence.

Further complexity is provided by the addition of lacustrine- or marine-feeding habits into the riverine migratory cycle (Fig. 3E). A sequence similar to that outlined above could occur in the movement of young from the reproductive habitat to the primary feeding habitat such as a lake. Movement within, and exit of migrants from, a lake may require a complex sequence of mechanisms and hierarchical use of them depending on diel timing of movement and weather. Recall, for example, the sequence of visual cues used by juvenile sockeye salmon migrants in lakes; each day migrants may shift from the sun to polarized light and perhaps to lunar cues, as well as to a magnetic compass if light or turbidity conditions did not permit them to orient visually. For anadromous species similar hierarchies and sequences of mechanisms may operate in guiding movement of growing subadults and maturing adults in the secondary feeding habitat, the ocean.

To this complexity may be added the interplay between environmental, physiological and genetic controls involved in the migration of fish through complex river systems or along the length of a single river where waterfalls or other barriers have acted as strong selective agents.

"MOTIVATIONS" FOR MIGRATION

Finally, perhaps the most intriguing and yet enigmatic aspect of the whole phenomenon of migration, why migrate? More explicitly, what advantages accrue to a riverine fish which adopts a migratory instead of resident life cycle? What selective pressures operate to favour one or the other life history, and what variability in migratory responses are available as the raw material for selection? Can we discern evolutionary steps in riverine migratory behaviour?

Primary Causes

Migration, as used here, consists of a cyclic movement between two or more habitats, most simply between a reproductive and a feeding habitat. In rivers and many other aquatic systems the most favourable areas for fish reproduction are not necessarily best for feeding. Indeed, as Mundie (1974) clearly pointed out, the minimal environmental unit of any stream, a riffle and a pool, are sequentially arranged and while the former is usually the site of maximum food production for salmonids, it is in the latter where much of the consumption often occurs. But few riverine salmonids spawn in pools. Thus there may be a fundamental structuring in stream habitats which promote at least short-distance longitudinal movement between spawning and feeding sites in salmonids, a group of freshwater fish exhibiting a high degree of migratory behaviour.

At the watershed level there are basic reasons for considerable temporal and spatial separation between optimal reproductive and feeding habitats. Changes in elevation, gradient and discharge between headwaters and lower reaches dictate major differences in stream size, substrate characteristics, temperature, oxygen, nutrients and other features.

The regular and generally predictable seasonality associated with summer and winter in temperate latitudes and with wet and dry periods in tropical regions also may promote development of migratory behaviour in riverine fishes. Shifts from summer feeding habitats to winter refuge habitats is a common pattern for fish populations in high arctic rivers. Likewise, in tropical rivers rich, wet-season feeding areas in flooded forests or inundated floodplains are vacated by mass movements of fish during low water of the dry season.

Adaptive Consequences

The suggestion that migration in fishes be viewed as an adaptation for increasing their abundance (or more precisely, for ensuring the highest number of viable offspring) has a long history (Gerbilsky 1958; Nikolsky 1963; Harden Jones 1968; Cushing 1969). Recently I considered various aspects of migration in freshwater fish which, by affecting their growth, fecundity, survival and abundance, might also enhance their production (Northcote 1978). Goulding (1980) noted that one of the major adaptations which has permitted most migratory characins of the Amazon River to maintain their relatively high biomasses has been their ability to exploit, by their well-timed movements, the allochthonous food available seasonally in the laterally flooded forests. Within a species individuals of migratory populations may attain much greater size than individuals of resident populations, a remarkable example being in some British Columbian streams where trout and char living in headwater reaches above a waterfall may weigh less than 50 g in weight, whereas lake migrants a short distance downstream may weigh several kilograms. The small resident females have fewer than a hundred eggs, whereas the large migratory ones have several thousands.

Migration can also increase survival by bringing about movement from temporarily unfavourable habitats into special refuge areas.

Riddell and Leggett (1981) suggest that parr in long-tributary stocks of Atlantic salmon migrate seaward in the autumn to avoid the higher energetic costs of overwintering as well as higher predation rates, whereas short-tributary stocks migrate downstream the following spring as smolts.

The migratory behaviour associated with reproductive homing promotes the orderly arrival on appropriate spawning grounds of mates in the correct physiological condition (Lindsey et al. 1959), and this must have considerable adaptive value. It also should serve to spread out the number of spawners going to a complex of localized, patchily distributed breeding sites in dendritic branches of watersheds and so ensure that in general all sites are optimally utilized. As Leggett and Carscadden (1978) commented concerning variation in spawning characteristics of American shad (*Alosa sapidissima*), "...fine tuning of reproductive strategies to local environmental conditions may be widespread among fish and may be the ultimate basis for evolution of homing."

Migrants can transport nutrients from highly productive feeding habitats to reproductive habitats where nutrients are in short supply. Decomposition of carcasses of migratory spawners adds greatly to nutrient inputs of low productivity streams and rivers (Mathisen 1972; Krogius 1973; Richey et al. 1975) and thereby enhances production of the offspring. Excretion by migrants which do not necessarily die after spawning may stimulate bacterial decomposition of leaf litter and thereby also enhance production in riverine ecosystems (Durbin et al. 1979).

Reproductive homing is probably never 100% accurate and indeed there may be selection against it being so. Lindsey et al. (1959) noted that, while "...homing in a species tends to balance the number of spawners using each stream against the reproductive capacity of the area," some straying may also have selective advantages, as individuals which invade new areas may obtain more resources and thereby have more viable offspring. Of course an important question is whether there are strays from all genotypes or just from certain genotypes (B. Jonsson, personal communication). For additional discussion of the adaptive role of migration in colonization of rivers see Northcote (1978) and Dingle (1980).

The more closely we look at the detailed aspects of migratory behaviour in riverine fish populations, the more evidence we uncover for marked local variation of a highly adaptive nature (Northcote 1969a, 1981; Hoar 1976; Bams 1976; Leggett and Carscadden 1978; Riddell and Leggett 1981; Riddell et al. 1981). Much of this variation seems to have a genetic basis so that fluctuating selective pressures even within habitats may operate quickly to shift responses in relatively few generations (Dingle et al. 1977; Dingle 1980).

ACKNOWLEDGMENTS

I am indebted to Bror Jonsson and Kjetil Hindar for reading the manuscript, making many useful suggestions and providing information on fish migration in Norwegian rivers. I am most grateful for the help of Linda Duncan in typing and of Linda Berg and Moira Greaven in literature searching and proofreading.

REFERENCES

Able, K.P. 1980. Mechanisms of orientation, navigation, and homing. Pages 284-373 *in* S.A. Gauthreaux, editor. Animal migration, orientation, and navigation. Academic Press, New York, New York, USA.

Alexander, D.R., and H.R. MacCrimmon. 1974. Production and movement of juvenile rainbow trout (*Salmo gairdneri*) in a headwater of Bothwell's Creek, Georgian Bay, Canada. Journal of the Fisheries Research Board of Canada 31:117-121.

Allen, K.R. 1944. Studies on the biology of the early stages of the salmon (*Salmo salar*). 4. The smolt migration in the Thurso River in 1938. Journal of Animal Ecology 13:63-85.

Alt, K.T. 1977. Inconnu, *Stenodus leucichthys*, migration studies in Alaska 1961-1974. Journal of the Fisheries Research Board of Canada 34:129-133.

Arawomo, G.A. 1981. Downstream movement of juvenile brown trout, *Salmo trutta* L., in the tributaries of Loch Leven, Kinross, Scotland. Hydrobiologia 77:129-131.

Armstrong, R.H. 1974. Migration of anadromous Dolly Varden (*Salvelinus malma*) in southeastern Alaska. Journal of the Fisheries Research Board of Canada 31:435-444.

Arnold, G.P. 1974. Rheotropism in fishes. Biological Reviews of the Cambridge Philosophical Society 49:515-576.

Asplund, C., and S. Sodergren. 1974. The spawning migration of river lampreys (*Lampetra fluviatilis*) in the River Rickleå. Zoologisk Revy 36:111-119.

Bagliniere, J.-L. 1976. Étude des populations de saumon atlantique (*Salmo salar* L., 1766) en Bretagne-Basse-Normandie. II. Activité de devalaison des smolts sur L'Elle. Annales d'Hydrobiologie 7:159-177.

Bagliniere, J.-L. 1979. Devalaison de truites (*Salmo trutta*) sur la rivière Elle. Bulletin Francais de Pisciculture 275:49-60.

Bailey, J.K., R.L. Saunders, and M.I. Buzeta. 1980. Influence of parental smolt age and sea age on growth and smolting of hatchery-reared Atlantic salmon (*Salmo salar*). Canadian Journal of Fisheries and Aquatic Sciences 37:1379-1386.

Bakshtanskiy, E.L., V.D. Nesterov, and M.N. Neklyudov. 1980. The behavior of young Atlantic salmon, *Salmo salar*, during downstream migration. Journal of Ichthyology 20:93-100.

Bams, R.A. 1976. Survival and propensity for homing as affected by presence or absence of locally adapted parental genes in two transplanted populations of pink salmon (*Oncorhynchus gorbuscha*) Journal of the Fisheries Research Board of Canada 33:2716-2725.

Banks, J.W. 1969. A review of the literature on the upstream migration of adult salmonids. Journal of Fish Biology 1:85-136.

Barnett, C. 1977. Aspects of chemical communication with special reference to fish. Biosciences Communications 3:331-392.

Barton, B.A. 1980. Spawning migrations, age and growth, and summer feeding of white and longnose suckers in an irrigation reservoir. Canadian Field-Naturalist 94:300-304.

Bayley, P.B. 1973. Studies on the migratory characin, *Prochilodus platensis* Holmberg 1889 (Pisces, Characoidei) in the river Pilcomago, South America. Journal of Fish Biology 5:25-40.

Bjornn, T.C. 1971. Trout and salmon movements in two Idaho streams as related to temperature, food, stream flow, cover, and population density. Transactions of the American Fisheries Society 100:423-438.

Bjornn, T.C., and J. Mallet. 1964. Movements of planted and wild trout in an Idaho river system. Transactions of the American Fisheries Society 93:70-76.

Bodznick, D. 1975. The relationship of the olfactory EEG evoked by naturally occurring stream waters to the homing behaviour of sockeye salmon (*Oncorhynchus nerka* Walbaum). Comparative Biochemistry and Physiology A. Comparative Physiology 52A:487-495.

Bodznick, D. 1978a. Water source preference and lakeward migration of sockeye salmon fry (*Oncorhynchus nerka*). Journal of Comparative Physiology A. Sensory, Neural, and Behavioral Physiology 127A:139-146.

Bodznick, D. 1978b. Characterization of olfactory bulb units of sockeye salmon with behaviorally relevant stimuli. Journal of Comparative Physiology A. Sensory, Neural, and Behavioral Physiology 127A:147-155.

Bodznick, D. 1978c. Calcium ion: An odorant for natural water discriminations and the migratory behavior of sockeye salmon. Journal of Comparative Physiology A. Sensory, Neural, and Behavioral Physiology 127A:157-166.

Bonetto, A.A., W. Dioni, and C. Pignalberi. 1969. Limnological investigations on biotic communities in the Middle Parana River Valley. Verhandlungen Internationale Vereinigung für Theoretische und Angewandte Limnologie 17:1035-1050.

Brannon, E.L. 1967. Genetic control of migrating behavior of newly emerged sockeye salmon fry. International Pacific Salmon Fisheries Commission Progress Report 16.

Brannon, E.L. 1972. Mechanisms controlling migration of sockeye salmon fry. International Pacific Salmon Fisheries Commission Bulletin 21.

Brannon, E.L. 1982. Orientation mechanisms of homing salmonids. Pages 219-227 *in* E.L. Brannon and E.O. Salo, editors. Salmon and trout migratory behavior symposium. University of

Washington, College of Fisheries, Seattle, Washington, USA.

Brannon, E.L., T.P. Quinn, G.T. Zucchette, and B.D. Ross. 1981. Compass orientation of sockeye salmon from a complex river system. Canadian Journal of Zoology 59:1548-1553.

Buckland, F. 1880. Natural history of British fishes. Univin, London, England.

Bull, H.O. 1931. The smolt descent on the river Tyne 1931. Report of the Dove Marine Laboratory 1931:32-43.

Bustard, D.R., and D.W. Narver. 1975. Aspects of the winter ecology of juvenile coho salmon (*Oncorhynchus kisutch*) and steelhead trout (*Salmo gairdneri*). Journal of the Fisheries Research Board of Canada 32:667-680.

Byrne, J.E. 1971. Photoperiodic activity changes in juvenile sockeye salmon (*Oncorhynchus nerka*). Canadian Journal of Zoology 49:1155-1158.

Campbell, J.S. 1977. Spawning characteristics of brown trout and sea trout (*Salmo trutta* L.) in Kirk Burn, River Tween, Scotland. Journal of Fish Biology 11:217-229.

Chapman, D.W. 1962. Aggressive behaviour in juvenile coho salmon as a cause of emigration. Jounrnal of the Fisheries Research Board of Canada 19:1047-1080.

Chavez, H. 1981. Marcado de robalo prieto, *Centropomus poeyi*, en la cuenca del papaloapan. Ciéncias Pesquisas 1:17-26.

Churikov, A.A. 1975. Features of the downstream migration of young salmon of the genus *Oncorhynchus* from the rivers of the northeast coast of Sakhalin. Journal of Ichthyology 15:963-970.

Claridge, P.N., and D.C. Gardner. 1978. Growth and movements of the twiate shad, *Alosa fallax* (Lacepede) in Severn Estuary. Journal of Fish Biology 12:203-211.

Cooper, J. 1982. Comment on electroencephalographic responses to morpholine and their relationship to homing. Canadian Journal of Fisheries and Aquatic Sciences 39:1544-1546.

Cooper, J.C., and A.D. Hasler. 1973. II. An electrophysiological approach to salmon homing. Fisheries Research Board of Canada Technical Report 415.

Cooper, J.C., and A.D. Hasler. 1976. Electrophysiological studies of morpholine imprinted coho salmon (*Oncorhynchus kisutch*) and rainbow trout (*Salmo gairdneri*). Journal of the Fisheries Research Board of Canada 33:688-694.

Cooper, J.C., and A.T. Scholz. 1976. Homing of artificially imprinted steelhead trout, *Salmo gairdneri*. Journal of the Fisheries Research Board of Canada 33:826-829.

Cooper, J.C., A.T. Scholz, P.M. Horrall, A.D. Hasler, and D.M. Madison. 1976. Experimental confirmation of the olfactory hypothesis with homing, artificially imprinted coho salmon (*Oncorhynchus kisutch*). Journal of the Fisheries Research Board of Canada 33:703-710.

Craig, P.C. 1978. Movements of stream-resident and anadromous arctic char (*Salvelinus alpinus*) in a perennial spring on the Canning River, Alaska. Journal of the Fisheries Research Board of Canada 35:48-52.

Craig, P.C., and V.A. Poulin. 1975. Movements and growth of Arctic grayling (*Thymallus arcticus*) and juvenile arctic char (*Salvelinus alpinus*) in a small arctic stream, Alaska. Journal of the Fisheries Research Board of Canada 32:689-697.

Craigie, E.H. 1926. A preliminary experiment on the relation of the olfactory sense to the migration of the sockeye salmon (*Oncorhynchus nerka* Wal.). Transactions of the Royal Society of Canada 5:215-224.

Cuinat, R., and M. Heland. 1979. Observations sur la devalaison d'alevins de truite commune (*Salmo trutta* L.) dans le Lissuraga. Bulletin Francais de Pisciculture 274:1-17.

Cushing, D.H. 1969. Migration and abundance. Pages 207-212 *in* Perspectives in fisheries oceanography. Special Bulletin of the Japanese Society of Fisheries and Oceanography.

Davies, R.W., and G.W. Thompson. 1976. Movements of mountain whitefish (*Prosopium williamsoni*) in the Sheep River watershed, Alberta. Journal of the Fisheries Research Board of Canada 33:2395-2401.

Dingle, H. 1980. Ecology and evolution of migration. Pages 1-101 *in* S.A. Gauthreaux, editor. Animal migration, orientation, and navigation. Academic Press, New York, New York, USA.

Dingle, H., C.K. Brown, and I.P. Hegmann. 1977. The nature of genetic variance influencing photoperiodic diapause in a migrant insect, *Ancopeltus fasciatus*. American Naturalist 111:1047-1059.

Dodson, J.J., and J.C. Young. 1977. Temperature and photoperiod regulation of rheotropic behavior in prespawning common shiners, *Notropis cornutus*. Journal of the Fisheries Research Board of Canada 34:341-346.

Donaldson, L.E., and G.H. Allen. 1957. Return of silver salmon, *Oncorhynchus kisutch* (Walbaum) to point of release. Transactions of the American Fisheries Society 87:13-22.

Døving, K.B., P.S. Enger, and H. Nordeng. 1973. Electrophysiological studies on the olfactory sense of char (*Salmo alpinus* L.). Comparative Biochemistry and Physiology A. Comparative Physiology 45A:21-24.

Døving, K.B., H. Nordeng, and B. Oakley. 1974. Single unit discrimination of fish odours released by char (*Salmo alpinus* L.) populations. Comparative Biochemistry and Physiology A. Comparative Physiology 47A:1051-1063.

Døving, K.B., R. Selset, and G. Thommesen. 1980. Olfactory sensitivity to bile acids in salmonid fishes. Acta Physiologica Scandinavica 108:123-131.

Durbin, A.G., S.W. Nixon, and C.A. Oviatt. 1979. Effects of the spawning migration of the alewife, *Alosa pseudoharengus*, on freshwater ecosystems. Ecology 60:8-17.

Ebel, W.J. 1980. Transportation of chinook salmon, *Oncorhynchus tshawytscha*, and steelhead, *Salmo gairdneri*, smolts in the Columbia River and its effects on adult returns. US National

Marine Fisheries Service Fishery Bulletin 78:491-505.

Eddy, S.B., and M.W. Lankester. 1978. Feeding and migratory habits of Arctic char, *Salvelinus alpinus*, indicated by the presence of the swimbladder nematode *Cystidicola cristivomneri* White. Journal of the Fisheries Research Board of Canada 35:1488-1491.

Edwards, R.J. 1977. Seasonal migrations of *Astyanax mexicanus* as an adaptation to novel environment. Copeia 1977:770-771.

Elliott, J.M. 1966. Downstream movements of trout fry *Salmo trutta* in a Dartmoor stream. Journal of the Fisheries Research Board of Canada 23:157-159.

Ellis, D.V. 1962. Preliminary studies on the visible migration of adult salmon. Journal of the Fisheries Research Board of Canada 19:137-148.

Ellis, D.V. 1966. Swimming speeds of sockeye and coho salmon on spawning migration. Journal of the Fisheries Research Board of Canada 23:181-187.

Ewing, R.D., C.A. Fustish, S.L. Johnson, and H.J. Pribble. 1980. Seaward migration of juvenile chinook salmon without elevated gill (Na+K)-ATPase activities. Transactions of the American Fisheries Society 109:349-356.

Ewing, R.D., S.L. Johnson, H.J. Pribble, and J.A. Lichatowich. 1979. Temperature and photoperiod effects on gill (Na+K)-ATPase activity in chinook salmon *Oncorhynchus tshawytscha*. Journal of the Fisheries Research Board of Canada 36:1347-1353.

Finnigan, R.J. 1978. A study of fish movement stimulated by a sudden reduction in rate of flow. Fisheries Research Board of Canada Data Report 98.

Fontaine, M. 1975. Physiological mechanisms in the migration of marine and amphihaline fish. Pages 241-255 *in* F.S. Russell and M. Yonge, editors. Advances in marine biology, volume 13. Academic Press, London, England.

Fried, S.M., J.D. McCleave, and G.W. LaBar. 1978. Seaward migration of hatchery-reared Atlantic salmon, *Salmo salar*, smolts in the Penobscot River estuary, Maine: riverine movements. Journal of the Fisheries Research Board of Canada 35:76-87.

Galaktionov, G.Z. 1978. Migration speed and the system of searching reactions in *Anguilla anguilla*. Soviet Journal of Ecology 1:88-91.

Ganapati, S.V. 1973. Ecological problems of man-made lakes of south India. Archiv für Hydrobiologie 71:363-380.

Geen, G.H., T.G. Northcote, G.F. Hartman, and C.C. Lindsey. 1966. Life histories of two species of catostomid fishes in Sixteenmile Lake, British Columbia, with particular reference to inlet stream spawning. Journal of the Fisheries Research Board of Canada 23:1761-1788.

Gerbilsky, N.L. 1958. The question of the migratory impulse in connection with the analysis of intraspecific biological groups. Trudy Soveshchanii Ikhtiologicheskoi Kimissii Academii Nauk SSSR 8:142-152.

Giles, M.A., and W.E. Vanstone. 1976. Changes in ouabain-sensitive adenosine triphosphatase activity in gills of coho salmon (*Oncorhynchus kisutch*) during parr-smolt transformation. Journal of the Fisheries Research Board of Canada 33:54-62.

Godin, J.-G. 1980. Temporal aspects of juvenile pink salmon (*Oncorhynchus gorbuscha*, Walbaum) emergence from a simulated gravel redd. Canadian Journal of Zoology 58:735-744.

Godin, J.-G. 1982. Migrations of salmonid fishes during early life history phases: daily and annual timing. Pages 22-50 *in* E.L. Brannon and E.O. Salo, editors. Salmon and trout migratory behavior. University of Washington College of Fisheries, Seattle, Washington, USA.

Godin, J.-G., P.A. Dill, and D.E. Drury. 1974. Effects of thyroid hormones on behavior of yearling Atlantic salmon (*Salmo salar*). Journal of the Fisheries Research Board of Canada 31:1787-1790.

Goulding, M. 1980. The fishes and the forest: explorations in Amazonian natural history. University of California Press, Los Angeles, California, USA.

Grau, E.G. 1982. Is the lunar cycle a factor timing the onset of salmon migration? Pages 184-189 *in* E.L. Brannon and E.O. Salo, editors. Salmon and trout migratory behavior. University of Washington College of Fisheries, Seattle, Washington, USA.

Gray, R.H., and J.M. Haynes. 1979. Spawning migration of adult chinook salmon (*Oncorhynchus tshawytscha*) carrying external and internal radio transmitters. Journal of the Fisheries Research Board of Canada 36:1060-1064.

Groot, C. 1965. On the orientation of young sockeye salmon (*Oncorhynchus nerka*) during their seaward migration out of lakes. Behaviour Supplement 14:1-198.

Hain, J.H.W. 1975. Migratory orientation in the American eel, *Anguilla rostrata*. Doctoral dissertation. University of Rhode Island, Kingston, Rhode Island, USA.

Hara, T.J. 1970. An electrophysiological basis for olfactory discrimination in homing salmon: a review. Journal of the Fisheries Research Board of Canada 27:565-586.

Hara, T.J. 1971. Chemoreception. Pages 79-120 *in* W.S. Hoar and D.J. Randall, editors. Fish physiology, volume 5. Academic Press, New York, New York, USA.

Hara, T.J. 1974. Is morpholine an effective olfactory stimulant in fish? Journal of the Fisheries Research Board of Canada 31:1547-1550.

Hara, T.J. 1975. Olfaction in fish. Progress in Neurobiology (Oxford) 5:271-335.

Hara, T.J. 1977. Further studies on the structure-activity relationships of amino acids in fish olfaction. Comparative Biochemistry and Physiology A. Comparative Physiology 56A:559-565.

Hara, T.J., and S.B. Brown. 1979. Olfactory bulbar electrical responses of rainbow trout (*Salmo gairdneri*) exposed to morpholine during smoltification. Journal of the Fisheries Research Board of Canada 36:1186-1190.

Hara, T.J., and S.B. Brown. 1982. Reply to comment on electro-encephalographic responses to morpholine and their relationship to homing. Canadian Journal of Fisheries and Aquatic Sciences 39:1546-1548.

Hara, T.J., and S. Macdonald. 1975. Olfactory responses to skin mucous substances in rainbow trout *Salmo gairdneri*. Comparative Biochemistry and Physiology A. Comparative Physiology 54A:41-44.

Hara, T.J., S. Macdonald, R.E. Evans, T. Marui, and S. Arai. 1984. Morpholine, bile acids and skin mucus as possible chemical cues in salmonid homing: electrophysiological re-evalation. Pages 363-378 *in* J.D. McCleave, G.P. Arnold, J.J. Dodson and W.H. Neill, editors. Mechanisms of migration in fishes. Plenum Press, New York, New York, USA.

Hara, T.J., K. Ueda, and A. Gorbman. 1965. Electroencephalograhic studies of homing salmon. Science (Washington DC) 149:884-885.

Harden Jones, F.R. 1968. Fish migration. St. Martin's Press. New York, New York, USA.

Hart, C.E., G. Concannon, C.A. Fustish, and R.D. Ewing. 1981. Seaward migration and gill (Na+K)-ATPase activity of spring chinook salmon in an artificial stream. Transactions of the American Fisheries Society 110:44-50.

Hartman, G.F. 1965. The role of behavior in the ecology and interaction of underyearling coho salmon (*Oncorhynchus kisutch*) and steelhead trout (*Salmo gairdneri*). Journal of the Fisheries Research Board of Canada 22:1035-1081.

Hartman, G.F., B.C. Andersen, and J.C. Scrivener. 1982. Seaward movement of coho salmon (*Oncorhynchus kisutch*) fry in Carnation Creek, an unstable coastal stream in British Columbia. Canadian Journal of Fisheries and Aquatic Sciences 39:588-597.

Hartman, G.F., T.G. Northcote, and C.C. Lindsey. 1962. Comparison of inlet and outlet spawning runs of rainbow trout in Loon Lake, British Columbia. Journal of the Fisheries Research Board of Canada 19:173-200.

Hasler, A.D. 1966. Underwater guideposts. University of Wisconsin Press, Madison, Wisconsin, USA.

Hasler, A.D., and A.T. Scholz. 1978. Olfactory imprinting in coho salmon. Pages 356-369 *in* K. Schmidt-Koenig and W.T. Keeton, editors. Animal migration, navigation and homing. Springer Verlag, New York, New York, USA.

Hasler, A.D., and W.J. Wisby. 1951. Discrimination of stream odors by fishes and relation to parent stream behavior. American Naturalist 85:223-238.

Hasler, A.D., A.T. Scholz, and R.M. Horrall. 1978. Olfactory imprinting and homing in salmon. American Scientist 66:347-355.

Hasler, A.D., R.M. Horrall, W.J. Wisby, and W. Braemer. 1958. Sun orientation and homing in fishes. Limnology and Oceanography 3:353-361.

Hayes, F.R. 1953. Artificial freshets and other factors controlling the ascent and population of salmon in the Le Have River, Nova Scotia. Fisheries Research Board of Canada Bulletin 99.

Haynes, J.M., and R.H. Gray. 1981. Diel and seasonal movements of white sturgeon, *Acipenser transmontanus*, in the mid-Columbia River. U S National Marine Fisheries Service Fish Bulletin 79:367-370.

Haynes, J.M., R.H. Gray, and J.C. Montgomery. 1978. Seasonal movements of white sturgeon (*Acipenser transmontanus*) in the mid-Columbia River. Transactions of the American Fisheries Society 107:275-280.

Heland, M. 1980a. La devalaison des alevins de truite commune *Salmo trutta* L. I. Characterisation en milieu artificiel. Annales de Limnologie 16:233-245.

Heland, M. 1980b. La devalaison des alevins de truite commune, *Salmo trutta* L. II. Activité des alevins "devalants" compares aux sedentaires. Annales de Limnologie 16:247-254.

Hellawell, J.M. 1976. River management and the migratory behaviour of salmonids. Fisheries Management 7:57-60.

Hoar, W.S. 1958. The evolution of migratory behaviour among juvenile salmon of the genus *Oncorhynchus*. Journal of the Fisheries Research Board of Canada 15:391-428.

Hoar, W.S. 1976. Smolt transformation: evolution, behavior, and physiology. Journal of the Fisheries Research Board of Canada 33:1234-1252.

Höglund, L.B., A. Bohman, and N.-A. Nillson. 1975. Possible odour responses of juvenile Arctic char (*Salvelinus alpinus* (L.)) to three other species of subarctic fish. Institute of Freshwater Research Drottningholm Report 54:21-35.

Huet, M., and J.A. Timmermans. 1979. Fonctionnement et rôle d'un ruisseau frayère a truites. Travaux du Station de Recherches des Eaux et Fôrets, Belgique 48.

Hynes, H.B.N. 1972. The ecology of running waters. University of Toronto Press, Toronto, Ontario, Canada.

Jellyman, D.J. 1979. Upstream migration of glass-eels (*Anguilla* spp.) in the Waikato River. New Zealand Journal of Marine and Freshwater Research 13:13-22.

Jensen, A.L., and R.N. Duncan. 1971. Homing of transplanted coho salmon. Progressive Fish-Culturist 33:216-218.

Johnsen, P.B. 1978. Contributions on the movements of fish. I. Behavioral mechanisms of upstream migration and homestream selection in coho salmon. Doctoral dissertation. University of Wisconsin, Madison, Wisconsin, USA.

Johnsen, P.B. 1982. A behavioral control model for homestream selection in migratory salmonids. Pages 266-273 *in* E.L. Brannon and E.O. Salo, editors. Proceedings of the salmon and trout migratory behavior symposium. School of Fisheries, University of Washington, Seattle, Washington, USA.

Johnsen, P.B. 1984. Establishing the physiological and behavioral determinates of chemosensory orientation. Pages 379-385 *in* J.D. McCleave, G.P. Arnold, J.J. Dodson and W.H. Neill, editors. Mechanisms of migration in fishes. Plenum Press, New York, New York, USA.

Johnsen, P.B., and A.D. Hasler. 1980. The use of chemical cues in the upstream migration of coho salmon, *Oncorhynchus kisutch* Walbaum. Journal of Fish Biology 17:67-73.

Johnson, W.E., and C. Groot. 1963. Observations on the migration of young sockeye salmon (*Oncorhynchus nerka*) through a large, complex lake system. Journal of the Fisheries Research Board of Canada 20:919-938.

Johnson, T., and K. Müller. 1978. Migration of juvenile pike, *Esox lucius* L., from a coastal stream to the northern part of the Bothnian Sea. Aquilo Ser Zoologica 18:57-61.

Jonsson, B. 1981. Life history strategies of trout (*Salmo trutta* L.). Doctoral dissertation. University of Oslo, Oslo, Norway.

Jonsson, B. 1982. Diadromous and resident trout *Salmo trutta*: Is their difference due to genetics? Oikos 38:297-300.

Kaeriyama, M., and S. Sato. 1979. Studies on the growth and feeding habit of the chum salmon fry during seaward migration in the Tokachi River system. III. Relationships between migration time and the growth or the feeding behavior of the fry during 1977. Scientific Reports of the Hokkaido Salmon Hatchery 33:47-73.

Kelso, B.W., and T.G. Northcote. 1981. Current response of young rainbow trout from inlet and outlet spawning stocks of a British Columbia Lake. Verhandlungen Internationale Vereinigung für Theoretische und Angewandte Limnologie 21:1214-1221.

Kelso, B.W., T.G. Northcote, and C.F. Wehrhahn. 1981. Genetic and environmental aspects of the response to water current by rainbow trout (*Salmo gairdneri*) originating from inlet and outlet streams of two lakes. Canadian Journal of Zoology 59:2177-2185.

Kowtal, G.V. 1972. Observations on the breeding and larval development of Chilka "Sahai", *Eleutheronema tetradactyium* (Shaw). Indian Journal of Fisheries 19:70-75.

Krogius, F.V. 1973. Population dynamics of growth of young sockeye salmon in Lake Dalnee. Hydrobiologia 43:45-51.

Leggett, W.C. 1977. The ecology of fish migrations. Annual Review of Ecology and Systematics 8:285-308.

Leggett, W.C., and J.E. Carscadden. 1978. Latitudinal variation in reproductive characteristics of American shad (*Alosa sapidissima*) evidence for population specific life history strategies in fish. Journal of the Fisheries Research Board of Canada 35:1469-1478.

Libosvarsky, J. 1976. On the ecology of spawning migration of brown trout. Zoologicke Listy. Folia zoologica 25:175-182.

Liley, N.R. 1982. Chemical communication in fish. Canadian Journal of Fisheries and Aquatic Sciences 39:22-35.

Lindsey, C.C., and T.G. Northcote. 1963. Life history of redside shiners, *Richardsonius balteatus* with particular reference to movements in and out of Sixteenmile Lake streams. Journal of the Fisheries Research Board of Canada 20:1001-1030.

Lindsey, C.C., T.G. Northcote, and G.F. Hartman. 1959. Homing of rainbow trout to inlet and outlet spawning streams at Loon Lake,

British Columbia. Journal of the Fisheries Research Board of Canada 16:695-719.

Lorz, H.W., and B.P. McPherson. 1976. Effects of copper or zinc in fresh water on the adaptation to sea water and ATPase activity, and the effects of copper on migratory disposition of coho salmon (*Oncorhyncus kisutsch*). Journal of the Fisheries Research Board of Canada 33:2023-2030.

Lorz, H.W., and T.G. Northcote. 1965. Factors affecting stream location and timing and intensity of entry by spawning kokanee (*Oncorhynchus nerka*) into an inlet of Nicola Lake, British Columbia. Journal of the Fisheries Research Board of Canada 22:665-687.

Malmquist, B. 1980. The spawning migration of the brook lamprey, *Lampetra planeri* Bloch, in a south Swedish stream. Journal of Fish Biology 16:105-114.

Mason, J.C. 1976. Response of underyearling coho salmon to supplemental feeding in a natural stream. Journal of Wildlife Management 40:775-788.

Mathisen, O.A. 1972. Biogenic enrichment of sockeye salmon lakes and stock productivity. Verhandlungen Internationale Vereinigung für Theoretische und Angewandte Limnologie 18:1089-1095.

Mayama, H. 1978. Ecological observation on the adult salmon. II. Diurnal variation of upstream migration of the adult chum salmon in the Chitose River. Scientific Reports of the Hokkaido Salmon Hatchery 32:9-18.

Mayama, H., and T. Takahashi. 1977. Ecological observation of the adult salmon. I. Diurnal variation of upstream migration of the adult chum salmon in the Chitose River. Scientific Reports of the Hokkaido Salmon Hatchery 31:21-28.

Meier, A.H., and A.J. Fivizzani. 1980. Physiology of migration. Pages 225-281 *in* S.A. Gauthreaux, editor. Animal migration, orientation, and navigation. Academic Press, New York, New York, USA.

Miles, R.G. 1968. Rheotaxis of elvers of the American eel (*Anguilla rostrata*), in the laboratory to water from different streams in Nova Scotia. Journal of Fisheries Research Board of Canada 25:1591-1602.

Morin, R., J.J. Dodson, and G. Power. 1981. The migrations of anadromous cisco (*Coregonus artidii*) and lake whitefish (*C. clupeaformis*) in estuaries of eastern James Bay. Canadian Journal of Zoology 59:1600-1607.

Mottley, C. McC. 1938. Fluctuations in the intensity of the spawning runs of rainbow trout at Paul Lake. Journal of the Fisheries Research Board of Canada 4:69-87.

Müller, K. 1982. Jungfischwanderungen zur Bottensee. Archiv fur Hydrobiologie 95:271-282.

Mundie, J.H. 1974. Optimization of the salmonid nursery stream. Journal of the Fisheries Research Board of Canada 31:1827-1837.

McBride, D.N. 1980. Homing of arctic char, *Salvelinus alpinus*

(Linnaeus) to feeding and spawning sites in the Wood River Lake system, Alaska. Alaska Department of Fish and Game Information Leaflet 184.

McDonald, J. 1960. The behaviour of Pacific salmon fry during their downstream migration to freshwater and saltwater nursery areas. Journal of the Fisheries Research Board of Canada 17:655-676.

McDowall, R.M. 1978. New Zealand freshwater fishes. Heinemann Educational Books Ltd., Auckland, New Zealand.

McDowall, R.M., and G.A. Eldon. 1980. The ecology of whitebait migrations (Galaxidae: *Galaxias* spp.). New Zealand Ministry of Agriculture and Fisheries. Fisheries Research Division Fisheries Research Bulletin 20.

McPhail, J.D., and C.C. Lindsey. 1970. Freshwater fishes of northwestern Canada and Alaska. Fisheries Research Board of Canada Bulletin 173.

Nihouarn, A. 1976. Les saumons juveniles dans la riviere Allier et leur devalaison en 1976. Conseil superieur de la Peche, Region piscicole Auvergne-Limousin, Clermond-Ferrand, France.

Nikolsky, G.V. 1963. The ecology of fishes. Academic Press, London, England.

Nordeng, H. 1971. Is the local orientation of anadromous fishes determined by pheromones? Nature (London) 233:411-413.

Nordeng, H. 1977. A pheromone hypothesis for homeward migration in anadromous salmonids. Oikos 28:155-159.

Northcote, T.G. 1958. Effect of photoperiodism on response of juvenile trout to water currents. Nature (London) 181:1283-1284.

Northcote, T.G. 1962. Migratory behaviour of juvenile rainbow trout, *Salmo gairdneri*, in outlet and inlet streams of Loon Lake, British Columbia. Journal of the Fisheries Research Board of Canada 19:201-270.

Northcote, T.G. 1969a. Patterns and mechanisms in the lakeward migratory behaviour of juvenile trout. Pages 183-203 *in* T.G. Northcote, editor. Symposium on salmon and trout in streams. H.R. MacMillan Lectures in Fisheries, University of British Columbia, Vancouver, British Columbia, Canada.

Northcote, T.G. 1969b. Lakeward migration of young rainbow trout (*Salmo gairdneri*) in the Upper Lardeau River, British Columbia. Journal of the Fisheries Research Board of Canada 26:33-45.

Northcote, T.G. 1978. Migratory strategies and production in freshwater fishes. Pages 326-359 *in* S.D. Gerking, editor. Ecology of freshwater fish production. Blackwell Scientific Publications, Oxford, England.

Northcote, T.G. 1981. Juvenile current response, growth and maturity of above and below waterfall stocks of rainbow trout, *Salmo gairdneri*. Journal of Fish Biology 18:741-751.

Northcote, T.G., and B.W. Kelso. 1981. Differential response to water current by two homozygous LDH phenotypes of young rainbow trout (*Salmo gairdneri*). Canadian Journal of Fisheries and Aquatic Sciences 38:348-352.

Northcote, T.G., S.N. Williscroft, and H. Tsuyuki. 1970. Meristic and lactate dehydrogenase genotype differences in stream populations of rainbow trout below and above a waterfall. Journal of the Fisheries Research Board of Canada 27:1987-1995.

Okada, Y. 1955. Fishes of Japan. Maruzen Co., Ltd., Tokyo, Japan.

Okazaki, T. 1978. Genetic differences of two chum salmon (*Oncorhynchus keta*) populations returning to the Tokachi River. Bulletin Far Seas Fisheries Research Laboratory (Shimizu) 16:121-128.

Oshima, K., W.E. Hahn, and A. Gorbman. 1969a. Olfactory discrimination of natural waters by salmon. Journal of the Fisheries Research Board of Canada 26:2111-2121.

Oshima, K., W.E. Hahn, and A. Gorbman. 1969b. Electroencephalographic olfactory responses in adult salmon to waters traversed in the homing migration. Journal of the Fisheries Research Board of Canada 26:2123-2133.

Osterdahl, L. 1969. The smolt run of a small Swedish River. Pages 205-215 *in* T.G. Northcote, editor. Symposium on salmon and trout in streams, H.R. MacMillan Lectures in Fisheries, University of British Columbia, Vancouver, British Columbia, Canada.

Ottaway, E.M., and A. Clarke. 1981. A preliminary investigation into the vulnerability of young trout (*Salmo trutta* L.) and Atlantic salmon (*S. salar* L.) to downstream displacement by high water velocities. Journal of Fish Biology 19:135-145.

Pavlov, D.S., A.M. Pakhorukov, G.N. Kuragina, V.K. Nezdoliy, N.P. Nekrasova, D.A. Brodskiy, and A.L. Ersler. 1977. Some features of the downstream migrations of juvenile fishes in the Volga and Kuban rivers. Journal of Ichthyology 17:363-374.

Pemberton, R. 1976. Sea trout in North Argyll sea lochs: population, distribution and movement. Journal of Fish Biology 9:157-179.

Potter, I.C. 1980. Ecology of larval and metamorphosing lampreys. Canadian Journal of Fisheries and Aquatic Sciences 37:1641-1657.

Power, J.H., and J.D. McCleave. 1980. Riverine movements of hatchery-reared Atlantic salmon (*Salmo salar*) upon return as adults. Environmental Biology of Fishes 5:3-13.

Quinn, T.P. 1980. Evidence for celestial and magnetic compass orientation in lake migrating sockeye salmon fry. Journal of Comparative Physiology A. Sensory, Neural, and Behavioral Physiology 137A:243-248.

Quinn, T.P., and E.L. Brannon. 1982. The use of celestial and magnetic cues by orienting sockeye salmon smolts. Journal of Comparative Physiology A. Sensory, Neural, and Behavioral Physiology 147A:547-552.

Quinn, T.P., R. Merrill, and E.L. Brannon. 1981. Magnetic field detection in sockeye salmon. Journal of Experimental Zoology 217:137-142.

Rajyalakshmi, T. 1973. The population characteristics of the Godavary *Hilsa* over the years 1963-1967. Indian Journal of Fisheries 20:78-94.

Raleigh, R.F. 1967. Genetic control in the lakeward migrations of sockeye salmon (*Oncorhynchus nerka*) fry. Journal of the Fisheries Research Board of Canada 24:2613-2622.

Raleigh, R.F., and D.W. Chapman. 1971. Genetic control in lakeward migrations of cutthroat trout fry. Transactions of the American Fisheries Society 100:33-40.

Raymond, H.L. 1979. Effects of dams and impoundments on migrations of juvenile chinook salmon and steelhead from the Snake River, 1966 to 1975. Transactions of the American Fisheries Society 108:505-529.

Reynolds, J.D. 1971. Biology of the small pelagic fishes in the new Volta Lake in Ghana. II. Schooling and migrations. Hydrobiologia 38:79-91.

Richey, J.E., M.A. Perkins, and C.R. Goldman. 1975. Effects of kokanee salmon (*Oncorhynchus nerka*) decomposition on the ecology of a sub-alpine stream. Journal of the Fisheries Research Board of Canada 32:817-820.

Riddell, B.E., and W.C. Leggett. 1981. Evidence of an adaptive basis for geographic variation in body morphology and time of downstream migration of juvenile Atlantic salmon (*Salmo salar*). Canadian Journal of Fisheries and Aquatic Sciences 38:308-320.

Riddell, B.E., W.C. Leggett, and R.L. Saunders. 1981. Evidence of adaptive polygenic variation between two populations of Atlantic salmon (*Salmo salar*) native to tributaries of the S.W. Miramichi River, N.B. Canadian Journal of Fisheries and Aquatic Sciences 38:321-333.

Robbins, W.H., and H.R. MacCrimmon. 1977. Vital statistics and migratory patterns of a potamodromous stock of smallmouth bass, *Micropterous dolomieui*. Journal of the Fisheries Research Board of Canada 34:142-147.

Roslyj, Y.S. 1975. The biology and the census of young Pacific salmons during their downstream migration in the Amur channel. Izvestiya TIRNO 98:113-128.

Ruggles, C.P. 1966. Depth and velocity as a factor in stream rearing and production of juvenile coho salmon. Canadian Fish Culturist 38:37-53.

Ruggles, C.P. 1980. A review of the downstream migration of Atlantic salmon. Canada Department of Fisheries and Oceans Technical Report 952.

Santos, U. de M. 1979. Observacões limnologicas sobre a asfixia e migracão de peixes na Amazonia Central. Ciéncia e Cultura 31:1034-1039.

Scholz, A.T., R.M. Horrall, J.C. Cooper, and A.D. Hasler. 1976. Imprinting to chemical cues: the basis for homestream selection in salmon. Science (Washington DC) 192:1247-1249.

Scholz, A.T., C.K. Gross, J.C. Cooper, R.M. Horrall, A.D. Hasler, R.I. Daly, and R.J. Poff. 1978a. Homing of rainbow trout transplanted in Lake Michigan: a comparison of three procedures used for imprinting and stocking. Transactions of the American

Fisheries Society 107:439-443.

Scholz, A.T., J.C. Cooper, R.M. Horrall, and A.D. Hasler. 1978b. Homing of morpholine imprinted brown trout, *Salmo trutta*. U S National Marine Fisheries Service Fishery Bulletin 76:293-295.

Scott, W.B., and E.J. Crossman. 1973. Freshwater fishes of Canada. Fisheries Research Board of Canada Bulletin 184.

Selset, R., and K.B. Døving. 1980. Behavior of mature anadromous char (*Salmo alpinus* L.) towards odorants produced by smolts of their own population. Acta Physiologica Scandinavica 108:113-122.

Shapley, S.P. 1961. Factors that influence the distribution and movement of Yellowstone cutthroat trout (*Salmo clarki lewisii*) fry in Kiakho Lake outlet, British Columbia. Master's thesis. University of British Columbia, Vancouver, British Columbia, Canada.

Shershnev, A.P., and A.I. Zhulkov. 1979. Features of the downstream migration of young pink salmon and some indices of the efficiency of reproduction of the pink salmon, *Oncorhynchus gorbuscha*, from Pritornaya River. Journal of Ichthyology 19:114-119.

Skeesick, D.G. 1970. The fall immigration of juvenile coho salmon into a small tributary. Research Report of the Fish Commission of Oregon 21:90-95.

Slaney, P.A., and T.G. Northcote. 1974. Effects of prey abundance on density and territorial behavior of young rainbow trout (*Salmo gairdneri*) in laboratory stream channels. Journal of the Fisheries Research Board of Canada 31:1201-1209.

Slivka, A.P., and G.F. Dovgopol. 1979. Qualitative characteristics of the Volga River stellate sturgeon and biological principles of its rational exploitation. Pages 188-200 *in* L.S. Berdichevskij, editor. Biological basis of sturgeon culture development in the USSR. Nauka, Moscow, USSR.

Smith, M.W., and J.W. Saunders. 1958. Movements of brook trout, *Salvelinus fontinalis* (Mitchill), between and within fresh and salt water. Journal of the Fisheries Research Board of Canada 15:1403-1449.

Solomon, D.J. 1973. Evidence for pheromone-influenced homing by migrating Atlantic salmon, *Salmo salar* (L.). Nature (London) 244:231-232.

Solomon, D.J. 1977. A review of chemical communication in freshwater fish. Journal of Fish Biology 11:363-376.

Solomon, D.J. 1978a. Some observations on salmon smolt migration in a chalkstream. Journal of Fish Biology 12:571-574.

Solomon, D.J. 1978b. Migration of smolts of Atlantic salmon (*Salmo salar* L.) and sea trout (*Salmo trutta* L.) in a chalkstream. Environmental Biology of Fishes 3:223-229.

Solomon, D.J. 1982. Migration and dispersion of juvenile brown and sea trout. Pages 136-145 *in* E.L. Brannon and E.O. Salo, editors. Salmon and trout migratory behaviour symposium.

University of Washington College of Fisheries, Seattle, Washington, USA.

Solomon, D.J., and R.G. Templeton. 1976. Movements of brown trout *Salmo trutta* in a chalkstream. Journal of Fish Biology 9:411-423.

Sopuck, R.D. 1978. Emigration of juvenile rainbow trout in Cayuga inlet, New York. New York Fish and Game Journal 25:108-120.

Spence, C.R. 1980. Radio telemetry investigation of the instream distribution and movement of adult Chilcotin River steelhead trout. British Columbia Fish and Wildlife Branch Technical Report F-80-2.

Stabell, O.B. 1982. Detection of natural odorants by Atlantic salmon parr using positive rheotaxis olfactometry. Pages 71-78 *in* E.L. Brannon and E.O. Salo, editors. Salmon and trout migratory behaviour symposium. University of Washington College of Fisheries, Seattle, Washington, USA.

Stauffer, T.M. 1972. Age, growth, and downstream migration of juvenile rainbow trout in a Lake Michigan tributary. Transactions of the American Fisheries Society 101:18-28.

Stepanov, A.S., A.V. Churmasov, and S.A. Cherkoshin. 1979. Sun orientation of pink salmon during their migration. Marine Biology (Vladivostok) 2:20-27.

Stuart, T.A. 1957. The migration and homing behaviour of brown trout (*Salmo trutta* L.). Freshwater Salmon Fisheries Research Scotland, Her Majesty's Stationery Office, Edinburgh, Scotland.

Suzuki, N. 1978. Effects of different ionic environments on the responses of single olfactory receptors in the lamprey. Comparative Biochemistry and Physiology A. Comparative Physiology 61A:461-467.

Svardson, G. 1966. Oringen. Fiske 66:2-31.

Teeter, J. 1980. Pheromone communication in sea lampreys (*Petromyzon marinus*): implications for population management. Canadian Journal of Fisheries and Aquatic Sciences 37:2123-2132.

Thorpe, J.E., and R.I.G. Morgan. 1978. Periodicity in Atlantic salmon *Salmo salar* L. smolt migration. Journal of Fish Biology 12:541-548.

Thorpe, J.E., L.G. Ross, G. Struthers, and W. Watts. 1981. Tracking Atlantic salmon smolts, *Salmo salar* L., through Loch Voil, Scotland. Journal of Fish Biology 19:519-537.

Thorpe, J.E., C. Talbot, and C. Villarreal. 1982. Bimodality of growth and smolting in Atlantic salmon, *Salmo salar* L. Aquaculture 28:123-132.

Todd, I.S. 1966. A technique for the enumeration of chum salmon fry in the Fraser River, British Columbia. Canadian Fish Culturist 38:3-35.

Tutty, B.D., and F.Y.E. Yole. 1978. Overwintering chinook salmon in the upper Fraser River system. Canada Department of Fisheries and Oceans, Manuscript Report 1460.

Vernon, E.H. 1966. Enumeration of migrant pink salmon fry in the Fraser River estuary. International Pacific Salmon Fisheries Commission Bulletin 19.

Welcomme, R.L. 1969. The biology and ecology of the fishes of a small tropical stream. Journal of Zoology (London) 158:485-529.
Welcomme, R.L. 1975. The fisheries ecology of African floodplains. Food and Agriculture Organization of the United Nations, Commission for Inland Fisheries, Africa, Technical Paper 3.
Werner, R.G. 1979. Homing mechanism of spawning white suckers in Wolf Lake, New York. New York Fish and Game Journal 26:48-58.
Westin, L. 1977. Temperature as orientation cue in migrating silver eels, *Anguilla anguilla* (L.). Contribution number 17 Askö Laboratory, University of Stockholm, Stockholm, Sweden.
Westin, L., and L. Nyman. 1977. The migrations of silver eels - when, where and how. Zoologisk Revy 39:2-11.
White, H.C. 1934. Some facts and theories concerning the Atlantic salmon. Transactions of the American Fisheries Society 64:360-362.
White, H.C. 1939. Factors influencing descent of Atlantic salmon smolts. Journal of the Fisheries Research Board of Canada 4:323-326.
Whitehead, P.J.P. 1959. The anadromous fishes of Lake Victoria. Revue de Zoologie et Botanique Africaines 59:329-363.
Woodhead, A.D. 1975. Endocrine physiology of fish migration. Oceanography and Marine Biology Annual Review 13:287-382.
Wrede, W.L. 1932. Versuche über den Artduft der Elritzen. Zeitschrift für Vergleichende Physiologie 17:510-519.
Zaugg, W.S. 1981. Advanced photoperiod and water temperature effects on gill $Na^+ - K^+$ adenosine triphosphatase activity and migration of juvenile steelhead (*Salmo gairdneri*). Canadian Journal of Fisheries and Aquatic Sciences 38:758-764.
Zaugg, W.S., and L.R. McLain. 1972. Changes in gill adenosinetriphosphatase activity associated with parr-smolt transformation in steelhead trout, coho and spring chinook salmon. Journal of the Fisheries Research Board of Canada 29:167-171.

HOMING AND STRAYING IN PACIFIC SALMON

Thomas P. Quinn

Department of Fisheries and Oceans
Fisheries Research Branch
Pacific Biological Station
Nanaimo, British Columbia V9R 5K6 Canada

ABSTRACT

The remarkable ability of Pacific salmon (genus *Oncorhynchus*) to home to their natal stream to spawn has tended to obscure the fact that a small proportion of the spawners stray to non-natal streams. It is hypothesized that straying is an evolutionary alternative to homing and that these two life-history strategies are in dynamic equilibrium. Straying should be relatively common in populations spawning in unstable streams (high annual variation in juvenile survival), and in species and populations spawning in geographically simple streams with similar nearby streams. Straying should also be relatively common in species with little variation in age at maturity.

INTRODUCTION

It is now well established that most Pacific salmon (genus *Oncorhynchus*) that survive to maturity spawn in the river they departed as juveniles years earlier. Once a matter of debate, the parent stream theory is now generally accepted on the basis of numerous marking studies. While it is clear that most salmon spawn in their natal river, some do spawn elsewhere and are known as strays. Strays have received relatively little attention, partly because their numbers are difficult to estimate. It is rarely possible to completely survey all streams to which marked fish could stray. Marks are generally not conspicuous, and some strays would escape detection under even the best of circumstances. On the other hand, entrance into a non-natal river (generally taken as evidence of straying) does not prove that the salmon would have spawned there. One conventional explanation for straying is that such salmon are lost. That is, all salmon are presumed to have the same general homing tendency, but sensory or memory failures or fatigue prevent some from locating their natal stream. This paper proposes another explanation for straying and suggests ways to evaluate predictions about straying rates generated from this explanation.

In contrast to the notion that natural selection favors optimal behavioral patterns within species, recent studies of reproductive biology have indicated that some individuals may successfully adopt a distinctly different alternative life-history strategy (Gadgil 1972; Dominey 1980; Gross and Charnov 1980). By application, this principle suggests that straying may be a viable alternative to homing under certain circumstances and that these two behavior patterns exist in dynamic equilibrium.

In a river whose characteristics are stable over a long period lineages of fish that home will be able to evolve physiological and behavioral specializations for that river (Ricker 1972). Such fish would be more successful than fish straying from other rivers, because the strays would either be generalists or specialists for another river. However, if one year all eggs in the river were lost due to a natural disaster, the only survivors of that population that year would be the progeny of salmon that strayed and spawned elsewhere. In stable rivers such disasters are by definition rare, and straying should not be favored. However, in unstable rivers with high annual variation in juvenile survival more strays should occur. Preliminary experimental evidence supporting this prediction comes from McCart's (1970) study of sockeye salmon (*O. nerka*) spawning in the tributaries of Babine Lake, British Columbia. Mature salmon were taken from spawning areas and released off the mouths of the streams or were displaced 16 km away in the lake. All of the salmon from a large stable river released at the mouth were recaptured in that river, and 93% of those displaced returned as well. By comparison, 87% of the salmon released at the mouths of early unstable streams were recaptured on the spawning grounds of those streams, and only 53% of the displaced sockeye returned to the unstable stream where they were first found.

Straying rates of coho salmon (*O. kisutch*) vary between populations as well. Donaldson and Allen (1958) reported that transplanted coho salmon returned almost exclusively to release sites (194 out of 195 fish). However, Shapovalov and Taft (1954) reported much lower homing rates (85% and 73%) in a 6-year study of wild coho salmon in two adjacent, small, unstable, coastal streams in California.

A second factor which may influence straying is the degree of similarity among spawning streams utilized by the species. The more diverse the streams, the greater the advantage conferred by homing. For example, the migrations of juvenile sockeye salmon in rivers and lakes frequently require complex responses to light, water flow, odors and compass-orienting cues (Groot 1965; Brannon 1972; Quinn 1980). Interpopulation differences in responses to these environmental stimuli are largely innate, and strays would be at a significant disadvantage. In contrast, pink salmon (*O. gorbuscha*) tend to spawn close to the ocean, where more generalized responses would be sufficient for migration to sea. In spite of the difficulties in assessing straying rates there are indications that sockeye salmon have a relatively strong homing tendency (Foerster 1936) and that pink salmon have a greater tendency to stray Helle 1966; Parker 1967). Pink salmon recolonized the Fraser River much faster than sockeye salmon did after removal of the rock slide obstruction at Hell's Gate (Vernon 1962), and pink salmon have also rapidly colonized the Great Lakes (Kwain and Lawrie 1981). Semko (1954) also discussed the relationship between homing and specialization for a river and presented preliminary observations indicating that pink salmon stray more than sockeye salmon.

Chum salmon (*O. keta*) often spawn in small, unstable, coastal streams similar to those used by pink salmon. While this might suggest that their straying rates should also be similar, an important difference in their life histories suggests otherwise. Pink salmon all spawn after 2 years of life, while chum salmon generally spawn after 3, 4 or 5 years (Neave 1966). Thus while any natural disaster on a pink salmon stream would eliminate the whole brood year from a given mating, such a disaster would only eliminate one age class from a chum salmon brood. Straying, therefore, has a temporal as well as a spatial dimension, and the lack of temporal straying by pink salmon may be balanced by spatial straying.

The age at which chinook salmon (*O. tshawytscha*) mature varies greatly, and they tend to spawn in relatively large, stable rivers compared to the other species of salmon. These two factors tend to favor homing over straying, and data from Cowlitz River Hatchery in the Columbia River system indicate that chinook salmon homing rates can be high. From 1978 to 1981, 98.6% of the marked Cowlitz River chinook salmon were recovered in Cowlitz River, despite extensive sampling for strays (40,503 out of 41,085 recoveries from brood years 1974-1977; Quinn and Fresh, unpublished data). Such a high homing rate is not universal, however, as 13% and 10% straying rates were reported for chinook salmon in California by Snyder (1931) and Sholes and Hallock (1979), respectively.

THE HYPOTHESIS

It is proposed that straying is under direct or indirect genetic control and is an alternative life-history strategy in dynamic balance with homing, the dominant strategy. According to this hypothesis, large stable rivers should have higher proportions of homing salmon than smaller, less stable ones. Further, straying rates should be lower in species and populations that have become highly specialized for their freshwater environment and should also be lower in species with variable age at maturity, because this constitutes a form of temporal straying. Natural disasters would temporarily shift the population structure to contain relatively more strays, but homing fish would subsequently re-establish dominance. In addition, the spatial distribution of strays should be nonrandom with rivers geographically close and hydrographically similar to the natal stream receiving most strays.

The hypothesized role of natural selection in the evolution of straying as an alternative strategy to homing does not exclude the influence of proximate environmental factors on straying. There is evidence that when conditions in the natal stream are highly unfavorable, significant straying may occur (IPSFC 1965; Whitman et al. 1982). Straying salmon may also be influenced by the odors of adult or juvenile conspecifics in selecting their spawning stream (Pete 1977; Quinn et al. 1983).

TESTING THE HYPOTHESIS

In order to evaluate the hypothesized roles of stream stability, age class variation and stream complexity in homing, accurate estimates of homing and straying must be established for wild salmon under three conditions.

1) Straying rates of sockeye salmon from stable and unstable tributaries of a lake should be determined. Because age structure and freshwater migratory behavior would be similar, the prediction of greater straying in the less stable tributaries would not be confounded by the other two variables.

2) Straying rates of pink and chum salmon from a given river should be determined. Stream stability and complexity would be the same, and greater straying by pink salmon is predicted due to their lack of variation in age at maturity.

3) Straying rates of chum and sockeye salmon from a given river should be determined. It is difficult to keep age structure and stream stability constant and to vary only complexity of freshwater habitat. However, chum and sockeye salmon often have similar age structures, and sympatric populations would experience the same degree of stream stability. Because chum salmon fry migrate directly to sea after they emerge from gravel incubation areas and sockeye salmon have complex lacustrine migrations, they might vary in the degree of specialization for the freshwater environment. Greater straying by chum salmon is predicted. Sympatric chum and sockeye salmon populations are uncommon, but they do exist, e.g. in Weaver Creek, British Columbia.

In summary, it is important to view migration and homing as integral parts of the reproductive biology of fish species. By considering homing from this evolutionary perspective, new insights can be gained and testable hypotheses can be generated.

ACKNOWLEDGMENTS

I thank the following individuals for valuable discussions and criticisms of the manuscript: R. Bams, C. Busack, C. Groot, M. Gross, B. Riddell and F. Withler.

REFERENCES

Brannon, E.L. 1972. Mechanisms controlling migration of sockeye salmon fry. International Pacific Salmon Fisheries Commission Bulletin 21:1-86.

Dominey, W.J. 1980. Female mimicry in male bluegill sunfish - a genetic polymorphism? Nature (London) 284:546-548.

Donaldson, L.R., and G.H. Allen. 1958. Return of silver salmon *Oncorhynchus kisutch* (Walbaum) to point of release. Transactions of the American Fisheries Society 87:13-22.

Foerster, R.E. 1936. The return from the sea of sockeye salmon (*Oncorhynchus nerka*) with special reference to percentage survival, sex proportions and progress of migration. Journal of the Biological Board of Canada 3:26-42.

Gadgil, M. 1972. Male dimorphism as a consequence of sexual selection. American Naturalist 106:574-580.

Groot, C. 1965. On the orientation of young sockeye salmon (*Oncorhynchus nerka*) during their seaward migration out of lakes. Behaviour, Supplement 14:1-198.

Gross, M.R., and E.L. Charnov. 1980. Alternative male life histories in bluegill sunfish. Proceedings of the National Academy of Sciences of the USA 77:6937-6940.

Helle, J.H. 1966. Behavior of displaced adult pink salmon. Transactions of the American Fisheries Society 95:188-195.

International Pacific Salmon Fisheries Commission (IPSFC). 1965. Annual report for 1964.

Kwain, W.-h., and A.H. Lawrie. 1981. Pink salmon in the Great Lakes. Fisheries (Bethesda) 6(2):2-6.

McCart, P.J. 1970. A polymorphic population of *Oncorhynchus nerka* at Babine Lake, involving anadromous (sockeye) and non-anadromous (kokanee) forms. Doctoral dissertation. University of British Columbia, Vancouver, British Columbia, Canada.

Neave, F. 1966. Salmon of the North Pacific Ocean, Part III. A review of the life history of North Pacific salmon. 6. Chum salmon in British Columbia. International North Pacific Fisheries Commission Bulletin 18:81-86.

Parker, R.R. 1967. Contributions of the 1964 brood year of Bella

Coola pink salmon to fisheries and escapements of the central British Columbia area. Fisheries Research Board of Canada Manuscript Report 935:1-19.

Pete, K. 1977. Species specific odor as a guiding mechanism for local orientation in homing chinook (*Oncorhynchus tshawytscha*) and coho (*O. kisutch*) salmon. Master's thesis. University of Washington, Seattle, Washington, USA.

Quinn, T.P. 1980. Evidence for celestial and magnetic compass orientation in lake migrating sockeye salmon fry. Journal of Comparative Physiology A. Sensory, Neural, and Behavioral Physiology 137A:243-248.

Quinn, T.P., E.L. Brannon, and R.P. Whitman. 1983. Pheromones and the water source preferences of adult coho salmon (*Oncorhynchus kisutch*). Journal of Fish Biology 22:677-684.

Ricker, W.E. 1972. Heredity and environmental factors affecting certain salmonid populations. Pages 19-160 *in* R.C. Simon and P.A. Larkin, editors. The stock concept in Pacific salmon. H.R. MacMillan Lectures in Fisheries, University of British Columbia, Vancouver, British Columbia, Canada.

Semko, R.S. 1954. The stocks of west Kamchatka salmon and their commercialization. Izvestiya Tikhookeanskogo Nauchno-issledovatelskogo Instituta Rybnogo Khozyaistva i Okeanografii 41:3-109. (Fisheries Research Board of Canada Translation Series number 288.)

Shapovalov, L., and A.C. Taft. 1954. The life histories of the steelhead rainbow trout (*Salmo gairdneri gairdneri*) and silver salmon (*Oncorhynchus kisutch*). California Department of Fish and Game, Fish Bulletin 98:1-375.

Sholes, W.H., and R.J. Hallock. 1979. An evaluation of rearing fall-run chinook salmon (*Oncorhynchus tshawytscha*) to yearlings at Feather River Hatchery, with a comparison of returns from hatchery and downstream releases. California Fish and Game 65: 239-255.

Snyder, J.O. 1931. Salmon of the Klamath River California. California Department of Fish and Game, Fish Bulletin 34:1-130.

Vernon, E.H. 1962. Pink salmon populations of the Fraser River system. Pages 53-58 *in* N.J. Wilimovsky, editor. Symposium on pink salmon. H.R. MacMillan Lectures in Fisheries. University of British Columbia, Vancouver, British Columbia, Canada.

Whitman, R.P., T.P. Quinn, and E.L. Brannon. 1982. Influence of suspended volcanic ash on homing behavior of adult chinook salmon. Transactions of the American Fisheries Society 111: 63-69.

MORPHOLINE, BILE ACIDS AND SKIN MUCUS AS POSSIBLE CHEMICAL CUES IN SALMONID HOMING: ELECTROPHYSIOLOGICAL RE-EVALUATION

Toshiaki J. Hara, S. Macdonald, Robert E. Evans, Takayuki Marui* and S. Arai**

Department of Fisheries and Oceans, Freshwater Institute
Winnipeg, Manitoba R3T 2N6 Canada

*Department of Oral Physiology
Kagoshima University Dental School
Kagoshima 890, Japan

**National Research Institute of Aquaculture
Tamaki, Mie 519-04, Japan

ABSTRACT

The olfactory-imprinting and pheromone hypotheses of salmon homing recognize the involvement of olfaction in the recognition of the homestream. However, physiological basis for olfactory recognition and the nature of homestream odors have not yet been established. In this paper the state of knowledge and advances in the study of chemical cues relevant to salmonid homing are reviewed, with special emphasis on 1) imprinting to morpholine, 2) skin mucus as a chemical signal, and 3) chemoreceptor responses to bile acids. Although homing of salmonids artificially imprinted to morpholine appears evident from behavioral studies, the olfactory detection of morpholine has not been adequately demonstrated. The skin mucus has been shown to be a potent olfactory stimulus for salmonids. Chemical characterization revealed that free amino acids present in the mucus were primarily responsible for olfactory stimulation; a synthetic mucus, a mixture of amino acids based on the analysis data, induced olfactory response indistinguishable from that induced by the original mucus. Electrophysiological studies showed that bile acids, especially taurine conjugates, were not only potent olfactory stimulants, but also highly-specific taste stimuli for rainbow trout. The threshold concentration for taurolithocholic acid, the most potent bile acid tested, was estimated at 10^{-12} M, nearly 4 log units lower than that for L-proline, the most potent taste stimulant reported for this species. In the olfactory system bile acids were as stimulatory as amino acids, with the threshold being almost 1,000 times higher than those for trout taste receptors. Because high sensitivity of the salmonid gustatory system to certain chemicals has now been demonstrated, it is no longer appropriate to consider olfaction to be the sensory modality for chemical detection only on the basis of its high sensitivity. In the light of these findings some important issues for future study are discussed.

INTRODUCTION

The olfactory hypothesis of salmon homing proposes that salmonids imprint to certain distinctive odors of the home stream during the early period of residence (Hasler 1966). As adults they use this information to locate the home stream tributary at least during the final stages of the homeward migration. Because of local differences in soil and vegetation of the drainage basin, each stream has a unique chemical composition and, thus, a distinctive odor to which the juvenile salmon imprint before they leave the home stream (Hasler and Wisby 1951). This hypothesis is supported by considerable circumstantial and direct experimental evidence (cf. Hasler et al. 1978).

This imprinting hypothesis has been challenged by Nordeng (1971, 1977), who proposes that homeward navigation is an inherited response to population-specific pheromone trails released by descending smolts. During the downstream migration smolts are present almost continuously in each river system. Thus the migrating smolts and nonmigrating young provide a constant source of population odor to which the adults may respond. The pheromones are thought to be released from the skin mucus. This pheromone hypothesis is also supported by various observations and experiments which suggest that migrating adults are attracted to streams containing conspecifics and that they are able to discriminate between populations on the basis of olfaction (cf. Nordeng 1977; Døving et al. 1980).

The hypotheses of Hasler and Nordeng may not necessarily be mutually exclusive. Under natural conditions, population-specific odor may be just one component in the chemical environment to which a salmon imprints as a juvenile and responds as an adult. More importantly, both hypotheses recognize the involvement of olfaction in the recognition of the homestream. Despite extensive investigations (cf. Døving et al. 1980; Cooper and Hirsch 1982), the physiological basis for olfactory imprinting and recognition and the nature of homestream odors has not yet been firmly established. In this paper the state of knowledge and major advances in the study of chemical cues relevant to salmonid homing are reviewed, with special emphasis on skin mucus and bile acids, and some important issues for future study are discussed in the light of recent electrophysiological findings.

OLFACTORY IMPRINTING TO MORPHOLINE: REVIEW

The homing of salmonids artificially imprinted or exposed to low concentrations of morpholine appears evident from extensive behavioral studies by Hasler and his colleagues (for review, see Hasler et al. 1978). Further support is provided by electrophysiological studies in which significant differences were obtained in the magnitude of the olfactory bulbar responses to 1% (1.1×10^{-1} M) morpholine of fish exposed to 5×10^{-5} mg/liter (5.7×10^{-10} M) morpholine as fingerlings as compared to unexposed fish (Dizon et al. 1973; Cooper and Hasler 1974, 1976). However, the validity of electrophysiological analyses of olfactory bulbar responses in salmonids imprinted to morpholine was later questioned (Hara 1974; Hara and Macdonald 1975; Hara and Brown 1979, 1982). One major criticism has been the use of high concentrations (10^{-2} - 10^{-1} M) of morpholine for electrophysiological studies. At these concentrations morpholine solutions have a pH greater than 10 and induce nonspecific, irritational responses in the fish olfactory system, which are not directly associated with normal olfactory function (Hara 1974). In all reported behavioral studies (Dizon et al. 1973; Scholz et al. 1973, 1976, 1978; Cooper and Hasler 1974, 1976; Cooper et al. 1976; Cooper and Scholz 1976; Johnsen and Hasler 1980) salmonids were exposed to morpholine at 5.7×10^{-10} M and decoyed to streams scented with the same chemical at equivalent or slightly higher concentrations. Electrophysiologically, however, no homing coho salmon (*Oncorhynchus kisutch*) responded to morpholine at concentrations lower than 1.1×10^{-2} M. In our studies with rainbow trout (*Salmo gairdneri*) morpholine produced no olfactory response until the concentration reached 10^{-3} M where it caused inhibition of background activity (Hara and Brown 1979). Furthermore, at 10^{-3} M, morpholine is nonstimulatory either for the olfactory system of Atlantic salmon (*Salmo salar*) (Sutterlin and Sutterlin 1971) or the gustatory system of rainbow trout (Marui, Evans, Zielinski and Hara unpublished data). Heart-rate conditioning studies by Hirsch (1977) show that coho salmon were unable to detect morpholine olfactorily at concentration lower than 2.3×10^{-8} M. This does not necessarily mean that fish are unable to detect the chemical. Electrophysiological and cardiac conditioning techniques may not be sensitive enough to record a response, or morpholine, instead of reacting with olfactory receptors, may have modified the characteristics of stream odors. In any event the involvement of olfaction in imprinting to morpholine at 5.7×10^{-10} M has not been adequately established. Alternatively, chemical imprinting and homing may have occurred via sensory channels other than olfaction. The gustatory systems of salmonids are highly sensitive to certain chemical stimuli (see below). Cauterization experiments will be required to prove that the behavioral responses attributable to morpholine are mediated by olfaction.

SKIN MUCUS AS A CHEMICAL SIGNAL

The skin mucus of fish, besides serving primarily as a protector, has been demonstrated to play an important role in schooling behavior, recognition of individuals and homing migration (Wrede 1932; Hemmings 1966; Todd et al. 1967; Nordeng 1971; Solomon 1973; Höglund and Astrand 1973; Höglund et al. 1975; Kinosita 1975; Selset and Døving 1980). Olfaction has been suggested to be the mediator of these behaviors. Electrophysiological studies have shown that skin mucus is an effective olfactory stimulus for rainbow trout and that the active components responsible for olfactory stimulation are heat-stable and nonvolatile, and have molecular weight less than 1,000 (Hara and Macdonald 1976). Mucus from migratory populations of Arctic char (*Salvelinus alpinus*) and Atlantic salmon produced differential neuronal responses in the olfactory bulb (Døving et al. 1973, 1974; Fisknes and Døving 1982). The results of further chemical characterization of mucous substances and the effectiveness of mucous substances as olfactory stimulants in rainbow trout and lake whitefish (*Coregonus clupeaformis*) will be described below. Preliminary results have been published (Hara 1977).

Analysis of Free Amino Acids in Mucus

Mucus was collected by washing the body surface of a fish with a jet of distilled water from a polyethylene wash bottle (Hara and Macdonald 1976). The vent and head regions of the fish were wrapped with tissues to avoid contamination with intestinal contents (cf. Stabell and Selset 1980). The mucous water thus collected was lyophilized and chromatographed (Dowex 2x8 Cl^- ion exchange resin) after deproteinization with 1% picric acid. The eluate (with 0.02 N HCl) was then analyzed for amino acids with an amino acid analyzer (Hitachi KLA-3B), employing the two-column system.

Figure 1 illustrates the concentration of amino acids and related compounds in the skin mucus of rainbow trout and lake whitefish. With few exceptions the pattern of free amino acids was similar in both species. Taurine was the most abundant amino acid, totalling 20.52 and 30.27 μmoles/100 mg mucus in the two species. α-Amino-butyric acid, galactosamine and anserine were not detected in whitefish mucus. All amino acids, except β-alanine in rainbow trout and phosphoserine in whitefish, were present in larger quantities in males than females. Furthermore, the total amount of free amino acids present in the mucus was significantly greater for males than females of both species (Table 1). Stabell and Selset (1980) also found a considerable amount of taurine plus 19 common amino acids in mucus collected from Atlantic salmon and Arctic char.

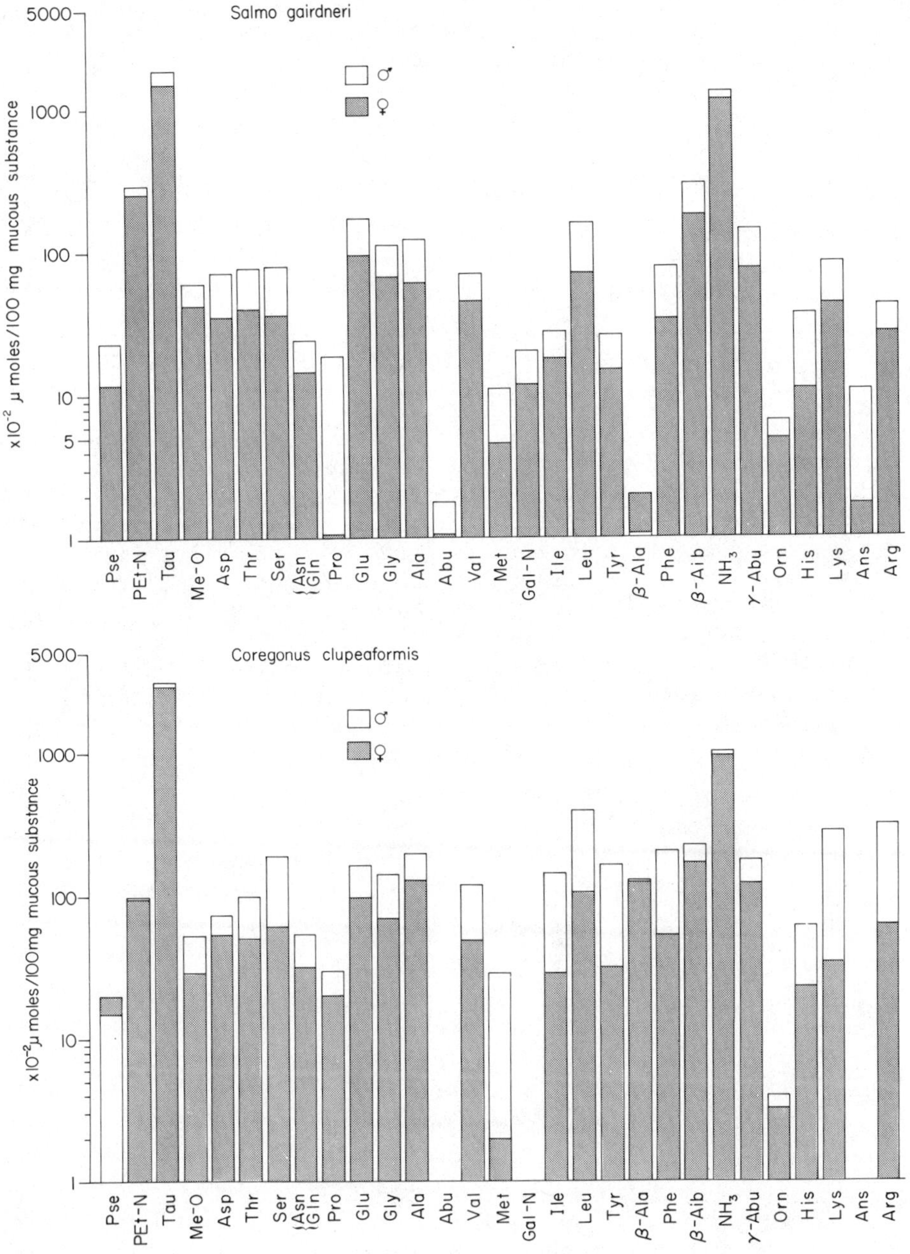

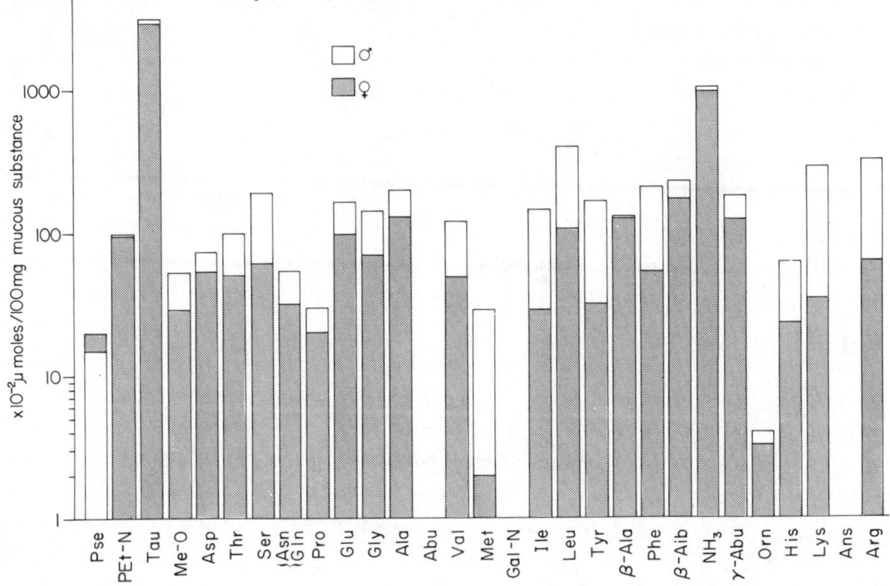

Figure 1. Free amino acids and related compounds in the skin mucus (µmoles/100 mg mucous substance). Results are means of five males and seven females for rainbow trout (*Salmo gairdneri*) (upper panel) and seven males and 13 females for lake whitefish (*Coregonus clupeaformis*) (lower panel).

Table 1. Total free amino acids in the skin mucus of rainbow trout, *Salmo gairdneri* and lake whitefish, *Coregonus clupeaformis* (μmoles/100 mg mucous substance). Numbers of fish analyzed in parentheses.

	Male	Female
Salmo gairdneri	36.55 ± 7.96 (5)	24.45 ± 8.94 (7)
Coregonus clupeaformis	55.18 ± 7.88 (7)	39.38 ± 8.85 (13)

Olfactory Responses to Skin Mucus

The original mucous water, the eluate of chromatography (called chromatographed mucous water) and a mixture of amino acids based on the above analysis data (called synthetic mucous water) were tested for their effectiveness as olfactory stimulants. The olfactory responses were measured by recording the electrical activities induced in the olfactory bulb when the nares were stimulated with mucous waters, according to methods described previously (Hara 1973; Hara et al. 1973). Typical olfactory responses of rainbow trout to the original and synthetic mucous waters are illustrated in Figure 2. A dose-response relationship for the synthetic mucous water was indistinguishable from that for the original mucus water (Fig. 3).

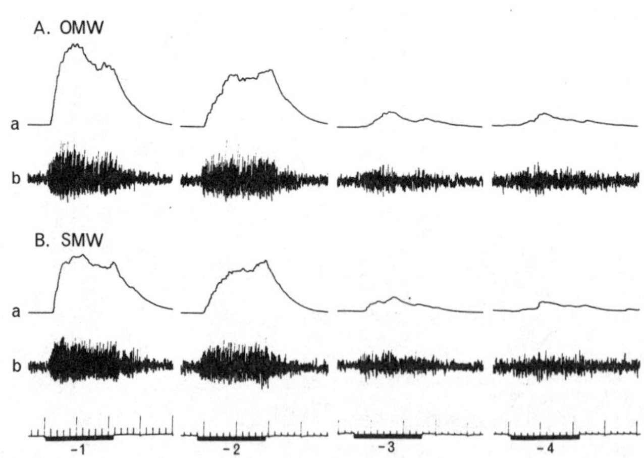

Figure 2. Electrical responses recorded from the olfactory bulb of rainbow trout (*Salmo gairdneri*) when the olfactory organ was stimulated with the original (OMW) and synthetic (SMW) mucous waters. In each pair of records (a) shows the integration of (b). Concentrations of mucous waters are indicated in logarithmic dilution at the bottom. Time scale, 1 division = 1 sec. Bold lines along time scale indicate duration of stimuli.

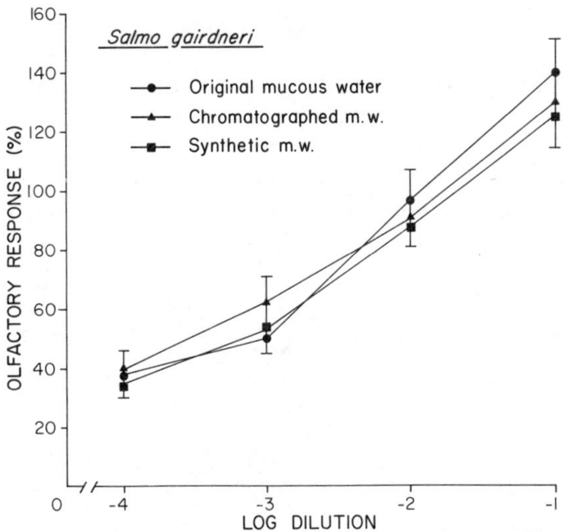

Figure 3. Dose-response relationships of the olfactory bulbar responses to the original, chromatographed and synthetic mucous waters in rainbow trout (*Salmo gairdneri*).

These data indicate that free amino acids present in the mucus are fully responsible for olfactory stimulation in rainbow trout. Essentially the same results were obtained with lake whitefish. The role of skin mucus in the behavior of these species is not well understood. Preliminary experiments showed that whitefish were either attracted or repelled by mucus collected from conspecifics suggesting a role in individual or sexual recognition. Skin mucus may be responsible for intraspecific attraction among individuals of juvenile Arctic char (Höglund and Åstrand 1973).

HIGH SENSITIVITY OF RAINBOW TROUT CHEMORECEPTORS TO BILE ACIDS

Døving et al. (1980) demonstrated that some bile acids, especially taurine and sulfate conjugates, were potent olfactory stimuli for Arctic grayling (*Thymallus thymallus*) and Arctic char with mean thresholds ranging between 6.3×10^{-9} and 8.0×10^{-8} M. High sensitivity of the olfactory receptors was implicated in their role as specific chemical signals in homing migration. Our recent electrophysiological findings that bile acids are not only potent olfactory stimuli but also highly specific gustatory stimuli for rainbow trout will be discussed below.

Olfactory Responses to Bile Acids

Olfactory responses were tested by recording the electro-olfactogram (EOG) using Ag-AgCl electrodes via Ringer-gelatin-filled capillary pipettes positioned directly on the surface of an olfactory

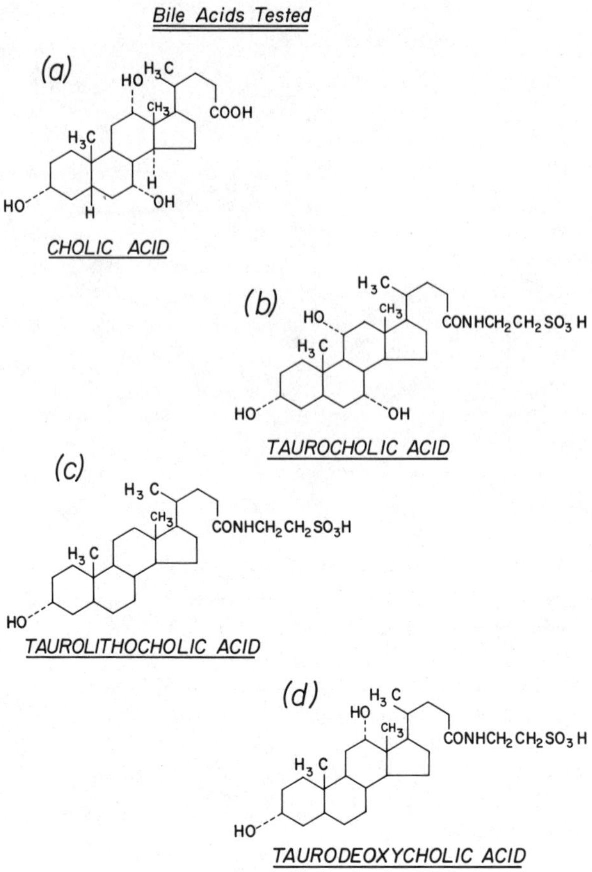

Bile Acids Tested

(a) CHOLIC ACID

(b) TAUROCHOLIC ACID

(c) TAUROLITHOCHOLIC ACID

(d) TAURODEOXYCHOLIC ACID

Figure 4. Structural formulae of bile acids tested for olfactory and gustatory responses.

lamella (Evans and Hara unpublished data). The EOG is the population average of receptor potentials responsible for the initiation of nerve impulses (Ottoson 1971). Bile acids tested are shown in Figure 4. Typical EOG responses to various concentrations of taurocholic acid, the most effective bile acid tested, are illustrated in Figure 5. The EOG responses to taurocholic acid remained unaffected during cross-adaptation to an amino acid, L-serine at 10^{-2} M, suggesting the existence of separate receptor sites for these two groups of chemicals (Figs. 5B and 6). All bile acids tested were equally or more stimulatory than L-serine, one of the most potent olfactory stimulants reported for this species (Fig. 6; cf. Hara 1982). The threshold concentration for taurocholic acid was estimated at 10^{-9} M or lower, which is comparable to those for stimulatory amino acids. The taurine conjugates were generally more stimulatory than the

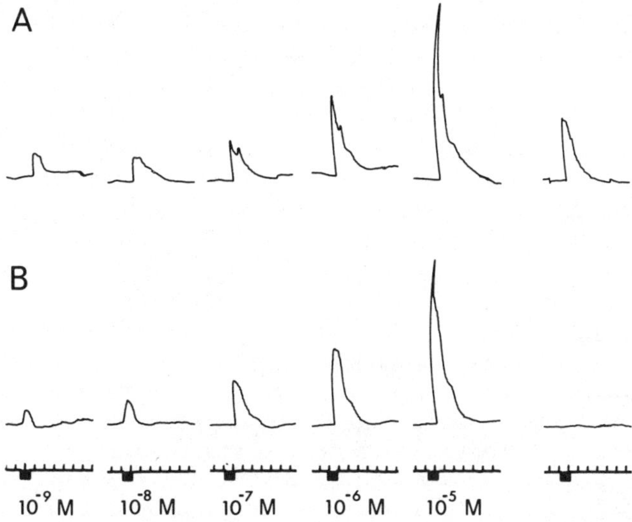

Figure 5. Electro-olfactogram (EOG) responses of the rainbow trout (*Salmo gairdneri*) to a bile acid, taurocholic acid, before (A) and during (B) cross-adaptation to 10^{-2} M L-serine. EOG response to a standard stimulus 10^{-5} M L-serine is shown at the right. Time scale, 1 division = 5 sec. Bold lines along time scale indicate duration of stimuli.

nonconjugate, and taurine itself was inactive. These results are in general agreement with the findings by Døving et al. (1980), who studied stimulatory effectiveness of bile acids by recording the DC

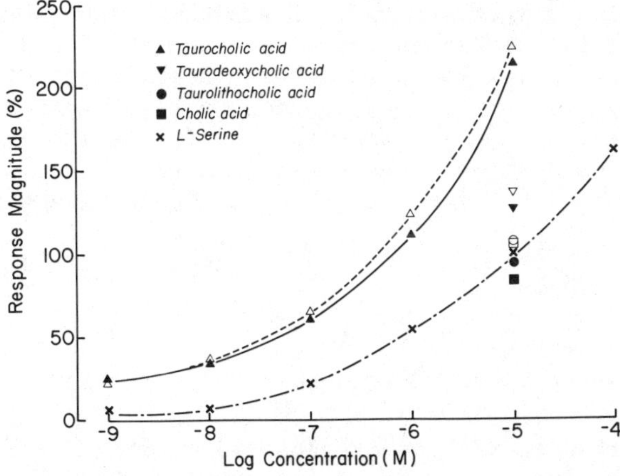

Figure 6. Dose-response relationships of EOG responses to bile acids and a standard stimulant L-serine. In each pair open symbols indicate responses recorded while the olfactory receptors were cross-adapted to 10^{-2} M L-serine.

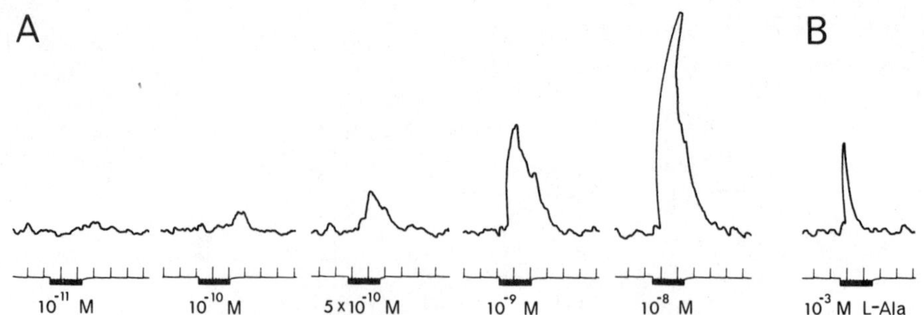

Figure 7. (A) Gustatory responses to a bile acid, taurolithocholic acid, recorded from palatine nerves innervating the upper lip and the palate in rainbow trout (*Salmo gairdneri*). (B) Gustatory response to a standard stimulus 10^{-3} M L-alanine. Time scale, 1 division = 5 sec. Bold lines along time scale indicate duration of stimuli.

potential shifts in the olfactory bulb of migratory salmonids.

Gustatory Responses to Bile Acids

Stimulatory effectiveness of bile acids on rainbow trout gustatory receptors were tested by recording the integrated electrical activity of palatine nerves innervating the upper lip and the palate (Marui, Evans, Zielinski and Hara unpublished data). Figure 7 shows typical responses to increasing concentration of taurolithocholic acid, the most effective bile acid tested. The response magnitude increased with logarithmic increase in concentration; the threshold ranged between 10^{-12} and 10^{-11} M (Fig. 8). The threshold for taurolithocholic acid is almost 4 log units lower than that for L-proline, the most potent taste stimulant recorded under the same experimental conditions (Marui, Evans, Zielinski and Hara unpublished data). Taurodeoxycholic acid and cholic acid were less effective than the other bile acids tested but still far exceeding L-proline (Fig. 8). The results suggest the existence of a large number of taste receptor sites with an extremely high affinity for this group of chemicals.

CONCLUDING REMARKS: FUTURE RESEARCH

The importance of olfactory cues in guiding the final stages of the salmonid homing migration is demonstrated by the significant reduction in homing success of olfactory-occluded fishes (Wisby and Hasler 1954; Hiyama et al. 1967; Groves et al. 1968; DeLacy et al. 1969). However, the nature of the olfactory cues has been variously characterized as volatile, nonvolatile, organic and nonorganic

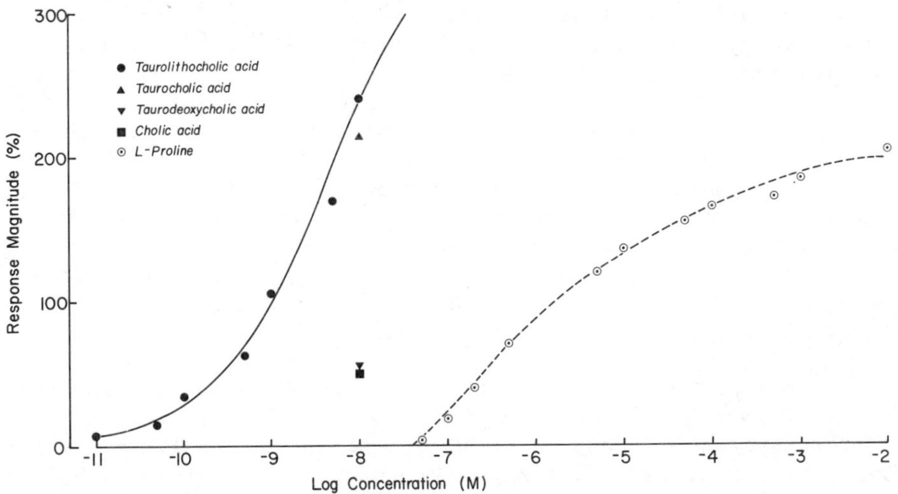

Figure 8. Dose-response relationships of gustatory responses to bile acids and a standard stimulant L-proline.

(Hasler and Wisby 1951; Idler et al. 1961; Fagerlund et al. 1963; McBride et al. 1964; Miles 1968). If odors distinctive to streams are an often-suggested mixture of a fingerprint type, as opposed to a single chemical, and fish discriminate subtle differences in their chemical composites, the identification of the olfactory cues by conventional methods seems extremely difficult. Pheromones, more appropriately substances such as skin mucus and intestinal contents emanating from conspecifics or other species, provide the same problems; no single chemical would seem to make population-specific odorants characteristic (Stabell and Selset 1980; Selset 1980; Stabell et al. 1982). The contribution of electrophysiology may be limited during this stage of the bioassay, when the effects of biological materials on particular neural patterns are determined (Døving et al. 1974; Fisknes and Døving 1982). Once the chemical substances responsible for the behavioral reaction have been identified, electrophysiological approaches would play a major role in investigating the specificity of the chemoreceptors. It should be noted that amino acids and bile acids identified in the skin mucus and intestinal contents represent the two major groups of chemicals responsible for effective olfactory and gustatory stimulation in salmonids. If mucus is used in chemical communication, including homestream recognition, it could certainly provide the specific clues needed for a successful identification.

Do salmonids imprint to morpholine olfactorily? It is evident from our studies that the olfactory bulbar responses induced by 1% (1.1×10^{-1} M) morpholine are nonspecific and irritational, and are caused by a mechanism not directly associated with normal olfactory function. All attempts have so far failed to provide direct evidence indicating that morpholine, at the levels used for field experiments, is detected by fish either through olfaction or taste. Since high sensitivity of the gustatory system to certain chemicals has now been demonstrated for salmonids, it is no longer appropriate to consider olfaction to be the sensory modality for chemical detection only on the basis of its lower effective concentration (Cooper 1982). In order to determine indirectly whether olfaction is involved in the detection of morpholine, artificial-imprinting or tracking experiments may be conducted using fishes with their olfactory organs ablated or cauterized. Ablation of taste organs is impracticable in fish, because taste buds are distributed over wider areas of the body surface and are innervated by a complex cranial nerve system.

There appears to be a relationship between nasal and genital functions, whereby sex hormones influence the function of the olfactory system. In the goldfish (*Carassius auratus*), for instance, administration of estradiol markedly enhances the olfactory bulbar responses to chemical and electrical stimulation of the olfactory organ (Hara 1967). The plasma concentrations of the steroid hormones of salmon show seasonal fluctuations coinciding with various physiological states during their life cycle (cf. Cedard et al. 1961), and enhanced secretion of gonadal steroids has sometimes been implicated as a triggering factor in spawning migration. Contrary to a notion (Cooper 1982), however, little or no information is available indicating that salmonids in spawning condition respond differently from nonspawning fish. It is thus encouraged to investigate whether the sensitivity of the chemoreceptor systems of salmonids varies with hormonal conditions during their life cycles.

REFERENCES

Cedard, L., M. Fontaine, and T. Nomura. 1961. Sur la teneur en oestrogènes du sang du saumon adulte (*Salmo salar* L.) en eau douce. Comptes Rendus hebdomadaires des Séances de l'Académie des Sciences, Paris, France 252:2656-2657.

Cooper, J. 1982. Comment on electroencephalographic responses to morpholine and their relationship to homing. Canadian Journal of Fisheries and Aquatic Sciences 39:1544-1546.

Cooper, J.C., and A.D. Hasler. 1974. Electroencephalographic evidence for retention of olfactory cues in homing coho salmon. Science (Washington DC) 183:336-338.

Cooper, J.C., and A.D. Hasler. 1976. Electrophysiological studies of morpholine-imprinted coho salmon (*Oncorhynchus kisutch*) and rainbow trout (*Salmo gairdneri*). Journal of the Fisheries Research Board of Canada 33:688-694.

Cooper, J.C., and P.J. Hirsch. 1982. The role of chemoreception in salmonid homing. Pages 343-362 *in* T.J. Hara, editor. Chemoreception in fishes. Elsevier Scientific Publishing Company, Amsterdam, Holland.

Cooper, J.C., and A.T. Scholz. 1976. Homing of artificially imprinted steelhead (rainbow) trout, *Salmo gairdneri*. Journal of the Fisheries Research Board of Canada 33:826-829.

Cooper, J.C., A.T. Scholz, R.M. Horrall, A.D. Hasler, and D.M. Madison. 1976. Experimental confirmation of the olfactory hypothesis with homing, artificially imprinted coho salmon (*Oncorhynchus kisutch*). Journal of the Fisheries Research Board of Canada 33:703-710.

DeLacy, A.C., L.R. Donaldson, and E.L. Brannon. 1969. Homing behavior of chinook salmon. Research in Fisheries 1968, College of Fisheries, University of Washington, Seattle, Washington, USA.

Dizon, A.E., R.M. Horrall, and A.D. Hasler. 1973. Olfactory electroencephalographic responses of homing coho salmon, *Oncorhynchus kisutch*, to water conditioned by conspecifics. US National Marine Fisheries Service Fishery Bulletin 71:893-896.

Døving, K.B., P.S. Engen, and H. Nordeng. 1973. Electrophysiological studies on the olfactory sense in char (*Salmo alpinus* L.). Comparative Biochemistry and Physiology A. Comparative Physiology 45A:21-24.

Døving, K.B., H. Nordeng, and B. Oakley. 1974. Single unit discrimination of fish odours released by char (*Salmo alpinus* L.) populations. Comparative Biochemistry and Physiology A. Comparative Physiology 47A:1051-1063.

Døving, K.B., R. Selset, and G. Thommesen. 1980. Olfactory sensitivity to bile acids in salmonid fishes. Acta Physiologica Scandinavica 108:123-131.

Fagerlund, U.H.M., J.R. McBride, M. Smith, and N. Tomlinson. 1963. Olfactory perception in migrating salmon. III. Stimulants for adult sockeye salmon (*Oncorhynchus nerka*) in home stream waters. Journal of the Fisheries Research Board of Canada 20:1457-1463.

Fisknes, B., and K.B. Døving. 1982. Olfactory sensitivity to group-specific substances in Atlantic salmon (*Salmo salar* L.). Journal of Chemical Ecology 8:1083-1092.

Groves, A.B., G.B. Collins, and P.S. Trefethen. 1968. Roles of olfaction and vision in choice of spawning site by homing adult chinook salmon (*Oncorhynchus tsawytscha*). Journal of the Fisheries Research Board of Canada 25:867-876.

Hara, T.J. 1967. Electrophysiological studies of the olfactory system of the goldfish, *Carassius auratus* L. III. Effects of sex hormone on the electrical activity of the olfactory bulb. Comparative Biochemistry and Physiology 22:209-226.

Hara, T.J. 1973. Olfactory responses to amino acids in rainbow trout, *Salmo gairdneri*. Comparative Biochemistry and Physiology A. Comparative Physiology 44A:407-416.

Hara, T.J. 1974. Is morpholine an effective olfactory stimulant in fish? Journal of the Fisheries Research Board of Canada 31:1547-1550.

Hara, T.J. 1977. Amino acids as olfactory stimuli in fish. Pages 157-164 *in* J. LeMagnen and P. MacLeod, editors. Olfaction and Taste VI. Information Retrieval, London, England.

Hara, T.J., editor. 1982. Chemoreception in Fishes. Elsevier Scientific Publishing Company, Amsterdam, Holland.

Hara, T.J., and S.B. Brown. 1979. Olfactory bulbar electrical responses of rainbow trout (*Salmo gairdneri*) exposed to morpholine during smoltification. Journal of the Fisheries Research Board of Canada 36:1186-1190.

Hara, T.J., and S.B. Brown. 1982. Comment on electroencephalographic responses to morpholine and their relationship to homing: reply. Canadian Journal of Fisheries and Aquatic Sciences 39: 1546-1548.

Hara, T.J., and S. Macdonald. 1975. Morpholine as olfactory stimulus in fish. Science (Washington, DC) 187:81-82.

Hara, T.J., and S. Macdonald. 1976. Olfactory responses to skin mucous substances in rainbow trout (*Salmo gairdneri*). Comparative Biochemistry and Physiology A. Comparative Physiology 54A:41-44.

Hara, T.J., Y.M.C. Law, and E. van der Veen. 1973. A stimulatory apparatus for studying the olfactory activity in fishes. Journal of the Fisheries Research Board of Canada 30:283-285.

Hasler, A.D. 1966. Underwater guideposts - Homing of salmon. University of Wisconsin Press, Madison, Wisconsin, USA.

Hasler, A.D., and W.J. Wisby. 1951. Discrimination of stream odors by fishes and its relation to parent stream behavior. American Naturalist 85:223-238.

Hasler, A.D., A.T. Scholz, and R.M. Horrall. 1978. Olfactory imprinting and homing in salmon. American Scientist 66:347-355.

Hemmings, C.C. 1966. Olfaction and vision in fish schooling. Journal of Experimental Biology 45:449-464.

Hirsch, P.J. 1977. Conditioning of heart rate in coho salmon (*Oncorhynchus kisutch*) to odors. Doctoral dissertation. University of Wisconsin, Madison, Wisconsin, USA.

Hiyama, Y., T. Taniuchi, K. Suyama, K. Ishioka, R. Sato, T. Kajihara, and T. Maiwa. 1967. A preliminary experiment on the return of tagged chum salmon to the Otsuchi River, Japan. Bulletin of the Japanese Society of Scientific Fisheries 33:18-19.

Höglund, L.B., and M. Åstrand. 1973. Preferences among juvenile char (*Salvelinus alpinus* L.) to intraspecific odours and water currents studied with the fluviarium technique. Institute of Freshwater Research Drottningholm Report 53:21-30.

Höglund, L.B., A. Bohman, and N.-A. Nilsson. 1975. Possible odour responses of juvenile Arctic char (*Salvelinus alpinus* L.) to three other species of subarctic fish. Institute of Freshwater Research Drottningholm Report 54:21-35.

Idler, D.R., J.R. McBride, R.E.E. Jonas, and N. Tomlinson. 1961. Olfactory perception in migrating salmon. II. Studies on a laboratory bio-assay for homestream water and mammalian repellent. Canadian Journal of Biochemistry and Physiology 39: 1575-1584.

Johnsen, P.B., and A.D. Hasler. 1980. The use of chemical cues in the upstream migration of coho salmon, *Oncorhynchus kisutch* Walbaum.

Journal of Fish Biology 17:67-73.

Kinosita, H. 1975. Schooling behavior of marine catfish eel (*Plotosus anguillaris*). Pages 135-154 *in* A. Okajima and K. Maruyama, editors. Modern biological sciences, volume 9, Locomotion and behavior, Iwanami, Tokyo, Japan.

McBride, J.R., U.H.M. Fagerlund, M. Smith, and N. Tomlinson. 1964. Olfactory perception in juvenile salmon. II. Conditioned response of juvenile sockeye salmon (*Oncorhynchus nerka*) to lake waters. Canadian Journal of Zoology 42:245-248.

Miles, S.G. 1968. Rheotaxis of elvers of the American eel (*Anguilla rostrata*) in the laboratory to water from different streams in Nova Scotia. Journal of the Fisheries Research Board of Canada 25:1591-1602.

Nordeng, H. 1971. Is the local orientation of anadromous fishes determined by pheromones? Nature (London) 233:411-413.

Nordeng, H. 1977. A pheromone hypothesis for homeward migration in anadromous salmonids. Oikos 28:155-159.

Ottson, D. 1971. The electro-olfactogram. Pages 95-131 *in* L.M. Beidler, editor. Handbook of sensory physiology, volume IV. Springer-Verlag, Heidelberg, Federal Republic of Germany.

Scholz, A.T., R.M. Horrall, J.C. Cooper, and A.D. Hasler. 1976. Imprinting to chemical cues: the basis for home stream selection in salmon. Science (Washington, DC) 192:1247-1249.

Scholz, A.T., J.C. Cooper, D.M. Madison, R.M. Horrall, A.D. Hasler, A.E. Dizon, and R.J. Poff. 1973. Olfactory imprinting in coho salmon: behavioral and electrophysiological evidence. Proceedings of 16th Conference of Great Lakes Research 16: 143-153.

Scholz, A.T., C.K. Gosse, J.C. Cooper, R.M. Horrall, A.D. Hasler, R.I. Daly, and R.J. Poff. 1978. Homing of rainbow trout transplanted in Lake Michigan: a comparison of three procedures used for imprinting and stocking. Transactions of the American Fisheries Society 107:439-443.

Selset, R. 1980. Chemical methods for fractionation of odorants produced by char smolts and tentative suggestions for pheromone origins. Acta Physiologica Scandinavica 108:97-103.

Selset, R., and K.B. Døving. 1980. Behaviour of mature anadromous char (*Salmo alpinus* L.) toward odorants produced by smolts of their own population. Acta Physiologica Scandinavica 108: 113-122.

Solomon, D.J. 1973. Evidence for pheromone-influenced homing by migrating Atlantic salmon, *Salmo salar* (L.). Nature (London) 244:231-232.

Stabell, O.B., and R. Selset. 1980. Comparison of mucus collecting methods in fish olfaction. Acta Physiologica Scandinavica 108: 91-96.

Stabell, O.B., R. Selset, and K. Sletten. 1982. A comparative chemical study on population specific odorants from Atlantic salmon. Journal of Chemical Ecology 8:201-217.

Sutterlin, A.M., and N. Sutterlin. 1971. Electrical responses of the olfactory epithelium of Atlantic salmon (*Salmo salar*). Journal of the Fisheries Research Board of Canada 28:565-572.

Todd, J.H., J. Atema, and J.E. Bardach. 1967. Chemical communication in social behavior of a fish, the yellow bullhead (*Ictalurus natalis*). Science (Washington, DC) 158:672-673.

Wisby, W.J., and A.D. Hasler. 1954. Effect of olfactory occlusion on migrating silver salmon (*O. kisutch*). Journal of the Fisheries Research Board of Canada 11:472-478.

Wrede, W. 1932. Versuche über den Artduft der Elritzen. Zeitschrift für vergleichende Physiologie 17:510-519.

ESTABLISHING THE PHYSIOLOGICAL AND BEHAVIORAL DETERMINANTS OF CHEMOSENSORY ORIENTATION

Peter B. Johnsen

Monell Chemical Senses Center
University of Pennsylvania
Philadelphia, Pennsylvania 19104 USA

ABSTRACT

Fishes make use of a variety of mechanisms in chemosensory-mediated orientation. Behavioral experiments with Pacific salmon (*Oncorhynchus* spp.) have led to the formulation of a behavioral control model which describes the upstream movements of homing salmonids. The model makes use of two feedback loops which represent swimming behaviors dependent on the spatial patterns of odor distribution encountered during the homeward migration. The possible sensory controls of these behaviors are discussed. In addition, stereotypic swimming patterns released by chemical cues rather than movements under continuous control of stimulus distribution are described.

INTRODUCTION

The directed movements of fishes have been attributed in many cases to chemical cues. Considerable experimental evidence demonstrates that chemical signals may be used by many fishes to facilitate directional locomotion between two points. However, little work has been published on the actual mechanisms of chemosensory orientation.

Unlike other environmental signals, such as light and sound, chemical signals have no directive component and vary only in intensity. This lack of directivity therefore imposes several restrictions on possible mechanisms by which an animal might orient to a stimulus source. Chemical signals may serve as sign-stimulus releasers. Fish, after detecting the signal, orient to some other directive cue such as water currents. Fish may also respond to gradients in the environment by both klinotaxis and tropotaxis. To measure the axis of a chemical gradient, fish may discern concentration differences by repeated sampling in time or space. Some species have morphological adaptations which permit them to compare concentrations between paired receptors at the same time, thereby perceiving the axis of the gradient across their body. A fourth and little understood mechanism involves the release of some stereotypic swimming pattern which has a certain probability of the animal's net displacement in the direction of the stimulus source.

The migratory movements of many fishes are, in fact, the results of various combinations of these general categories of orientation mechanisms. A specific example is the upstream migration of salmonids. These fishes may make use of several mechanisms depending on the spatial distribution of stimuli. To learn more of the details of how salmon make use of chemical cues in their upstream movements, experiments were conducted in streams with controlled distributions of the homestream (imprinting) odor, which simulated conditions found in nature.

In a dendritic river system patterns of odor distribution exist in two general forms, which have particular importance in the homeward migration of salmon. Downstream from a source, odors are distributed fully across the width of the stream. They may not be present in uniform concentration owing to discontinuous mixing, but are nevertheless present at suprathreshold concentrations. A second pattern of distribution exists at the confluence of two streams. The stream carrying the odor of interest is scented from bank to bank, while the other tributary does not carry any of this scent. Where these two channels unite there exists for a distance downstream two discrete plumes, one with the odor and one without. As these plumes of water move downstream they become mixed. This pattern of distribution and mixing is repeated throughout the river system as the water flows to the ocean.

This paper describes experimental work on chemosensory orientation and a control model for homestream migration in salmonids. One aspect of the movement patterns of salmon will be examined in detail illustrating the value of this approach in the study of mechanisms of migration.

EXPERIMENTAL OBSERVATIONS

By controlling the distribution of odor within the stream and in separate experiments verifying the distribution with dye, the movements of individual fish which had been imprinted to synthetic compounds were recorded in relation to the actual spatial distribution of the odor (Johnsen and Hasler 1980). When imprinting odors were distributed across the full width of the stream, fish imprinted to that compound exhibited positive rheotaxis. If the homestream (imprinting) compound was absent, fish exhibited negative rheotaxis and moved downstream until they again detected the homescent. Thus, the segregation of fish imprinted to different odors from different homesites is based on differential rheotactic responses released in the presence or absence of the imprinting odor.

To simulate the conditions found at the confluence of two streams, an imprinting chemical was released into the stream so that its distribution was confined to one half of the stream channel, thus creating adjacent scented and unscented water masses. We observed that fish confined their movements to the side of the stream which contained the imprinting odor (Johnsen and Hasler 1980). Detailed analysis of the fish's position with respect to the odor within the stream revealed that the fish made use of the interface between the scented and unscented masses of water to a greater extent than the odor corridor itself (Johnsen 1978). By zig-zagging along the edge of the trail, fish made sure and steady progress up the trail.

These observations of fish movements in relation to controlled distributions of homestream odor allowed me to develop a behavioral control model which describes the movements of homing salmonids as they make their way from the sea to their homestream in a dendritic river system. The model, previously described (Johnsen 1982), makes use of two feedback loops which represent swimming behaviors dependent on the spatial patterns of odor distribution encountered along the way (Fig. 1).

The two major patterns of upstream movement include straight positive rheotaxis in water scented from bank to bank and zig-zag swimming along the interface of the scented plume at the confluence of two tributaries. An additional feature includes the negative rheotaxis in the absence of the homestream odor. I believe that these three simple behaviors, positive and negative rheotaxis and zig-zagging at the confluence of two streams, are sufficient to guide fish to the homestream.

While this model may not be complete, it does provide a framework on which to base critical experiments and observations. Obviously, each behavior has a physiological basis which must be understood. The model then allows us to isolate and examine in more detail specific mechanisms of orientation.

Because the physical-chemical characteristics of an interface boundary may be simulated and manipulated in a variety of ways, I believe it is possible to obtain a greater understanding of the physiological and behavioral principles of chemosensory orientation under these particular conditions. This will aid in a general understanding of the process as a whole.

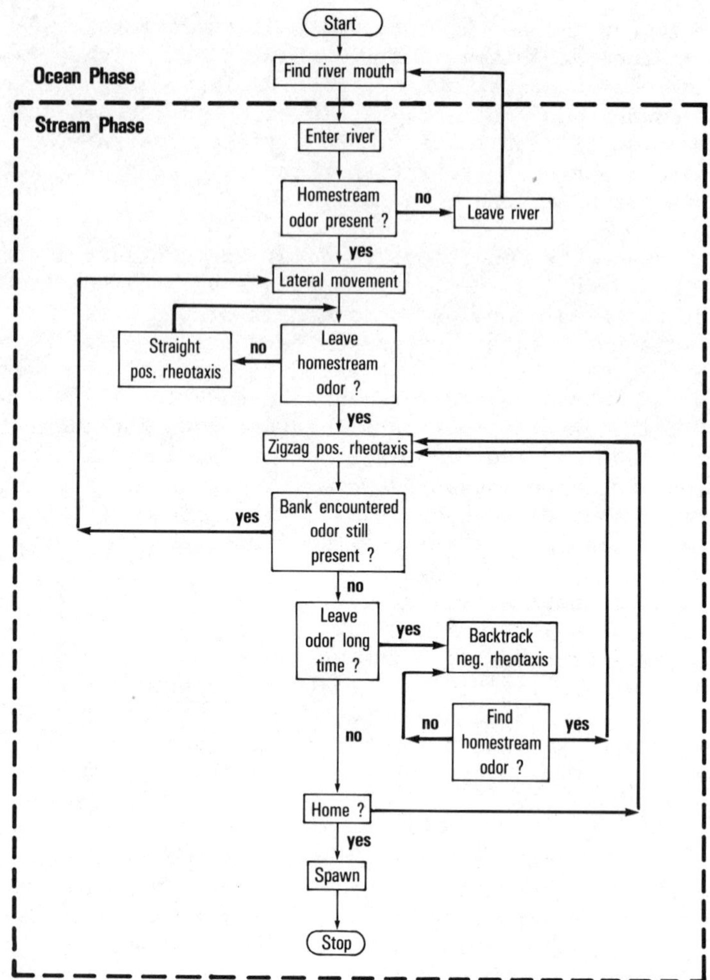

Figure 1. Proposed behavioral control model for upstream movement and homestream selection in migrating salmonids (from Johnsen 1982).

ZIG-ZAG BEHAVIOR

To begin consideration of the problem, what are the functional advantages of movement at the interface or edge of the odor plume as compared to swimming within the plume and turning back into the center of the plume when encountering either of the two edges? The first suggested advantage is that moving in and out of the stimulus limits the animal's exposure time and thus reduces the effects of sensory adaptation and the concomitant loss of important sensory input during the critical situation of choosing the correct tributary at the confluence of two stream channels. Additionally, at the edge of the scent trail the greatest contrast of signal is present. Here, in a relatively short distance, the animal is able to distinguish between stimulus-laden water and stimulus-free water thus reducing the need to detect precisely differences in stimulus concentration. The fish may rely on simple presence or absence information.

What initiates the turning in zig-zagging behavior? I have hypothesized that when a fish traveling within a scented watermass encounters a rapidly changing stimulus concentration, thus indicating an interface, the behavior of swimming at an angle to the current rather than against the current is released. This movement allows the fish to move in the axis of greatest probable concentration change. After establishing the presence of the interface, the fish starts the zig-zag swimming behavior. To determine the factors which are responsible for each turn, one can examine the stimulus intensity at each of the turn points during the behavior. One turn occurs at a relatively low concentration and the other at a relatively higher one. Two different strategies may be employed by the fish to control the turning. Under one scheme, with the animal turning back from the unscented water and swimming at an angle to the current, the fish enters the scented plume (below threshold concentration) and continues until it encounters a threshold concentration. It then turns and swims on the opposite tack traveling through the plume until the concentration goes below threshold, thus initiating another turn. A second scheme involves the animal turning at some critical concentration difference with respect to the lower and higher concentrations. This is different from the first strategy in that the animal can perceive the presence of the stimulus but does not turn until some specific concentration is detected. Computer simulation models of zig-zag behavior generate the same path using both schemes indicating that further experimental work is required to resolve the actual mechanism.

In both fishes and insects the amplitude of zig-zags seems to be related to the slope of the gradient across the edge of the plume. In cases where the gradient is quite shallow the animals show larger amplitudes in their zigs and zags. As the gradient steepens the turning rate increases and the amplitude decreases. This is consistent with both schemes and again suggests the need for further experimentation.

Effects of adaptation, or the decrement of sensitivity after prolonged exposure to a stimulus, would greatly influence the behavioral responses of the fish. In both the situation of the fish searching for some absolute difference in concentration between the lower and upper concentrations and the fish turning when the concentration reaches threshold, turns would become less frequent due to the depressed sensitivities of the fish. The fish, as it enters the plume, is not at threshold concentration, yet the olfactory system is exposed to stimuli and the process of sensory adaptation is underway. As a result the threshold sensitivity of the animal slides upward. The longer the animal remains in the subthreshold concentration, the higher the required concentration for threshold stimulation. The same thing would occur for an animal seeking some concentration difference. Because the apparent concentration is different than the actual, the animal would move to higher and lower concentrations before initiating a turn.

I have conducted laboratory experiments to learn more about the problem of sensory adaptation. Cardiac-conditioning experiments are being conducted to determine the consequences of prolonged exposure to chemical stimuli. There is a proportional increase in the concentration of stimulus required for detection as the background concentration rises. More studies on the effects of prolonged subthreshold-stimulus exposure are required.

Another important aspect of the zig-zag behavior is the detection of the edge of the plume and the absence of stimuli. We know from peripheral nerve recordings that the response of olfactory nerves does not totally adapt after prolonged exposure. Thus the animal is presumably able to sense the off-set of a stimulus and would be capable of detecting movement out of the stimulus. However, there are other explanations of the animal's ability to detect the edge of the plume. These may include restimulation after rinsing out by subthreshold water and a restimulation by suprathreshold water found at the discontinuously mixed interface. Experiments on the fish's ability to detect decreasing concentrations will be required to determine the actual mechanism of the release of zig-zag behavior.

RELEASED BEHAVIORS

While these explanations of how the animal performs its zigs and zags include a degree of behavioral control by the stimulus, Johnsen (1982) proposed that the actual pattern of zig-zagging may be generated by some internal gating mechanism. I investigated gating in fish through the use of directed stimulation (Johnsen, unpublished data). The technique makes use of a headmount and tubes which deliver stimuli directly into the nares of a fish. With this technique I can simulate many spatial distributions of stimuli. The responses of bonnethead (*Sphyrna tiburo*) sharks to large volume stimulations applied directly into the water were observed in the laboratory. These fish demonstrated the usual circling and turning behavior which has been described for sharks for almost seventy years (Parker 1914). The proposed mechanism by which a shark remains in the vicinity of the stimulus involves the animal reversing its swimming direction when it swims beyond the water containing suprathreshold quantities of stimuli. This can occur in both stagnant and moving water. However, the headmount allowed for a more precise delivery and distribution of stimuli of small volume and short duration to the sharks. After the stimulus passed through the nasal structure, the concentration fell below threshold by dilution. Yet the fish showed the same pattern of circling and turning at the stimulation site as with the large volume stimulations. Even more strikingly, sharks stimulated in the presence of a current circle in the projected site which was moving down current even though the stimulus had diluted to below threshold levels. This behavior indicated that the turning behavior exhibited by sharks is an internally generated one released by a chemical signal. In this way chemical stimuli may serve only to modulate the released stereotypic behaviors which result in the animal's displacement in the direction of the stimulus source.

CONCLUSION

I believe that the development of these simple sorts of models greatly aids in the organization and structuring of research into orientation mechanisms. For example, our model has led to the realization that we need to know the physiological limits of detecting the decrease in concentration of a stimulus and the role of long term stimulation on sensory acuity. One of the most important things that we have learned recently is that many behaviors which were previously thought to be tightly controlled by chemical stimuli may in fact be released by the signal and are part of some internally driven stereotypic behavior. The ability to control precisely the concentration and distribution of stimuli is greatly increasing our understanding of the process of orientation.

REFERENCES

Johnsen, P.B. 1978. Contributions on the movements of fish: I. Behavioral mechanisms of upstream migration and homestream selection in coho salmon (*Oncorhynchus kisutch*) II. Winter aggregations of carp (*Cyprinus carpio*) as revealed by ultrasonic telemetry. Doctoral dissertation. University of Wisconsin, Madison, Wisconsin, USA.

Johnsen, P.B. 1982. A behavioral control model for homestream selection in migratory salmonids. Pages 266-273 *in* E.L. Brannon and E.O. Salo, editors. Proceedings of the salmon and trout migratory behavior symposium. School of Fisheries, University of Washington, Seattle, Washington, USA.

Johnsen, P.B., and A.D. Hasler. 1980. The use of chemical cues in the upstream migration of coho salmon, *Oncorhynchus kisutch* Walbaum. Journal of Fish Biology 17:67-73.

Parker, G.H. 1914. The directive influence on the sense of smell in the dogfish. US National Marine Fisheries Service Fishery Bulletin 33:63-68.

DOWNSTREAM MOVEMENTS OF JUVENILE SALMONIDS: A FORWARD SPECULATIVE VIEW

J. E. Thorpe

Department of Agriculture and Fisheries for Scotland
Freshwater Fisheries Laboratory
Pitlochry, Perthshire PH16 5LB Scotland

ABSTRACT

Site attachment and the ability to maintain hydromineral balance in a hypotonic medium permit juvenile salmonids to live in a riverine environment. Abandonment of such adaptations results in downstream emigration. Numerous morphological and physiological changes have been associated with this seasonal emigratory stage (collectively called smoltification); however, the motivation for the migration is not understood. The seasonality of the migration implies developmental synchrony achieved through perception of photoperiod or thermoperiod change, via the hypothalamo-pituitary target-organ axies. As regulators of hydromineral balance prolactin and corticosteroid hormones are likely to play a prominent role. In this context Meier's two-oscillator model of migratory behaviour regulation (Meier and Fivizzani 1980), expressed in the seasonal relative phase-changes of the circadian rhythms of these two hormones, is a hypothesis worthy of testing with young salmon. Intraspecific stock discreteness implies genetic control over such phase relationships and needs investigation. Finally, inhibition of emigration through physiological competition by sexual maturation should throw further light on the regulatory mechanism.

INTRODUCTION

In a recent paper (Thorpe 1982) I argued that downstream movements of juvenile salmonids could be considered most simply in terms of Balchen's (1976) model of migration as a "...simple-minded local process of maximising 'comfort'..." Taylor and Taylor (1977) had much the same conception as Balchen in saying that "...movement is a fundamental biological response to adversity," or more obscurely, Slobodkin and Rappaport (1974) in seeing migration as the "...quintessential response..." in the evolutionary "...existential game..." In essence my argument was that behavioural specialisations at successive growth stages of juvenile salmonids represented compromises by the species consequent on long-term genetic optimisation. Interpretations of the nature of these compromises depended on assumptions about the evolutionary history of salmonids, which I reviewed briefly. On the basis of the supposed antiquity of the salmoniform design (Greenwood et al. 1966) and Balon's recent discussions of evidence for neotenic development with the salmonine group (Balon 1968, 1980), I took as a working hypothesis the view that salmonid fishes were derived from a marine ancestor and that specialisations during early life history were adaptations of a marine animal to a freshwater habitat. In Balchen's terms the implications of this hypothesis were that movements in freshwater were the consequences of a set of discomforts experienced by a marine animal as each successive behavioural compromise was replaced by another. Emiliani (1982) suggested that the coastal environment, being oceanically marginal, represents one in which evolutionary radiation is more likely to have taken place than in the less diverse habitats of the open ocean. Dingle (1980) pointed out that inshore-offshore breeding migrations are common among crustaceans and whales as well as among fishes. It may be more realistic, then, to consider the protosalmoniform as a coastal evolutionary product, which has radiated both outward and inward, evolving a simple and highly successful body design for utilising open-oceanic resources and various degrees of adaptation for protected reproduction and early juvenile development in the specialised, restricted, riverine environment. Those degrees of adaptation form a modern comparative series from wholly marine fish such as the capelin (*Mallotus villosus*) (an osmerid relative of salmonids, which demonstrates spawning behaviour ranging from broadcasting eggs on the sea bed to 60 m depth to burying eggs in beaches at the tide-margins [Tempelman 1947]), through other osmerids and galaxiids which enter freshwater and spawn by distributing unprotected eggs on the shore-line or stream bed (McDowall 1968), to salmonids which themselves range from *Brachymystax lenok*, which broadcasts its eggs openly on the stream bed (Smolanov 1961), to the majority of salmonines which bury their few large eggs under 15-20 cm of gravel. This paper focuses on the possible mechanisms promoting movement rather than on the evolution of the behaviour.

LIFE-HISTORY STRATEGY

However, in accordance with Tinbergen's (1963) first requirement for understanding behaviour, it is necessary first to consider movement and migration in the overall context of life-history strategy of the species that moves. Salmonid fishes are relatively primitive anadromous teleosts which enter freshwater to spawn, the majority of species depositing their large yolky eggs in depressions they excavate in the gravel beds of rivers, which they cover with more gravel excavated immediately upstream. The embryonic stage is completed in this habitat, and the emergent juveniles are comparatively large (c. 2-4 cm long), morphologically and behaviourally equipped to take external food immediately and depend on it totally. Residence time in the riverine habitat varies between and within species, but the prime advantage of that residence appears to be a protective one during a developmental period when young fish in general are highly vulnerable. (This idea was also suggested by Jackson [1961] for stream spawning among African Lake fishes.) This assertion is supported by the evidence of relatively low fecundity (less than 100 to a few thousand eggs per female), coupled with large egg size leading to large first-feeding fry, in a relatively unproductive environment incapable of supporting many large predators (cf. Dingle 1980). But the very fact of low productivity in a riverine environment as compared to the sea limits its advantages as a growth environment for the developing salmonid too. Movement away to the sea can therefore be viewed as a balanced risk of greater growth opportunity traded for greater exposure to predators (cf. Schaffer and Elson 1975). For most species more than 99% of weight growth of survivors to adulthood is achieved at sea before the maturing fishes return to their natal streams to spawn.

ADAPTATIONS TO RIVERINE ENVIRONMENTS

Whatever the actual evolutionary routes, use of both the freshwater and oceanic environments implies the possession of different sets of adaptations for each. The principal problems that face such a group of fishes in freshwater are both behavioural and physiological. Behaviourally, the questions are how to maintain station in a relatively high-speed, unidirectional flow and how to obtain adequate nourishment to meet that cost as well as to grow. Physiologically, the primary question is how to maintain hydromineral balance in a hypotonic medium. Effective adaptations meeting these needs would ensure retention and successful development there. The fact that in many salmonid species movement away from this environment occurs, implies that either the adaptations become inadequate or the development needs change. It was suggested above that in order to achieve full growth potential the fish must move to a more productive environment. In this case it is the developmental needs that are not being met. However, in order for the movement to occur the adaptations which have served so well to retain the animal on a particular site in a healthy functional condition must themselves be abandoned. Evidence that such abandonment does occur has been obtained experimentally for behavioural adaptations in Atlantic salmon (*Salmo salar*) (Thorpe and Morgan 1978; Tytler et al. 1978; Thorpe et al. 1981) and was reviewed together with evidence from other salmonids in Thorpe (1982). The outcome of such abandonment when in a strong flow must necessarily be downstream displacement, ultimately to the sea. That the passage downstream is consistent with passive displacement has been established by acoustic tracking by several workers (Fried et al. 1978; LaBar et al. 1978; McCleave 1978; Tytler et al. 1978; Thorpe et al. 1981).

REGULATION OF MOVEMENTS

Downstream movement by progressive abandonment of riverine behavioural adaptations implies the progressive abandonment of physiological adaptations also. That this occurs is suggested from work on masu salmon *Oncorhynchus masou* (Kubo 1955) and coho salmon *O. kisutch* (Folmar and Dickhoff 1981) in which progressive reductions of blood electrolytes were recorded during downstream passage of smolts of these species. More importantly it raises questions about neural and hormonal regulation of both behavioural and physiological processes in these fishes, about the environmental events which synchronise and influence these regulators, and about the sensory systems and pathways by which these events are perceived and information transferred to the hypothalamus, pituitary and target tissues (cf. Groot 1982). Because morphological changes also occur at the time of smolting in salmonids, e.g. changes of body shape (Wankowski and Thorpe 1979), silvering (Johnston and Eales 1967), and dentition (Gorbman et al. 1982), and increase of salt-secreting cells in the gills (Hoar 1951), these and other likely structural changes, such as size of paired fins and kidney structure, should be investigated in relation to habitat change. In short, what motivates movement, and how may indices of this be measured environmentally, morphologically, behaviourally and physiologically?

MOTIVATION FOR MOVEMENT: FUTURE QUESTIONS

Hoar (1965) drew attention to the seasonality of physiological changes in salmonids, and various authors have found evidence of endogenous developmental rhythms (Clarke et al. 1978; Lundqvist 1980; Eriksson et al. 1982). It has also been shown that such rhythms are responsive to entrainment by photoperiod (Saunders and Henderson 1970; Wagner 1974; Clarke et al. 1978; Eriksson and Lundqvist 1980; Lundqvist 1980; Groot 1982). Identification of a major environmental oscillator which may act to synchronise physiological processes is an important first step, but little has been done to identify the pathways by which information on changing daylength is perceived and processed by salmonids. Eriksson (1972) found evidence of pineal photoresponsiveness in brook charr (*Salvelinus fontinalis*) and in brown trout (*Salmo trutta*). Weber and Smith (1980) suggested that melatonin synthesis by the pineal might provide the neurohormonal transmission of photoperiod information to the brain, and thus, via hypothalamus and pituitary, the prolactin rhythm of steelhead trout (*Salmo gairdneri*) would be modulated. Prolactin has been regarded as the primary hormone maintaining hydromineral balance of fish in freshwater, reducing sodium efflux and water uptake (Bentley 1971). As melatonin was synthesised in the dark, increasing photoperiod would lead to reduced melatonin output and ultimately reduced prolactin secretion. Such a reduction would become evident physiologically through lowered serum electrolyte levels and lethargy (Pickford et al. 1966). It was noted above that masu and coho salmon show reductions in electrolyte level during their downstream migration and that Atlantic (and coho salmon, Smith 1982) become unwilling to swim at more than 2 body lengths per second at this time. The ontogeny of prolactin activity in juvenile salmonids has not been fully explored, and it deserves particular attention as the reduction in function of this hormone system could provide a major component of the internal motivation for downstream movement in spring.

Recently a theoretical model of the regulation of migration in diadromous fishes has been proposed by Meier and Fivizzani (1980). In the gulf killifish (*Fundulus grandis*), which lives in the warm saline waters of the Gulf of Mexico in winter and migrates into freshwater to spawn in spring, both prolactin and cortisol (which promotes osmoregulatory functions in seawater) showed diel rhythms of serum concentration. Whereas the cortisol rhythm remained at the same phase at 20 C (winter marine temperature) and 28 C (summer freshwater temperature), the prolactin rhythm shifted by 16 h from the lower to the higher temperature. The daily peaks of the two hormones coincided at 8 h after dawn in winter, but prolactin peaked at dawn and cortisol 8 h later in summer. Under experimental conditions by injecting the hormones simultaneously they could reproduce physiological responses in the fish characteristic of winter (fattening, gonad growth, high salinity preference), and by injecting cortisol 8 h after prolactin they could reproduce responses characteristic of summer (fat loss, gonad regression, low salinity preference). They regarded the diel oscillations of the hormones as expressions of the effects of two separate neural oscillators, whose temporal relations varied with temperature. Such a model suggests phase setting of oscillators by temperature rather than by light and encourages the testing of hypotheses about major environmental regulators. But, as importantly, it also points to the need for more careful investigation of the diel rhythms of serum hormone titres and of the desirability of multivariate approaches in interpretation of hormonal effects. It would be profitable to examine the physiological motivation for salmonid movement from such a viewpoint. The Meier and Fivizzani (1980) model is worthy of much more attention by fish biologists, because Meier and his co-workers have demonstrated similar mechanisms occurring in bird migration, involving the same major hormone systems and similar phase shifts (Martin and Meier 1973; Meier 1975, 1976).

It has also been shown that salmonid species are usually composed of large numbers of genetically discrete stocks, largely isolated from one another through precision of homing to spawn at geographically discrete sites (see STOCS symposium papers 1981). Because seasonal arrival in a local sea area by migrant smolts is compressed into a few weeks at most, initiation of movement from differing parts of major river systems is likely to occur at different times under differing light and temperature conditions. If the physiological responses leading to movement are mediated by mechanisms of the Meier and Fivizzani (1980) type, then the phasing of those oscillators must be under precise genetic control. Such a possibility also requires testing.

There are also apparent differences between species in the nature of downstream movement. It has been suggested that two groups exist: those that orient to their conspecifics within a few days of emergence from the gravel; and those that orient to physical cues on the stream bed (the latter group becoming territorial). Thorpe (1982) argued that the former group were those which had not evolved away from the pelagic ancestral model to the same extent as the latter group and had not evolved the same degree of site attachment. Among the former group is the sockeye salmon (*Oncorhynchus nerka*). Several workers have shown evidence for navigation in the active downstream migration of this species (Groot 1965; Brannon 1972; Quinn 1980; Brannon et al. 1981; Quinn and Brannon 1982), whereas no evidence has been found for such navigation in the Atlantic salmon smolt migration (a member of the latter group). This discrepancy deserves closer scrutiny.

Finally, considering life-history strategies, there is evidence among many salmonid species of early maturation, particularly of male fish, while still in the freshwater phase (Thorpe 1982). Among these and other species there are populations which are facultatively "landlocked." In such circumstances it would appear that the motivation for downstream movement, characteristic of other populations within the species, has been overcome by motivation to reproduce. The ecological, physiological and behavioural basis of this flexibility of life-history strategy which avoids long-range movement should also be explored. A start has been made in this direction in the Pitlochry laboratory (Buck and Youngson 1982).

REFERENCES

Balchen, J.G. 1976. Principles of migration in fishes. Foundation for Scientific and Industrial Research (SINTEF), Trondheim, Norway. Technical note number 81.

Balon, E.K. 1968. Notes to the origin and evolution of trouts and salmon with special reference to the Danubian trouts. Vestnik Ceskoslovenske Spolecnosti Zoologicke 32:1-21.

Balon, E.K. 1980. Early ontogeny of the lake charr, *Salvelinus (Cristivomer) namaycush*. Pages 485-562 *in* E.K. Balon, editor. Charrs. W. Junk, The Hague, The Netherlands.

Bentley, P.J. 1971. Endocrines and Osmoregulation. Springer-Verlag, Berlin, Federal Republic of Germany.

Brannon, E.L. 1972. Mechanisms controlling migration of sockeye salmon fry. International Pacific Salmon Fisheries Commission Bulletin 21.

Brannon, E.L., T.P. Quinn, G.L. Lucchetti, and B.D. Ross. 1981. Compass orientation of sockeye salmon fry from a complex river system. Canadian Journal of Zoology 59:1548-1553.

Buck, R.J.G., and A.F. Youngson. 1982. The downstream migration of precociously mature Atlantic salmon, *Salmo salar* L., parr in autumn; its relation to the spawning migration of mature adult fish. Journal of Fish Biology 20:279-288.

Clarke, W.C., J.E. Shelbourn, and J.R. Brett. 1978. Growth and adaption to seawater in underyearling sockeye (*Oncorhynchus nerka*) and coho (*O. kisutch*) salmon subjected to regimes of constant or changing temperature and daylength. Canadian Journal of Zoology 56:2413-2421.

Dingle, H. 1980. Ecology and evolution of migration. Pages 1-101 *in* S.A. Gauthreaux, Jr., editor. Animal migration, orientation, and navigation. Academic Press, New York, New York, USA.

Emiliani, C. 1982. Extinctive evolution: extinctive and competitive evolution combine into a unified model of evolution. Journal of Theoretical Biology 97:13-33.

Eriksson, L.-O. 1972. Die Jahresperiodik augen- und pinealorganlöser Bachsaiblinge *Salvelinus fontinalis* Mitchell. Aquilo Ser. Zoologica 13:8-12.

Eriksson, L.-O., and H. Lundqvist. 1980. Photoperiod entrains ripening by its differential effect in salmon. Naturwissenschaften 67:202.

Eriksson, L.-O., H. Lundqvist, E. Brännäs, and T. Eriksson. 1982. Annual periodicity of activity and migration in the Baltic salmon, *Salmo salar* L. Pages 415-430 *in* K. Müller, editor. Coastal research in the Gulf of Bothnia. W. Junk, The Hague, The Netherlands.

Folmar, L.C., and W.W. Dickhoff. 1981. Evaluation of some physiological parameters as predictive indices of smoltification. Aquaculture 23:309-324.

Fried, S.M., J.D. McCleave, and G.W. LaBar. 1978. Seaward migration of hatchery reared Atlantic salmon, *Salmo salar*, smolts in the Penobscot River estuary: riverine movements. Journal of the Fisheries Research Board of Canada 35:76-87.

Gorbman, A., W.W. Dickhoff, J.L. Mighell, E.F. Prentice, and F.W. Waknitz. 1982. Morphological indices of developmental progress in the parr-smolt coho salmon, *Oncorhynchus kisutch*. Aquaculture 23:1-19.

Greenwood, P.H., D.E. Rosen, S.H. Weitzmann, and G.S. Myers. 1966. Phyletic studies of teleostean fishes with a provisional classification of living forms. Bulletin of the American Museum of Natural History 131:339-456.

Groot, C. 1965. On the orientation of young sockeye salmon (*Oncorhynchus nerka*) during the seaward migration out of lakes. Behaviour Supplement 14:1-198.

Groot, C. 1982. Modifications on a theme--A perspective on migratory behavior of Pacific salmon. Pages 1-21 *in* E.L. Brannon and E.O. Salo, editors. Salmon and Trout Migratory Behaviour Symposium. School of Fisheries, University of Washington, Seattle, Washington, USA.

Hoar, W.S. 1951. Hormones in fish. Publications of the Ontario Fisheries Research Laboratory 71:1-51.

Hoar, W.S. 1965. The endocrine system as a chemical link between the organism and its environment. Transactions of the Royal Society of Canada, Series IV 3:175-200.

Jackson, P.B.N. 1961. The impact of predation, especially by the Tigerfish (*Hydrocyon vittatus* Cast.) on African freshwater fish. Proceedings of the Zoological Society of London 136:603-622.

Johnston, C.E., and J.G. Eales. 1967. Purines in the integument of the Atlantic salmon (*Salmo salar*) during parr-smolt transformation. Journal of the Fisheries Research Board of Canada 24:953-964.

Kubo, T. 1955. Changes of some characteristics of blood of smolts of *Oncorhynchus masou* during seaward migration. Bulletin of the Faculty of Fisheries Hokkaido University 6:201-207.

LaBar, G.W., J.D. McCleave, and S.M. Fried. 1978. Seaward migration of hatchery-reared Atlantic salmon (*Salmo salar*) smolts in the Penobscot River estuary, Maine: open-water movements. Journal du Conseil Conseil International pour l'Exploration de la Mer 38:251-269.

Lundqvist, H. 1980. Influence of photoperiod on growth in Baltic salmon parr (*Salmo salar* L.) with special reference to the effect of precocious sexual maturation. Canadian Journal of Zoology 58:940-944.

McCleave, J.D. 1978. Rhythmic aspects of estuarine migration of hatchery-reared Atlantic salmon (*Salmo salar*) smolts. Journal of Fish Biology 12:559-570.

McDowall, R.M. 1968. *Galaxias maculatus* (Jenyns) - the New Zealand Whitebait. New Zealand Marine Department Fisheries Research Bulletin 2:1-84.

Martin, D.D., and A.H. Meier. 1973. Temporal synergisms of corticosterone and prolactin in regulating orientation in the migratory white-throated sparrow (*Zonotrichia albicollis*). Condor 75:369-374.

Meier, A.H. 1975. Chronoendocrinology of vertebrates. Pages 469-549 *in* B.E. Eleftheriou and R.L. Sprott, editors. Hormonal correlates of behaviour. Plenum Press, New York, New York, USA.

Meier, A.H. 1976. Chronoendocrinology of the White-Throated Sparrow. Proceedings of the 16th International Ornithological Congress 1974:355-368.

Meier, A.H., and A.J. Fivizzani. 1980. Physiology of migration. Pages 225-282 *in* S.A. Gauthreaux, Jr., editor. Animal migration, orientation, and navigation. Academic Press, New York, New York, USA.

Pickford, G.E., P.K.T. Pang, and W.H. Sawyer. 1966. Prolactin and serum osmolality of hypophysectomised killifish, *Fundulus heteroclitus*, in freshwater. Nature (London) 206:1040-1041.

Quinn, T.P. 1980. Evidence for celestial and magnetic compass orientation in lake migrating sockeye salmon fry. Journal of Comparative Physiology A. Sensory, Neural, and Behavioral Physiology 137A:243-248.

Quinn, T.P., and E.L. Brannon. 1982. The use of celestial and magnetic cues by orienting sockeye salmon smolts. Journal of Comparative Physiology A. Sensory, Neural, and Behavioral Physiology 147A:547-552.

Saunders, R.L., and E.B. Henderson. 1970. Influence of photoperiod on smolt development and growth of Atlantic salmon (*Salmo salar*). Journal of the Fisheries Research Board of Canada 27:1295-1311.

Schaffer, W.M., and P.F. Elson. 1975. The adaptive significance of variations in life history among local populations of Atlantic salmon in North America. Ecology 56:577-590.

Slobodkin, L.B., and A. Rappaport. 1974. An optimal strategy of evolution. Quarterly Review of Biology 49:181-200.

Smith, L.S. 1982. Decreased swimming performance as a necessary component of the smolt migration in salmon in the Columbia River. Aquaculture 28:153-161.

Smolanov, I.I. 1961. Razvitiye lenka *Brachymystax lenok* (Pallas). Voprosy Ikhtiologii 1:136-148.

STOCS symposium papers. 1981. Canadian Journal of Fisheries and Aquatic Sciences 38:1457-1921.

Taylor, L.R., and R.A.J. Taylor. 1977. Aggregation, migration, and population mechanics. Nature (London) 265:415-421.

Templeman, W. 1947. Life history of the capelin (*Mallotus villosus* O. F. Müller) in Newfoundland waters. Bulletin of the Newfoundland Government Laboratory, Newfoundland Research Bulletin 17.

Thorpe, J.E. 1982. Migration in salmonids with special reference to juvenile movements in freshwater. Pages 86-97 *in* E.L. Brannon and E.O. Salo, editors. Salmon and Trout Migratory Behavior Symposium. School of Fisheries, University of Washington, Seattle, Washington, USA.

Thorpe, J.E., and R.I.G. Morgan. 1978. Periodicity in Atlantic salmon (*Salmo salar* L.) smolt migration. Journal of Fish Biology 12:541-548.

Thorpe, J.E., L.G. Ross, G. Struthers, and W. Watts. 1981. Tracking Atlantic salmon smolts (*Salmo salar* L.), through Loch Voil, Scotland. Journal of Fish Biology 19:519-537.

Tinbergen, N. 1963. On the aims and methods of ethology. Zeitschrift für Tierpsychologie 20:410-433.

Tytler, P., J.E. Thorpe, and W.M. Shearer. 1978. Ultrasonic tracking of the movements of Atlantic salmon smolts (*Salmo salar* L.) in the estuaries of two Scottish rivers. Journal of Fish Biology 12:575-586.

Wagner, H.H. 1974. Photoperiod and temperature regulation of smolting in steelhead trout (*Salmo gairdneri*). Canadian Journal of Zoology 52:219-234.

Wankowski, J.W.J., and J.E. Thorpe. 1979. The role of food particle size in the growth of juvenile Atlantic salmon (*Salmo salar* L.). Journal of Fish Biology 14:351-370.

Weber, L.J., and J.R. Smith. 1980. Possible role of the pineal gland in migratory behaviour of salmonids. Pages 313-320 *in* W.J. McNeil and D.C. Himsworth, editors. Salmonid ecosystems of the North Pacific. Oregon State University Press, Corvallis, Oregon, USA.

MIGRATION AND LEARNING IN FISHES

M. E. Bitterman

Bekesy Laboratory of Neurobiology
Pacific Biomedical Research Center
University of Hawaii
Honolulu, Hawaii 96822 USA

ABSTRACT

In the first part of this paper, some laboratory experiments on discriminative learning in fishes are reviewed--experiments on the discrimination of food signals, reward and nonreward, temporal context, compound stimuli, and the components of compound stimuli. The results, which closely resemble those obtained with birds and mammals in analogous experiments and point to the operation of common vertebrate mechanisms, should be of interest to students of migration in fishes who are concerned with the role of discriminative learning in migratory behavior. Comparative experiments on partial reinforcement, on successive acquisition and extinction, and on incentive contrast are reviewed in the second part of this paper. The results suggest that there has been some evolutionary divergence in vertebrate learning and that generalizations to fishes from work with birds and mammals (about whose learning a great deal is known) should be made with care.

INTRODUCTION

The title of this paper may be misleading because I have little to say about migration, which only provides the context for a discussion of learning, but I can find no simple way to express the idea that it is a paper on learning for students of migration. There seem to me to be two closely related reasons why students of migration in fishes might want to know something about learning in fishes (Hasler 1966): one is that learning methods are used to study the discriminative capabilities thought to play a role in migration, and the second is that migratory behavior may itself depend on discriminative learning. My plan, therefore, is first to review some experiments on discrimination which, while methodologically interesting, serve mainly to direct your attention beyond simple questions of sensory capacity to some more advanced questions of sensory organization and control that may deserve a place in your thinking about migration. Then I shall have something to say about similarities and differences between learning in fishes and learning in other vertebrates. It would be fortunate for your purposes if there were no differences, because most of what is known about learning comes from work with other vertebrates--and there do indeed seem to be some broad commonalities--but there are indications as well of evolutionary divergence that should be considered carefully.

DISCRIMINATIVE LEARNING IN FISHES

We know that an animal can discriminate between two stimuli if it responds differently to them, although it may discriminate without responding differently. (Detection is a special case of discrimination--we know that an animal can detect a stimulus if it responds differently in the presence than in the absence of the stimulus.) Differential responding may occur without special training, as when a naive animal shows a preference for one rather than another food. The point is perhaps too obvious to require illustration, but I ask you to look at some relevant data (as yet unpublished) out of interest in the method employed.

Food Preferences

Carp (*Cyprinus carpio*) and goldfish (*Carassius auratus*),--six of each--were tested individually in a tank at one end of which, side by side, were two illuminated Plexiglas targets. At the center of each target (Fig. 1) was a small food cup baited with a drop (10 μl) of liquid food that was replaced automatically as it was taken (a pump was activated by the animal's contact with the cup). In each of six daily 20-min sessions, two of four foods were available--one that was low both in energy and in protein, a second that was high in energy but low in protein, a third that was low in energy but high in protein, and a fourth that was high both in energy and in protein. Each subject was exposed to all six possible pairs of the four foods, with the order of pairs and the assignment of foods to the two food cups balanced systematically over subjects and sessions. The proportion of responses to each of the four foods in all sessions combined is plotted in Figure 2, which shows that high protein was clearly preferred to low protein, while energy-level made no difference. The results for carp and goldfish were very much the same, as were the results for goldfish with transected olfactory tracts, supporting the suggestion that "taste" rather than "smell" may be responsible for food selection by fishes (Bardach and Atema

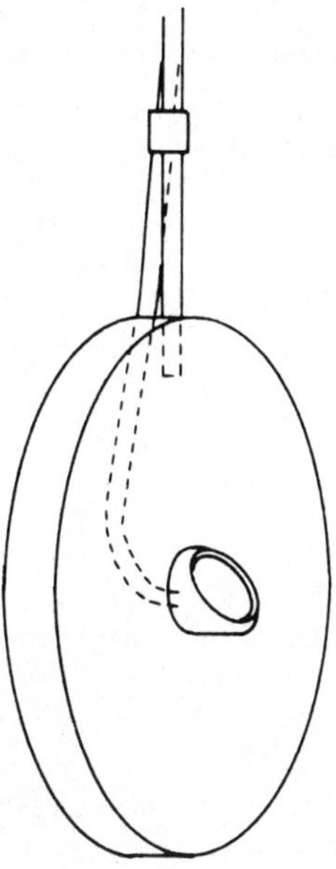

Figure 1. Plexiglas target (4 cm in diameter) with a food cup into which liquid food can be pumped through plastic tubing indicated by curved broken lines. Colored lights and patterns, projected on the target from the rear, serve as conditioned stimuli. The target is mounted on a rod connected to a strain-gauge circuit, the output of which pulses a response-relay when the animal pushes against the target.

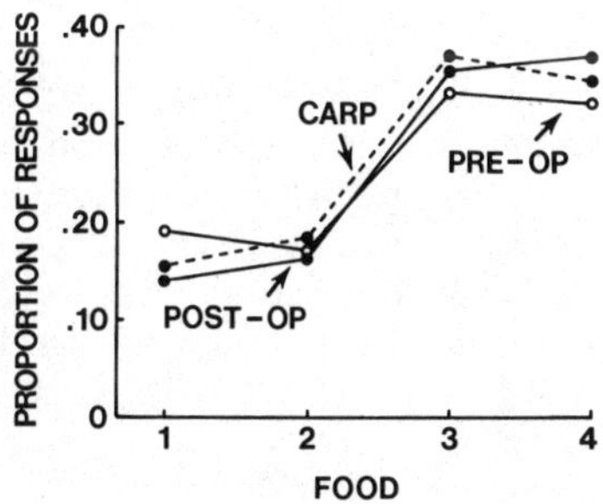

Figure 2. Food-preferences in normal carp and in goldfish before (pre-op) and after (post-op) transection of the olfactory tracts.

1971). The technique employed for this work, developed by Holmes and Bitterman (1969), is an extremely efficient one, but note that seemingly minor variations either in the apparatus or in the way it is used may lead to marked reduction in efficiency. I was surprised recently to read that goldfish studied by Zippel and Voigt (1982) in an apparatus much like ours in conception took 4-5 weeks to learn to associate food with the tip of a funnel through which it was delivered. Most of our animals take food from the cup in the first training session, although the rate of feeding increases gradually over the next six or seven sessions.

Discrimination of Food Signals

If, instead of two targets illuminated with light of the same color and providing two different foods, we use two different colors and the same food, the animals may not respond differently to the targets because they have no color preference, but it is easy to demonstrate discrimination of color by appropriate training. Our aim in discriminative training is to produce differential response to stimuli by subjecting the stimuli to different treatments that encourage the animals to learn different things about them. The treatments fall into two main categories called "classical" and "instrumental." In classical conditioning the stimulus signals some impending event which the animal can only anticipate, and we look for evidence of anticipation in its response to the signal; for example, the general activity of goldfish soon comes to be heightened in the presence of a light that always is followed by electric shock (Horner et al. 1960). In instrumental conditioning the stimulus signals a response-contingent event--an event that the animal can produce or prevent by making some defined response; for example, light may signal an impending shock that the goldfish can avoid by swimming from one end of a tank to the other (Horner et al. 1961). The target shown in Figure 1 can be used both for classical and for instrumental conditioning. For example, red light projected on the target may be used as a signal that food soon will be delivered (independently of the animal's behavior) to the empty food cup, which quickly produces anticipatory responding to the food cup (Woodard and Bitterman 1974); or the light may be used to signal that food will be delivered only if the animal strikes--or only if it fails to strike (Brandon and Bitterman 1979)--the target. Let me show you as an example some unpublished data of an instrumental conditioning experiment in which two different colors signaled two different foods. Eight goldfish were trained in 20-trial sessions to strike a target that was red on some trials and green on the rest in quasi-random order. Each trial began with illumination of the target for a 20-sec period during which responses to the target were recorded but not rewarded; in the next 10 sec, each response produced a drop of food, after which the target light was turned off and the trial ended. The foods signaled by the colors were a highly attractive, standard diet and the same diet made somewhat less attractive by the addition of bitters--the

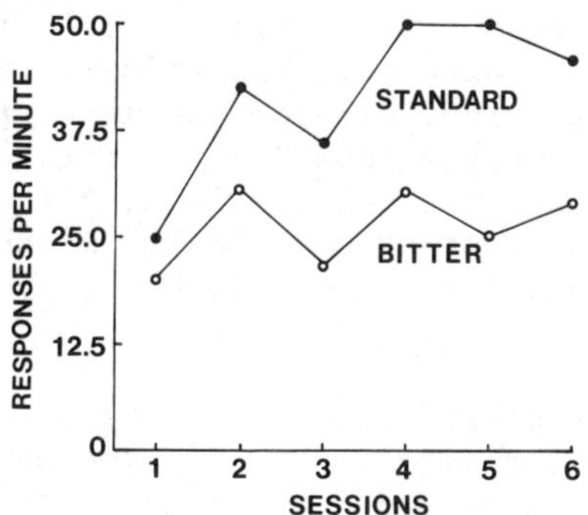

Figure 3. Discrimination of color by goldfish under conditions in which response to one color was rewarded with a preferred (standard) diet and response to the other with the same diet adulterated with bitters.

standard diet signaled by red and the bitter by green for half the animals, and the opposite for the rest. As Figure 3 shows, discrimination developed quickly, with more rapid responding to the signal for the preferred food than to the signal for the less-preferred food.

Discrimination of Reward and Nonreward

Differential responding to stimuli that are treated differently is, of course, to be expected only if the animals are differently affected both by the stimuli and by the treatments. When we are interested in the discriminability of the stimuli, we choose treatments whose differential effectiveness can be taken for granted; to find out whether goldfish are able to discriminate red and green, we would not pair one of the colors with a food of reduced palatability, but with no food at all. When we are interested in treatments, we use stimuli whose discriminability can be taken for granted, and yet it may happen in such experiments that we come upon discriminative capabilities about which we might not otherwise have thought to inquire. Consider, for example, the so-called "partial reinforcement effect," long known in laboratory rats, which occurs also in goldfish (Gonzalez and Bitterman 1967). Two groups of animals are rewarded with food for making some simple response to a stimulus, one group consistently (on every trial), the other partially (which means intermittently, as, for example, on a random 50% of the trials). Then reward is withheld for both groups, and the persistence of the animals in making the response (their "resistance to extinction") is measured. Since reward tends in general to strengthen the tendency to make a response, while nonreward tends to weaken it, we may be surprised to find greater resistance to extinction in the partially rewarded animals. A simple explanation of this phenomenon is that the consequences of response on any given trial--reward and nonreward--are themselves discriminable stimuli to which the animal may learn to respond differentially on the next trial. Since the partial animals are frequently rewarded in training for responding after nonreward, they show greater persistence when reward is no longer given than do the consistent animals, which never have been rewarded for responding after nonreward, but only after reward (Hull 1952).

A simple test of the hypothesis that the reward or nonreward given on one trial may acquire discriminative control of performance on the following trial is provided by "single-alternation training"--that is, training in which rewarded and nonrewarded trials are regularly alternated (Couvillon and Bitterman 1981). Eight goldfish were trained to strike a lighted target for food in a series of daily 40-trial sessions with an interval in darkness of 10 sec between trials. At the start of each trial the target was illuminated with blue light for a period of 20 sec in which response to the target was recorded but not rewarded. Then there was a 10-sec period in which, on odd-numbered trials, each response was rewarded with a drop of food; on even-numbered trials, no food was given. Asymptotic performance (pooled over the last four of 15 training sessions) is plotted in the left-hand portion of Figure 4. The curves show clearly that the animals learned to respond less rapidly on nonrewarded (B-) trials (previous reward predicted nonreward) than on rewarded (B+) trials (previous nonreward predicted reward). The discrimination seemed to be lost between sessions--responding on the first nonrewarded trial of a session was no less rapid than on the first rewarded trial--but recovered quickly as the training in the session continued. The curves in the right-hand portion of Figure 4

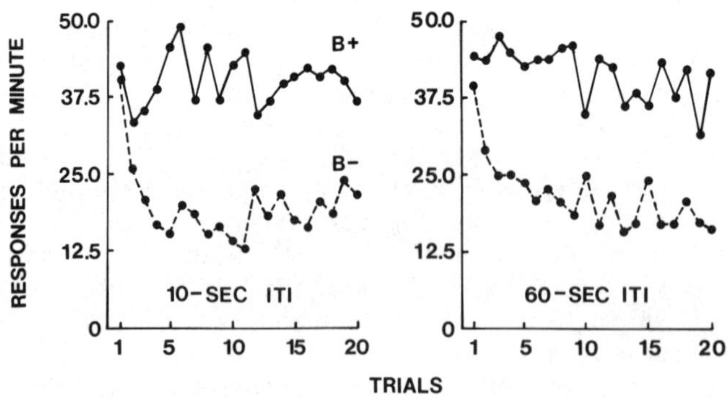

Figure 4. Single-alteration patterning in goldfish. Response to a blue target was rewarded on odd-numbered (B+) trials and not on even-number (B-) trials. The intertrial interval (ITI) was 10 sec to begin with and later was increased to 60 sec.

show good discrimination also in subsequent sessions when the interval between trials was increased to 60 sec.

Discrimination of Trial Stimuli

You may be interested in one further example of the subtle discriminations that we sometimes come upon in the course of work with readily discriminable stimuli subjected to different treatments out of interest only in the way in which the treatments work. In this case 14 goldfish were trained with a target that was red on some trials and green on others; in each session, there were 10 trials with each color in quasi-random order (Couvillon 1982). Response to one of the colors (the RN color) was rewarded on its first five presentations but not on its last five presentations, while response to the alternative color (the NR color) was nonrewarded on its first five presentations and rewarded on its last five presentations. This training produced the patterned responding shown in the left-hand portion of Figure 5. Consider especially the NR curve. The rate of responding to the NR stimulus was low at the onset of each session and then increased over the next few trials *despite nonreward*, suggesting that the animals had learned that response to the stimulus soon would be rewarded, although (being unable to tell time very well or to count) they did not quite know when. The simplest interpretation is that the animals were discriminating what we call "trial stimuli"--stimuli correlated with time or number of trials since the beginning of each training session; response to the NR stimulus was nonrewarded in the context of early-trial stimuli but rewarded in the context of later-trial stimuli. Before presenting some evidence in support of this interpretation, I must tell you that the 20 trials with the NR and RN colors were intermixed with 20 trials on which two other colors were presented, response to one of them (S+) always rewarded and response to the other (S-) never. The left-hand portion of Figure 5 shows also the rates of responding to these colors, which were used to make it impossible for the animals to predict the outcome of response to the NR or to the RN color on any trial from the events of the immediately preceding trial. In the sessions that provided the data for the right-hand portion of Figure 5, the procedure was exactly the same except that 20 additional trials with S+ and S-, 10 of each, were given either just before (Preceding) or just after (Following) the usual 40-trial mixed sequence of all four stimuli. In the Preceding condition, the first presentation of the NR stimulus occurred in the context of later-trial stimuli, and the animals (as expected) responded rapidly.

Conditional Discrimination

These results are interesting because they demonstrate sensitivity, not only to trial stimuli, but to a conditional relationship between trial stimuli and color--early-trial stimuli suppressed responding to the NR color but not to the RN color. A clearer

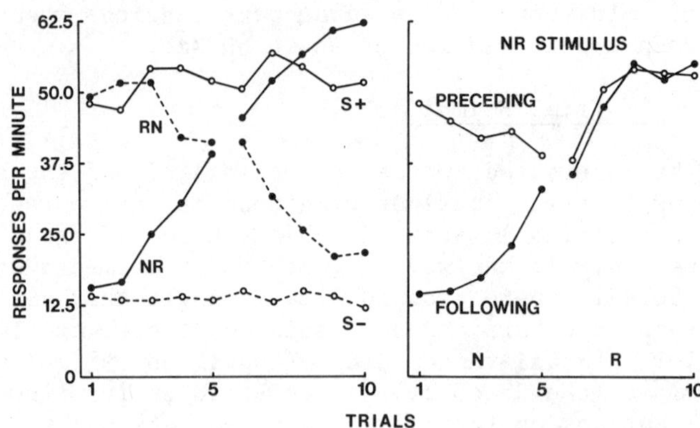

Figure 5. (Left) Response of goldfish to four stimuli presented 10 times each in each session of 40 trials. One (S+) was rewarded on all trials, another (S-) never rewarded, a third (RN) rewarded on the first five trials but not on the next five, and a fourth (NR) nonrewarded on the first five trials but rewarded on the next five. (Right) Response to the NR stimulus under the same conditions except for the addition of 20 trials with S+ and S- either before (preceding) or after (following) the block of 40 trials with all four stimuli.

example of conditional discrimination in goldfish is provided by some unpublished work with color-tone compounds. Goldfish were trained with a target sometimes red and sometimes green, and with a 200-Hz tone sometimes on and sometimes off, yielding four different conditions of stimulation--red-tone, red-no tone, green-tone, and green-no tone--that were presented in quasi-random order five times each in each session. Some animals were rewarded for responding only to red-tone and green-no tone, while the rest were rewarded for responding only to red-no tone and green-tone. In neither case, then, could food be predicted from color alone or from the presence or absence of tone alone, but only from the two in combination. Figure 6 shows samples of performance in the tenth training session, by which time the animals had mastered the problem. The curves on the left are for an animal rewarded for responding to red-tone and green-no tone, while those on the right are for an animal rewarded for responding to green-tone and red-no tone. (The results for the first trial show that there was some loss of the discrimination in the interval between sessions, but that the discrimination was rapidly restored as training in the session continued.) While it is convenient to speak in such circumstances of "conditional" discrimination, we are not, of course, required by the results to assume any conceptual ability on the part of the animals. A simpler explanation is that the afferent processes triggered by the components of a

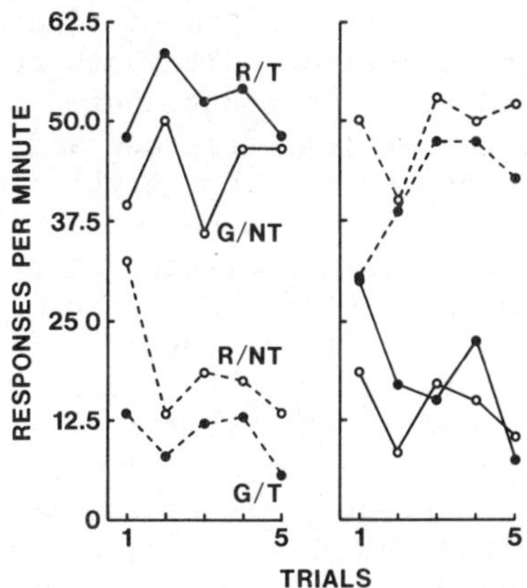

Figure 6. Conditional discrimination in goldfish. The data at left are for an animal that was rewarded for response to red-tone (R/T) and green-no tone (G/NT). The data at right are for an animal rewarded for response to red-no tone (R/NT) and green-tone G/T).

compound stimulus interact to produce compound-unique properties (Hull 1952). A whole may be quite different from the sum of its parts.

Within-Compound Association

It must not be supposed, however, that the separate identities of the parts are lost entirely in the whole, as we know from experiments showing that the components of a compound stimulus may be associated *with each other* in the course of experience with the compound. This phenomenon--called "within-compound association"--is evident in some as yet unpublished goldfish data. In the first stage of training 16 goldfish were fed for striking a target that displayed a white vertical line against a red background on some trials and a white horizontal line against a green background on other trials. In each session there were 10 rewarded trials with each compound in quasi-random order. In the second stage of the experiment some animals were trained to discriminate between the white lines alone (against a dark background), while the rest were trained to discriminate between the red and green backgrounds alone (without the lines). In the third stage of the experiment the animals which had been trained in the second stage with the lines were required to discriminate between the colors, while the animals which had been trained with the colors were required to discriminate between the lines.

All stimuli were properly balanced over subjects, but it may be well in order to better understand the design of this rather complicated experiment to consider the treatment of a particular pair of animals. After rewarded experience with both compounds in the first stage, they were trained to discriminate between the lines in the second stage, response to vertical rewarded and response to horizontal nonrewarded. In the third stage both animals were trained to discriminate between the colors. One of them, a member of the Nonreversal Group (which contained eight subjects), was rewarded for response to red, the background of the vertical line in the first stage of the experiment, but not for response to green, the background of the horizontal line in the first stage. If an association had been formed in the first stage between red and vertical, and between green and horizontal, this animal should have an advantage in the third stage. The second animal, a member of the Reversal Group (also with eight subjects), was rewarded in the third stage for response to green rather than red, which--by the same logic--should place it at a disadvantage. The results for the two groups are plotted in Figure 7 in terms of mean rates of responding to the rewarded and unrewarded stimuli in each of the nine sessions of the third stage. They bear out the expectation (based on the concept of within-compound association) that there should be a preference in the third stage for the component associated in the first stage with the rewarded component of the second stage. The interpretation implies, of course, that the components do largely retain their identities in the compound--for example, the red of the third stage must be recognized as the red of the first stage. Within-compound association between autonomous components may, in fact, be one source of the compound-uniqueness inferred from conditional discrimination.

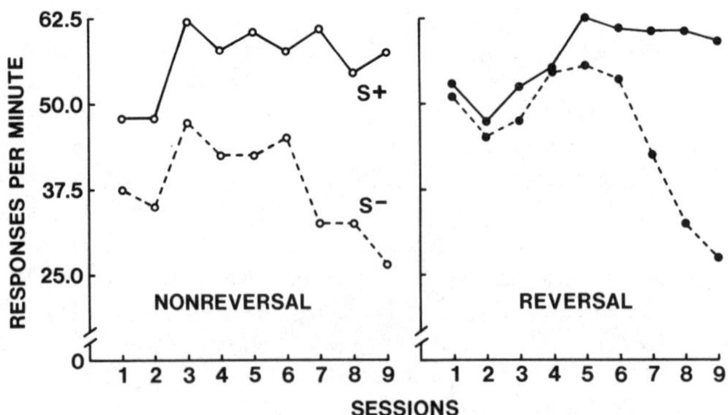

Figure 7. Performance of two groups of goldfish previously trained in such a way that within-compound associations would facilitate (nonreversal) or impair (reversal) discrimination between the rewarded (S+) and nonrewarded (S−) stimuli.

Overshadowing

To complicate matters further, let me illustrate yet another kind of interaction between the components of a compound--a competitive interaction that produces what is called "overshadowing." In one experiment (Tennant and Bitterman 1975a) two groups of carp (six subjects in each) were trained to discriminate between lines tilted 30° and 60° from the vertical. In each training session the lines were presented six times each in quasi-random order with half the animals in each group rewarded for response to the 30° line and the rest rewarded for response to the 60° line. The Control Group was trained with lines that were of the same color--both red for half the animals in the group or both green for the rest--and discrimination therefore was possible only on the basis of angle. For the Overshadowing Group, the two lines were different in color--the positive line was red for half the animals in the group and green for the rest--and discrimination therefore was possible either on the basis of color, or on the basis of angle, or the two together. Both groups learned rapidly, their acquisition functions nearing asymptote by the fifth or the sixth session, but the training was continued for 20 sessions in order to provide extensive experience with the stimuli. In the 21st session the animals were tested to determine what they had learned about angle, the procedure being the same as in training except that both lines were yellow and response to neither was rewarded. The response of the two groups to each of the stimuli in the last training session and in the test is plotted in Figure 8. The training curves show that the performance of the Overshadowing Group was somewhat better than that of the Control Group, while the test curves suggest that it was based exclusively on color. The Control Group responded significantly more in the test to the previously rewarded angle than to the previously nonrewarded angle, but the Overshadowing Group did not--the difference in color had somehow prevented learning about angle. Angle was "overshadowed," as we say, by color.

Attention

There is no place here for consideration of the various explanations of overshadowing that have been proposed (Bitterman et al. 1979), but the larger significance of the phenomenon should be touched upon briefly. I have said that an animal will come to respond differently to stimuli that are treated differently only if it can discriminate the stimuli and if the treatments are differentially effective, but now we see that these two conditions, although necessary, are not sufficient. There seems also to be some sort of selective process at work. In the example just considered animals trained with lines differing both in angle and color learned about color (showing the treatments to be effective) but nothing about angle (shown by the performance of the Control Group to be readily discriminable). There have been some efforts to train

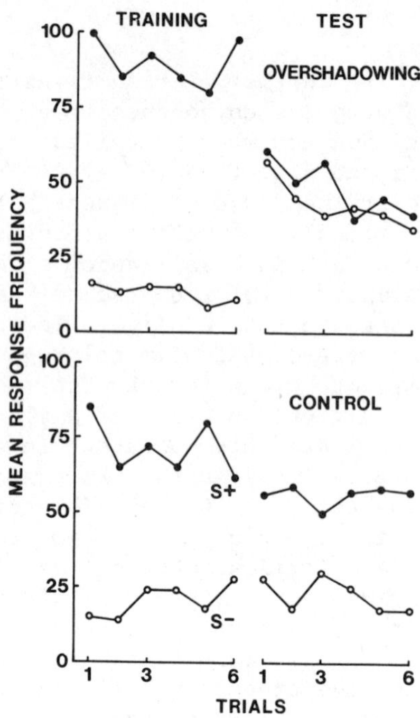

Figure 8. Performance of two groups of carp after training to discriminate two lines (S+ and S-) differing both in angle and color (overshadowing) or only in angle (control). (Left) Data of the last training session. (Right) Data of a nonrewarded test with two lines of a third color differing only in angle. After Bitterman (1975). Copyright (1975) American Association for the Advancement of Science.

goldfish to attend to one rather than another feature of a compound stimulus (Tennant and Bitterman 1973). While those efforts have not been carried far, I do want you to consider evidence that shows at least a negative influence--evidence that prior training can reduce the control of behavior acquired by the properties of a subsequently-encountered novel stimulus (Tennant and Bitterman 1975b). The results are interesting as well because they illustrate the generalization method of studying discrimination, which is basically the method used in experiments on imprinting in upstream migration (Cooper and Hirsch 1982). Instead of being trained with two stimuli that are treated differently, an animal is trained only with one, after which some property of the stimulus is changed and the effect of the change on responding is studied. If, for example, a goldfish is rewarded in the presence of a 200-Hz tone and then tested with a 400-Hz tone which it encounters for the first time, it may respond more than an untrained animal to the 400-Hz tone, although less than to the 200-Hz tone. Responding to the 400-Hz tone is referred to as "generalization" and provides an index of the similarity of the two tones. Less responding to the 400-Hz tone than to the 200-Hz tone--"generalization decrement"-- shows both that the animal can discriminate between the two frequencies and that it has in fact learned something about frequency. In the experiment now to be considered animals were given prior experience with colors that was intended to influence their subsequent learning about frequency.

In the first stage of the experiment two groups of five goldfish each were trained with a target that was green on some trials and blue on others in quasi-random order. For the True Discrimination (TD) Group the two colors were differentially rewarded (response to one of them always followed by food and response to the other never). For the Pseudodiscrimination (PD) Group the colors were nondifferentially rewarded, which is to say that half the trials with each color were rewarded and the rest nonrewarded. The TD Group mastered the problem rapidly, but training continued for 10 sessions during which the TD animals continued to respond differentially to the colors while the PD animals responded at an intermediate rate to both. The second stage of the experiment consisted of five sessions in which there were no trials with the blue and green targets, but both groups were rewarded consistently for response to a red target in the presence of a tone (200 Hz for half the animals in each group and 800 Hz for the rest). In the third stage there was a generalization test to determine what the animals had learned about the frequency of the tone in the second stage. It consisted of a series of unrewarded trials with the red key together with tones of

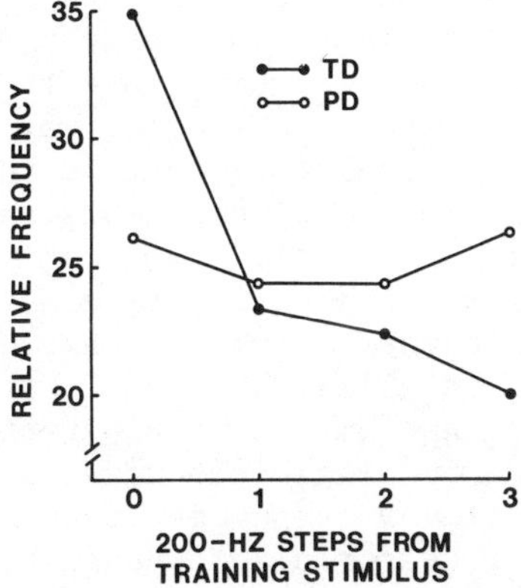

Figure 9. The performance of two groups of goldfish in a nonrewarded test with four tones after rewarded training with one of them. The TD group had still earlier been trained to discriminate between two colors with which the PD group was given nondifferentially rewarded experience.

200, 400, 600, or 800 Hz presented in quasi-random order. The results are plotted in Figure 9 in terms of the relative frequency of response to the four tones--such curves are called "relative generalization gradients." The TD curve shows that the animals are capable of distinguishing the training frequency from the others and that the TD animals did in fact learn something about frequency in the second stage of the experiment. By contrast, the PD animals were unaffected by variation in frequency, perhaps because their nondifferential training in the first stage had made them generally less attentive to changes in the experimental environment.

LEARNING IN FISHES AND OTHER VERTEBRATES

Some of you, at least the more pragmatic, may be skeptical about the relevance of such phenomena for the problem of migration. You may well be asking whether they are more than laboratory curiosities. There is, after all, no very good reason to believe that every property of an animal--structural or functional--must play some important role in its normal life. Harlow (1958) pointed out, for example, that monkeys tested in laboratory situations are able to solve problems far more complex than ever they meet in the natural environment. You may be put off, too, by the extent of my reliance on data for goldfish, asking whether like results would be obtained in experiments with more interesting fishes, such as salmon, if indeed such experiments were at all feasible. We can begin to deal with these questions in the context of a larger and more ambitious one, which has to do with the relation between learning in fishes and learning in other vertebrates.

Although there have been demonstrations of learning, especially of discriminative learning, in a wide variety of fishes (Herter 1953), the analytical work on learning in fishes has been done primarily with goldfish. The same specialization is evident in the work with other vertebrate classes: what we know about learning in birds comes primarily from work with pigeons, and what we know about learning in mammals comes primarily from work with rats and dogs. Is any meaningful generalization possible from so narrow a data base? One might think not. Consider, however, that every phenomenon of discriminative learning in goldfish which I have described for you is found also in analogous experiments with pigeons and rats and dogs. The goldfish experiments on single-alternation patterning and on within-compound association were suggested to begin with by work with rats (Tyler et al. 1953; Rescorla and Cunningham 1978); the experiments on the role of trial stimuli in NR patterning and on pseudodiscrimination were suggested by work with pigeons (Thomas et al. 1970; Couvillon et al. 1980); and the experiments on compound-uniqueness and on overshadowing by work with dogs (Pavlov 1927). Since the phenomena are found in markedly divergent animals trained under necessarily diverse conditions, the most parsimonious assumption (however bold) is that they are common to vertebrates, reflecting common mechanisms of substantial biological importance, and it follows from this assumption that the same phenomena would be found in a wide range of other fishes. You may be interested to know that all of the phenomena for which we have looked thus far appear also in free-flying honeybees trained after the fashion of von Frisch (Couvillon and Bitterman 1980, 1982)--compound-uniqueness, within-compound association, overshadowing, and even (in experiments as yet unpublished) poor discrimination after PD as compared with TD training. Common phenomena certainly do not require us to assume common mechanisms, and it may seem more reasonable in view of the remoteness of the evolutionary relationship between honeybees and vertebrates to look for different mechanisms shaped independently by similar adaptive constraints. To assume convergence rather than homology is not, however, to diminish the biological significance of the phenomena or of the underlying mechanisms, but the contrary.

The Partial Reinforcement Effect

To test the hypothesis that phenomena found in different species are based on common mechanisms, we may turn to detailed analysis of the conditions under which the phenomena occur, and it may interest you to consider a case in which parametric variation leads us to suspect the operation of different mechanisms (Bitterman 1975). Goldfish show the partial reinforcement effect--greater resistance to extinction after intermittent than after consistent reward--in "massed" trials (that is, when the intertrial interval is relatively brief) but not in widely spaced trials (when the intertrial interval is 24 hr). The same is true of African mouthbreeders (*Tilapia macrocephala*) and of painted turtles (*Chrysemys picta picta*), although pigeons and rats show the effect in widely spaced as well as in massed trials. What can the difference mean? If greater resistance to extinction after intermittent reward is due to the fact that the animals have been rewarded for responding after nonreward, information about preceding nonreward must be available on subsequent rewarded trials, and two ways in which that might happen are schematized in Figure 10. Let S_1 be the conditioned stimulus, R_1 the required response to S_1, and S_2 the consequence of response, reward or nonreward. One idea is that, where trials are massed, certain components of S_2 on one trial (T) *carry over* to the next trial (T+1), as, for example, the taste and smell of lingering particles of food, or feedback from a persisting emotional response to nonreward. Another idea is that S_1 and S_2 are associated, with the result that S_2 is *associatively reinstated* when S_1 is presented again on trial T+1. The two ideas are not contradictory, and since the first is indisputable, we are left with the question of whether the second serves any useful purpose. The answer is that it does: we need it to account for the partial reinforcement effect (where the effect occurs) in widely spaced trials. Animals that do not show the effect in spaced trials may not form the S_1-S_2 associa-

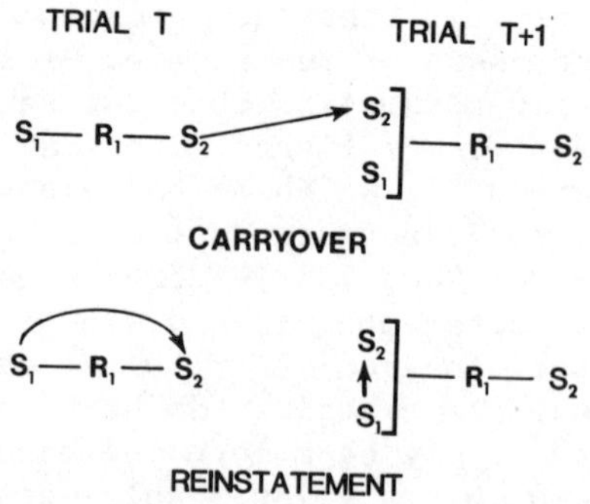

TRIAL T TRIAL T+1

$S_1 - R_1 - S_2 \rightarrow S_2, S_1 - R_1 - S_2$

CARRYOVER

$S_1 - R_1 - S_2$ $S_1 \rightarrow S_2 - R_1 - S_2$

REINSTATEMENT

Figure 10. Two ways in which the consequence of response on one trial (T) can exercise discriminative control of responding on the next trial (T+1). S_1, conditioned stimulus; R_1, conditioned response; S_2, consequence of response. In the carryover case, S_2 persists during the intertrial interval; in the reinstatement case, S_2 is evoked by S_1 with which it has been associated.

tion (may not remember the consequences of their actions), which seems unlikely, although it is well to note that the effects of the consequences of a response on subsequent performance of the response can be understood without reference to memory of the consequences. A second possibility is that the S_1-S_2 association is formed, but that it is temporally unstable. A third is the association is stable, but does not for some reason acquire discriminative control of responding.

The second possibility, which implies only a quantitative difference in mechanism, can be tested by giving massed trials under conditions designed to control for the effects of carryover. Couvillon (1982) rewarded two groups of eight goldfish each for striking a colored target (S_1), one group (CRF) consistently and the other (PRF) intermittently. Interpolated between successive S_1 trials were trials with two other colors, of which one (S+) was always rewarded and the other (S-) never. For both groups, therefore, rewarded S_1 trials sometimes followed reward and sometimes followed nonreward. A like experiment had previously been done with pigeons (Couvillon et al. 1980), and the difference in the results for the two species, which are plotted in Figure 11, parallels the difference found in experiments with widely spaced trials: the pigeons showed the partial reinforcement effect, but the goldfish did not. The results for pigeons can be understood on the assumption of associative reinstatement, which is not required by the results for goldfish.

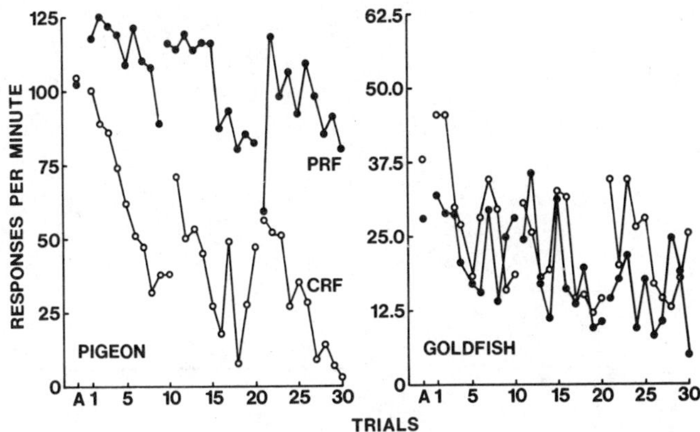

Figure 11. Resistance to extinction in pigeons and goldfish trained with intermittent (PRF) or consistent (CRF) reward. Performance in the last training session shown at A. In extinction there were 10 nonrewarded trials with the previously rewarded stimulus in each of three daily sessions. Interpolated trials with other stimuli to control for carryover were used throughout. Copyright (1980) American Psychological Association. Adapted by permission.

Successive Acquisition and Extinction

The point is important enough perhaps to warrant a look at the results of another goldfish-pigeon comparison afforded by the same series of experiments. In what is called "SAE training" (successive acquisition and extinction), response to a stimulus (S_1) is consistently rewarded for a time, then consistently nonrewarded, then consistently rewarded again, and so forth. Here the training (of 16 goldfish) was done in two ways: in the Between case, there were two sessions with response to S_1 rewarded, followed by two sessions with response to S_1 nonrewarded, then two rewarded sessions, then two nonrewarded sessions, and so forth; in the Within case, everything was the same, except that the transitions from reward to nonreward, and from nonreward to reward, occurred in midsession instead of at the beginning of a session. In both cases there were 20 trials with S_1 in each session and 20 interpolated trials with two other colors (S+ and S−) to control for carryover. The asymptotic resistance to extinction produced by the two kinds of training is plotted in Figure 12; the data used are those of sessions in which S_1 was never rewarded. A striking feature of the results for pigeons is the much greater resistance to extinction produced in the Within than in the Between case, which can be explained on the assumption that at within-sessions transitions from nonreward to reward the animals remember nonreward on immediately-preceding S_1 trials and therefore are rewarded for response to the memory of

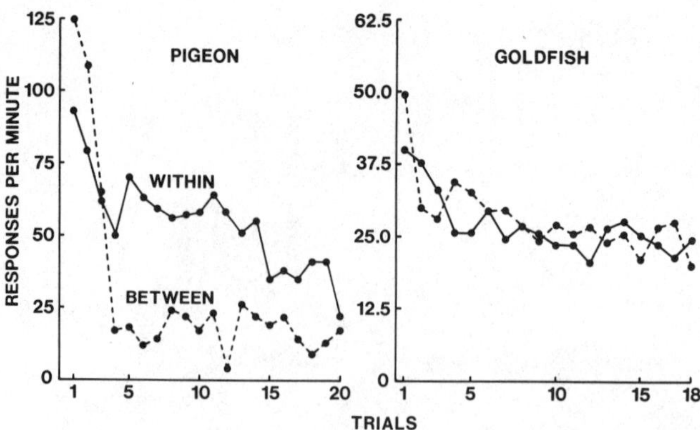

Figure 12. Resistance to extinction in pigeons and goldfish given SAE training (successive acquisition and extinction) with transitions from reward to nonreward, and from nonreward to reward, either within or between sessions. Interpolated trials with other stimuli were used to control for carryover. Copyright (1980) American Psychological Association. Adapted by permission.

nonreward. At corresponding between-sessions transitions pigeons presumably are unable to remember the outcome of response on the preceding S_1 trials, which in this case were given 24 hr earlier. That pigeons show the partial reinforcement effect at a 24-hr intertrial interval suggests that they can remember nonreward at least that long, but a distinction should be made between the ability to remember *that* they have been nonrewarded and the ability to remember *when* they were nonrewarded (Mackintosh 1974); the spaced-trials partial reinforcement effect requires only the former ability. Goldfish show no difference in resistance to extinction and make it unnecessary to assume any control of performance by memory of nonreward.

Incentive Contrast

The results for goldfish are negative also in experiments designed to show control of performance by memory of reward (Bitterman 1975). Consider the case in which two groups of rats are trained in widely spaced trials to make some simple response, one group for a preferred reward and the other group for a less-preferred reward. The animals work reasonably well for the less-preferred reward, although not as enthusiastically as for the preferred reward, but when the animals trained with the preferred reward are shifted suddenly to the less-preferred reward, their performance is disrupted markedly for many trials--they do not work as well for the less-preferred reward as do the animals that have known nothing better. This phenomenon, which is called "successive negative contrast" and seems to depend on memory of the preferred reward, has not been found in goldfish. The performance of goldfish shifted from a preferred to a less-preferred reward may decline gradually, but it becomes no poorer than the performance of goldfish trained from the outset with the less-preferred reward. Nor are these results for goldfish unique. Successive negative contrast has failed also to appear in painted turtles and, more recently, in very young rats (Chen et al. 1981): after a shift from large to small reward, the performance of 20-21 day old rats falls to the level of the small-reward control group, but the performance of 25-26 day old rats falls precipitously to a considerably lower level.

As Brookshire (1976) noted in his admirably balanced review of the rather disputatious comparative literature, there are some conditions in which the contrast effect fails also to appear in (adult) rats, and the possibility must be considered that in our work with other subjects we have not looked for it in the right place. Accordingly, we have been making further efforts to find successive contrast in goldfish, although still without success. Let me show you some recent, unpublished data. Two groups of nine goldfish each were trained in massed trials to strike a target for food reward. At the beginning of each trial the target-light was turned on for a 20-sec period during which responses were recorded but not rewarded, after which there was a reward period in which each response earned a drop of food. For the experimental group the reward period was 10 sec; for the control group it was 1 sec; and, as shown in Figure 13, this difference in amount of reward produced a marked difference in the rate of anticipatory response to the target. When the reward period of the 10-sec group was reduced to 1 sec, there was no disruption in performance, but only a gradual decline over many

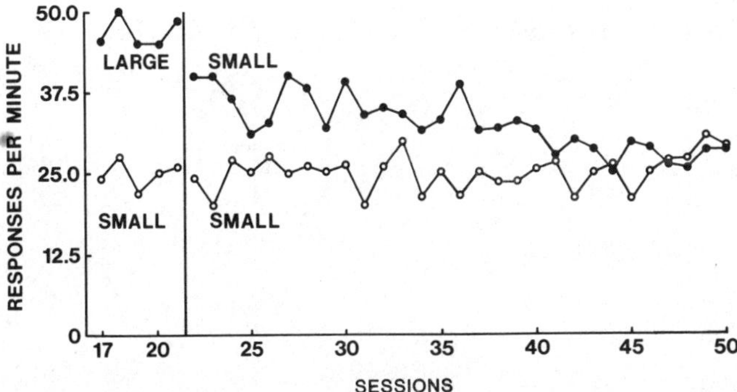

Figure 13. Instrumental performance of two groups of goldfish, one trained throughout with a 1-sec reward (small), the other trained for 20 sessions with a 10-sec reward (large) and then shifted to the 1-sec reward.

sessions to the level of the l-sec group.

In another experiment we varied quality rather than quantity of reward and measured consummatory rather than anticipatory responding. The model is provided by an experiment in which rats were given access to sucrose solution for a period of 5 min each day, and the number of licks at the drinking tube during that time was measured (Flaherty et al. 1978). For one group a 4% solution was used throughout, while a second group began with 32% and was shifted after 11 sessions to 4%. In the twelfth session the shifted group took much less of the 4% solution than did the unshifted group and continued to do so in the next five sessions, at which point the experiment was terminated. In our experiment with goldfish (two groups of 12 each) consummatory responding was measured for 30 min in each session, and the two foods were our attractive standard diet and the same diet adulterated with quinine. One group was given the standard diet and the other the quinine diet for 24 sessions. The rate of responding for the standard diet increased rapidly over eight sessions and then stabilized at about 55 per min, while the rate of responding for the quinine diet increased much more slowly, stabilizing only after some 19 sessions at about 25 per min. In sessions 25-35 both groups were given the quinine diet.

In Figure 14 the within-sessions performance of the two groups is plotted for sessions just before the shift, immediately after the shift, and at the end of the experiment. The preshift curves show that the quinine diet was taken somewhat less readily than the

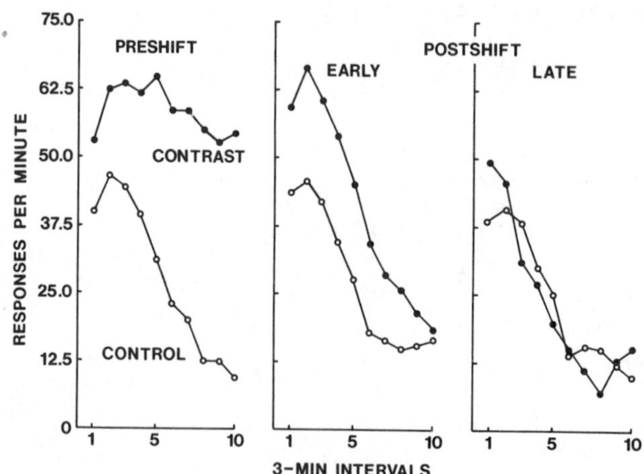

Figure 14. Consummatory behavior of two groups of goldfish, one (contrast) group shifted from a preferred to a less-preferred food, and a second (control) group given the less-preferred food throughout.

standard from the outset of each session, although the main difference between the curves is that response to the quinine fell off sharply in the course of the session while response to the standard diet did not. (I should say parenthetically that goldfish tested under these conditions continue to take the standard food for hours, not swallowing it, of course, but expelling it through the gill covers. In these animals, as in mammals, there seems to be a "mouth factor" as well as a "stomach factor" in feeding, responding for the diet adulterated with quinine being instigated by the stomach factor but not by the mouth factor. To prolong the sham feeding expelled food must be carried away quickly in the outflow; if permitted to accumulate in the tank, it can be enjoyed in the course of normal respiration without need for response to the target. The "satiation" functions obtained in our first work with this consummatory technique [Holmes and Bitterman 1969] probably can be understood in terms of our use of a small, closed system with relatively inefficient filtration.) The postshift curves plotted in Figure 14 give no indication of contrast. In fact the shifted animals took significantly more of the adulterated diet than did the unshifted animals, at least in the first few postshift sessions, after which their rate of responding declined to the level of the unshifted group. If the shifted animals remembered the preshift food that must in some way have been responsible for their more enthusiastic behavior, the memory did not serve to diminish their appreciation of the postshift food.

Interpretation

The meaning of such sharp differences in the results of analogous experiments with vertebrates of different classes still is unclear. They may point to some fundamental divergence in the mechanisms of learning, or perhaps to mere quantitative change--as, for example, in the temporal stability of memory of reward and nonreward, or in its susceptibility to interference; widely discrepant outcomes of comparative experiments may sometimes be derived from a common set of equations that differ only in their parametric values (Bitterman 1975). Whatever the truth of the matter, it is evident that care must be exercised in generalizing to fishes from the results of experiments on learning in other vertebrates, although there do seem to be important commonalities of which we might take predictive advantage if we had a better understanding of the reasons for the differences. As to whether the results of experiments on learning in goldfish and other readily available species convenient for laboratory work permit us to say anything about fishes in general, it is possible, I think, to be somewhat more optimistic, because agreement of the results with those for vertebrates of other classes suggests that we are dealing with some rather general mechanisms. Even where there are differences (as between the results for goldfish and rats in contrast experiments), there also are agreements (as between the results for goldfish and painted turtles in the same experiments) which rule out an interpretation in terms of the idiosyncracy of goldfish. The question of whether our laboratory findings, even if assumed to hold for fishes in general, have any bearing on the problem of migration, I must leave for you to answer. You may have been interested to hear about them only out of a broad concern with the nature of fishes, although I should think that in considering the role of learning in migration you might do well to keep them in mind.

REFERENCES

Bardach, J.E., and J. Atema. 1971. The sense of taste in fishes. Pages 293-336 *in* L.M. Beidler, editor. Handbook of sensory physiology, volume IV, part 2, Taste. Springer-Verlag, Berlin, Federal Republic of Germany.

Bitterman, M.E. 1975. The comparative analysis of learning. Science (Washington, DC) 188:699-709.

Bitterman, M.E., V. LoLordo, J.B. Overmier, and M.E. Rashotte. 1979. Animal learning: Survey and analysis. Plenum Press, New York, New York, USA.

Brandon, S.E., and M.E. Bitterman. 1979. Analysis of autoshaping in goldfish. Animal learning and behavior 7:57-62.

Brookshire, K.H. 1976. Vertebrate learning: Evolutionary divergences. Pages 191-216 *in* R.B. Masterton, M.E. Bitterman, C.B.G. Campbell, and N. Hotton, editors. Evolution of brain and behavior and vertebrates. Lawrence Erlbaum Associates, Hillsdale, New Jersey, USA.

Chen, J.-S., K. Gross, and A. Amsel. 1981. Ontogeny of successive negative contrast and its dissociation from other paradoxical reward effects in preweanling rats. Journal of Comparative and Physiological Psychology 95:146-159.

Cooper, J.C., and P.J. Hirsch. 1982. The role of chemoreception in salmonid homing. Pages 343-362 *in* T.J. Hara, editor. Chemoreception in fishes. Elsevier Scientific Publishing Company, Amsterdam, The Netherlands.

Couvillon, P.A. 1982. The performance of goldfish in patterned sequences of rewarded and nonrewarded trials. Doctoral dissertation, University of Hawaii, Honolulu, Hawaii, USA.

Couvillon, P.A., and M.E. Bitterman. 1980. Some phenomena of associative learning in honeybees. Journal of Comparative and Physiological Psychology 94:878-885.

Couvillon, P.A., and M.E. Bitterman. 1981. Analysis of alternation patterning in goldfish. Animal learning and behavior 9:169-172.

Couvillon, P.A., and M.E. Bitterman. 1982. Compound conditioning in honeybees. Journal of Comparative and Physiological Psychology 96:192-199.

Couvillon, P.A., S.E. Brandon, W.T. Woodard, and M.E. Bitterman. 1980. The performance of pigeons in patterned sequences of rewarded and nonrewarded trials. Journal of Experimental

Psychology: Animal Behavior Processes 6:137-154.

Flaherty, C.F., R. Blitzer, and G.H. Collier. 1978. Open-field behaviors elicited by reward reduction. American Journal of Psychology 91:429-443.

Gonzalez, R.C., and M.E. Bitterman. 1967. Partial reinforcement effect in the goldfish as a function of amount of reward. Journal of Comparative and Physiological Psychology 64:163-167.

Harlow, H.F. 1958. The evolution of learning. Pages 269-288 *in* A. Roe and G.G. Simpson, editors. Behavior and evolution. Yale University Press, New Haven, Connecticut, USA.

Hasler, A.D. 1966. Underwater guideposts: Homing of salmon. University of Wisconsin Press, Madison-Milwaukee, Wisconsin, USA.

Herter, K. 1953. Die Fisch Dressuren und ihr Sinnesphysiologische Grundlagen. Akademie Verlag, Berlin, German Democratic Republic.

Holmes, N.K., and M.E. Bitterman. 1969. Measurement of consummatory behavior in the fish. Journal of the Experimental Analysis of Behavior 12:39-41.

Horner, J.L., N. Longo, and M.E. Bitterman. 1960. A classical conditioning technique for small aquatic animals. American Journal of Psychology 73:623-626.

Horner, J.L., N. Longo, and M.E. Bitterman. 1961. A shuttle box for fish and a control circuit of general applicability. American Journal of Psychology 74:114-120.

Hull, C.L. 1952. A behavior system. Yale University Press, New Haven, Connecticut, USA.

Mackintosh, N.J. 1974. The psychology of animal learning. Academic Press, London, England.

Pavlov, I.P. 1927. Conditioned reflexes. Oxford University Press, Oxford, England.

Rescorla, R.A., and C.L. Cunningham. 1978. Within-compound flavor associations. Journal of Experimental Psychology: Animal Behavior Processes 4:267-275.

Tennant, W.A., and M.E. Bitterman. 1973. Some comparisons of intra- and extradimensional transfer in goldfish. Journal of Comparative and Physiological Psychology 83:134-139.

Tennant, W.A., and M.E. Bitterman. 1975(a). Blocking and overshadowing in two species of fish. Journal of Experimental Psychology: Animal Behavior Processes 1:22-29.

Tennant, W.A., and M.E. Bitterman. 1975(b). Extradimensional transfer in the discriminative learning of goldfish. Animal Learning and Behavior 3:201-204.

Thomas, D.R., F. Freeman, J.G. Svinicki, D.E.S. Burr, and J. Lyons. 1970. Effects of extradimensional training on stimulus generalization. Journal of Experimental Psychology Monograph 83, number 1, part 2.

Tyler, D.W., E.C. Wortz, and M.E. Bitterman. 1953. The effect of random and alternating partial reinforcement on resistance to extinction in the rat. American Journal of Psychology 66:57-65.

Woodard, W.T., and M.E. Bitterman. 1974. Autoshaping in the goldfish.

Behavior Research Methods and Instrumentation 6:409-410.

Zippel, H.P., and R. Voigt. 1982. Neuronal correlates of olfactory behavior in goldfish. Pages 181-199 *in* T.J. Hara, editor. Chemoreception in fishes. Elsevier Scientific Publishing Company, Amsterdam, The Netherlands.

METHODOLOGICAL ASPECTS OF MIGRATION AND ORIENTATION IN FISHES

Wim J. van der Steen

Department of Biology
Free University
P.O. Box 7161
1007 MC Amsterdam, The Netherlands

ABSTRACT

General methodology, defined as a branch of philosophy of science, could be a powerful tool in the development of programs for research on fish migration and orientation. The explication of concepts should be central in a methodological approach. Current concepts for various aspects of orientation and migration in fishes are unsatisfactory because they are loaded with theory. Commonly used classifications of methods of migration and orientation are incomplete. They have generated confusion due to a conflation of different classification criteria (e.g., tracks, stimulus-response connections, physiological mechanisms). The impact of concepts on research is considerable. Methodology can reveal how much they steer investigations. General methodology, statistics and model building can be applied integratively in the planning of new research. The ensuing benefits are substantial.

INTRODUCTION

To begin with, I want to plead guilty. Fish biology was beyond my horizon until recently, so my paper is likely to contain inaccuracies, distortions and omissions. Moreover, my approach is a philosophical one, and philosophers mostly think that they should not set the course of future studies by scientists (the main purpose of this NATO Advanced Research Institute). Nonetheless, I have gladly accepted the invitation to contribute to the Institute. A brief exposition of my views on the philosophy of science, especially biological science, will explain my motives. The term "philosophy" is used here in a restricted sense, mostly for a branch of philosophy called "(general) methodology."

It is obvious that philosophers of science and scientists should often investigate similar problems, especially when general methodology is at issue. Biology, however, is an exception among the sciences. Biologists mostly do not consider topics belonging to the area of philosophy, and philososphers are often not acquainted with real biology.

Evolutionary biology is among the few fields where biologists and philosophers do meet (cf. many recent papers in Systematic Zoology; Ruse 1973; Hull 1974), but, as yet, the interactions are not very fertile. The biologists have indeed adopted methodological principles which are obsolete in the eyes of most philosophers. Operationism and falsificationism are among the examples.

Operationism is the thesis that theoretical concepts must be defined in terms of fixed procedures at the level of observations. It must not be confused with the weaker thesis that concepts must be operational in the sense that there be links with the level of observations. Operationism arose half a century ago as an attempt to give physics a solid footing. Both philosophers and physicists have recognized afterwards that it cannot be effectuated, because the links between theories and observations are extremely involved and indirect (Carnap 1966). This development clearly has implications for biology (Hull 1968). Nonetheless, remnants of operationism are common in biology (e.g., Sneath and Sokal 1973).

Falsificationism is the view that hypotheses--and laws, theories --must be testable in a specific sense; it must be possible to show that they are false, if they are false. That is, they must be falsifiable.

In the first few decades of this century another paradigm of testability called verificationism was widely defended. In simplistic terms it implied that hypotheses must be testable in the very strong sense of provable. This paradigm was soon abandoned. Philosophers, especially Popper (1959; original, German edition 1935), have rightly argued that hypotheses are often universal statements. Such statements are unverifiable because they refer to infinitely many instances. The development of falsificationism as an alternative to verificationism was mainly Popper's work. Unfortunately, many philosophers have meanwhile shown--decisively so in my opinion--that falsificationism is as untenable as verificationism (Harding 1976). We therefore need a more liberal testability criterion.

In biology, however, falsificationism has recently been welcomed as a salutory device for excising disputable theories (see recent issues of Systematic Zoology).

The philosophical (methodological) techniques recommended by biologists are obviously not the best ones available in the philosopher's stock. Moreover, the use of these techniques is mostly limited to evolutionary biology. Even research involving application of evolutionary biology is hardly affected by general methodology. For example, current theories about temperature adaptation are seemingly untestable, but the issue of testability is never discussed by biologists elaborating such theories (van der Steen 1981).

Such matters deserve attention, especially because the philosopher's kit contains many utensils that are useful in any field of science. Philosophers, unfortunately, hardly stimulate the application of their theories at an elementary level.

Any application of philosophy should begin with conceptual matters. Problems concerning hypothesis testing, explanation, and the like, simply cannot be solved unless conceptual clarity is first attained. So I shall emphasize concepts in the analysis of orientation and migration in fishes. The main purpose is to show that the impact of concepts on research strategies is great. Strategies can be improved once this is recognized.

My approach is somewhat idiosyncratic. I have explained it in a recent book (van der Steen 1982) which is only available in Dutch. For this reason methodological themes will be considered here in some detail. Biological topics are developed at the same time, such that methodology and biology become intertwined. If this makes difficult reading, my only excuse is that I am trying to fill a gap in current research.

THE PUZZLE OF ADAPTATION

Theoretical concepts in biology, even those with a central position, often remain undefined. This may hamper research, because many of the terms used for such concepts are ambiguous.

With some hesitation I have chosen *adaptation* to introduce this topic. It is the most untractable term of evolutionary biology known to me. (The term has different meanings outside evolutionary biology, but in the evolutionary sense it also has a tremendous import in almost any area of biology.)

My aim can be but modest. I want to sketch a strategy that may prepare the ground for new research, with implications for the study of fish migration. Conceptual matters discussed in later sections will result in more substantial proposals for new research.

The current proliferation of discussions on the status of evolutionary theory should indicate that something is wrong at the conceptual level. However, this point is seldom considered. Testability rather than conceptual clarity is the subject of disputes. The controversy between Lewontin (1978) and Maynard Smith (1978), discussed by Brady (1979), is an example. The disagreement concerns the testability of "the general hypothesis of adaptation" (roughly: "all characters are adaptive") and more specific hypotheses, for example, in the context of optimal foraging theory.

The authors do not discuss the question of whether the falsifiability principle is appropriate. As argued above, it is not, but this point is of minor importance. From a methodological perspective the salient feature of the controversy is the absence of clear definitions for any of the central concepts (adaptation, natural selection, fitness, function). This is unsatisfactory, because attempts to apply a testability criterion are simply fruitless unless meanings of concepts are first specified.

The scarceness of definitions in evolutionary biology wrongly allows simple notions to take the place of complex concepts. The following tentative definition of "adaptation" in one particular sense illustrates the complexity: "Within taxon T, character X_i is adaptive relative to character X_j, and relative to a set of overall phenotypes S, in environment E" $=_{DF}$ "Within taxon T, individuals with X_i have better chances of survival and reproduction than individuals with X_j in environment E, when individuals with X_i or X_j have phenotypes in set S" (" $=_{DF}$" is a conventional symbol for "is to mean the same as"). The definition is meant to illustrate that the ascription of adaptiveness to characters involves many comparisons. Adaptation in the intended sense is indeed essentially a relative notion in evolutionary biology. As a consequence statements to the effect that some character is adaptive, without qualification, are meaningless. At the very least the context should provide qualifications.

For example, the statement that some migration pattern A_1 is adaptive in a particular species of fish needs expansion. It may take the following form: "Migration pattern A_1 (in the relevant species) is adaptive in comparison with pattern A_2, in environment E, when characters in the set S (e.g., large body size) are regarded as given." It is possible that migration pattern A_1 should not be called adaptive in comparison with a third pattern, A_3, or relative to another environment, or on the assumption that body size may change. So the truth of our statement depends on various standards of comparison. The standards may refer either to actual or to hypothetical entities.

Specifications of this kind are seldom spelt out in the literature; the excellent discussion of the concept of function by Hinde (1982) is among the exceptions. It is indeed inevitable that the results of science are mostly rendered less precisely. When our concepts and theories would be fully specified, booklets couched in mathematical terms could take the place of brief expositions, and that is not always helpful. However, overly informal approaches will likewise impede the progress of research. A middle course is obviously called for. The above definition of adaptation is meant to exemplify the style of such a course.

Analysis of concepts may put problems continuously resisting solution in a new perspective. The problem discussed in the following passage taken from Harden Jones (1981, page 140) is presumably a candidate for analysis:

> Migratory behaviour is one of several features in the life histories of fish directed towards reproductive success. Here the Grand Strategy appears to be that of producing a sufficient number of viable offspring to maintain the population up to the limit - in numbers or in weight - that can be fed. Nikolsky ... regards migration as "an adaptation towards increasing the abundance of a species" and this is a view with which I have some sympathy. But Cushing ... puts the question as to "whether fish are abundant because they migrate or vice versa". I do not know of any simple argument to resolve this circularity and when others have tried to do so, the results have not been very constructive ... .

The migration-abundance problem has developed into a rather untractable one, and precisely this may indicate that the distinction of factual and conceptual matters has been obliterated. I shall not venture to propose a solution of the problem, but only mention two items that may illuminate it.

1. Explication of various concepts of adaptation. Adaptation is used for characters (behaviours among them), strategies, individuals and populations. I am convinced that a single concept cannot cover all these cases. At present, it is difficult to extract relations between the meanings involved from the literature. This point is important here, because abundance and migration belong to different categories of features. Abundance, as opposed to migration, necessarily involves populations rather than individuals.

2. Specification of standards of comparison, subsequent to explication of concepts. Standards are needed because populations showing migration may be compared with other actual populations (not showing migration), or with purely hypothetical populations, or with hypothesized ancestral populations. Different comparisons may yield different acceptable statements about the adaptive value of migration.

Elaboration of such items would involve some methodological research. The dissolution rather than the solution of the original problem may be the final outcome. I am not sure whether Harden Jones agrees that the problem must be dissolved. The comments that follow the passage cited above (Harden Jones 1981, page 140) suggest it:

> Perhaps it is sufficient to make two points. Firstly, the food resources of different areas would not be available to stocks unless they moved between them, and secondly, those groups which are the more important commercially - cods, herrings, tunas and salmons - are strongly migratory.

I shall conclude this section with another, more specific example. Beitinger and Fitzpatrick (1979, page 320), in a study of preferred temperatures in fish, comment on Fry's final temperature preferendum paradigm, which is said to imply three *falsifiable* hypotheses (my italics). One of their hypotheses is formulated as follows:

> Taking into account circadian rhythms ..., the final thermal preferendum will coincide with temperatures at which key physiological, biochemical and life-history processes are optimized.

In an elaboration of the hypothesis, the authors state:

> A promising approach to resolving the adaptive value of a precise thermal preferendum in these species is to examine physiological and life-history correlates of preferred temperatures, and to compare optima of various rate functions (e.g., metabolism, growth, swimming speed) with thermal preferenda. If rate-function optima coincided with thermal preferenda, the adaptive value should be clear.

The meanings of "optimize" and "optimum" are not specified in the text, but the presentation of results indicates that the terms stand for "maximize" and "maximum."

Beitinger and Fitzpatrick would apparently endorse the following specification of the general hypothesis rendered above: "Maximization of swimming speed at the final thermal preferendum is adaptive." I shall try to show, by a succinct methodological analysis, that this hypothesis is problematic for conceptual reasons.

Consider a migrating fish moving in a horizontal temperature gradient, for example, in an ocean gyre. Let the fish select a preferred temperature (in the sense of final preferendum), and suppose that swimming speed is maximal at this temperature. The behaviour of the fish in this situation should serve as a test of the above hypothesis. But it is obvious that we need more information to make the test operative, and this is where standards of comparison (related to "adaptation") come in. One item will suffice to show what I mean. Swimming speed will not be the only aspect of locomotion involved in temperature effects. So the adaptiveness of some relation between temperature and speed will depend on how temperature influences other characteristics of locomotion. Let us assume that such characteristics remain unaffected (unless, of course, they are connected with speed). In this case, the chances are that the fish will leave regions with the preferred temperature, even when other factors do not interfere with temperature preference.

This conclusion is strange. My comments have apparently ended in incoherence, because the concept of preferred temperature has broken down. Maximization of swimming speed at preferred temperatures may indeed become *logically* impossible upon further specification of our hypothesis. The implication is that it would make sense to ascribe adaptive value to behaviours which are impossible. You will agree, I think, that this is an unpalatable conclusion.

Problems of this kind can only be solved by an adequate explication of concepts. If we consider testability and testing before explication, we are simply putting the cart before the horse

HOW TO LOCATE MISSING LINKS: ORIENTATION

The previous section touched but one methodological prescript for concept formation, namely that theoretical concepts must be clarified by appropriate explications. The basic concept of adaptation emerged as problematic under this criterion. The relevant methodology and the allied view on biology are both expanded in this section by an analysis of more specific concepts relating to orientation. The ensuing results will be applied to the study of orientation in fishes.

The recognition of ties between concept formation and classification is often essential for an adequate appraisal of theories. Such ties are preeminently visible in the field of animal orientation. For example, the conceptual distinction of taxes and kineses is tightly connected to a classification of patterns or mechanisms of orientation.

Distinctions of this kind are adequate only if principles of classification are satisfied. Three principles are presupposed in most contexts of research: 1) Classifications should be exclusive, i.e. classes must not share elements; 2) Classifications should be exhaustive, i.e. each element in the relevant domain must belong to a class; and 3) Criteria for classifying entities must be used consistently.

My main purpose in this section is to locate gaps in research, especially those resulting from incomplete classifications (cf. the second principle). I shall develop this theme against the background of previous research (van der Steen and ter Maat 1979). To set the stage for a discussion of orientation in fishes, some of the ideas developed in our 1979 paper are summarized below, together with new points of view.

Animal orientation is primarily a behavioural phenomenon. An important issue in research on this phenomenon is that one wants to explain observed patterns of locomotion. Various kinds of questions can be asked about a given pattern, e.g.:

1. Is the pattern adaptive (is it functional; is it effective)?

2. Which features characterize the tracks covered in locomotion?

3. Which stimulus-response (SR) connections generate the tracks?

4. Which anatomical structures and physiological mechanisms mediate responses to stimuli?

Such questions are associated with possible classification criteria for orientation. The following, simplistic elaboration of criteria illustrates this:

1. *Adaptiveness* (or functionality, or effectiveness): adaptive; nonadaptive (two classes).

2. *Track*: straight; not straight (two classes, admittedly very unequal ones).

3. *Operative feature of stimulus*: stimulus intensity (I); stimulus direction, i.e. angle between stimulus direction and body axis (A). (I stands for the intensity directly registered by the experimenter, not for intensities at receptor sites of animals.)
Aspects of locomotion affected: velocity (V); amount of turning (T); direction of turning (D).
Resulting *SR-connection* ("*behavioral mechanism*"): IV; IT; ID; AV; AT; AD (six classes).

4. *Anatomical or physiological aspects*: one receptor organ; two or more receptor organs (two classes). (Data for "I" at receptor sites would also relate to this criterion.)

The four criteria yield four classifications that satisfy the principles of classification discussed above. (The characterizations are somewhat vague; see, for example, comments on adaptation in the previous section.) You will have noticed that I have kept numbers of classes as low as possible by the choice of general descriptions. I have also disregarded various classification criteria, e.g., physiological processes; time lags in SR-connections, that is, memory or physiological adaptation. In spite of these restrictions, a combination of the four criteria would result in no less than 48 logically possible classes for orientation at the lowest level of complexity that is conceivable.

The classifications available in the literature mostly involve many criteria, but none of them even remotely approaches the sophistication that should be expected. Fraenkel and Gunn (1961), whose classification is fairly involved, distinguished only eight classes. The discrepancy could be explained by the assumption that various possible kinds of orientation have been disregarded on the basis of empirical evidence. A careful inspection of the literature has indeed convinced me that this cannot be the whole story. My diagnosis, instead, is as follows.

Various kinds of orientation have been discovered and subsequently described in terms of adaptiveness, behaviour, physiology, etc. The descriptions have gradually attained the status of definitions. In practice, factual matters (descriptions) and conceptual matters (definitions) are mostly intermingled. Undiscovered kinds of orientation (which I presume to be numerous) naturally remained undefined. The resulting terminologies now block the road to new discoveries. For example, if an animal follows an erratic course, it is normally assumed that the intensity rather than the direction of some stimulus may be effective. Orientation involving erratic courses and effects of stimulus direction is not covered by any concept, and for this reason alone it could remain undetected.

All current classifications are obviously incomplete. Moreover, classification criteria are not used consistently. For example, straight tracks are mostly emphasized in the identification of taxes, whereas SR-connections feature prominently in the identification of kineses. To make things worse, various schools are now engaged in orientation studies (see van der Steen and ter Maat 1979; some new developments are covered by Bean 1979, McNab 1979, Schone 1980, and Bell and Tobin 1982). They have different classifications (with similar terms for different concepts), and communication between them is limited. I shall briefly comment on Fraenkel and Gunn's (1961) classification, which is widely used in fish biology (e.g., Harden Jones 1968).

Fraenkel and Gunn (1961) distinguished various major kinds of orientation which may be summarily described as follows in terms of SR-connections: 1) kineses - stimulus intensity effects an undirected response; 2) taxes - stimulus direction effects a directed response; 3) transverse orientations: idem, but with the body axis not in line with the stimulus direction. Their subdivision of the major categories fits our distinctions as follows: orthokinesis = IV; klinokinesis = IT; klinotaxis = combination of ID and AD; tropotaxis, telotaxis, all transverse orientations = AD. (AD, as defined, may be accompanied with AT in practice.) The incompleteness of Fraenkel and

Gunn's classification is visible even when SR-connections are the only criterion considered; AV and AT are disregarded, and ID occurs only in combination with AD.

I recommend that terminologies for adaptiveness, tracks, behavioural mechanisms (SR-connections), and anatomy cum physiology be kept separate to prevent needless bias in research. Fraenkel and Gunn's (1961) terminology can be retained, but it is wise to narrow the meanings of their concepts by the use of a single classification criterion. SR-connections are my choice.

The limitations of the terminology (and any other current terminology) need continual attention. Firstly, some SR-connections are not defined.

Secondly, the relevant features of stimuli and responses (I, A, V, T, D) represent unspecified quantities. Any specification will complicate the situation (see also THE PROS AND CONS OF MODELS below). A given stimulus or stimulus feature can be expressed by more than one stimulus variable (e.g., variables may have either levels or changes in levels of some stimulus as values, and various scales of measurement can be used in either case). Analogous remarks apply to responses. So the envisioned classifications have a provisional character.

Thirdly, the concepts involved only apply to simple systems. The introduction of a separate term for each complex mechanism, however, would lead to an undesirable proliferation. Flow diagrams may provide an alternative (e.g., Bell and Tobin 1982).

The following terminology may facilitate the distinction of some special cases that involve more than one stimulus:

Orienter: stimulus (S) considered in the initial classification - in terms of SR-connections - of (behavioural) orientation mechanisms.

Releaser: stimulus that triggers some mechanism of orientation.

Modifier: stimulus that modifies effects of S on R when the SR-connection is already operative.

Terminator: stimulus that inactivates some mechanism of orientation. These terms stand for concepts that pervade much fish biology. Many other terms, e.g., sign stimulus (releaser, modifier, or both?) are now commonly used. So it would be wise to adopt a convention. Otherwise, the concepts are easily confused.

The specification of adequate variables for stimuli and responses is indispensable, if research is to move beyond the stage of exploration. The study of relations between variables should involve appropriate experimental designs, mathematical models and data anlysis. My approach will remain untechnical, but some comments on the use of mathematics and statistics are given later.

The methodology introduced so far is sufficient as a basis for an informal discussion of orientation in fishes. *Rheotropism* will serve as an example. I shall begin with a concise presentation of some common views, mostly based on the excellent survey by Arnold (1974) and then proceed with methodological comments.

Rheotropism is a broad category of reactions to currents. It is meant to include rheotaxis, rheokinesis and more complex reactions. According to Arnold (1974), pure rheokineses have never been demonstrated, whereas rheotaxes are common in migrating fish. Rheotaxis is essentially a directed response. Fish performing it are either headed upcurrent (positive rheotaxis) or downcurrent (negative rheotaxis). Responses occur only above rheotactic thresholds in current or background velocity. Current may be detected by visual or by tactile cues. Much information on visual responses has been obtained through an indirect approach, in experiments with fish in an optomotor apparatus (usually a circular tank with vertical stripes on a cylindrical background). Optomotor reactions are taxes, but they may include an orthokinetic component. Apparently, the direction and the speed of swimming can be affected at the same time in a rheotropism. The sign of the taxis and the intensity of the kinetic response (i.e., the swimming speed) can be altered by environmental factors such as temperature and chemical stimuli and by internal (e.g. endocrine) factors.

These views call for the following comments:

1. The concepts used involve both "tracks" and SR-connections as classification criteria. This is simply a consequence of adopting Fraenkel and Gunn's (1961) classification. As argued above, a choice had better be made. At the very least, the demarcation of definitions and factual statements could be made more explicit.

2. The distinction between rheotropism and other patterns of orientation is less clear than the literature suggests. Suppose, for example, that some form of rheotropism is induced by visual cues. The stimulus in this case is not some feature of the current. That is, the reactions to current are only a result of a mechanism operating on other information. For this reason orientation patterns like thermotaxis (or -kinesis) and chemotaxis (or -kinesis) should also deserve the label "rheotaxis" (or "-kinesis") in many cases.

The current usage of the "rheo-" prefix might be justified by two arguments. Firstly, currents are more strongly associated with the implicated stimuli than with factors such as temperature. (With chemical stimuli this is somewhat doubtful.) Secondly, we have no convenient terms (such as "temperature") to represent the stimuli. These arguments have some force from a practical standpoint, but the danger remains that the very concept of rheotropism leads to biased interpretations of experimental data and field observations. Thus one could be tempted to give temperature a minor role, as a releaser or a modifier rather than an orienter, on a priori grounds.

3. Suppose again, by way of an example, that visual cues (VC) are operative as stimuli, in some form of rheotaxis. Two obvious aspects of VC may affect orientation: the velocity of displacement relative to the background (example of I); and the angle between the body axis and the direction of this displacement (example of A). The intensity of VC--VC(I)--is supposedly effective when VC acts as a releaser (cf. rheotactic thresholds). When VC acts as an orienter, both VC(I) and VC(A) could play a role. As far as I know, this point has never been investigated. Quite possibly the problem has been masked by available concepts.

4. The absence of data pointing to (pure) rheokinesis is intriguing. I suspect that the following implicit line of reasoning may underlie the conclusion that rheokinesis is at best a rare phenomenon.

> Rheotaxis is a common phenomenon. Many migrating fishes indeed show orientation to currents: they are headed upstream or downstream, depending on the phase of migration. Erratic swimming patterns have also been observed in fish migrating in currents, but their effectiveness is unknown. Moreover, little is known about the nature of the stimuli and the responses involved. So it cannot be assumed that rheokinesis is possible as a form of orientation.

This argumentation may have misleading results because orientation is probably used in two senses, in the normal wide sense and for taxis (and transverse orientation). This ambiguity is common in literature on fishes; the source is Fraenkel and Gunn (1961).

The argumentation also involves the use of various classification criteria (SR-connection, track, effectiveness), and this is even more misleading. The presence of taxes is postulated with reference to tracks, and kineses are treated, less kindly, with reference to SR-connections. (Effectiveness is another complicating factor which I disregard now.) The interpretation of the data will change drastically when this asymmetry is abolished. Both rheotaxis and rheokinesis are then well-known phenomena (if tracks are made defining) or little is known about their prevalence (if SR-connections are made defining).

This hypothetical example shows that conceptual matters can definitely steer research.

5. Experimental equipment is also a source of bias in this case. The construction of a normal optomotor apparatus simply does not permit the detection of rheo-klinokinesis. This may explain the common identification of rheokinesis (or the rheokinetic component) with rheo-orthokinesis (e.g., Dodson and Young 1977; Emanuel and

Dodson 1979). An apparatus with a moving underground instead of background, and with dots instead of stripes, could yield interesting results.

Field research on upstream migration in salmon by Johnson and Hasler (1980) does suggest that rheo-klinokinesis (combined with chemo-orientation) may be involved in orientation. Tracks of fish in a river with morpholine at one side are an indication. The authors do not mention this interpretation.

The foregoing comments, taken as a whole, show that the number of possible mechanisms that may underly even simple kinds of rheotropism is vastly underestimated. Of course, it may not be sensible to investigate all possibilities, but it is almost certain that many mechanisms that could make sense have never been investigated.

LOADED CONCEPTS AND MIGRATION

The theoretical load carried by concepts has a great impact on research. That much became clear in the analysis of adaptation and orientation. The implications for migration studies were marginally touched. They will have a central position in the present section. For this purpose various basic concepts relating to migration will be analysed in some detail. Some mixing of different terminologies cannot be avoided since there is no single, generally accepted conceptual framework. The views of Harden Jones (1968) and Baker (1978) will serve as the main background. Various common notions from the literature are first introduced to facilitate the analysis.

The concept of *migration* itself has been defined in many ways. Baker's (1978, page 23) definition is presumably the widest, "... the act of moving from one spatial unit to another." I shall also use migration in a wide sense without trying to improve upon the definition. Harden Jones uses "migration" in a narrower sense. A *lifetime track* (Baker 1978, page 27) is "... a path traced out in space by an individual between birth and death." Many authors (e.g., Harden Jones 1968, 1981) use *homing* as defined by Gerking (1959) for "... the return to a place formerly occupied instead of going to other equally probable places." This concept is similar to Baker's (1978, page 26) *return migration*, "... migration to a spatial unit that has been visited previously." Return migration is distinguished from exploratory migration and removal migration. Processes involved in migration are referred to by terms such as drift, search, orientation and navigation. With the exception of drift, these terms are so ambiguous that it would be unwise to give provisional definitions. This is explained by the analysis given below.

Harden Jones' (1968) book did not become a classic by accident. The essentials of the views it contains remain representative of research on fish migration, so I feel some hesitation in suggesting that his conceptual distinctions need improvement.

Harden Jones (1968, Chapters 11 and 12) distinguishes two major problems concerning migration, "Firstly, how does a returning migrant get home? Secondly, how does the migrant recognize the home when it has reached it?" The second problem is discussed with reference to local landmarks (topographical features, chemical landmarks). I shall concentrate on the first problem, which involves "... movement without reference to local landmarks."

Four methods which may be used, singly or in combination, to reach the home are distinguished by Harden Jones.

1. Passive drift (a relatively unproblematic notion).

2. Search. a) Random (undirected) search: search random with respect to direction. b) Systematic (orientated, directed) search. (Harden Jones [1968] does not use the term "directed" here, but "directed" and "orientated" are used interchangeably in other cases.) Search presupposes that the environment does not affect movement.

3. Kineses. Movement random with respect to direction (undirected). Stimulus intensity affects speed (orthokinesis) or frequency of turning or angle of turn (klinokinesis). "Here the fish does respond to changes in the environmental field before the home is reached, and this is the basis of the distinction between search and kinesis" (Harden Jones 1968, page 199).

4. Directed (orientated) movements: taxes, transverse orientations.

Some aspects of this classification are commented on in the following point-by-point survey.

1. The concept of randomness (cf. search, kinesis) may be misleading. "Random" is a relative notion, i.e., a phenomenon can be random only with respect to some variable and a probability distribution for the variable. Harden Jones (1968) realises this (cf. "locomotion random with respect to direction"), but perhaps not sufficiently. For example, the fact that kineses can result in aggregation shows that they are not random in direction in all respects. Similar remarks apply to the term directed (systematic, orientated). As a consequence, the distinction of random and directed methods is fuzzy. I think that it should be abandoned in advanced research.

2. The terms "random," "systematic," "undirected" and "orientated" refer both to tracks and to SR-connections. Harden Jones (1968, cf. Chapter 10) follows Fraenkel and Gunn (1961) rather

closely here. The resulting classification is incomplete even if SR-connections are considered alone. One of Harden Jones' own examples, an alternative for Saila and Shappy's (1963) well-known model of Pacific salmon migration, illustrates this (pages 216-217). It involves rheotropism with visual cues in a uniform horizontal field. The implied mechanism is labeled as an orthokinesis because the response is undirected. But direction could well be the effective feature of the stimulus. The mechanism then shares properties of taxes and kineses. This is not to say that it is a combination of a taxis and a kinesis; Harden Jones explicity recognized such combinations.

The incompleteness of the classification may restrict the scope of research. Any kind of directional cue could indeed steer migration through effects on swimming velocity or turning, even in homogeneous environments. One mechanism in this category is indeed embedded, by assumption, in Saila and Shappy's (1963) model. The explicit recognition of such possibilities in the classification of migration methods should enhance the discovery of new mechanisms.

3. The involvement of various classification criteria and other problems with terms ("random," etc.) will not make the recognition of search very easy. The concept of search is indeed problematic for other reasons as well. Searching is necessarily searching for something, and the something in the present context is a home. So the identification of some aspects of migration as search presupposes that a home is identified. Baker's (1978, page 871) comments on this point reveal interesting pitfalls:

> A number of difficulties arise during the execution of displacement experiments. In particular, the true test of navigation is whether the animal can find its way to the location of the required resource. The experimenter, however, can only guess where this location might be, and usually, in fact always, assumes that it is the site at which the animal was originally captured. Whether or not this is a valid assumption depends on whether the animal's threshold for removal migration from that site has been exceeded since the time it was captured there. Thus an animal captured at a transient home range may, by the time of release, require the resources offered by its summer home range. The test of navigation in this case is not whether the animal returns to its site of capture but whether it finds its way to the summer home range.

The context in which we consider a given pattern of locomotion apparently determines whether it involves search or something else. It may represent search with respect to a transient home, although there is no search for the home at the end of a lifetime track.

Analogous comments apply to the description, in terms of effectiveness, of any method of migration whatsoever. Statements to the effect that a method works always presuppose that some goal is known. Presuppositions of this kind are often reasonable, as in the case of homing anadromous fish. But even then the ascription of (in)effectiveness to migration methods remains dangerous. The possible existence of additional transient goals implies that migration methods cannot be effective in an absolute sense. In more accurate words, effectiveness cannot be absolute for conceptual reasons because "effectiveness" is a relative notion.

The difficulties hinted at by Baker (1978) obviously relate to the applicability of concepts and thereby to the testability of hypotheses. Testability is seldom considered in migration studies, with possible detriments to research. My attention will remain focused on conceptual matters, but the implications for testability should be kept in mind.

Orientation and *navigation* are key words in the study of return migration. Two meanings of "orientation" were mentioned so far: a general meaning covering kineses, taxes and transverse orientations; and a specific one excluding kineses. "Orientation" in a third, even more specific sense, and "navigation" are mostly used outside fish biology (e.g., Schmidt-Koenig 1979, for birds). Baker (1978) applies the concepts to fish migration as well. He gives the following explications (page 28):

> Orientation: the mechanism by which an animal moves in a given plane or compass direction. Animals using an orientation mechanism, when displaced laterally, continue to move in their original plane or compass direction.
>
> Navigation: the mechanism by which an animal determines the position of a given point in space. Animals using a navigation mechanism to migrate to point A, when displaced laterally (within certain limits) can still migrate to point A.

These definitions are somewhat misleading because they suggest that the concepts have an extremely narrow meaning. The above comments on search show that this is not Baker's intention. The concepts have a relative nature. Thus navigation is navigation to a reference point, and the reference may change.

At first sight, then, "navigation" sensu Baker seems synonymous with Griffin's (1952) type III "(true) navigation (or homing)," but actually it also covers part of his type I "navigation" (which includes navigation with reference to local landmarks). "Orientation" sensu Baker is a mixture of Griffin's type I and type II "navigation."

Griffin's (1952) concepts have been used, with modifications, by several authors (Harden Jones 1968, page 4). Harden Jones thinks that they can be useful for the study of birds, but not fish. However, Baker (1978) maintains that his own concepts (which after all are in the spirit of Griffin) suit the description even of fish migration involving passive transport in ocean gyres. The subject is conceptually muddled indeed, the more so because many authors (e.g., Schmidt-Koenig 1979) do not give clear definitions.

In my opinion the concepts carry too heavy a load. (This will sound like a litany by now.) They involve mechanisms, tracks, and relations with a goal. I propose, again, that mechanisms and tracks be described with separate terminologies to prevent bias.

Reference to goal-like entities in definitions leads to additional problems of a different kind (see also comments on search). It easily brings mentalistic connotations in the description of behaviour. Apparently mentalistic concepts are indeed common in migration studies ("search," "exploration," "recognition," etc.), and limitations of our present knowledge make mentalism a potent source of untestability. I do agree with Griffin (1978, 1981, 1982) and Hinde (1982) that the study of the mental should be resumed in biology.

Baker (1978, Chapter 33) classifies navigation models--without being faithful to his own navigation concept--as follows:

> 1. Non-map models (random search models, sensory-contact models, downcurrent models, and inertial navigation); 2. Map models, with two categories, depending on the kind of map. *a* Grid map (sun-arc map, night-sky map, geomagnetic map, other maps of this kind). *b* Maps involving past experience, i.e., familiar-area maps. The term map is here used in a very wide sense, which I adopt for convenience.

The classification is anything but complete. The following hypothetical example, which covers views common in fish biology, illustrates this.

> The lifetime track of some anadromous fish consists of a number of sub-tracks, each charaterized by one or more mechanisms of orientation (in the general sense) and drift. At the end of each sub-track, terminators and releasers change the mechanism of orientation. A releaser (or a modifier) may set a particular compass course--not necessarily in the direction of the ultimate home--but there are many other possibilities. Terminators and releasers involve many kinds of factors, e.g., chemicals emanating from local faunas, physical characteristics. The sequence of orientation patterns ensures the return to the breeding site. The mechanisms involved could even allow for return after artificial displacements.

This example does not belong to any of Baker's (1978) categories. It closely resembles the description of a familiar-area map, but experience need not be involved.

I suspect that the incompleteness of Baker's (1978) classification (and many other, less articulated classifications) results from a preoccupation of navigation biologists with a mentalistic, anthropocentric perspective. Sophisticated bicoordinate navigation and coursing along tracks known by experience are indeed the paradigms of conscious navigation in man. The scene may drastically change when mechanisms rather than goals are emphasized in the definition of concepts.

For lack of new terminologies, I shall use "navigation" roughly in Baker's (1978) wide sense in some concluding remarks bearing on fish migration.

Baker's (1978) ultimate preference is the familiar area map as *the* tool for navigation. The form of his argument is, in my words, "A, B, C ... and X are the available methods; A, B, C ... do not work; navigation occurs; so X is the only workable method." This is a fallacy because the first premise is false (ultimately for conceptual reasons: the underlying classification is logically incomplete). Naturally, I do not want to imply that familiar area maps are unimportant; that would also be a fallacy.

Neave (1964) is among the few authors supposing that navigation may explain fish migration. His reasoning is that passive transport, sun-compass orientation and orientation to gradients do not suffice to explain the return migration of Pacific salmon. Therefore, "it is difficult to avoid the conclusion that throughout the period of ocean life some awareness of position in relation to the place of origin is maintained" (Neave 1964, page 1242). That is, a sophisticated form of bicoordinate navigation may be involved. Harden Jones (1968) has rightly argued that this is not the only possibility, although he did not qualify the argument as a fallacy. (Neave's argument resembles Baker's argument.) Harden Jones would agree, I think, that the hypothetical example presented above is an alternative which is acceptable in fish biology.

If this is true, the mechanisms implied in the example deserve more attention. Experimental work dealing with effects of releasers and modifiers on orientation mechanisms is then of primary importance. Such effects may take many forms that were never investigated. For example, I would not be surprised if food quality (e.g. chemicals emanating from a local fauna) were shown to affect the course of fish that use a magnetic or an electric field for orientation.

If these ideas are sound, possible orienters should be studied under many different experimental conditions, and this will necessitate changes in the prevailing strategy of research.

THE PROS AND CONS OF MODELS

Methodological analysis, as exemplified in previous sections, is inherently ambivalent. My purpose was the removal of ambiguities from theories by articulation of concepts. But I have not attained any definite preciseness, because the language to be improved was also used in the analysis. It would be unwise to expect much more as long as one remains in the domain of a natural language.

Preciseness can indeed be reached when artificial languages enter the scene, e.g., in the garbs of mathematical models, but applied mathematics ultimately has similar limitations. A formal approach amounts to a temporary departure from the natural context in which any resulting models should come to be embedded, and all-pervasive preciseness in the domain of the natural is simply a fiction. I shall not dwell on such limitations, but their presence behind the scenes should be effectual.

Our research group is presently trying to combine the benefits of formal and informal approaches in theoretical biology. Against this background, we have tackled animal orientation from the philosophy of science angle stressed in this paper and from a mathematical angle. My colleague, Paul Doucet, and co-workers are now preparing two papers representing the latter approach. I shall briefly mention some of their results before proceeding with general comments on mathematics in the study of fish migration.

Doucet and co-workers have studied kineses of a hypothetical animal through computer simulations. The animal is a simple one. It follows a step-angle track when it moves, i.e., linear movements are separated by sharp angles. Adaptation does not occur. Locomotion of the animal was simulated with reference to four variables: speed (by assumption constant within steps), step length, step time (= time between angles) and angle size. Three independent variables are actually sufficient for a complete specification of tracks (speed = step length/step time). In the simulations the variables were made stimulus-dependent in various combinations. Thus results were obtained for orientation resembling orthokinesis and klinokinesis.

The outcomes of the simulations were most interesting. Orthokinesis gave results in agreement with informal studies in the literature. The results of klinokinesis were also in line with the literature, in the sense that they explain persisting controversies about its effectiveness. The spatial patterns resulting from slightly different forms of klinokinesis were indeed quite different. Moreover, the nature of the difference depended on the gradient being bounded or unbounded.

Obviously, slight differences in the orientation mechanism of a simple model animal can substantially affect the ensuing behaviour. The moral of this conclusion will be clear. Firstly, it again shows that it is unwise to conflate different classification criteria (mechanism, track, effectiveness) in the construction of a conceptual framework for animal orientation. Secondly, casual interpretation of data concerning orientation (by far the most common species of interpretation) is dangerous. It must certainly be discouraged when more complex mechanisms of orientation are involved (which is mostly the case).

Doucet et al. have also developed an analytical model for the hypothetical animal by simplifying Patlak's (1953) rather untractable model. The results of the analytical approach tended to confirm the simulation studies. In the development and the application of the model, Doucet et al. distinguished three logically independent forms of kinesis. The distinction of two categories (orthokinesis, klinokinesis) is indeed inadequate for describing the behaviour of the model animal. Thus the general distinctions used in the third section are indeed not very precise. However, it would be rash to conclude that the distinction, for example, of three rather than any other number of kineses is always preferable. It is probable that the tools needed for a mathematical approach to orientation will be highly context-dependent. Animals to be modeled can have quite dissimilar properties, and this will affect the choice of variables and thereby the classification of mechanisms.

Methodological demands such as generality, realism and precision simply have different implications (Levins 1968), and this is why we have to compromise.

Simulation and the building of analytical models do not exhaust the opportunities for applying mathematics. Research on animal orientation and migration could also benefit from the use of statistical methods hitherto disregarded in experimental work. Multivariate statistics for factorial designs and response surface exploration deserve special attention. Such methods are indispensable for the solution of problems discussed in previous sections, which mostly involve interactions between environmental factors (cf. the distinction of categories for stimuli: orienters, releasers, modifiers, terminators). They have been used in very few cases until now.

It is true that standard methods of circular statistics (Batschelet 1981) are widely used in orientation research. Such methods perfectly suit the aim of showing that animals do orientate. But detailed information on orientation can only be acquired by means of more advanced statistics, circular or non-circular.

The study of fish migration should benefit from a joint application of general methodology, model building and statistics. To illustrate this, I shall comment on a paper by Neill (1979) on behavioural thermoregulation. His work is important for understanding orientation of fish staying in ocean gyres during migration.

Neill (1979, page 306) distinguishes two categories of behavioural thermoregulation, denoted as predictive and reactive: There would seem to exist the possibility of two fundamentally different mechanisms of behavioural thermoregulation in fishes. The critical distinction between the two lies in whether or not the fish has and can use predictive information about the thermal structure of its environment. Requisite information may be obtained through either the individual fish's own experience or that of its ancestors; that is, the information may be based on either learning or instinct. We may call the effective use of such prior knowledge of environmental structure "predictive behavioural thermoregulation" and reserve "reactive behavioural thermoregulation" for the remaining class of thermoregulatory mechanisms.

Neill presents various hypothetical models for either kind of thermoregulation. Simulations showed that one particular model can effectively generate reactive thermoregulation in the form of klinokinetic avoidance behaviour (klinokinesis involving a form of adaptation). Simulations of other models (not specified by Neill) had shown that klinokinesis without adaptation and orthokinesis are not sufficiently effective. Predictive thermoregulation was not simulated. I have the impression that it is held to involve taxes or transverse orientations.

The following itemized comments on Neill's (1979) work may contribute to an elaboration of his approach.

1. The terms "(un)predictable" and "(un)predictive" are widely used in ecology. I think that they are somewhat hazy because they presuppose too much theory. Seemingly innocuous mentalistic language (cf. "knowledge") associated with these terms might actually hide such theory.

Concepts such as "learning" and "instinct" are not very simple either. "Instinct" refers to something determined genetically, and it is either uninformative or senseless to call some property of an organism genetically determined. Any proper reconstruction of genetics will reveal that "genetic determination" and "environmental determination" refer to *differences* between organisms or populations. So we need a standard of comparison (see discussion of adaptation in the second section).

I shall not continue this analysis but merely suggest that an appraisal of concepts could improve our understanding of behavioural thermoregulation.

2. The distinction of predictive and reactive thermoregulation involves many classification criteria (cf. properties of the environment; learning and instinct; mechanism of orientation; involvement of other factors besides temperature). An explicit

formulation of these criteria would be helpful since conceptual matters (definitions) and factual matters (descriptions) are intermingled here. Anyhow, the fact that many criteria are involved ensures that the number of conceivable mechanisms underlying behavioural thermoregulation must be very large indeed. So the a priori distinction of just two categories with a heavy theoretical load could impart undetected, though unwanted, bias on research.

It would be fruitful to make an inventory of possible mechanisms, for example, through the construction of a logically complete set of flow diagrams for a given level of complexity. (Neill [1979] gives two examples of diagrams.) This would strengthen the ties between model building and conceptualization. The detection of biased research steered by concepts should thus be facilitated.

3. There is an unfortunate tendency in the literature not to dwell on negative results. Neill's (1979) brief comments on klinokinesis without adaptation and orthokinesis are an illustration. Data concerning the relevant simulations should have some importance in view of the results obtained by Doucet et al., namely that minor changes in a simulation program can visibly alter the pattern of orientation.

4. Simulation studies should involve appropriate designs, and the results must be subjected to statistical analysis. But statistics is even more important in research that may precede model building. Experiments with a factorial design could serve to identify stimuli affecting behavioural thermoregulation and interactions between them. The ensuing results will permit the selection of subsets of flow diagrams making empirical sense. Simulation studies should focus on diagrams selected in this way.

Various other types of mathematical models for fish migration recently became available. Those developed by DeAngelis (1978) and Balchen (1979) are already well-known. The fact that the models differ in many respects shows that the number of feasible models must be staggering. The development of criteria for making the right choices is therefore badly needed. A word of caution, serving as a final comment, is appropriate here. There is no such thing as the right model for a particular phenomenon. Rightness, you will guess it by now, is a relative notion. Things can be right indeed, but only in relation to a given purpose. So purposes will need some attention in modeling work.

ACKNOWLEDGMENTS

Gratefully I thank my colleagues Paul Doucet, Hans Jager and Bart Voorzanger for helpful comments, and Thea Laan for typing the manuscript.

REFERENCES

Arnold, G.P. 1974. Rheotropism in fishes. Biological Reviews of the Cambridge Philosophical Society 49:515-576.

Baker, R.R. 1978. The evolutionary ecology of animal migration. Hodder and Stoughton, London, England.

Balchen, J.G. 1979. Modeling, prediction, and control of fish behavior. Pages 99-146 *in* C.T. Leondes, editor. Control and dynamic systems, volume 15. Academic Press, New York, New York, USA.

Batschelet, E. 1981. Circular statistics in biology. Academic Press, New York, New York, USA.

Bean, B. 1979. Chemotaxis in unicellular eukaryotes. Pages 335-354 *in* W. Haupt and M.E. Feinleib, editors. Encyclopedia of plant physiology, new series, volume 7. Springer-Verlag, Berlin, Federal Republic of Germany.

Beitinger, T.L., and L.C. Fitzpatrick. 1979. Physiological and ecological correlates of preferred temperature in fish. American Zoologist 19:319-329.

Bell, W.J., and T.R. Tobin. 1982. Chemo-orientation. Biological Reviews of the Cambridge Philosophical Society 57:219-260.

Brady, R.H. 1979. Natural selection and the criteria by which a theory is judged. Systematic Zoology 28:600-621.

Carnap, R. 1966. Philosophical foundations of physics. Basic Books, New York, New York, USA.

DeAngelis, D.L. 1978. A model for the movement and distribution of fish in a body of water. ORNL/TM-6310. Oak Ridge National Laboratory, Oak Ridge, Tennessee, USA.

Dodson, J.J., and J.C. Young. 1977. Temperature and photoperiod regulation of rheotropic behavior in prespawning common shiners, *Notropus cornutus*. Journal of the Fisheries Research Board of Canada 34:341-346.

Emanuel, M.E., and J.J. Dodson. 1979. Modification of the rheotropic behavior of male rainbow trout (*Salmo gairdneri*) by ovarian fluid. Journal of the Fisheries Research Board of Canada 36: 63-68.

Fraenkel, G.S., and D.L. Gunn. 1961. The orientation of animals. Dover Publications, New York, New York, USA.

Gerking, S.D. 1959. The restricted movements of fish populations. Biological Reviews of the Cambridge Philosophical Society 34: 221-242.

Griffin, D.R. 1952. Bird migration. Biological Reviews of the Cambridge Philosophical Society 27:359-393.

Griffin, D.R. 1978. Prospects for a cognitive ethology. Behavioral and Brain Sciences 4:527-538.

Griffin, D.R. 1981. The question of animal awareness. Rockefeller University Press, New York, New York, USA.

Griffin, D.R., editor. 1982. Animal mind - human mind. Springer-Verlag, Berlin, Federal Republic of Germany.

Harden Jones, F.R. 1968. Fish migration. Arnold, London, England.

Harden Jones, F.R. 1981. Fish migration: strategy and tactics. Pages 139-165 *in* D.J. Aidley, editor. Animal migration. Cambridge University Press, Cambridge, England.

Harding, S.G., editor. 1976. Can theories be refuted? Reidel, Dordrecht, The Netherlands.

Hinde, R.A. 1982. Ethology. Collins, Glasgow, Scotland.

Hull, D.L. 1968. The operational imperative: sense and nonsense in operationism. Systematic Zoology 17:438-457.

Hull, D.L. 1974. Philosophy of biological science. Prentice-Hall, Englewood Cliffs, New Jersey, USA.

Johnson, P.B., and A.D. Hasler. 1980. The use of chemical cues in the upstream migration of coho salmon, *Oncorhynchus kisutch* Walbaum. Journal of Fish Biology 17:67-73.

Levins, R. 1968. Evolution in changing environments. Princeton University Press, Princeton, New Jersey, USA.

Lewontin, R.C. 1978. Adaptation. Scientific American 239:157-168.

Maynard Smith, J. 1978. Optimization theory in evolution. Annual Review of Ecology and Systematics 9:31-56.

McNab, R.M. 1979. Chemotaxis in bacteria. Pages 310-334 *in* W. Haupt and M.E. Feinleib, editors. Encyclopedia of plant physiology, new series, volume 7. Springer-Verlag, Berlin, Federal Republic of Germany.

Neave, F. 1964. Ocean migrations of Pacific salmon. Journal of the Fisheries Research Board of Canada 21:1227-1244.

Neill, W.H. 1979. Mechanisms of fish distribution in heterothermal environments. American Zoologist 19:305-317.

Patlak, C.S. 1953. A mathematical contribution to the study of orientation of organisms. Bulletin of Mathematical Biophysics 15:431-476.

Popper, K.R. 1959. The logic of scientific discovery. Basic Books, New York, New York, USA.

Ruse, M. 1973. The philosophy of biology. Hutchinson, London, England.

Saila, S.B. and R.A. Shappy. 1963. Random movement and orientation in salmon migration. Journal du Conseil Conseil Permanent International pour l'Exploration de la Mer 28:153-166.

Schmidt-Koenig, K. 1979. Avian orientation and migration. Academic Press, New York, New York, USA.

Schone, H. 1980. Orientierung im Raum. MBH, Stuttgart, Federal Republic of Germany.

Sneath, P.H.A., and R.R. Sokal. 1973. Numerical taxonomy. Freeman, San Francisco, California, USA.

Steen, W.J. van der. 1981. Testability and temperature adaptation. Oikos 37:123-125.

Steen, W.J. van der. 1982. Algemene methodologie voor biologen. Bohn, Scheltema and Holkema, Utrecht, The Netherlands.

Steen, W.J. van der, and A. ter Maat. 1979. Theoretical studies on animal orientation. I. Methodological appraisal of classifications. Journal of Theoretical Biology 79:223-234.

AN INTRODUCTION TO MODELING MIGRATORY BEHAVIOR OF FISHES

D. L. DeAngelis and G. T. Yeh

Environmental Sciences Division
Oak Ridge National Laboratory
Oak Ridge, Tennessee 37830 USA

ABSTRACT

This paper reviews models of basic fish movement patterns and their incorporation into models of fish migration. Computer simulations of various types of kineses and taxes have led researchers to the conclusion that only certain types of turning response are capable of producing directed movement of organisms. Klinokinesis-with-adaptation is possibly the most realistic model of the movement of fish along a stimulus gradient. In modeling fish migration, however, simpler models of movement, such as biased random walk models, are more convenient to use. A biased random walk model of the migration of a group of fish in a hypothetical ocean-coastal region is described. Environmental heterogeneity is built into the model to simulate a realistic situation. It is shown that such a model is equivalent to an advection-diffusion partial differential equation model. The results of the advection-diffusion model are compared with those of the biased random walk model.

INTRODUCTION

Concepts of the "animal as a mechanism" go back at least to the time of Descartes, as do attempts to simulate animal movement with mechanical devices. The cybernetic revolution of the twentieth century further stimulated the construction of mechanical models to help in understanding "purposeful" movement.

The Philips "dog," developed during the 1920s, had two photo-electric cells for eyes. When one of these was exposed to light, it switched on one of two motors that turned the dog toward the light until both cells received equal fluxes of light. Then both motors began to move the dog straight toward the light. A more complicated artificial animal was the *Machina speculatrix* of W. Grey Walter (1951), which would search randomly, maneuver around obstacles, and approach a light source, but move away from it if it were too intense. In addition to these machines, a "mouse" that could find its way out of a labyrinth, a "squirrel" that gathered "nuts," animals that could display conditioned reflexes, and numerous other of models had already been built by 1960 (see Nemes 1970). Many of these could behave in a way that is uncannily realistic.

These cybernetic models have much in common with the mathematical models of migration that we shall describe. Both the mechanical models and the mathematical models involve abstractions or simplifications of their biological prototypes. In fact, if a mechanical model can be designed, a corresponding exact mathematical model is possible, and vice versa.

Like the mechanical models, mathematical models can make wonderful toys for anyone with access to a computer. But can they be useful in understanding animal movement--migration in particular? Saila and Flowers (1969) remark: "Biologists have often considered mathematical models as inappropriate or worthless for their work and have not made much use of them." We feel that this attitude has been changing rapidly during the last decade, in part due to the efforts of Saila and others like him. In the study of animal movement researchers have recognized that mathematical models, with all their limitations and oversimplifications, have a useful role to play. While the various mechanisms of movement may certainly be described verbally, the implications of these mechanisms for spatial displacement and distributions are not intuitive. Models that allow quantitative computation of these implications are indispensable.

A few examples should confirm this point. Rohlf and Davenport (1969) used computer simulations to verify Ullyott's (1936) conclusion, based on empirical observations, that simple klinokinetic (see definition later) movement of animals has no aggregating effect over the long term. In a revealing comment the authors admit that they undertook their work because one of them had long had intuitive doubts about Ullyott's conjecture. Neill's (1979) computer model confirmed Ivlev's (1960) and Sullivan's (1954) suggestions that orthokinetic (see definition later) responses would be poor mechanisms for moving fish or other organisms, using behavioral thermoregulation, toward desired temperature ranges. Saila and Shappy (1963) used a model of migration to show that the open-ocean phase of salmons' return to their natal streams could be achieved with much less orientation than had commonly been supposed.

In all three cases verbal reasoning was unable to predict the implications of a particular mechanism of movement. There was, to be sure, relevant empirical evidence in two of these cases. But empirical studies of animal behavior can seldom be done under conditions that control unobserved factors. Mathematical models, on the other hand, can be designed to test each postulated mechanism unconfounded by other factors.

The judicious use of mathematical models, then, may play an important role in unraveling the physiological and environmental factors involved in fish movement in general and fish migration in particular. Models can disprove hypotheses, corroborate them, or, as Neill (1979) has so clearly shown, suggest new hypotheses. Neill's warning comment on his own model of thermoregulation, however, should be assimilated by all who might be tempted to put too much trust in model results without adequate empirical validation: "But the extent to which biological reality is reflected in this or any other model of reactive thermoregulation can only be judged from research yet to be performed."

It might be added, too, that the biological reality of fish swimming in the vast ocean is far different from the reality of fish swimming in an experimental tank. The modeling of fish under experimental conditions has been quite successful, and there is nothing that we can add to that work. However, there is still room for developing models that can accurately simulate and predict movements of fish in their actual environments over long temporal and spatial ranges. Again, the groundwork for this was laid some time ago by the pioneering work of Skellam (1951), Saila and Shappy (1963), Patten (1964), Saila and Flowers (1969), and others. Additional steps can still be taken, nevertheless, in the development of models of migration that incorporate environmental complexity and yet are still convenient to use.

MECHANISMS OF MICROSCALE MOVEMENTS

Some of the things that a modeler, like any scientist, first thinks about in describing a phenomenon are the appropriate spatial and temporal scales for observing it. The typical spatial scale of fish migration is in tens, hundreds, or perhaps even thousands of kilometers. The typical temporal scale is in tens or hundreds of days. It would seem a gross incongruity, therefore, to try to describe fish migrations in terms of the fish movements, or "darts," observed under experimental conditions and having the spatial scale of meters and temporal scale of seconds. Yet science proceeds by breaking observable phenomena into simple constituent elements that can be quantified and understood. The hope is that the constituent elements can be added together to describe the complexity occurring in nature.

The phenomenon of active animal movement in space can be decomposed into several basic types. Fraenkel and Gunn (1961, original edition 1940) describe a useful set of such types that they derived partly from earlier work by Loeb (1918), Kuhn (1929), and others. Here we list these types, relying also on definitions by Rohlf and Davenport (1969) and Okubo (1980). We also describe some of the inferences that have been drawn from mathematical models of these basic mechanisms. In a later section the actual details of the mathematical models will be described, and these simple models will lead to a general migration model.

Simple and Weighted Random Walk

Suppose an animal is initially situated at a point (i,j) in a two-dimensional grid of uniformly spaced points (Fig. 1). Here i represents a position at distance Y_i along the y-axis, and j represents a position at distance X_j along the x-axis. During each time step it is assumed that the organism can move with equal probability (P = 0.25) in any of four directions, to the points (i-1,j), (i+1,j), (i,j-1), or (i,j+1).

A question of key interest, the probability of the animal arriving at a point (k,m) after n steps, can be answered analytically for this case of simple random walk. The answer is especially interesting in the limit that this discrete random walk approaches continuous random walk. If we let time $t = n\tau$, where τ is the time step size, and $X_k = k\lambda$ and $Y_m = m$, where λ is the interpoint distance on the grid, then the probability that the animal will be in the area bounded by the lines $X_k, X_k + \Delta X, Y_m$, and $Y_m + \Delta Y$ at time t is

$$P(X_k,Y_m;t) = \frac{\Delta X \Delta Y}{4\pi Dt} \exp\{-\frac{X_k^2 + Y_m^2}{4Dt}\}, \quad (1)$$

where D is defined as the limit of $\lambda^2/2\tau$ as λ and τ become small (that is, in which n, k, and m $\rightarrow \infty$ for fixed t, X_k and Y_m). This is a bivariate Gaussian distribution about the origin. The significant fact revealed by this probability distribution is that the probability for arrival at a given radial distance from the starting point is the same in all directions; that is, there is radial symmetry. Thus a popultion of fish starting from a common initial point (0,0) and moving according to pure random walk shows no average spatial displacement in any direction.

An important generalization of the simple random walk model is the weighted or biased random walk in which the probability of a move is not the same in all four directions. If the probability of movement in, for example, the positive x-direction is increased relative to the probability of movement in the negative x-direction (say, $P_{+x} = 0.4$ and $P_{-x} = 0.1$), while the probabilities of movement in the positive and negative y-directions remain the same (0.25), then the probability of reaching the interval $\{X_k, Y_m; X_k + \Delta X, Y_m + \Delta Y\}$

after n steps ($n \to \infty$) is

$$P(X_k, Y_m; t) = \frac{\Delta X \Delta Y}{4\pi D t} \exp - \left\{ \frac{[X_k - 0.3(\frac{\lambda}{\tau})t]^2 + Y_m^{\ 2}}{4Dt} \right\}. \tag{2}$$

This is simply a two-dimensional Gaussian distribution that has been translated at a distance $0.3(\lambda/\tau)t$ along the x-axis.

Both the simple and weighted random walk models violate some of our intuitive notions concerning movement. First, real movement is spatially continuous, not discrete as in the model, and real incremental movements can occur in any angular direction, not just along the positive or negative x- and y-axes. Hence, this "lattice walk" model, as it is called, would seem to be very unrealistic. There are, of course, many other types of random walk models. In Pearson's walk (Pearson 1905), for example, an animal moves a distance s, then turns through some randomly chosen angle anywhere between 0° and 360°. Continuous-time random walk models have also been formulated. We will, nevertheless, continue to use the lattice walk model, because it is conceptually simple and because it is not a bad approximation in describing movement over large distances compared to the grid scale (interpoint distance). Many physical and biological problems are successfully analyzed by this convenient fiction (see, for example, Barber and Ninham 1970).

A second lack of realism in the present model is that it does not reflect the typical movement patterns of animals with bilateral symmetry, such as fish. A more realistic model in this respect is presented next.

Random Walk with Effects of Bilateral Symmetry

An animal with bilateral symmetry (that is, having a front and back end) tends to move forward in the direction it is facing or to turn to face left or right and then to proceed in that direction, but not to move backward. A particular model incorporating this effect of bilateral symmetry, but still resembling the simple random walk in other respects, can be summarized succinctly by the following properties: 1) The animal at (i,j) has a probability of 0.5 of taking the next step in the same direction as the last, e.g. if the animal was previously at point (i-1,j), then it has a probability of 0.5 of being next at (i+1,j); and 2) The animal has probabilities of 0.25 of turning left (counterclockwise) and 0.25 of turning right (clockwise).

Rohlf and Davenport (1969) simulated on the computer the movements of 500 animals starting at the point (0,0) and following the above mechanism of movement. They found that the mean position along the x-axis after 100 steps was $\overline{X} = 0.058$, with a standard deviation (SD) of X = 12.60, assuming that the nearest-point distance on the lattice was 1.0. Hence, the displacement was not significant. The distribution of animals about (0,0) appeared to be indistinguishable from the two-dimensional Gaussian (Eq. 1).

The resultant Gaussian distributions for the simple random walk and the random walk adjusted for bilateral symmetry are not very surprising. Neither of these types of model would have been expected prima facie to produce net displacement or aggregation of a population. However, we now come to types of movement where intuition is not a reliable guide to predicting the final disposition of the population. Two types of movement, kinesis and taxis, have been proposed as mechanisms that result in directed or purposive movement. Kinesis has been defined by Rohlf and Davenport (1969) as, "...undirected locomotory reactions, in which the speed of movement or frequency of turning depend upon the intensity of stimulation...," while a taxis is motion that is directed along a gradient of a stimulus. The difference is that in a taxis the animal is assumed able to detect the direction of the gradient and direct its movement accordingly. In kinesis, however, the directed movement of the animal, if it occurs, is assumed to result not from directed responses to the stimulus gradient per se but from different speeds or turning frequencies in different stimulus intensities. We shall first consider some possible types of kineses.

Orthokinesis

Orthokinesis is defined as movement in which step length is a function of the stimulus intensity. Because this stimulus intensity is in general a function of spatial position, one can directly express this step length as a function of position. For example, Rohlf and Davenport (1969) considered a plane whose x-axis ranged from X = -50.0 to X = 100.0, where the step size varied as

$$\Delta X = 1.0 + 0.02\ X, \tag{3}$$

that is, from 0.0 to 3.0. Along the y-axis step size remained constant at $\Delta Y = 1.0$. The lattice can be used to represent this type of movement simply by making the interpoint spacing along the x-axis vary with X_i;

$$X_{i+1} = X_i + 1.0 + 0.02\ X_i. \tag{4}$$

The simulations of Rohlf and Davenport (1969) for this type of movement yielded a skewed distribution after 100 steps, with a higher density of organisms where the step size was small (to the left of the $X_i = 0$ axis). However, there was little net displacement ($\overline{X}$ = 1.654, SD of X = 13.386). Hence, orthokinesis alone can result in local aggregation but not in significant net movement in a direction (some slight net movement not being excluded, however).

Simple Klinokinesis

Suppose now that it is not the size of the step but the probability of an animal changing its direction that is a function of its position in space. For example, the probability of turning left or right rather than continuing forward could increase along the positive x-direction. Rohlf and Davenport (1969) set the probability of turning, P_{turn}, equal to

$$P_{turn} = 0.5 + 0.005\ X_i, \tag{5}$$

so that on a range from X = -50.0 to X = 50.0, P_{turn} ranged from 0.0 to 1.0. Simulations yielded $\bar{X} = 0.262$, SD of $\bar{X} = 12.087$, again not a significant displacement. Simple klinokinesis, besides showing no net displacement, also indicated no aggregating effect.

Topotaxis

Taxis differs from kinesis in that the animal is assumed able to directly detect a stimulus gradient and actively orient its movement along it, rather than to simply respond to different intensities of stimulus by altering the intensities of various types of behavior such as turning or speed. Directed movement along a stimulus gradient, which Okubo (1980) calls topotactic movement (Rohlf and Davenport [1969] used the label "tropotactic" movement), requires spatially separated sensors, like the photocell "eyes" on the Philips "dog," that can immediately detect differences of intensity. The animal can then adjust its direction accordingly.

In their computer simulations Rohlf and Davenport (1969) assumed that the stimulus has a gradient along the x-axis. When the organism was facing perpendicular to the x-axis, it had a two-thirds chance of turning in the "correct" direction toward increasing stimulus. When it was facing along the x-axis, it had a one-half probability of continuing in the same direction. Rohlf and Davenport found that $\bar{X} = 17.492$ after 100 steps, a significant displacement, showing that topotaxis is an effective cause of net movement in animals.

A shortcoming of the topotaxis model is that it assumes at any instant that an animal can measure the gradient of a stimulus. It is not likely that many animals will have sensors that are capable of detecting typical spatial variations in concentrations of a chemical attractant on the scale length of their bodies. Another hypothesized mechanism, "klinokinesis-with-adaptation," discussed next, circumvents this difficulty.

Klinokinesis-with-adaptation and Klinokinetic Avoidance

Klinokinesis-with-adaptation refers to a type of behavior,

described by Fraenkel and Gunn (1961) and Rohlf and Davenport (1969), that is related to "klinokinetic avoidance" hypothesized by Neill (1979). The precise mechanisms are different but similar enough that we need not bother about the differences here. We describe Neill's model.

Neill's (1979) model specifically addresses the problem of orientation in a temperature gradient, though any other gradient could be substituted. Neill postulated that for negligible thermal stress a fish orients randomly through simple kinesis, but when the fish is far enough from its ideal temperature range in a sufficient gradient, it reacts through directed movement. This could be done through a clever incorporation of "memory" in the fish's physiology. Fish have temperature sensors in both the skin and deep in the core. The surface sensors register present ambient temperatures. The core temperature, on the other hand, represents an average of the water temperatures the fish has been swimming through in the recent past. The fish then has information not only about its current water temperature state but also about its earlier ambient temperatures.

It would be simple for the nervous system of the fish to compare present and earlier ambient temperatures to judge whether its present direction of motion is improving or worsening its situation. If the fish's skin sensors detect that the present temperature is not ideal, it can make the following four appropriate corrections to its movement:

Current state	Probability of changing direction
Too warm and warming	$P_{turn} > 0.5$
Too warm and cooling	$P_{turn} < 0.5$
Too cool and warming	$P_{turn} < 0.5$
Too cool and cooling	$P_{turn} > 0.5$

Neill (1979) conducted simulations that yielded interesting results. When the temperature gradients were steep enough that the fish could effectively sense the differences in core and ambient temperatures, they efficiently moved in the direction of their ideal range of temperatures, forming a density there. If the gradient were too weak, this spatial distribution of fish could become highly dispersed or even bimodal, with the peaks close to lethal temperatures.

Conclusions

Discussion of the above microscale movements is useful because it gives us an idea of what sorts of mechanisms give rise to directed movements, which is important to know when attempting to model migration. Of course, all sorts of combinations of the above mechanisms are also possible and are described by Rohlf and Davenport (1969), but we will not discuss them here.

In the next section we will do two things. First we will show how these mechanisms of movement are actually described mathematically. Second, we will show that for the study of large-scale movements a model combining the biased random walk with a topotactic mechanism is adequate, a fact that will make further advances in model analysis possible, as described later.

MATHEMATICAL DESCRIPTION OF MICROSCALE MOVEMENT

In the preceding section basic mechanisms of animal movements were reviewed. In the present section more precise mathematical descriptions are given for these mechanisms. This may seem somewhat redundant because the verbal descriptions were generally quite complete in themselves. Nevertheless, the more mathematical representations will give one a better flavor for how the computer programs for the simulation of movement actually work.

Consider again Figure 1, showing a small section of the two-dimensional grid. We define on this grid the discrete function $k(i,j \rightarrow i+\varepsilon,j+\eta)$, which is the transition probability from a point (i,j) to each of its neighbors, when $(\varepsilon,\eta) = (0,1), (1,0), (-1,0)$, or $(0,-1)$. We also define the probability $P(i,j;t)$ that at time t an animal is situated at the point (i,j). We must have

$$\sum_{\varepsilon,\eta} k(i,j \rightarrow i+\varepsilon,j+\eta) = 1, \qquad (6)$$

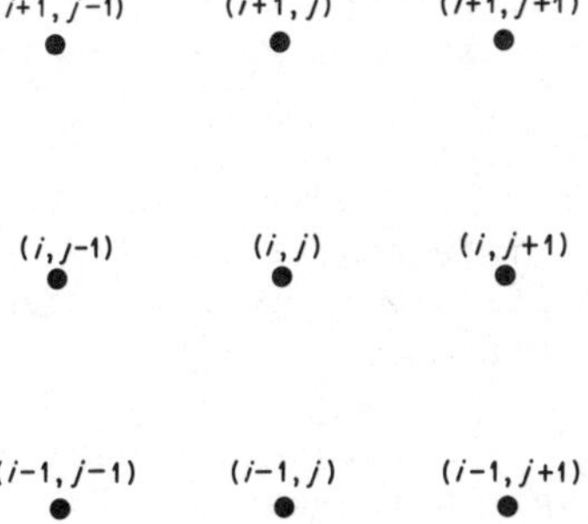

Figure 1. Portion of a two-dimensional lattice for modeling animal movement.

where the summation is over all four neighbors.

For *simple random walk* the form that the transition probability takes is very simple:

$$k(i,j \rightarrow i+\varepsilon,k+\eta) = 0.25 \qquad (\text{all } \varepsilon,\eta \text{ and all } i,j). \tag{7}$$

This is implemented in the computer program by using a pseudorandom number generator to choose a number, r, uniformly on the interval [0,1]. If $(0.0 < r \leq 0.25)$, then the transition to $(\varepsilon,\eta) = (0,1)$ is executed; if $(0.25 < r \leq 0.50)$ then the transition to $(\varepsilon,\eta) = (1,0)$ is executed, and so forth.

This simple model can easily be extended to a *biased random walk* model by increasing the probability of motion in a particular direction. For example,

$$\begin{aligned} k(i,j \rightarrow i+1,j) &= 0.25 \\ k(i,j \rightarrow i-1,j) &= 0.25 \\ k(i,j \rightarrow i,j+1) &= 0.40 \\ k(i,j \rightarrow i,j-1) &= 0.10 \end{aligned}$$

describes a pronounced average movement in the positive x-direction.

To describe *simple kinesis* a slightly more complex formulation, involving conditional probabilities, must be used. In particular,

$$k(i,j \rightarrow i+\varepsilon,j+\eta|\ i+\varepsilon',j+\eta')$$

represents the probability of a transition from (i,j) to $(i+\varepsilon,j+\eta)$, given that the preceding point was $(i+\varepsilon',j+\eta')$, where $(\varepsilon',\eta') = (0,1)$, (1,0), (0,1), or (-1,0). For example, the transition probability for movement from (i,j) to the particular point (i+1,j) is given by

$$k(i,j \rightarrow i+1,j) = \sum_{\varepsilon',\eta'} k(i,j \rightarrow i+1,j|\ i+\varepsilon',j+\eta'). \tag{8}$$

In Table 1 an example of a complete set of conditional transition probabilities for simple kinesis is exhibited.

The mathematical description of *orthokinetic movement* does not differ from that of simple kinesis except that, as discussed in the preceding section, the interpoint distances in the grid would vary, according to some specification, so that step size would vary as a function of position in space.

In *simple klinokinesis* the conditional probabilities, $k(i,j\ i+\varepsilon,j+\eta|\ i+\varepsilon',j+\eta')$, vary with spatial distance. For example, if the turning probability increases with X_i, we would have the transitional probability shown in Table 2, where the probabilities satisfy the conditions

$$0 \le g(X_j),\ f(X_j) \le 1.0$$

$$f(X_j) + 2\ g(X_j) = 1.0,$$

and where $g(X_j)$, the probability of turning, increases with X_j.

Table 1. Set of conditional transition probabilities as an example of simple kinesis.

k(i,j → i-1,j| i-1,j) = 0.0	k(i,j → i-1,j| i-1,j) = 0.25
k(i,j → i-1,j| i+1,j) = 0.5	k(i,j → i-1,j| i-1,j) = 0.25
k(i,j → i-1,j| i,j-1) = 0.25	k(i,j → i-1,j| i,j-1) = 0.0
k(i,j → i-1,j| i,j+1) = 0.25	k(i,j → i-1,j| i,j+1) = 0.5
k(i,j → i+1,j| i-1,j) = 0.5	k(i,j → i+1,j| i-1,j) = 0.25
k(i,j → i+1,j| i+1,j) = 0.0	k(i,j → i+1,j| i+1,j) = 0.25
k(i,j → i+1,j| i,j-1) = 0.25	k(i,j → i+1,j| i,j-1) = 0.5
k(i,j → i+1,j| i,j+1) = 0.25	k(i,j → i+1,j| i,j+1) = 0.0

Table 2. Set of conditional transition probabilities as an example of simple klinokinesis.

k(i,j → i-1,j| i-1,j) = 0	k(i,j → i-1,j| i-1,j) = $g(X_j)$
k(i,j → i-1,j| i+1,j) = $f(X_j)$	k(i,j → i-1,j| i+1,j) = $g(X_j)$
k(i,j → i-1,j| i,j-1) = $g(X_j)$	k(i,j → i-1,j| i,j-1) = 0
k(i,j → i-1,j| i,j+1) = $g(X_j)$	k(i,j → i-1,j| i,j+1) = $f(X_j)$
k(i,j → i+1,j| i-1,j) = $f(X_j)$	k(i,j → i+1,j| i-1,j) = $g(X_j)$
k(i,j → i+1,j| i+1,j) = 0	k(i,j → i+1,j| i+1,j) = $g(X_j)$
k(i,j → i+1,j| i,j-1) = $g(X_j)$	k(i,j → i+1,j| i,j-1) = $f(X_j)$
k(i,j → i+1,j| i,j+1) = $g(X_j)$	k(i,j → i+1,j| i,j+1) = 0

In *topotactic movement*, as pointed out in the preceding section, the probability of turning does not vary with X_j or Y_i. However, there will be a bias, constant spatially, for turning in the proper direction, toward the stimulus. Therefore, if the stimulus increases with increasing X_j, the conditional probabilities have the form shown in Table 3, where $0 < f_i,\ g_i \le 1.0$ (for all i,j), $f_1 + g_3 + g_4 = 1.0$, $2g_1 + f_3 = 1.0$, $2g_2 + f_2 = 1.0$; and where $f_3 \ge f_1 \ge f_2$, $g_1 > g_2$, and $g_4 > g_3$.

Klinokinetic-avoidance behavior uses the same set of conditional

Table 3. Set of conditional transition probabilities as an example of topotaxis when the stimulus increases in the positive x-direction.

$k(i,j \to i-1,j \mid i-1,j) = 0.0$	$k(i,j \to i-1,j \mid i-1,j) = g3$
$k(i,j \to i-1,j \mid i+1,j) = f_1$	$k(i,j \to i-1,j \mid i+1,j) = g3$
$k(i,j \to i-1,j \mid i,j-1) = g_1$	$k(i,j \to i-1,j \mid i,j-1) = 0.0$
$k(i,j \to i-1,j \mid i,j+1) = g_2$	$k(i,j \to i-1,j \mid i,j+1) = f_2$
$k(i,j \to i+1,j \mid i-1,j) = f_1$	$k(i,j \to i+1,j \mid i-1,j) = g_4$
$k(i,j \to i+1,j \mid i+1,j) = 0.0$	$k(i,j \to i+1,j \mid i+1,j) = g_4$
$k(i,j \to i+1,j \mid i,j-1) = g_1$	$k(i,j \to i+1,j \mid i,j-1) = f_3$
$k(i,j \to i+1,j \mid i,j+1) = g_2$	$k(i,j \to i+1,j \mid i,j+1) = 0.0$

probabilities as simple klinokinesis. But g, the probability of turning, now depends not on X_i or Y_i but on whether or not the fish is stressed and whether or not the fish detects itself moving in a direction that is reducing the stress. In the case of thermoregulation, the fish's core temperature might act as memory in deciding if the fish is currently improving its situation or not. A simple iteration scheme, such as

$$T_{core}(t) = \alpha T_{ambient}(t-1) + \beta T_{core}(t-1) \quad (\alpha,\beta > 0) \tag{9}$$

may be sufficient for estimating the core temperature at a given time t, though Fechhelm and Neill (1982) have developed a more precise model.

If the fish finds $T_{ambient}(t)$ too high and $T_{ambient}(t) > T_{core}(t)$, then it "chooses" a high value for g in Table 2; if $T_{ambient}(t)$ is too high and $T_{ambient}(t) < T_{core}(t)$, the fish chooses a low value of g, and so forth.

After this extended tour of the various proposed kinetic mechanisms, we argue that a biased random walk model with some features of topotaxis is the most suitable for representing long-range fish movements such as migration. Klinokinetic avoidance (or conversely, attractance) might be the most realistic model for simulating detailed movements of fish, since it explicitly incorporates the effects of bilateral symmetry and the mechanism of memory. However, in modeling migration it is not practical to keep track of every dart and turn of a fish. Convenient step sizes in a migration model will be of the order of kilometers, not meters. On this scale we probably do not have to explicitly incorporate memory into a model, because the fish can directly sense the stimulus gradient itself in the time scale involved. This suggests a topotaxis model. But even the degree of complexity in the topotaxis model may be unnecessary. Surely, the bilateral symmetry of the fish is not important on the scales of kilometers.

These arguments lead us to believe that the next simplest model, a biased random walk model with the bias in the appropriate direction of the stimulus gradient, may be best for modeling fish migration. One should not be led to think, however, that this simple model will automatically reproduce all the results of more complex models. For example, let us compare a model of topotaxis with the biased random walk model. Consider the example where the conditional transition probabilities of the topotaxis model are $f_1 = 0.2$, $f_2 = 0.5$, $f_3 = 0.6$, $g_1 = 0.4$, $g_2 = 0.2$, $g_3 = 0.1$, and $g_4 = 0.4$. The transition probabilities for an equivalent biased random walk model are $k(i,j \rightarrow i-1,j) = 1/2(1 - 9/17)$, $k(i,j \rightarrow i+1,j) = 1/2(1 - 9/17)$, $k(i,j \rightarrow i,j-1) = 1/17$, and $k(i,j \rightarrow i,j+1) = 8/17$.

We performed simulations with both of the above models and calculated the movements of 100 fish in 75 steps (Fig. 2). The two cases are compared below.

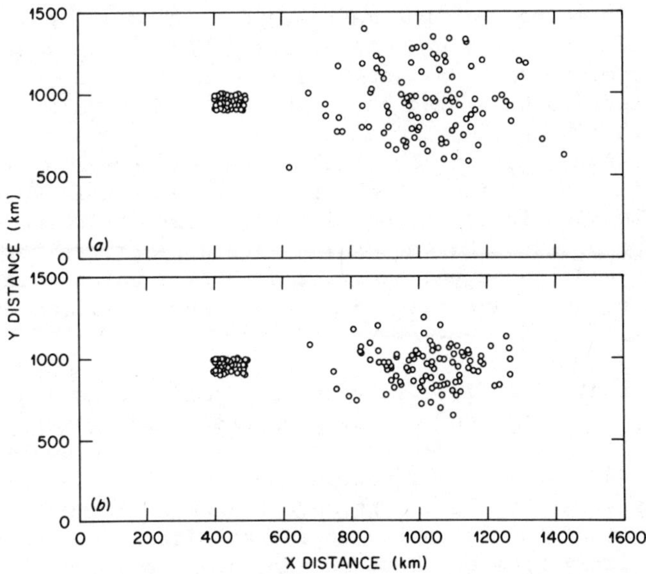

Figure 2. Initial tight clusters of fish (starting at a distance of about 400-500/km along the x-axis) and the spatial distributions of the fish after 75 steps: (a) for a topotaxis model with net movement in the positive x-direction, and (b) for a simpler biased random walk model, parameterized to imitate the topotaxis model. The biased random walk model is poor at simulating the increasing standard deviation of the spatial distribution of the topotaxis model for reasons given in the text.

	$\overline{X}$	SD(X)	$\overline{Y}$	SD(Y)
Topotaxis	1030.4	15.5	939.8	22.4
Biased random walk	1032.8	12.2	934.4	15.4

The comparison is reasonably good as far as average displacements in X and Y are concerned. However, the standard deviations, especially on the y-axis, are significantly different. The reason for this seems to be that in the topotaxis model--as in the models of simple kinesis, orthokinesis, and klinokinesis--there is a greater probability than in the biased random walk model of the fish exhibiting "spurts" of several steps in the same direction, even if it is not the "preferred" direction. The biased random walk model could possibly account for this if we scaled the x- and y-axes differently. Nevertheless, it is clear that specific mechanisms of short-range movement can lead to long-range effects that the biased random walk model, used in a naive way, could miss.

WEIGHTED RANDOM WALK MODEL OF FISH MIGRATION IN A HETEROGENEOUS ENVIRONMENT

Migration is movment on the macroscale. In modeling migration it is not only impractical but unnecessary to try to represent each incremental movement--each dart and turn--of a fish. At the end of the preceding section we argued that a biased random walk model was adequate for describing the general movement of a population of fish. In doing so we gave up trying to simulate the bilateral symmetry effects of a fish by letting it move in any direction in space during a time step. We also increase the distance between lattice points on the plane to the order of kilometers, a size appropriate for migration. The model does not simulate each dart but simulates motion as consisting of average "forays" of several kilometers at a time.

An aspect that comes into increasing prominence as spatial and temporal scales are increased is environmental heterogeneity. Simple models of migration have tended to ignore this. For example, Saila and Shappy's (1963) model of migration represents the coastline as an infinitely long straight line. The model takes no account of varying ocean currents or of the extension of the chemical attractant gradient irregularly from the home stream into the ocean. Also, as Patten (1964) pointed out, the model does not take into account the possibility of fish behavior changing as a function of its position in the ocean. Saila and Shappy's model and others like it are excellent for obtaining a general idea of the nature of migration but are not able to describe the particulars of a given situation.

How are the heterogeneous features of the real environment best taken into account? It is clear that models of a more complex nature are necessary and that some effort must be invested in making such models efficient and easy to use. In this and the next section we shall use computer models with these goals in mind.

Consider a hypothetical ocean coastline (Fig. 3). We have drawn arrows to represent the general direction that fish are likely to follow in the absence of olfactory attraction. This general direction may be the consequence of several factors--passive drift or active reaction to ocean currents (e.g., Arnold 1981), geomagnetic orientation, and so forth. An estuary is also pictured, the mouth of which is surrounded by isopleths of gradient strength of a chemical attractant. It is assumed that the probabilities of the fish being attracted up the gradient increase with gradient strength.

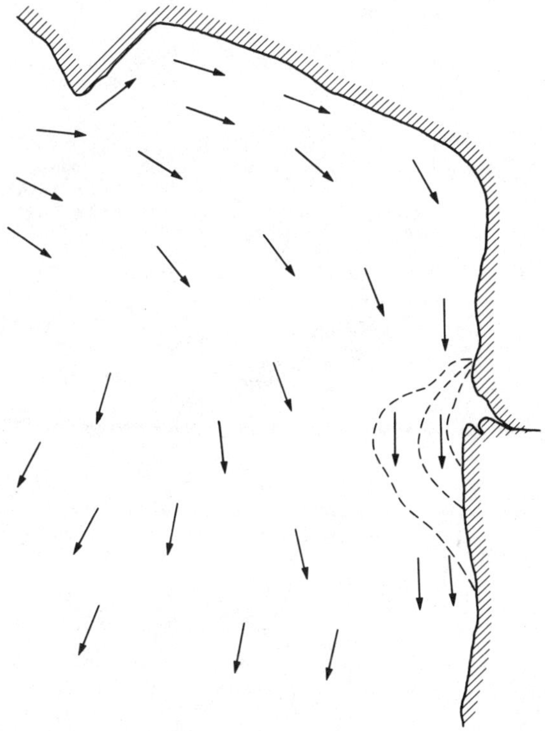

Figure 3. Hypothetical area of coastal ocean with estuary. The arrows represent instantaneous velocities that the fish would have in each part of the ocean region (due to a combination of drift with currents and active swimming) in the absence of random movements and movement directed up a chemical gradient (denoted by dotted isopleths of chemical strength near the estuary).

If one were to accurately represent this situation using a two-dimensional lattice, it would be necessary to assign different values of transition probabilities to every point on the lattice. This is not very practical. As a much more convenient alternative, the region was divided into subregions. The weighted random walk transition probabilities within each subregion were assumed identical for all points in the subregion. Perfect reflection was assumed at the coastline boundary (See Simms and Larkin [1977] for a similar type of model of sockeye salmon [*Oncorhynchus nerka*] migration in a lake.)

Consider a particular subregion, say region 3, as shown in Figure 4 (see Fig. 5a for a schematic of the assumed regions). There is a component of movement in the southerly direction (effects of ocean currents, etc.). Olfactory stimulation causes a component of

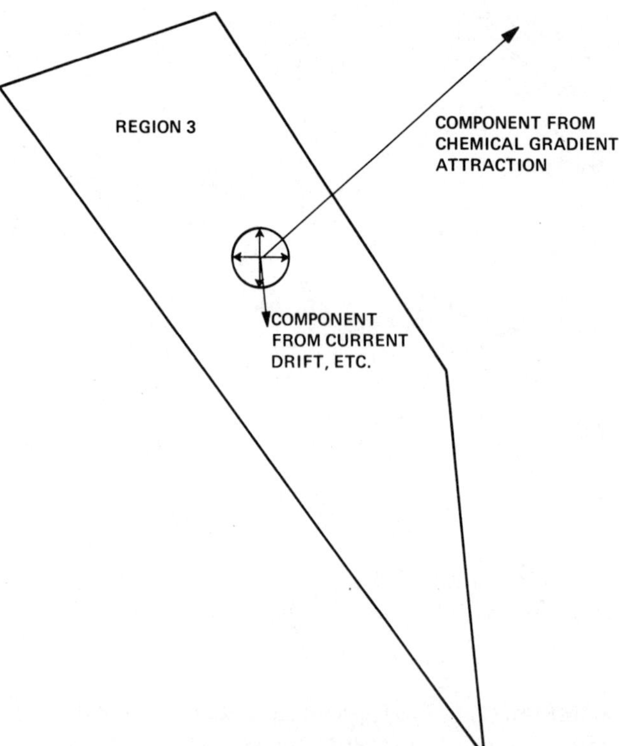

Figure 4. Components of movement of fish in subregion 3 of Figure 3. There is a southerly tendency due to ocean currents, a northeasterly tendency due to the chemical attractant, and a smaller random component that is the same in all directions. The first two components of motion must be projected along the x- and y-axes in the model.

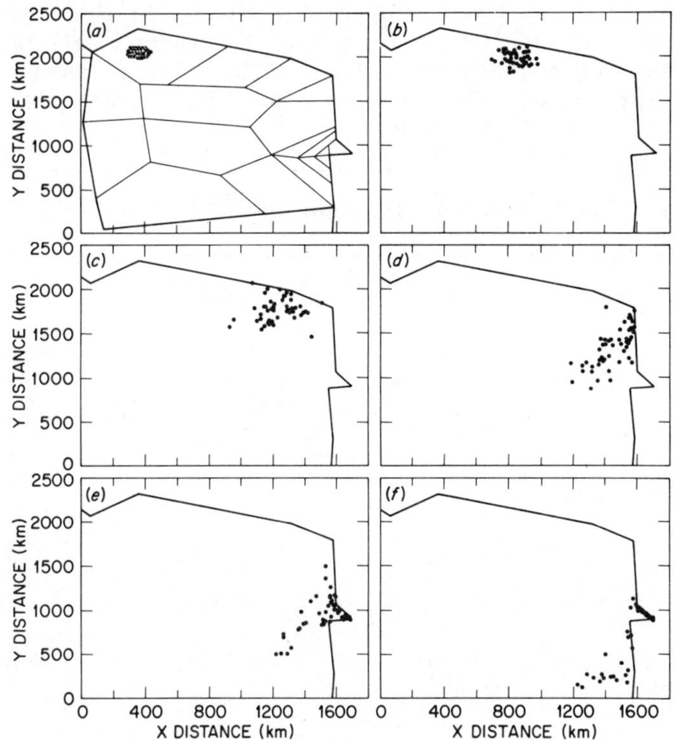

Figure 5. The figures show the movement through time of a cluster of fish modeled by a biased random walk. In (a) the initial cluster of fish is shown. This figure also shows the division of the total region into subregions, in each of which the average velocity of fish movement differs. The remaining figures show the simulated fish cluster after 30 (b), 50 (c), 60 (d), 80 (e), and 100 (f) days.

movement in the northeasterly direction. There will also be some degree of "random" movement. The problem, of course, is to estimate the relative importance of these effects in deciding on the transition probabilities of the weighted random walk model.

Here we look at the results of a given set of assumptions on the transition probabilities. Consider a cluster of 50 fish starting in a certain region of the ocean. The predicted patterns after various periods of time (days) are shown in Figure 5. Table 4 shows the numbers of fish that have successfully arrived in the estuary by certain times.

This particular illustration is only hypothetical. It is instructive nonetheless to experiment with it. For example, a small

Table 4. Accumulated numbers of fish arriving in the tributary as a function of time according to the random walk model. Two sets of starting positions of the fish in space are represented. In the second set the average fish starts 200 km further south.

Time (days)	1st starting distribution	2nd starting distribution
0	0	0
5	0	0
10	0	0
15	0	0
20	0	0
25	0	0
30	0	0
35	0	0
40	0	0
45	0	0
50	0	0
55	0	0
60	0	0
65	0	0
70	4	0
75	11	1
80	20	4
85	25	5
90	27	5
95	28	5
100	28	5
105	28	5
110	29	5
115	29	5
120	29	5

change in initial positions of the fish (an average of 200 km further south in Fig. 5a) may significantly alter the number of successes (Table 4).

THE EQUIVALENT ADVECTION-DIFFUSION MODEL OF MIGRATION

The weighted or biased random walk model is a highly flexible approach and may be suitable for a great variety of fish migration problems. However, one can visualize instances in which an environment may be so complicated that an inordinately large number of grid points might be required to represent it accurately. In some cases, where ocean current speeds vary with depth, for example, three spatial dimensions may be necessary in the model, in which case there would be very many lattice points and Monte Carlo methods of simulation described above would become inefficient.

For the above reasons it is useful to look for alternative formulations of the biased random walk model. Saila and Flowers (1969) advocated that generalized models of fish migration be based on advection-diffusion equations, where advection would represent the general directional movement of the fish and diffusion would represent random movements. These authors referred back to the pioneering work of Skellam (1951) in this field. There is, as Skellam showed, a close connection between the random walk type of model and advection-diffusion partial differential equation models. To describe this we shall rely heavily on Okubo's (1980) excellent text (see also Ahlstrom et al. 1977). Let us again consider the lattice (Fig. 1) for a region in which the transition probabilities, $k(i,j \rightarrow i+\varepsilon,j+\eta)$, are spatially and temporally constant. We will define $S_{ij}(t)$ as the number of fish at a particular point (i,j) at time t, $\lambda = X_{j+1} - X_j = Y_{i+1} - Y_1$ as the interval between adjacent lattice points, and τ as the time step such that $k(i,j \rightarrow i+e,\varepsilon j+\eta)$ represents the transitional probability of a fish during that time.

Therefore, during the time τ, a net number of fish, $k(i,j \rightarrow i,j+1)S_{ij}(t) - k(i,j+1 \rightarrow i,j)S_{i,j+1}(t)$ move from point j to point j+1. An analogous expression exists for movement along the y-axis. The current, or flow, of fish along the x-axis can be expressed as the velocity λ/τ times the number that moved during τ; or

$$J_x = \frac{\lambda^2}{\tau} \{\frac{k(i,j \rightarrow i,j+1)S_{ij}(t) - k(i,j+1 \rightarrow i,j)S_{i,j+1}(t)}{\lambda}\}. \qquad (10)$$

Let us now make use of the fact that the k's are the same at every point but differ for transitions in different directions. We will define k_e = probability of transition to the east (larger X), k_w = probability of transition to the west, k_n = probability of transition to the north, and k_s = probability of transition to the south. Hence

$$J_x = \frac{\lambda^2}{\tau} \{\frac{k_e S_{ij}(t) - k_w S_{i,j+1}(t)}{\lambda}\}. \qquad (11)$$

If the grid size λ is allowed to shrink, then it is possible to expand $S_{i,j+1}$ as the first two terms of a Taylor series:

$$S_{i,j+1}(t) = S_{ij}(t) + \lambda \frac{\partial S_{ij}}{\partial X}, \qquad (12)$$

so that J_x becomes

$$J_x = \frac{\lambda}{\tau} (k_e - k_w) S_{ij} - \frac{\lambda^2}{\tau} \frac{\partial S_{ij}}{\partial X} k_w. \tag{13}$$

A similar expression exists for the flow of fish in the y-direction:

$$J_y = \frac{\lambda}{\tau} (k_n - k_s) S_{ij} - \frac{\lambda^2}{\tau} \frac{\partial S_{ij}}{\partial X} k_s. \tag{14}$$

It is now necessary to shrink the time step, τ, λ, $k_e - k_w$, and $k_n - k_s$ proportionally. We define the limits

$$u_x \underset{\lambda,\tau,k_e-k_w \to 0}{=} \frac{\lambda}{\tau} (k_e - k_w) \qquad u_y \underset{\lambda,\tau,k_n-k_s \to 0}{=} \frac{\lambda}{\tau} (k_n - k_s)$$

$$D_x \underset{\lambda,\tau \to 0}{=} k_w g^2/t \qquad D_y \underset{\lambda,\tau \to 0}{} k_s \lambda^2/\tau$$

so that, finally, two expressions are found:

$$J_x = u_x S(X,Y,t) - D_x \frac{\partial S(X,Y,t)}{\partial X} \tag{14a}$$

$$J_y = u_y S(X,Y,t) - D_y \frac{\partial S(X,Y,t)}{\partial Y}. \tag{14b}$$

These expressions for the flow of animals may be combined with the equation for number conservation,

$$\frac{\partial S}{\partial t} + \nabla \cdot \underline{J} = 0, \tag{15}$$

to obtain

$$\frac{\partial S}{\partial t} + u_x \frac{\partial S}{\partial X} + u_y \frac{\partial S}{\partial Y} - D_x \frac{\partial^2 S}{\partial X^2} - D_y \frac{\partial^2 S}{\partial Y^2} = 0. \tag{16}$$

Equation (16) is the advection-diffusion equation, and its solution will tell us the time evolution of fish density through space. How this is solved is the subject of the next section.

USE OF THE INTEGRATED COMPARTMENTAL MODEL IN SOLVING THE ADVECTION-DIFFUSION MODEL OF FISH MIGRATION

It is not the purpose of this paper to go into the details of solving the advection-diffusion equations. The solution of partial

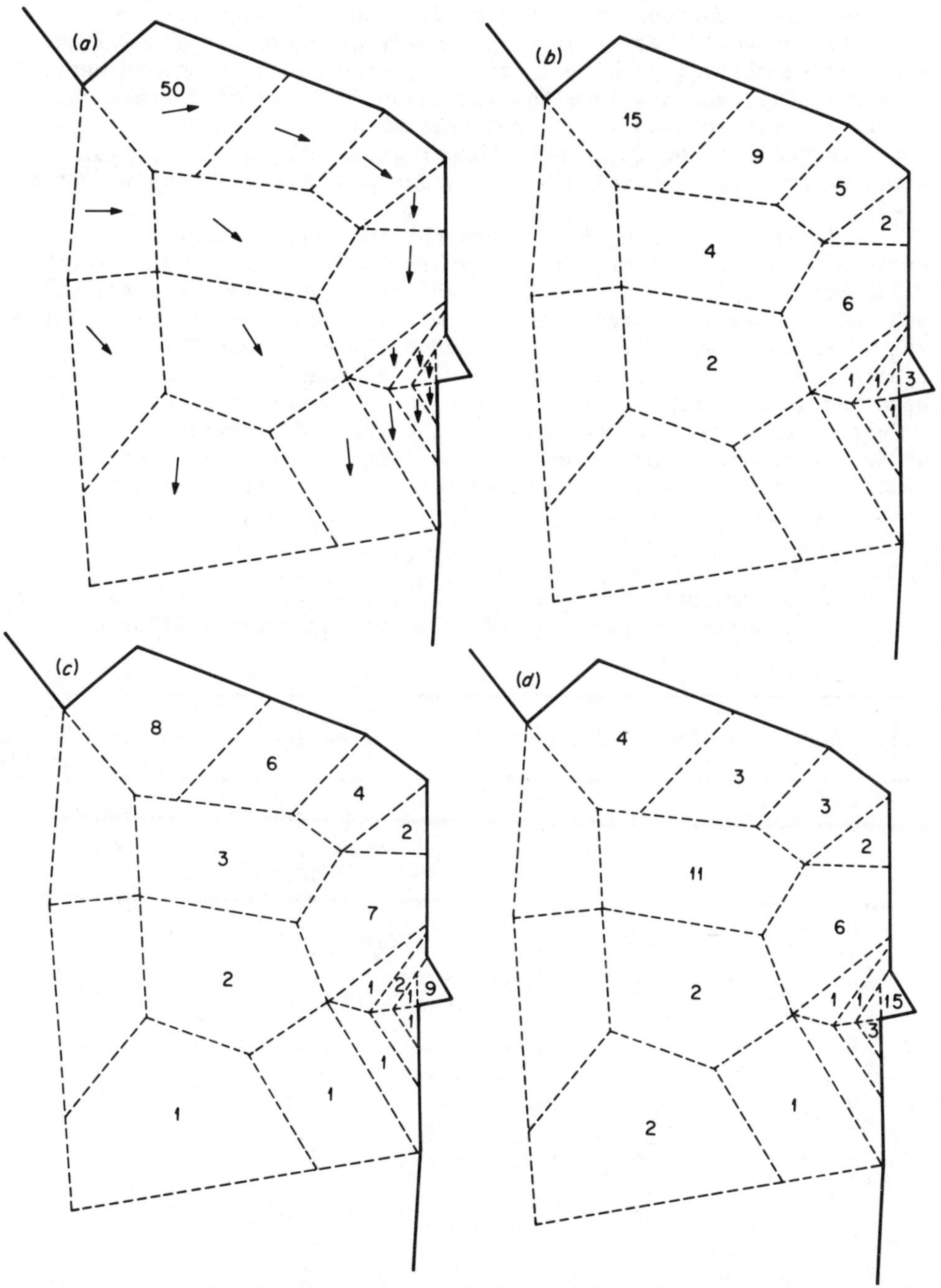

Figure 6. Simulation of advection-diffusion model analogous to the weighted random walk model of Figure 4. The distributions are those at the start (a) and after 40 (b), 60 (c), and 80 (d) days.

differential equations under the conditions of irregular boundaries and heterogeneous conditions is generally an extremely complex and difficult numerical problem. Recently, significant progress was made by Yeh (1981), who combined the traditional finite difference and finite element methods with compartmental analysis techniques. This approach, called the Integrated Compartmental Method (ICM), is suitable for highly complex problems and has been rigorously tested.

As in any other numerical approach, the spatial region is divided into a number of discrete subregions. In the ICM approach, unlike other approaches, the sizes and shapes of these subregions or subcompartments are largely arbitrary. In fact, the division of the region into subregions, which we used earlier in describing the application of the weighted random walk approach (Fig. 4a), can be used here. The ICM requires no information beyond that used in the weighted random walk approach. The only real difference is that now we do not consider individual fish on lattice points but total numbers of fish in the discrete subregions. A certain amount of resolution is thus lost.

Table 5. Accumulated numbers of fish arriving in the tributary as a function of time according to the advection-diffusion model.

Time (days)	Number of fish	Time (days)	Number of fish
0	0	90	17
5	0	95	18
10	0	100	19
15	0	105	20
20	0	110	21
25	0	115	22
30	1	120	23
35	2	125	24
40	3	130	25
45	4	135	25
50	5	140	26
55	7	145	26
60	8	150	27
65	10	155	27
70	11	160	27
75	13	165	28
80	14	170	28
85	16	175	28

We compared the ICM with the weighted random walk approach by duplicating, as far as possible, the simulations shown in Figure 4--this time, however, by solving Equation (16). It is assumed that there is no return flux of fish from the estuary, so when fish enter that subregion they remain there and accumulate.

Interestingly there is a greater spatial dispersion of fish in the advection-diffusion simulation (Fig. 6) than in the random walk model (Fig. 4). This is then reflected by a greater spread of times of arrival of fish according to the advection-diffusion equation (Table 5) than according to the weighted random walk model (Table 4). This spatial and temporal dispersion is characteristic of the compartmental approach taken in solving the equation and could be partially alleviated by dividing the region into more subregions. Note that the number of fish ultimately reaching the estuary is roughly the same in the two models.

CONCLUSIONS

We reviewed some of the basic modeling techniques appropriate to fish migration. The biased lattice random walk model and the advection-diffusion model have been discussed and illustrated in some depth because these seem to the authors to be both practical to use and capable of at least partially imitating the great complexity of real environments. It appears from our comparisons of these two types of models that advection-diffusion models may suffer numerical dispersion problems unless care is taken in choosing small enough sizes for the spatial compartments. This problem with advection-diffusion models, plus the greater conceptual simplicity of biased random walk models, may make the latter more attractive to most modelers of migration.

We have not discussed the techniques involved in fitting the migration models to data, a subject perhaps best kept separate from our simple account of available modeling methods. Application of the models is, of course, the most difficult aspect of the art of modeling, and it is here that creative ideas and the collaboration of empirical and theoretical scientists is needed. It may be that organisms as complex as fish will not be easy to fit into any simple modeling scheme. It is interesting, though somewhat depressing, that while kinetic models have been fit quite adequately to bacteria, attempts to fit such models to somewhat more complex organisms like paramecium have not been judged successful as yet (Lapidus and Levandowsky 1980). In the meantime, however, we can be encouraged that easily used computer models are becoming readily available to almost any potential user.

ACKNOWLEDGEMENTS

This research was sponsored by the Office of Health and Environmental Research, U.S. Department of Energy, under contract W-7405-eng-26 with Union Carbide Corporation. Publication number 2101, Environmental Sciences Division, Oak Ridge National Laboratory.

REFERENCES

Ahlstrom, S.W., H.P. Foote, R.C. Arnett, C.R. Cole, and R.J. Serne. 1977. Multicomponent mass transport model: Theory and numerical implementation (discrete-parcel-random walk version). BNWL-2127. Battelle, Pacific Northwest Laboratories, Richland, Washington, USA.

Arnold, G.P. 1981. Movements of fish in relation to water currents. Pages 55-79 *in* D.J. Aidley, editor. Animal migration. Cambridge University Press, Cambridge, England.

Barber, M.N., and B.W. Ninham. 1970. Random and restricted walks: Theory and applications. Gordon and Breach, New York, New York, USA.

Fechhelm, R.G., and W.H. Neill. 1982. Predicting body-core temperature in fish subjected to fluctuating ambient temperature. Physiological Zoology 55:229-239.

Fraenkel, G.S., and D.L. Gunn. 1961. The orientation of animals. Dover Publications, New York, New York, USA. (Original version 1940, Oxford University Press, London, England.)

Ivlev, V.S. 1960. Analysis of the mechanisms of distribution of fishes under the conditions of a temperature gradient. Zoologicheskii Zhurnal 39:494-499.

Kuhn, A. 1929. Phototropism und Phototaxis der Tiere. Bethes Handbook of Normal Pathological Physiology 12:17-35.

Lapidus, I.R., and M. Lvandowsky. 1980. Modeling chemosensory responses of swimming eukaryotes. Pages 388-396 *in* W. Jager, H. Rost, and P. Tautu, editors. Biological growth and spread: Mathematical theories and applications. Springer-Verlag, New York, New York, USA.

Loeb, J. 1918. Forced movements, tropisms and animal conduct. Philadelphia, Pennsylvania, USA.

Neill, W.H. 1979. Mechanisms of fish distribution in heterothermal environments. American Zoologist 19:305-317.

Nemes, T. 1970. Cybernetic machines. Gordon and Breach, New York, New York, USA.

Okubo, A. 1980. Diffusion and ecological problems: Mathematical models. Springer-Verlag, New York, New York, USA.

Patten, B.C. 1964. The rational decision process in salmon migration. Journal du Conseil Conseil International pour l'Exploration de la Mer 28:410-417.

Pearson, K. 1905. The problem of random walk. Nature (London) 72:294-342.

Rohlf, F.J., and D. Davenport. 1969. Simulations of simple models of animal behavior with a digital computer. Journal of Theoretical Biology 23:400-424.

Saila, S.B., and J.M. Flowers. 1969. Toward a generalized model of fish migration. Transactions of the American Fisheries Society 98:582-588.

Saila, S.B., and R.A. Shappy. 1963. Random movement and orientation in salmon migration. Journal du Conseil Conseil International pour l'Exploration de la Mer 23:153-166.

Simms, S.E., and P.A. Larkin. 1977. Simulation of dispersal of sockeye salmon (*Oncorhynchus nerka*) underyearlings in Babine Lake. Journal of the Fisheries Research Board of Canada 34:1379-1388.

Skellam, J.G. 1951. Random dispersal in theoretical populations. Biometrika 38:196-218.

Sullivan, C.M. 1954. Temperature reception and responses in fish. Journal of the Fisheries Research Board of Canada 11:153-170.

Ullyot, P. 1936. The behaviour of *Dendrocoelum lacteum* I. Responses at light-and-dark boundaries. Journal of Experimental Biology 13:253.

Walter, W.G. 1951. An imitation of life. Scientific American 182:42-45.

Yeh, G.S. 1981. ICM: An integrated compartment method for numerically solving partial differential equations. ORNL-5684. Oak Ridge National Laboratory, Oak Ridge, Tennessee, USA.

MEASURING PHYSICAL-OCEANOGRAPHIC FEATURES RELEVANT TO THE MIGRATION OF FISHES

Ronald Lynn

National Oceanic and Atmospheric Administration
National Marine Fisheries Service
Southwest Fisheries Center, La Jolla Laboratory
La Jolla, California 92038 USA

ABSTRACT

Knowledge of the physical marine environment is a prerequisite to understanding migration of fishes. Variations in currents, drifts, tides, upwelling and other forms of motion in the ocean have direct and indirect effects upon the timing and paths of migration and upon reproductive strategies in general. This paper reviews some of the approaches for studying the marine environment that should be considered in developing projects concerning the migration and distribution of fish stocks. The three principal sources of physical oceanographic data are 1) publications and archives, 2) agency monitoring programs, and 3) field measurements. Physical data taken contemporaneously with fish observations have particular value. All three sources are frequently combined in a well-planned project. Remote sensing of the marine environment is an important new method that is now being used to influence the deployment of research and industry vessels.

INTRODUCTION

An important factor in the migration of fishes is the fields of motion in the ocean environment. Water motion, as it may influence fish stocks, occurs across a broad range of temporal and spatial scales. It can impinge directly upon fish in the form of long-term drift or strong, short-term advection by currents or tidal flow. It also can act in an indirect manner by creating or destroying food aggregations. Upward motion enriches the euphotic zone. Olfactory clues for migrating fish may be carried with the flow, and proper spawning habitats may depend on flow patterns. Fish may move with the drift or swim against it. These fields of motion may be used selectively by fish. Current systems often undergo cyclical changes in response to tidal or seasonal events, and they usually show even larger aperiodic fluctuations.

The published literature contains some excellent studies in which some part of the cycle of fish migration is associated with water motion or other events in the physical environment. There are other studies, however, which appear to have been handicapped by a failure in their design to give adequate attention to the vagaries of the ocean. There is a definite need to take physical-oceanographic measurements that might assist in explaining observed biological phenomena. The objective of this presentation is to review some approaches for studying the physical environment that should be considered in developing projects concerning the migration or distribution of fish stocks. It is not my intent to outline complex and expensive research into physical oceanography, but rather to consider some relevant measurements of the ocean that are affordable by a migration research group with a modest budget and which might indicate some physical order to biological distributions and might cue or direct migratory behavior.

EDDIES AND OCEAN FRONTS

Two major features in the ocean that form strong patterns in the distributions of physical characteristics are fronts and eddies. Fronts and eddies are regions where ocean dynamics are intensified. They are regions of horizontal shear, convergence and divergence, and vertical convection. Owen (1981), in a review of these classes of motion and their biological consequences, noted that diverse marine life forms, from the very small to the very large, alter their distributions in the presence of such flow patterns.

Eddies may be classified as free or stationary. Free eddies are formed from flow instabilities and atmospheric forcing. Examples include the warm-core and cold-core eddies that are formed when meanders are pinched off major current systems such as the Gulf Stream and the Kuroshio. These eddy "rings" transport large amounts of heat, salt and biota. The effect of such an eddy is the injection of a community of organisms and a body of water into a foreign water mass. However, free eddies are unlikely to provide a reliable mechanism for fish migration because of their episodic nature.

Stationary eddies, on the other hand, are topographically controlled and therefore persistent. Their circulation patterns are determined by banks, islands and the configuration of coastlines. They usually are semipermanent features but may have a large range of fluctuation. The Southern California Eddy (reviewed by Owen 1980) is an example of a stationary cyclonic eddy in which vertical convection produces an upward transfer of nutrients into the euphotic zone. A large number of studies has shown the Southern California Eddy to be a refuge for a variety of organisms including pelagic fishes. Smith (1978) found that the region about the Southern California Eddy, which comprises only 12% of the spawning area of the northern anchovy's (*Engraulis mordax*) central subpopulation, contained, on average over 24 years, 48% of the spawned larvae.

Ocean fronts may be classified into five major categories. Fronts are formed 1) at the boundary of intense currents, 2) in regions of convergent surface flow driven by wind stress, 3) as a result of coastal upwelling, 4) at boundaries of estuarine discharge plumes, and 5) as a result of restratification of water from tidal or wind stirring in regions of shallow topography. In the major oceans large-scale fronts lie in zonal bands in response to the time-integrated effect of large-scale wind stress.

Well-studied examples of ocean fronts are the subarctic and subtropic fronts of the North Pacific Ocean. The subarctic front lies about the 40°N parallel, except near its eastern terminus, where it curves southward to the vicinity of 35°N. The subtropic front lies near the 32°N parallel. In the 1970's the distribution of albacore (*Thunnus alalunga*) as they migrated toward the North American summer troll fishery was investigated in relation to these fronts (Laurs and Lynn 1977). In the late spring of 1972 and 1973 the fronts were well developed. The fronts had spatial continuity and strong gradients of temperature and salinity. Albacore catches were distributed along and between the fronts (Fig. 1). Such aggregation may have occurred in response to the availability of forage. The forage, in turn, was dependent on lower trophic levels and ultimately on the high productivity created by frontal dynamics. The changing patterns in the size composition of the albacore in these offshore catches over 7 weeks were repeated, after a delay, in the nearshore fishery, indicating that albacore were actively migrating through this area during the period of the study.

A repeat of the survey in the late spring of 1974 revealed significant differences (Fig. 1). The frontal structure was poorly developed and water-mass boundaries were less distinct. Albacore catches were made over a larger region and did not tend to persist in any local area as they had in the previous 2 years. There was, however, one pocket of high catches that was sustained over several days. A frontal meander of modest proportions nearly encompassed these catches. A sequence of catches--first within the meander, then coastward of the meander, followed by a recurrence of catches within the meander--suggested a funneling of the migrating albacore related to the feature. Thus, spring distribution and migration of albacore seem strongly influenced, whether directly or indirectly, by the presence of these large-scale fronts, the degree of development of

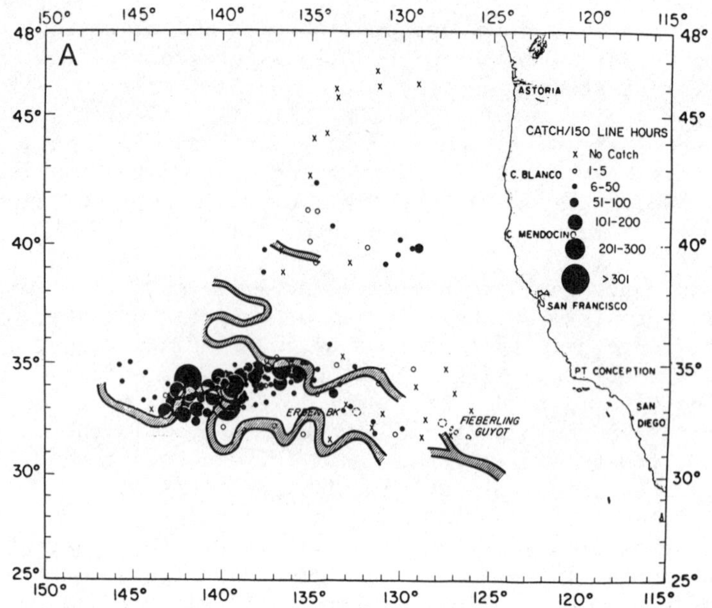

Figure 1. Albacore catch per 150 line-hours by 12 chartered fishing vessels during a preseason exploratory study and the subarctic (northern) and subtropic (southern) ocean fronts during (A) 1973 and (B) 1974 (from Laurs and Lynn 1977).

the fronts and the form of mesoscale features within the fronts.

OTHER CLASSES OF MOTION

Classes of motion which impact biological distributions, other than fronts and eddies, include coastal upwelling, current systems, long-term drift, turbulent mixing, tides, and river flow. In a comparative study of oceanographic regimes and spawning strategies within the California Current system, Parrish et al. (1981) described the seasonal cycle of upwelling, offshore transport and nearshore countercurrents. The reproductive strategies of pelagic fish in this highly productive region were shown to accommodate to these seasonal fluctuations in nearshore coastal dynamics. The strategies include seasonal timing, short planktonic stages and large numbers of spawn.

CONVENTIONAL OCEAN MEASUREMENTS

Methods of ocean measurement must be chosen with consideration for the characteristics of the species of fish studied as well as the availability of financial and manpower resources. The objective of

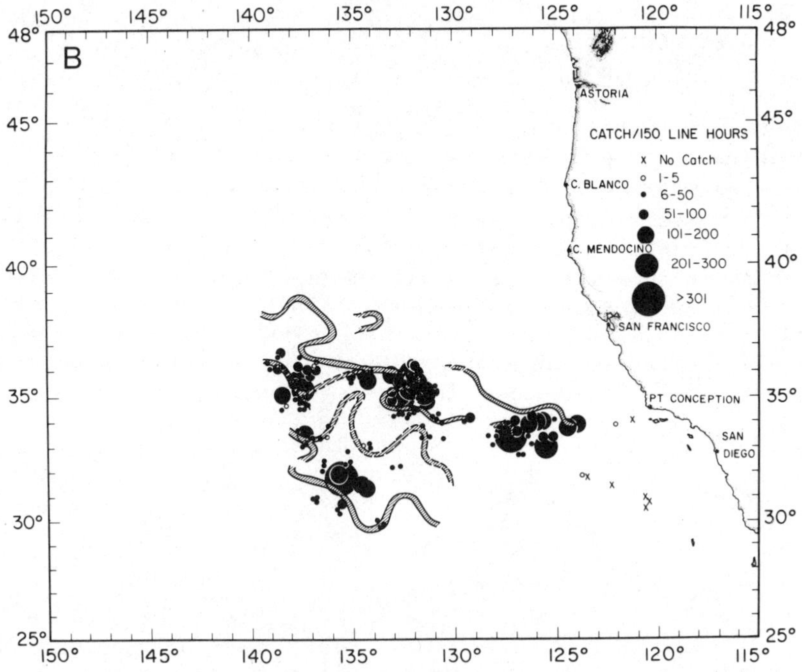

Figure 1 continued.

such measurements of ocean currents is to estimate the effective fields of motion over spatial and temporal time scales that are pertinent to the migrations of a particular population of fish. Often there is no obvious method to achieve the desired results directly. Moored current meters can provide relatively dense coverage in time. However, if one is to achieve reasonable coherence among current meters, a modest deployment can cover only a small part of the migratory domain. That still might represent a large investment in time, effort and money. Thus current meters are most effective in confined regions such as channels, embayments and continental shelves. In the open ocean current meters usually involve logistics of such proportions that they would dominate, in an operational sense, a modest project on fish migration.

Other direct methods of current measurement used in larger-scale regions are drogued drifters, ship's drift and, in coastal zones, drift cards. Recently, acoustic doppler logs and expendable temperature/velocity probes have been used successfully to measure velocity profiles from a moving vessel.

More commonly, we observe fields of motion indirectly by sampling the density (temperature and salinity) field in hydrographic Conductivity/Temperature/Depth (CTD) surveys from which geostrophic velocities can be computed and water masses and fronts located. Other variables such as water clarity, productivity and oxygen content can also be measured. Distributions of these variables may strongly influence the distribution and movement of fish. This method has limitations which arise from the assumption of synopticity, and it produces only one slice in time of a varying state. Repeated surveys (a costly endeavor), which are often done on monthly intervals, can be seriously biased. When data are too sparse to examine events within a season, an accumulation of measurements may permit calculation of long-term mean seasonal changes that may relate to mean seasonal distributions of fish. Other indirect methods include using Ekman drift to estimate time-varying aspects of surface currents over periods of weeks to months and using records of tide level to follow changes in coastal currents.

Sources of Ocean Data

The sources of data for marine environmental studies may be divided into three principal categories.

1. One source is historical files in the form of data archives, scientific literature, and summary publications including atlases.

2. A second source is ongoing ocean monitoring programs. These programs, often found in government agencies and universities, may provide ocean data in the planning stages or during the operations of a study project. Examples include maritime reports of weather and sea-surface temperature, remote-sensed satellite imagery, and sea level.

3. The third major source is field measurements conducted during fisheries studies. These data have particular value in being contemporaneous with observations of fish. They may come from measurements aboard a research vessel or gathered by arrangements aboard commercial fishing vessels or other ships-of-opportunity.

Perhaps a fourth source is model simulations of ocean dynamics. Most or all of these sources usually are brought to bear upon a problem.

Case Studies

For example, let us look at a project in progress involving albacore tuna; it uses all three principal data sources. A phase of the complex migration of North Pacific albacore is the movement following the seasonal decline of the United States-Canadian surface fishery in the fall to the subsurface longline fishery conducted in winter by Japan, South Korea and Taiwan. This is a southward and westward movement from a region offshore of the US west coast to a broad zone across much of the North Pacific. Catch records of the

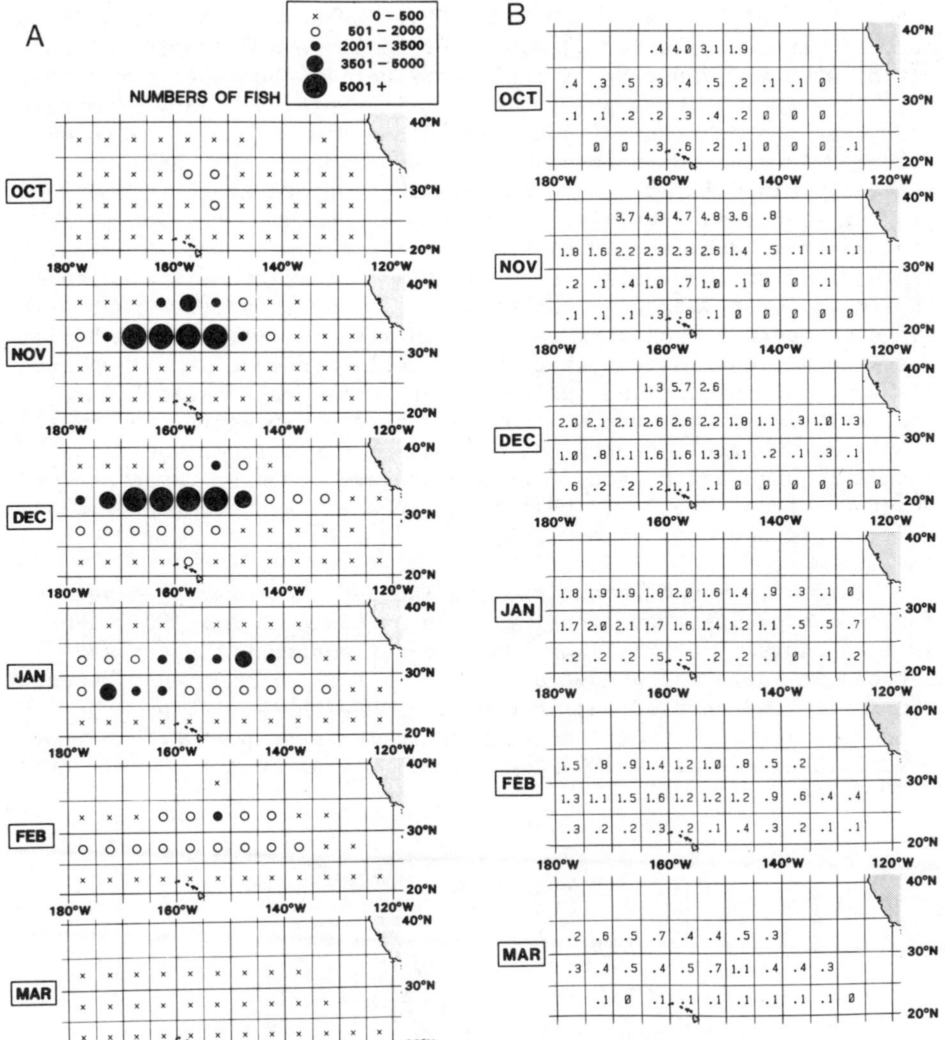

Figure 2. (A) Japanese longline albacore catches in the eastern north Pacific for October through March averaged for 1969-1979 (from Lynn and Bliss 1982). (B) The same data as catch per 100 hooks.

Japanese longline fishery for albacore east of the international dateline (Lynn and Bliss 1982) show that the major catches in the eastern North Pacific are taken between 30°N and 35°N during the months of November through February (Fig. 2a). In the early part of this period there are also some modest catches north of 35°N and between 145°W and 165°W. The interesting aspect is that the catch rates north of 35°N are high (Fig. 2b). Based upon a resolution of monthly catch totals for 5° blocks of latitude and longitude, the catch rates are close to five albacore per 100 hooks compared to no more than two to three per 100 hooks elsewhere. We know from earlier work that the subarctic front lies within the zone of 35°N to 40°N. November is a period of cooling and deepening of the shallow, seasonal, mixed layer. Beneath this layer there is the deeper frontal structure remaining from the previous winter. Oceanographic atlases (e.g. Robinson 1976) show that in November the 16 C to 18 C isotherms from the surface to 60 m fall in this zone (Fig. 3) and are thus coincident with the mean position of the subarctic front. This range of ocean temperature is characteristic of albacore fisheries. These conditions suggest that in the early phases of the longline fishery, as albacore and surface layer isotherms retreat from northerly latitudes, the albacore are once again aggregated by events about the subarctic front, thus producing the circumstances that result in high catch rates. By mid or late December temperatures about the front are lower than those apparently preferred by albacore and catches decrease accordingly.

This background information influenced the strategy developed for recent exploratory longline surveys (Laurs et al. 1981, 1982). In these surveys US fishing vessels experimented with gear and methods in a project conducted by the American Fishermen's Research Foundation and carried out in cooperation with the National Marine Fisheries Service, under the scientific leadership of R.M. Laurs. Five fishing vessels are participating in the recent survey. Prior

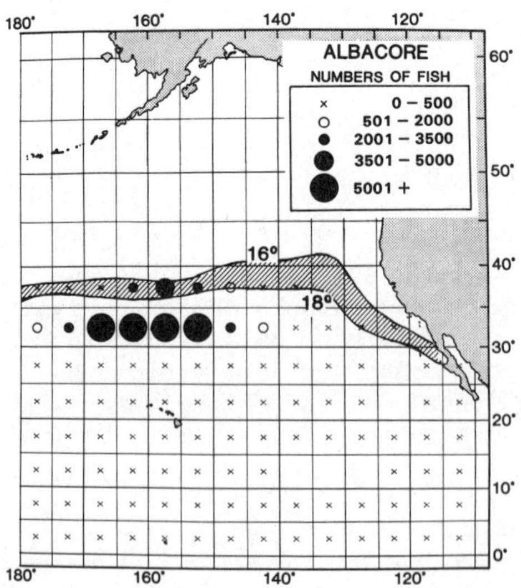

Figure 3. The albacore longline catch for November from Figure 2A and the monthly mean 16 C and 18 C isotherms.

to sailing, up-to-date expendable bathythermograph (XBT) data were obtained from the US Navy's Fleet Numerical Oceanographic Center. Plots of isotherm depth in the upper 150 m of the survey region revealed a zone of abrupt changes identifying a strong front. Thus, an ocean monitoring program provided input for tactics in deploying the vessels. Each of the five vessels was instructed to record twice-daily weather and sea conditions and sea-surface temperature. Also, each vessel made daily XBT casts. This measure of thermal structure, taken concurrently with fishing operations, will be combined with the background information to address the question of the association of migration with ocean fronts.

The 1972-74 albacore surveys (Laurs and Lynn 1977) in which the spring migration into the North American surface fishery was investigated, offer an example of a more comprehensible level of operations using physical oceanography in conjunction with fisheries research. The surveys combined the use of the R/V DAVID STARR JORDAN and chartered commercial-fishing vessels. The JORDAN conducted oceanographic observations on a planned grid. Traditional observations were made: CTD profiles; XBT and water bottle casts; continuous recording of surface temperature, salinity and chlorophyll concentration; and analysis of water bottle samples for O_2 and chlorophyll concentrations. Nutrient chemistry was desired but proved to be beyond our resources. Micronekton tows, to obtain potential forage organisms for albacore, were made using an Isaacs-Kidd midwater trawl. The JORDAN also trolled for albacore as time permitted. The catches made by the JORDAN, while not comparable in effort to those of the chartered fishing vessels, did provide confirmation that areas well beyond the regions of fronts were less productive or unproductive. The chartered fishing vessels recorded weather and sea conditions and made XBT casts. These additional physical observations extended our interpretation of the ocean conditions beyond the grid covered by the JORDAN.

An added factor in these surveys was the cooperation of unchartered fishing vessels. Daily broadcasts of findings from the JORDAN and chartered fishing vessels were given on radio frequencies used by fishermen. Many of the fishermen responded in kind by reporting their daily catches, thus assisting in the success of the operations, and many also voluntarily completed catch logs as part of an interstate and federal program (Laurs et al. 1975). This project combined a number of activities directed toward describing and understanding the migration of albacore of which the study of the physical environment was an integral part.

Interpreting and Using Ocean Measurements

The CTD, XBT and water-sample casts provide the basis for identifying water masses and fronts (Fig. 4) and for calculating the dynamics of stability, mixing and geostrophic currents. Meanders or

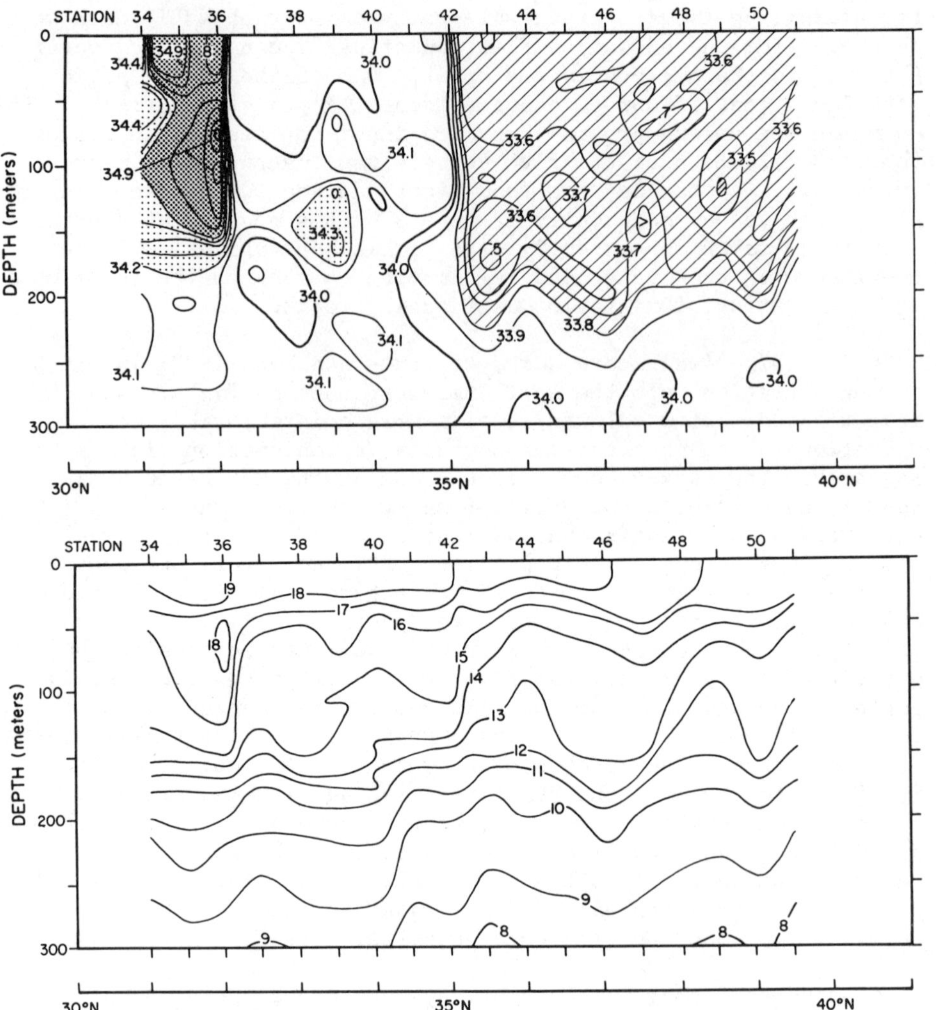

Figure 4. Vertical section of salinity (upper panel) and temperature (lower panel) along 137.5°W for June 1973. The subtropic front is seen at 32°N and the subarctic at 35°N. Hatching indicates salinities < 33.8 °/oo; light stippling indicates salinities between 34.2 and 34.6 °/oo and heavy stippling indicates salinities > 34.6 °/oo.

eddies spun off the fronts can be described. The nutrient chemistry, chlorophyll determinations and micronekton hauls provide the links between the physical and biological regimes. The thermosalinograph, which provides a continuous underway recording of surface layer temperature and salinity, reveals the presence of surface frontal gradients as they are crossed (Fig. 5). Thus, opportunity arises for

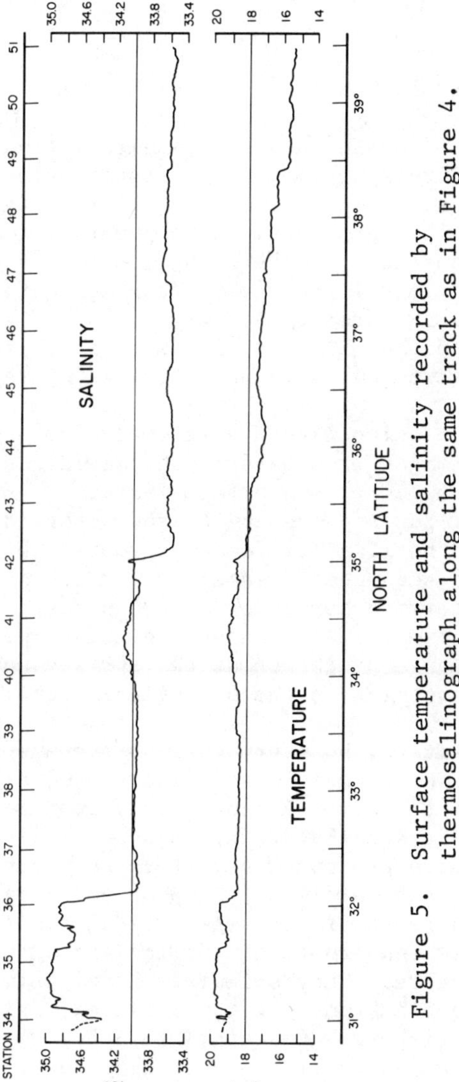

Figure 5. Surface temperature and salinity recorded by thermosalinograph along the same track as in Figure 4.

a tactical response in sampling procedures and fishing operations. The temperate zone frontal system of the eastern North Pacific Ocean often exhibits a temperature change on the order of only 0.7-1.0 C in 4-20 km. It is the abruptness of change in an otherwise flat signal that identifies these features. The salinity change at the front, 0.5-1.0 ^{o}/oo, is even greater relative to its ambient fluctuations in the surface layer.

APPLICATIONS OF REMOTE SENSING

Remotely sensed measurements from aircraft and satellites have been employed in directing commercial-fisheries and fisheries-research operations and in analysis of the results of biological- and physical-oceanographic projects. This powerful tool is growing in acceptance and importance in the fishery sciences and industries. The applications of satellite remote sensing in fisheries, including the use of thermal infrared radiometry, the Coastal Zone Color Scanner (CZCS), and the SEASAT-A microwave scatterometer, has recently been reviewed by Laurs and Brucks (in press).

For many ocean scientists the initial exposure a decade ago to thermal-infrared images from satellites raised questions as to relevance of the images to subsurface features. Experience has demonstrated that, for the most part, the observed plumes, fronts, eddies, wakes and the like are features rooted in the mixed layer or deeper. In a time sequence of satellite images some are found to evolve slowly in form; others show a strong response to major wind events in a short time. This imagery reveals expressions of the advection and exchange processes at the surface and potentially reflects related patterns in the biological distributions.

The major, western, boundary-current systems are monitored by satellite. Contemporary multidisciplinary studies of the warm- and cold-core rings in the North Atlantic rely upon remotely sensed data. Satellite imagery is equally useful in small-scale studies and in the lesser-gradient features found in eastern boundary currents. Using images from the NOAA-5 satellite, Simpson and Pingree (1978) confirmed the occurrence of convergent tidal fronts in the Celtic Sea. These shallow sea fronts divide regions of tidally mixed and well-stratified waters. They stimulate local primary productivity and concentrate zooplankton. At the National Marine Fisheries Service Southwest Fisheries Center (NMFS/SWFC) we have used imagery to direct effort in a detailed survey of an upwelling plume. Also the SWFC participated in a joint venture of industry, university and government to supply salmon and albacore fishermen on the US west coast with timely maps of gradient features (Montgomery 1981). Recently the development of processing techniques for data from the CZCS aboard the Nimbus 7 has added a new instrument to our arsenal. R.M. Laurs and colleagues are studying an application of the CZCS to the west coast albacore fishery, which has altered our perception as to the mechanism of response of these fish to upwelling fronts. Albacore, which are visual feeders, are now suspected to be responding to strong changes in turbidity which are seen as color fronts associated with upwelled waters rather than to the temperature fronts (R.M. Laurs personal communication).

R. Lasker and colleagues at the SWFC are applying satellite imagery to studies of northern anchovy spawning in the Southern California Bight. Anchovy eggs and larvae were sampled during March and April 1980 in a closely spaced grid of stations covering most of the Bight. The first-day larvae were well represented in net tows in some areas and absent from others. There was remarkable coincidence in the larval distribution's southern boundary with a strong gradient of chlorophyll concentration as measured by the blue-green ratio in a CZCS image taken in early April, 1980. Spawning appears to have occurred within the plankton-rich areas and was absent from the clear blue, and more sterile, waters offshore and to the south. The northern limit of egg distribution was coincident with the 14-C surface isotherm as determined by direct measurements and thermal IR satellite imagery. Temperatures less than 14 C, while not physiologically unsuitable for anchovy spawning, indicate recently upwelled waters. The 14-C isotherm may denote a zone of change in the succession of planktonic-community composition or in the vertical stratification of the water column, both of which are important factors in spawning (Lasker 1978). A repeated set of data for February 1981 did not confirm the correspondence with the surface-chlorophyll gradient. More observations are scheduled.

PRESENT DAY DEVELOPMENTS AND FUTURE EXPECTATIONS

I expect application of remote-sensing techniques to continue to advance our understanding of the oceans and to expand in usefulness in marine biology. There are several recent improvements in remote sensing, and more are anticipated. The present application of thermal IR imagery is to utilize the relative temperature-gradient patterns for interpretation of events. Until recently the accuracy of estimates of sea temperature from single channel IR data was no better than 4-5 C (or 2-3 C if sea surface measurements were available). Algorithms have now been developed for multichannel IR data, limiting error to 0.6-0.9 C (Bernstein 1982; McClain et al. 1983).

Overcast and atmospheric moisture continue to be the major problems in collecting useful IR and visual imagery. Bernstein (1982) has devised a routine which retrieves sea-surface temperatures even from small scattered openings in clouds. Using image-screening techniques the routine eliminates all but the best data. Applying the routine to a series of images collected over days to weeks produces a field of data not unlike that from maritime ship reports. The satellite data have a smaller bias than do maritime ship reports, however, and are not constrained to shipping lanes. This routine might be used to improve the present program for monitoring sea-surface temperature that relies on maritime reports and would likely improve the spatial resolution. It also could be applied regionally.

One experimental sensor that is unaffected by clouds is the microwave scatterometer used aboard the ill-fated SEASAT-A. It provides the basis for estimates of wind stress. Bakun and Parrish (1980) have found a spatial alternation in the pattern of wind-curl stress along the California and Baja California coastal zone. A lobe of positive curl between regions of negative curl appears to block the seasonal development of the countercurrent, resulting in two semi-indepedent gyres. These gyres correspond to separate subpopulations of pelagic fish. The wind stress was derived from a composite of many years of data. The wind-stress curl could not have been resolved by the methods available for synoptic sampling. Bakun and Parrish suggested that the design resolution of 50 km for the microwave scatterometer would allow synoptic viewing of wind-stress features on a size scale relevant to fish stocks.

Within the past decade we have seen substantial technological advances in computers and in data handling systems. These improvements have facilitated the archiving and use of voluminous time series of environmental data. Numerous resulting studies and papers have examined interactions between the oceans and atmosphere on scales from regional to global. The benefits to fisheries research, which are yet to be fully realized, should include the development of environmental indices and simulation models which have potential for interfacing with fisheries data sets and migration models.

CONCLUSION

The linkage between the distribution and movement of fish and fields of motion and related ocean features is complex and involves multiple trophic levels. It may not be possible to demonstrate a direct causal relationship, but evidence and intuition point toward such linkages. Both physical and biological data sets tend to be deficient because the number of samples is usually small, the medium is highly variable, and most data sets require considerable interpolation. Despite these deficiencies, evidence is overwhelming for the influence of spatial and temporal variations in the marine environment upon fish. Hypotheses so raised can lead to improved experiments and alternate approaches to confirm or dispel the ideas.

The technological advances also provide for rapid data retrieval and sophisticated graphic display of data from oceanographic archives--quite the equivalent of an electronic atlas. As these developments become available they improve our capabilities for accessing marine environmental data for development of project plans and in analysis. However, the advances in remote sensing and data technologies do not supplant the need for field measurements but rather supplement them. There is still the need for lowering instruments over the side to take the measure of the ocean environment concurrently with sampling fish.

REFERENCES

Bakun, A., and R.H. Parrish. 1980. Environmental inputs to fishery population models for eastern boundary current regimes. Pages 67-104 *in* G.D. Sharp, editor. Workshop on the effects of environmental variation on the survival of larval pelagic fishes. IOC Workshop Report 28, UNESCO, Paris, France.

Bernstein, R. 1982. Sea surface temperature estimation using the NOAA-6 satellite AVHRR radiometer. Journal of Geophysical Research 87(C-10):7865-7872.

Lasker, R. 1978. The relation between oceanographic conditions and larval anchovy food in the California Current: identification of factors leading to recruitment failure. Rapport et Procès Verbaux des Réunions Conseil International l'Exploration de la Mer 173:212-277.

Laurs, R.M., and J. Brucks. In press. Living marine resources applications. Chapter 13 *in* B. Saltsman, editor. Advances in geophysics, volume 83.

Laurs, R.M., C. Hooker, L. Hreha, and R. Lincoln. 1975. Marine Fisheries Review 37(11):14-21.

Laurs, R.M., and R.J. Lynn. 1977. Seasonal migration of north Pacific albacore, *Thunnus alalunga*, into North American coastal waters: distribution, relative abundance, and association with transition zone waters. US National Marine Fisheries Service Fishery Bulletin 75:795-822.

Laurs, R.M., R.J. Lynn, R. Dotson, R. Nishimoto, K. Bliss, and D. Holts. 1982. Exploratory albacore longline fishing in the eastern North Pacific during winter 1982. National Oceanic and Atmospheric Administration National Marine Fisheries Service, Southwest Fisheries Center Administrative Report LJ-82-06.

Laurs, R.M., R.J. Lynn, R. Nishimoto, and R. Dotson. 1981. Albacore trolling and longline exploration in eastern North Pacific waters during mid-winter 1981. National Oceanic and Atmospheric Administration Technical Report NOAA-TM-NMFS-10.

Lynn, R.J., and K.A. Bliss. 1982. Summary of Japanese longline catches of albacore, 1967-1979, in the eastern North Pacific Ocean. US National Marine Fisheries Service, Southwest Fisheries Center Administrative Report LJ-82-23.

McClain, E.P., W.G. Pichel, C.C. Walton, Z. Ahmad, and J. Sutton.

1983. Multi-channel improvements to satellite-derived global sea surface temperature. Advances in space research 2(6):43-47.

Montgomery, D.R. 1981. Commercial applications of satellite oceanography. Oceanus 24(3):56-65.

Owen, R.W. 1980. Eddies of the California Current system: physical and ecological characteristics. Pages 237-263 *in* D. Powers, editor. The California Islands. Museum of Natural History, Santa Barbara, California, USA.

Owen, R.W. 1981. Fronts and eddies in the sea: mechanisms, interactions, and biological effects. Pages 197-233 *in* A.R. Longhurst, editor. Analysis of marine ecosystems. Academic Press, London, England.

Parrish, R.H., C.S. Nelson, and A. Bakun. 1981. Transport mechanisms and reproductive success of fishes in the California Current. Biological Oceanography 1:175-203.

Robinson, M.K. 1976. Atlas of North Pacific Ocean monthly mean temperatures and mean salinities of the surface layer. US Department of Navy, Washington, DC, USA.

Simpson, J.H., and R.D. Pingree. 1978. Shallow sea fronts produced by tidal stirring. Pages 29-42 *in* M.J. Bowman and W.E. Esian, editors. Ocean fronts in coastal processes. Springer-Verlag, Berlin, Federal Republic of Germany.

Smith, P.E. 1978. Central subpopulation northern anchovy: first order corrections for missing region and month survey effort. US National Marine Fisheries Service, Southwest Fisheries Center Administrative Report LJ-78-2.

BIOENERGETIC CONSIDERATIONS IN FISH MIGRATION

D. Weihs

Department of Aeronautical Engineering
Technion - Israel Institute of Technology
Haifa, Israel

ABSTRACT

A hydromechanical approach to analysis of fish migratory behavior is presented. This mathetmatically-oriented method enables quantitative studies and development of predictive models of requirements and capabilities for migration, and it offers possible explanations for various behavioral and morphological adaptations. A review of existing work, including optimal cruising speeds, schooling and non-steady swimming is used as a basis for suggestions for future work, both theoretical and experimental.

INTRODUCTION

The engineer's task is often to design a machine or system that will perform some previously defined task in the most effective manner. This rather loose definition (specifically - what criteria are there for effectiveness?) serves as a basis also for the hydromechanical approach to fish migration I present here.

We said that the task that the system has to perform is previously defined. This translates into the migration that the fish (or school of fish) has to perform in order to achieve some specific goal, such as reproduction or improved feeding. The geographical locations of starting and endpoints of the migratory route are well defined in the sense that they are rather precisely located in most (especially reproductive) cases.

The available time interval is usually a constraint of the problem in the mathematical sense. This means that while the exact period for migration is not defined by outside requirements there are limits on the time, such as seasonal changes in the environment. Thus, the time spent while migrating is a degree of freedom which can be varied by the fish to obtain better results.

Migrations have been mentioned several times but still have to be defined in the present context. They will be separated into two types: long-range and short-range migrations. The long-range migrations are now defined as the movements from one geographical location to another, at least a few days, and up to a few months, distant in terms of the cruising speeds of the species involved. The distances in question can be from a few tens to several thousands of kilometers. Long-range migrations may include feeding periods, but in many cases performance and survival has to be based on the original internal energy store. Short-range migrations can be both horizontal and vertical (as in the diurnal vertical movements of scattering layers) and have in common directed movements of a few hours duration, usually repeated on a diurnal or semi-diurnal (tidal) basis. Alexander (1972) has summarized, in an excellent example of biomechanical analysis, the energetic costs and advantages of the various modes of swimming and buoyancy control used in the vertical migrations of pelagic and mesopelagic fish.

Returning to the definition given in terms of system design in the first paragraph, we now look at effectiveness. Determination of criteria for the effectiveness of migratory behavior is one of the main purposes of this paper, together with a description of the ways various fish have adapted to applying these. Such criteria will, among others, include efficient (energy-sparing) swimming behavior and time-saving techniques.

The hydromechanical approach serves to address certain problems relating to the understanding and analysis of migration. Can a certain species perform a given migration? How, in terms of trajectory, time and energy, should this be done in order to arrive at the endpoint with maximum fitness for feeding or reproduction? Can various morphological and behavioral adaptations observed in migratory species be explained in terms of evolutionary solutions to the questions above?

Quantitative answers to these questions have only recently begun to emerge. The methods of analysis require some steps which may not be typical for fisheries studies, such as defining a generalized fish as a point mass that produces, and is acted on, by various forces. These methods permit the formulation of general statements that hold for all fish at the price of losing details of specific adaptations. The purpose behind such generalizations is to allow mathematical analysis that shows the relations between various parameters, their relative importance and, thus, the possibility of reducing the number of measurements required by pointing out the most significant variables and their probable ranges.

As mentioned above this relatively new way of looking at problems of fish migration has only just begun to provide answers to some of the outstanding questions so that much of the present paper is devoted to suggesting future work.

BASIC HYDROMECHANICS

The common denominator for all types of migration is substantial, sustained motion relative to the water mass. Water is a relatively dense, slightly viscous fluid so that motion in and through it requires energy. While this is intuitively clear, a quantitative description of the forces encountered while moving through the water is useful. We begin our description by limiting the discussion to forward motion in still water, i.e. motion in which the time-average of side forces vanishes. Thus, we look at motions in a vertical plane (Fig. 1), for which one can discern six contributions to the time-averaged forces. The term *time-averaged forces* has just been used twice in a context in which constant (time-independent) forces are usually defined. However, all self-propelled motion of fish is essentially periodic because of the oscillatory nature of fish propulsion systems (Weihs and Webb in press), so that time-averages of the forces are required even when the forward speed is essentially constant. These forces are defined in Fig. 1, which differs from the usual description (Magnuson 1978, page 251, for example) in that the possibility of center of mass motions normal to the longitudinal axis and of nonhorizontal movements are included. The forces acting on the fish are written in co-ordinates parallel (x), and perpendicular to (z) the longitudinal axis of the fish, with

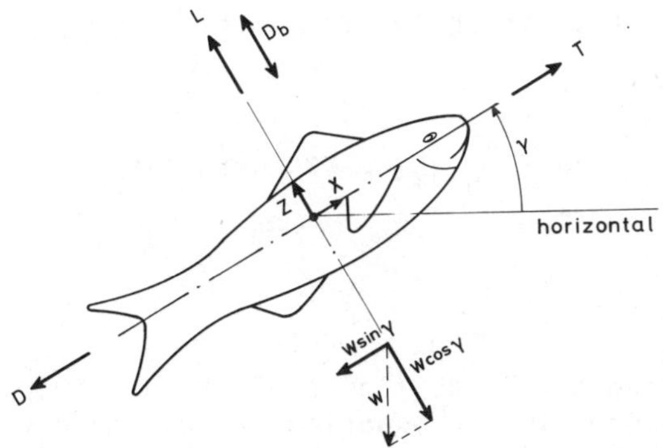

Figure 1. Schematic force balance on a fish moving at angle upward, relative to a horizontal plane. See Appendix for symbols.

the origin attached to the fish's center of mass (and thus moving with it). The force balances can then be written as:

$$T - D - W\sin\gamma = m\ddot{x}; \tag{1}$$

and

$$L - W\cos_{\gamma} \pm D_b = m\ddot{z}, \tag{2}$$

where m_x and m_z are the vertical masses (fish mass + added mass) in the x and z directions and $\ddot{x}$, $\ddot{z}$ are linear accelerations in the x,z directions respectively. γ is the instantaneous angle between the fish longitudinal axis and the horizon.

In the basic case of constant-speed horizontal motion, in which $\gamma = 0$, and $\ddot{x}$, $\ddot{z}$ also vanish, equations (1) and (2) reduce to:

$$T - D = 0; \tag{3a}$$

and

$$L - W = 0. \tag{3b}$$

Thus the thrust is exactly equal and opposite to the drag, and the lift precisely opposes the weight. However, this is just a special case, and when thrust and drag are not equal in magnitude, either acceleration or deceleration of horizontal motion results. When the lift and weight do not balance, motions in the z direction occur, giving rise to an additional hydrodynamic resistance D_b. The resultant of these motions is then at an angle γ, as in equations (1) and (2).

Hydrodynamics enters the problem as a means of defining the forces T, D, D_b and L as functions of geometrical and kinematic parameters. Thus, each of these forces is an independent function of the speed U relative to the water mass:

$$T = \tfrac{1}{2}\, \rho\, V^{2/3} U^2 C_T; \tag{4a}$$

$$D = \tfrac{1}{2}\, \rho\, V^{2/3} U^2 C_D; \; D_b = \tfrac{1}{2}\, \rho\, V^{2/3} U^2 \tan^2\gamma C_{D_b}; \tag{4b}$$

and

$$L = \tfrac{1}{2}\, \rho\, V^{2/3} U^2 C_L. \tag{4c}$$

The nondimensional coefficients[1] of thrust (C_T), drag (C_D) and lift (C_L) appearing in equation (4) can be estimated, with varying degrees of accuracy, by the techniques of fluid mechanics.

Calculation of these coefficients has been the goal of a large number of studies (see reviews by Webb 1975; Lighthill 1975; Yates in press). I shall not describe these analyses in detail, except to point out some general features and trends in the results.

Most migratory species produce thrust for swimming by means of oscillations of the body or the caudal fin. These motions cover the range from the anguilliform to the carangiform modes of motion (Breder 1926). The former is characterized by oscillations of the whole body with amplitude increasing progressively towards the tail. This mode is relatively inefficient from the hydromechanical point of view, as momentum transfer is produced by portions of the body "pushing" the surrounding water. Thus, to produce a backwards motion of the water relatively large sideways motions of the body are required producing a larger frontal area and resulting in higher drag. In a calculation of the drag of an anguilliform swimmer Lighthill (1971) showed that the drag on the fish, swimming actively at a speed in the cruising range, is four times that of the fish's body being towed (while stretched straight) at the same speed. The latter is defined as the gliding drag. Other anguilliform and subcarangiform swimmers showed increases in the drag by factors of three to five (Webb 1975). In the carangiform mode typically only the rear one-third of the fish's length is oscillated with large amplitudes (of one-tenth the fork length) obtained only at the caudal peduncle and fin. Thus, the drag increment due to active swimming is reduced to a minimum, which is less than one and one-half times the gliding drag. On the other hand, a smaller part of the body is used for thrust, so that more efficient propulsors are required. One such development is the so-called "lunate-tail," which is a caudal fin with large aspect ratio (i.e. the spanwise dimension of the fin is much larger than the fin chord). These fins, which are typical of scombroids and some sharks, are designed to produce large forces by causing circulation of flow around them as an aircraft wing produces lift. The ratio of lift to drag on a lunate tail may be between five and ten (Yates in press), so that a large increase in the effectiveness of the propulsive mechanism is obtained. This effectiveness is further improved (Katz and Weihs 1978) by stiffening the caudal fin, so that it can bear increased loading without bending. The stiffening is obtained by compaction and fusion of the fin rays and in selachians by thickening the fin.

For steady horizontal swimming the level of sustained thrust available T_a defines the swimming speed, as, from equation (3a):

---

[1] Note that the value for the characteristic area squared is taken to be the volume (V) raised to the two-thirds power, for all coefficients, and not as usually found in the literature where the frontal area is used for drag, and the lifting or propulsor area used for lift and thrust respectively.

$$U^2 = T_a/(\tfrac{1}{2}\rho V^{2/3} C_D) \quad (5)$$

For Reynolds numbers $R_e > 10^4$ the drag coefficient is insensitive to swimming speed (Arnold and Weihs 1978), so that the maximum sustained swimming speed is proportional to the square root of the available thrust. From equation (3b) we see that a lift, equal and opposite to the submerged weight, is also required. Neutrally-buoyant fish have this requirement fulfilled ab initio, but negatively buoyant fish need to produce this lift. In many cases this forces the latter to swim above a minimum speed (Magnuson 1978), thus defining a range of speeds possible for migration.

REDUCTION OF ENERGY CONSUMPTION IN MIGRATION BY BEHAVIORAL ADAPTATIONS

In both long-range and short-range migrations it is always advantageous to save energy while performing the predefined task. This has served as the starting point for most of the existing hydromechanical analyses of optimization of long-term sustained (see Weihs and Webb in press) locomotion of fishes. Active migrations, as opposed to planktonic motion following water movements, will belong to the periodic class of motion almost by definition. We now look at behavioral options such as speed adjustment and schooling.

The Optimal Swimming Speed

The easiest behavioral parameter for a fish to control is its swimming speed relative to the water surrounding it. In cases of swimming in shallow or confined waters, where an external spatial reference is available, the absolute speed can also be controlled. Looking initially at constant-speed swimming, the energy optimization principle can be stated as: find the constant swimming speed at which the range of energy expended per unit distance is minimal.

This approach first presented by Weihs (1973a) can be described simply by assuming that no feeding, evacuation or growth take place during the period of swimming analyzed. Thus, the only significant contributions to the energy flux are due to locomotion and resting metabolism. The energy per unit time P required for locomotion at constant speed (actually "almost constant" speed because of the small variations due to the periodic mode of propulsion) is

$$\eta P = TU, \quad (6)$$

where η is the propulsive efficiency, T the (average) thrust, which from equation (3a) is equal and opposite to the drag, and U the swimming speed relative to the water. The total rate of working is

$$E = P + M = (K/\beta)\, C_D\, U^{3-\alpha-\delta} + M, \quad (7)$$

where P and M are the rates of working attributable to propulsion and maintenance. The relation between drag and velocity, and the empirical relation between efficiency and aerobic swimming speed (Kliashtorin 1973) are substituted into P. The terms K, β, and C_{D_o} are constants for a given fish, where $K = \frac{1}{2} \rho V^{2/3}$, β is an empirical constant of proportionality relating efficiency and an empirical power of the swimming speed ($\eta = \beta U^{\alpha}$, Kliashtorin 1973), and C_{D_o} is the reduced drag coefficient which is independent of velocity. The drag coefficient $C_D = C_{D_o} U^{\delta}$, where δ has values found empirically to range from 0 to 0.3. The distance crossed by the fish per unit time is numerically equal to the swimming speed. The distance traversed per unit of energy used is the distance crossed per unit time divided by the energy expended in the same unit of time, i.e. U/E. The optimal speed U_o is obtained from the mathematical condition:

$$\frac{d}{dU} \frac{U}{E} = 0,$$

which leads to

$$U_o = [\beta M/KC_{D_o}(2-\alpha-\delta)]^{1/3-\alpha-\delta}. \tag{9}$$

By substituting U_o into (7) one sees that the ratio of metabolic to propulsive rates of working at the optimal speed is:

$$(M/P)_o = 2-\alpha-\delta. \tag{10}$$

The coefficient α has values of $0.7 \leq \alpha \leq 1.0$, so that the ratio $(M/P)_o$ takes values of $1.3 < (M/P)_o < 0.7$. The rate of energy consumption at optimal speed is thus predicted to be 2 ± 0.3 times the standard metabolic rate. This value is in good agreement with data from Winberg (1956), Kerr (1971) and Ware (1975, 1978).

This approach can be carried further to include size effects (Weihs 1977). Applying various empirical relationships for the scaling of the quantities appearing in (8) it was predicted that

$$U_o \simeq \text{constant } L_f^{0.43} \tag{11}$$

with the constant having a numerical value of approximately 0.5 when U_o is measured in meters per second and L_f is the fork length in meters. Ware (1978), by physiological arguments, independently obtained the same (to both significant figures) exponent of size dependence of the optimal speed; he used data for sockeye salmon (*Oncorhynchus nerka*). The empirically obtained constant was about 8% larger than the one quoted above in one case and about 40% larger in another. The constant in equation (11) was obtained essentially by using data on trout (Webb 1971a,b), salmon (Brett 1979) and some other fishes swimming in flumes. But all of these fish were of a limited size range (0.3-0.7 m in length), and to rigorously test the validity of the predicted size dependence, data on large, as well as small, fish are required. Fish that are much smaller than the above range (less than 0.05 m) are less likely to be involved in migrations, and so information on larger fish is of greater interest here. Few such data are available because it is difficult to study large fish under controlled conditions and at the same time make the appropriate metabolic measurements. A first attempt appeared recently (Weihs et al. 1981), in which voluntary swimming speeds of two species of carcharhinid sharks in a large tank were reported (Fig. 2). The fish were fully acclimated to the tank, having been in it for months, and even years. The relationship described in equation (11) seems to be surprisingly accurate taking into account the multiplicity of parameters and the assumptions involved. The conclusion is that migrating fish tend to move at speeds which minimize the energy expenditure per unit distance traversed.

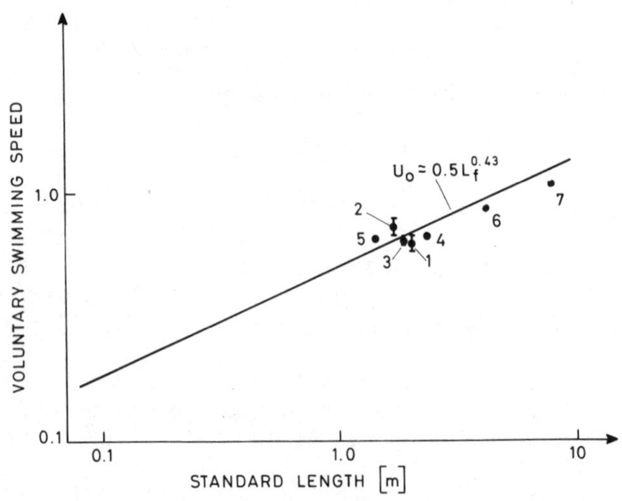

Figure 2. Voluntary swimming speeds for large sharks as a function of standard length, adapted and updated from Weihs et al. (1981). 1,2 - bull sharks (*Carcharhinus leucas*); 3 - sandbar shark (*C. plumbeus*); 4,5 - mako (*Isurus* sp.) (F.G. Carey, personal communication); 6 - Great white shark (*Carcharodon carcharias*) (Carey, personal communication); 7 - Basking shark (*Cetorhinus maximus*) (Q. Bone, personal communication). The line is the theoretical prediction of the optimal swimming speed (equation 11). Points 1-3 were measured under controlled conditions, while 4-7 were obtained from freely swimming fish.

The Effects of Currents

The optimal speeds discussed above were calculated and measured (as far as possible) for uniform relative motion between fish and water mass. Under natural conditions this corresponds to motion in still water. Many migrations, however, take place in areas where there are constant currents, or reversing tidal streams, with speeds of up to 1.5 m/sec to be traversed. In many cases migrating fish will adapt their path to make use of such currents (Harden Jones 1968).

A recent attempt to generalize the results of the previous section to include the effect of currents (Trump and Leggett 1980) predicts that the optimum speed relative to the water mass will still be the previous U_o (equations 9 or 11) plus the average current speed so that the absolute speed still stays U_o. Trump and Leggett's analysis is limited to slowly varying currents and thus does not yet include such problems as: what happens in eddies; how are the current speeds (especially in constant currents) estimated by the fish; and what course should the fish choose when the current is at an angle to the required direction of motion? The problem of optimization of motion in currents of a general nature is thus still unsolved.

But in certain specific cases patterns of behavior have been clearly shown to minimize the energy costs of migration. One such example is selective tidal stream transport (Harden Jones et al. 1978). In the North Sea the migration routes of flatfish, such as plaice (*Pleuronectes platessa*) and sole (*Solea solea*) (de Veen 1967), pass through zones of rapid tidal streams (0.7-1.2 m/sec equivalent to a swimming speed of 2-3 body lengths per second). Such speeds are much larger than U_o and gains of 50% or more are possible for fish that move up into midwater when the tide is flowing in the appropriate direction and rest on the seabed when it is flowing in the opposite direction (Weihs 1978).

Two-phase Periodic Locomotion

A common mode of locomotion among many migratory species is a two-phase periodic behavior, in which one phase includes one or more locomotory cycles of the body and fins causing acceleration (the beat phase), followed by a comparable period of coasting with no propulsive motions (the coast phase). At the end of the coasting period the speed has slowed down to the initial value and another burst and coast cycle begins.

This behavior, which seems rather complicated at first sight, produces large energy savings for sustained (aerobic) swimming over a long distance (Weihs 1974). Thus, the burst-and-coast swimming mode should be seen as an additional technique for minimization of energy consumption during migration. As mentioned previously, the hydrodynamic drag of the flexing fish body is three to five times larger than the stretched out fish body coasting at the same speed (Alexander 1967; Webb 1975), so that the coasting phase is economical. This is the key factor, leading to an overall saving of the energy required for total propulsion by this mode of behavior. Coasting must alternate with periods of acceleration to retain the required average forward speed.

Quantitative analysis of the benefits of this behavior showed that the gains in efficiency are functions only of the initial and final speeds of the beat phase, and the ratio ζ of drag when swimming actively to drag when coasting (Weihs 1974). The optimal conditions are obtained at relatively low average speeds. The increase in efficiency is greater when ζ is bigger, or alternatively when the burst phase becomes shorter. Thus, fish swimming in the anguilliform mode, for which ζ is larger, stand to gain more from this beat-and-coast mode of swimming behavior. The optimal beat phase includes one or two propulsive cycles only.

Recently, beat-and-coast swimming was shown theoretically to be energetically advantageous for anaerobic high-speed sprints also (Videler and Weihs 1982). There is experimental evidence for this behavior in gadoids. Here also, the greatest gains are obtained at the low end of the sprint speed-range (around 2-3 body lengths/sec for the individuals studied).

A variant of the energy-sparing two-phase swimming behavior, which could be useful for negatively buoyant fish, was analyzed by Weihs (1973b) (Fig. 3). This includes a phase of passive gliding at constant speed, utilizing the submerged weight of the fish to produce a component of force countering the hydrodynamic drag (see equations 1 and 2). As a result the fish sinks obliquely at the gliding angle γ to a greater depth. The second phase consists of the fish swimming diagonally upwards until it returns to its original depth. Then swimming oscillations stop and another glide phase starts. As with beat-and-coast swimming, the gains obtained here are due to the

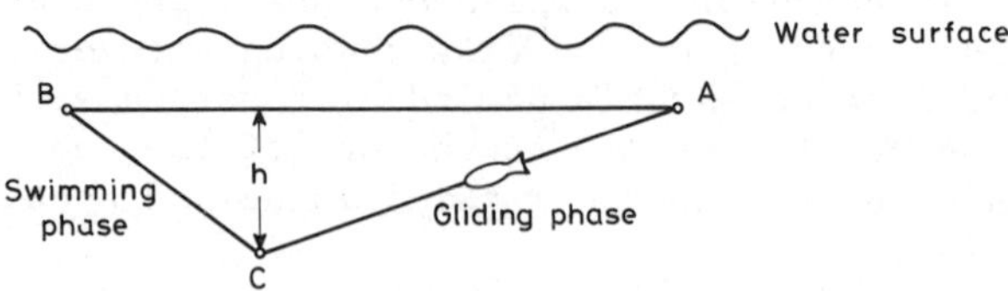

Figure 3. Schematic description of a two-phase glide-and-swim motion shown to be an efficient migratory strategy for negatively buoyant fish. Distances and angles are descriptive only and not to scale of fish drawn.

difference between drag when swimming and coasting. Optimal gains in range for a given energy store can be twofold. Anguilliform swimmers, such as flatfish, tend to gain more from the glide-and-swim technique than thunniform swimmers. However, an increase in saving is also possible for fish with low drag and relatively high submerged weight, the ratio of which at any speed determines the glide angle. Thus, there is an incentive for fish using carangiform movements applying this method to evolve low-drag shapes with high density. This combination of morphological adaptations is predominantly found in predatory, highly active scombrids and sharks (Magnuson 1978). Such glide-and-swim locomotion has been observed in flounders (Olla et al. 1972); indications of periodic depth changes of the required frequency and magnitude have been noted for dogfish (H. Westerberg, personal communication) and istiophorids (F.G. Carey, personal communication).

Schooling

A different type of adaptation for efficient long distance motion involves cooperative interactions of many individuals in a school. Many reasons for schooling have been suggested (Cushing and Harden Jones 1968; Shaw 1978): here the possible hydrodynamic advantages of schooling for migrating species are examined.

Synchronized motion in tight formation has been shown mathematically to reduce the thrust (and thus the energy) required to move at any given sustained speed (Weihs 1973c, 1975). When a fish swims by body and caudal fin oscillation, it sheds a wake consisting of a series of vortices in two staggered rows (Lighthill 1969). Each vortex induces velocities in the surrounding water in the same direction of rotation as the vortex and with strength inversely proportional to the distance from the vortex. As a result a mean velocity in the direction opposite to the fish motion is induced directly behind the fish, while diagonally behind it (outside the vortex "street") a forward velocity is induced (Fig. 4). A fish situated directly behind its predecessor would experience a higher relative velocity and would have to work harder. In the diagonal position, however, less effort is required than for a solitary fish moving at the same absolute speed. The best position has been predicted to be laterally midway between two fish of the preceding row (Weihs 1975). In this formation some fish help others with no direct gain to themselves, as only alternate horizontal rows of fish can make use of this bonus. However, by staying close to neighbors of the same lateral row and moving in antiphase synchrony (moving the tail in a direction opposite to the two closest neighbors) a channelling effect is produced that can increase the thrust produced by given bodily motions, again increasing the effectiveness of swimming in schools. Similar channelling effects can be obtained with neighbors vertically above and below, resulting in an elongate diamond shape as theoretically, the basic optimal structure of a fish

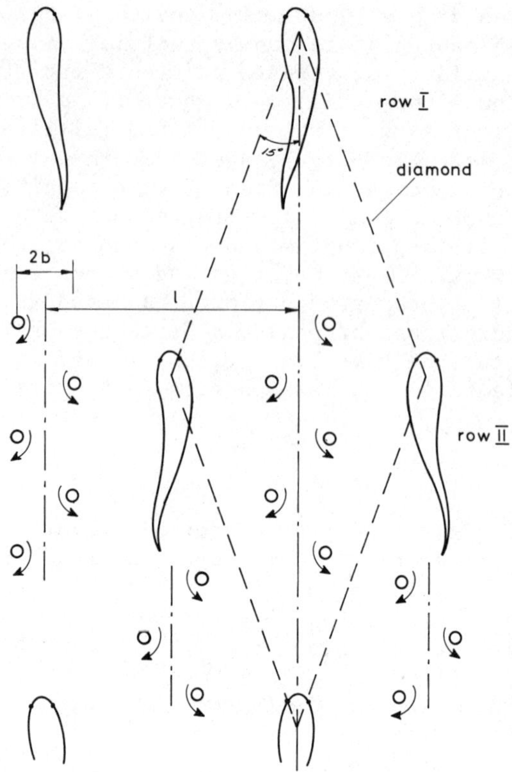

Figure 4. Idealised description of part of one horizontal layer of a large fish school. The unit formation (the diamond) has one member of row I, two members of row II and one member of row III. The maximum sideways amplitude of the body-caudal fin oscillation is indicated by b and the lateral distance between fish (less than one fish length for efficient interactions) by l.

school. Since this model was suggested, Breder (1976) reanalysed data on *Trachurus symmetricus* of Hunter (1966) and showed that the mean measured location of individuals approximates to the diamond shape. Partridge and Pitcher (1979) did not find such formations in saithe (*Pollachius virens*), herring (*Clupea harengus*) or cod (*Gadus morhua*), but more recently (B.L. Partridge personal communication) found that the lateral positioning of individuals in single row, hunting schools of Atlantic bluefin tuna (*Thunnus thynnus*) indicated a possible hydrodynamic advantage.

The various other behavioral mechanisms described above were all set in the framework of saving energy for moving at a prescribed speed or moving with a minimal expenditure of energy per unit distance crossed. Some of these same mechanisms, such as schooling, might equally be used to increase the average speed of migration to overcome local difficulties. Such difficulties can include regions of tidal currents which must be crossed in a given time so as to avoid being swept off course or areas in which predators abound. This may be the basis for the high speeds observed in migrating schools both in nature and in the laboratory (Zuyev and Belyayev 1970; Peterson 1976).

A Caveat

As a conclusion to this section, and before suggesting useful directions for future research, one potentially serious criticism of the entire hydromechanical approach as related to migration must be considered. Harden Jones (1981) pointed out that, while it may be both useful and attractive to present the results of mathematical analyses of potential energy saving mechanisms in terms of percentage savings, it is essential to examine the actual total energy saving before any conclusion can be reached about the relevance of any mechanism to observed or postulated patterns of behavior.

Harden Jones' (1981) example concerns the plaice which spawn in the Southern Bight of the North Sea and which make an annual reproductive migration of 280 km. The energy cost of this migration is reduced from the equivalent of about 10 g of fat - the cost of swimming the whole distance - to about 3.5 g by the use of selective tidal stream transport. Although in relative terms this is a considerable saving (by a factor of 3), in absolute terms it represents less than 1% of the total mass of the fish. Thus, at first sight the energetic advantage of migration by tidal stream transport might seem questionable, but if the saving in energy is used either to increase the total number of eggs in the gonad, or to increase the yolk reserve in each individual egg, a selective advantage may well accrue. While Harden Jones' (1981) criticism is well taken, this cautionary note does not reduce the utility or importance of the hydromechanical approach.

DIRECTIONS FOR FUTURE RESEARCH

It is a difficult task to predict worthwhile directions for future research. There are many interesting and important problems and unanswered questions awaiting - more than would occupy all of us for many years to come - and the decision of what problems to address becomes the cardinal one. I will consider possible generalizations of existing theoretical work, experimental studies required to confirm (or disprove) existing hypotheses and results and finally some possibilities for future theoretical work.

Extensions of Existing Theory

The main theoretical axiom, upon which almost all existing hydromechanical analyses of fish migration rest, is that the various adaptations and behavioral patterns observed tend to reduce energy requirements. This will probably serve as a starting point for more studies to include the effects of parameters not appearing in present analyses as well as more detailed work directed to improving present models.

It may be useful to look at the effects of temperature differences along migratory routes in an attempt to see whether choosing the ambient temperature can affect energy requirements. This seems reasonable as it is well known that aerobic muscular efficiency is highly temperature dependent in fish (Ware 1975; Brett 1979). It would thus be relatively straightforward to determine whether it is advantageous for a migrating school to choose its route through warmer regions. Such an analysis should include possible tradeoffs between the gains due to the effects of water temperature and the resulting changes in both vertical and horizontal trajectories. For example, one can imagine a situation in which in migrating towards warmer climes the fish may have the choice of moving along with a colder current or remaining in a warmer, and stationary, or even counterflowing, water mass. This effect may even limit the possible areas the fish may traverse (Barkley et al. 1978).

The arguments above have assumed that the fish in question were poikilothermic and in thermodynamical equilibrium. As a result water and fish temperature can be taken to be equal. In certain cases an additional thermal effect, heat transfer, may be of importance. Fish musculature, like any other engine, produces heat as a by-product of the propulsive motions, and this heat must be "lost" to the surroundings. When the ambient waters are cool, this usually does not consitute a problem; on the contrary, various species such as tunas have developed adaptations to slow down this heat loss and keep specific parts of the body warm. In warm waters, however, strenuous efforts may cause hyperthermia due to reduced heat loss, and it is of interest whether this can serve as a bound for migration. Such a bound does not have to be exclusive, but could cause a speed limit in certain regions.

Water viscosity changes with temperature; the viscosity of fresh water is 29% less at 20°C than at 10°C. This means that the drag that the fish has to overcome is lower by essentially the same ratio as the drag coefficient is directly proportional to the viscosity in most cases (Hoerner 1965). Less energy is thus required for motion in warmer waters, and the interaction of this effect with the changes in efficiency in defining an optimal ambient temperature for migration should be examined. It must be recalled, however, that all of the above effects may have influenced in varying amounts the in vivo experiments in which the relationships between muscular efficiency and temperature were established.

The viscosity of water also changes with salinity, which means that salinity may influence migratory behavior in a manner analogous to temperature. Less is known about the effects of salinity on aerobic efficiency, but it seems a reasonable assumption that such a dependence exists. Ambient salinity may also affect the body's energetic balance by changing ionic gradients.

The quantities discussed up to now in this section have been characteristic of the water mass. We now go on to consider interactions with external factors such as the surface, the bottom topography and various types of currents.

Surface effects are probably not very significant for migration, except for possible directional clues discussed by other authors in this volume. The main point is that swimming close to, or breaking, the water surface, is costly from an energetic point of view. The extra wave drag incurred would, therefore, probably be avoided except when feeding on surface-dwelling prey. Cetaceans with their requirement for a regular supply of atmospheric air are an exception, and it has been recently shown that for fast motion, dolphins can actually save energy by "running," a behavior that includes periodic leaping diagonally out of the water (Au and Weihs 1980).

Motion close to the bottom in both riverine and oceanic migrations, on the other hand, can be highly efficient for a number of reasons. It has been known for a long time (Osborne 1961) that salmon during their upstream migrations in rivers make use of the shape of the bottom topography to move through zones of lower local water velocity. They also move close to the bottom taking advantage of the bottom boundary layer and its reduced water speed. Quantitative analyses of the actual gains thus obtained, as opposed to the added complication of swimming close to obstacles that need to be avoided, would be of interest here. Ground effect on the hydrodynamic lift of the pectoral fins may also conceivably aid in keeping station in rest periods, by causing negative lift that pushes the fish against the ground and prevents it being swept downstream even without active propulsive motions. This behavior is well known in salmon parr (Kalleberg 1958) and is analogous to behavior observed in plaice (Arnold 1969; Arnold and Weihs 1978).

The only analysis of the effect of currents on migratory motions (Trump and Leggett 1980) is, as mentioned previously, limited to slowly varying currents. This type of study urgently needs to be generalized to include more rapid changes in current speed. Motion at a time varying angle to the current is also of interest as, for example, in the case of a fish moving through a circular eddy. One of the relevant questions is how should such an eddy be traversed, when its center is not necessarily along the migration path. Such oceanic eddies usually have a complicated, time dependent, three dimensional structure, so that such an analysis can become rather unwieldy mathematically.

Another aspect of having to deal with currents is that currents may have different, even opposite, velocities at different depths in the same geographic location. Thus, it may be advantageous from the energy standpoint to move at different depths. This has to be examined in conjunction with the temperature and salinity differences encountered.

Experimental Verification

The second category of research contains the experimental verification of the validity of the assumptions made in the course of existing and projected theoretical studies and the establishment of specific values for the various parameters appearing in the analyses. Here the situation is more difficult and the prospects less clear than for the theoretical part. First, as in all biological systems the natural variation between individuals makes any measurement, especially of behavioral and physiological features, extremely difficult and costly to obtain. Measurements from many individuals are required, and conditions closely approximating those in the wild have to be recreated and maintained. Success is not necessarily guaranteed even then as migratory species frequently do not adjust to captivity and rapidly lose their capabilities. Work in natural conditions may then be essential.

Physiological measurements such as oxygen consumption are difficult to obtain in any case, but are especially so for small larvae or large (>50 cm fork-length) adults. There is thus an urgent need for improved measurements of oxygen consumption, either by improving existing techniques or by the application of new methods, such as tracers. Neurophysiological measurements may be of use in conjunction with high-speed film records of fish motion (e.g. Videler and Wardle 1978) to help obtain independent estimates of thrust production.

Many migrating species are in the "large" category as defined above. This raises the possibility of measurements by telemetry from imbedded transducers. Temperature measurements have been taken by these means for several years now (Carey et al. 1971), and external pressure and acceleration have been added more recently (Ogilvy and Dubois 1981). It seems that this technique has tremendous potential for multiple parameter measurements in vivo, especially in view of the continuing trend to miniaturization in electronics. The external pressure distribution over the flexing fish body, measured continuously, or at least at a rate of around 100 Hz, is of great importance to the estimation of the hydrodynamic drag coefficient of the swimming fish. Estimates of this coefficient are derived at present from calculations based upon flow over rigid bodies. As mentioned previously these can be in error by a factor of five. Present drag estimates of swimming (flexing) fish are based on corrections originating from measurements of oxygen consumption or fish swimming capability. Both of these are indirect methods suffering from many inaccuracies (as partially discussed previously) and including additional effects. Thus, present estimates of the drag on a swimming fish are usually only accurate to within a factor of two. If hydromechanical analyses are to be of practical use in specific cases this situation has to be drastically improved.

Accurate position measurement by imbedded transponders at rates of 1 Hz or more can help to obtain actual instantaneous speeds and accelerations of swimming fish. These again will serve as independent measures of energy expenditure when taken together with physiological data.

All of these measurements are needed for each species individually, together with their variations during the various phases of life, if one wants to obtain information on specific adaptations or look for reasons for bodily changes (in shape, buoyancy, etc.) during the fish's life-span.

Other Possibilities

The two categories of future research outlined above are directly related to the concept of energy efficient migration. Other directions are less well defined at the present time and so I discuss them briefly, lumping together dissimilar subjects.

One of the more important questions bearing directly on the energetics of migration is that of the cues and clues (following Harden Jones 1981) by which prospective migrants initiate, carry out and terminate their actions. These include the meteorological, chemical and biological indicators that the time has come to start and end migration and the measures and data required by individuals or schools that the migration is going according to plan, or is reasonably well adjusted to unexpected changes in conditions en route.

Thus it is important to establish how accurately fish can measure the different kinds of information required, giving us an idea of the feedback mechanism by which the actual motions are adjusted to the (assumed) optimum. It can be shown that decision making under conditions of uncertainty and noise in the data may lead to solutions that are far from the optimum. An understanding of the data inputs to the fish will give limits on the possible correlation between observed behavior and the hydromechanically obtained prediction. For example, if the cue leading to beginning of the migration is "fuzzy," resulting in a delay in starting, the average speed may end up being significantly higher than predicted as the fish belatedly realise their "mistake."

The effects of possible predator-prey interactions on deviations from optimal trajectories should also be taken into account. These may cause changes resulting from the habits and locations of the migrant's predators or prey. For example, migrant fish may have to dash through predator-infested waters (Peterson 1976). Obviously, such interactions influence the energy cost and timing of migration. Also, it must be recalled that feeding itself may be the reason for certain migratory behavior, especially of the short-range type. Thus, the level of starvation or alternatively the abundance of food may define the constraints on locomotion.

The connection between external and internal morphological features, such as fin placement, distribution of red and white musculature and body shape, and migratory habits seems to be another useful avenue for research. Of special utility here are cases of closely related species, some of which migrate. Thus, changes in the features mentioned could be correlated with the requirements of migration.

ACKNOWLEDGEMENTS

I thank Paul W. Webb for stimulating discussion and the U.S. National Science Foundation for supporting part of this work during my visit to Ann Arbor, Michigan, through grant number PCM-8006469 to Paul W. Webb.

REFERENCES

Alexander, R.McN. 1967. Functional design in fishes. Hutchinson University Library, London, England.

Alexander, R.McN. 1972. The energetics of vertical migration by fishes. Pages 273-294 *in* M.A. Sleigh and A.G. Macdonald, editors. The effect of pressure on organisms. Cambridge University Press, Cambridge, England.

Arnold, G.P. 1969. The reactions of the plaice (*Pleuronectes platessa* L.) to water currents. Journal of Experimental Biology 51:681-697.

Arnold, G.P., and D. Weihs. 1978. The hydrodynamics of rheotaxis in the plaice (*Pleuronectes platessa* L.). Journal of Experimental Biology 75:147-169.

Au, D., and D. Weihs. 1980. At high speeds dolphins save energy by leaping. Nature (London) 284:548-550.

Barkley, R.A., W.H. Neill, and R.M. Gooding. 1978. Skipjack tuna, *Katsuwonus pelamis*, habitat based on temperature and oxygen requirements. US National Marine Fisheries Service Fishery Bulletin 76:653-662.

Breder, C.M. Jr. 1926. The locomotion of fishes. Zoologica (New York) 4:159-256.

Breder, C.M. Jr. 1976. Fish schools as operational structures. US National Marine Fisheries Service Fishery Bulletin 74:471-502.

Brett, J.R. 1979. Environmental factors and growth. Pages 599-675 *in* W.S. Hoar and D.J. Randall, editors. Fish physiology, volume 8. Academic Press, New York, New York, USA.

Carey, F.G., J.M. Teal, J.W. Kanwisher, K.D. Lawson, and J.S. Beckett. 1971. Warm-bodied fish. American Zoologist 11:135-145.

Cushing, D.H., and F.R. Harden Jones. 1968. Why do fish school? Nature (London) 218:918-920.

Harden Jones, F.R. 1968. Fish migration. Arnold, London, England.

Harden Jones, F.R. 1981. Fish migration: strategy and tactics. Pages 139-165 *in* D.J. Aidley, editor. Animal migration. Cambridge University Press, Cambridge, England.

Harden Jones, F.R., M. Greer Walker, and G.P. Arnold. 1978. Tactics of fish movement in relation to migration strategy and water circulation. Pages 185-207 *in* H. Charnock and G. Deacon, editors. Advances in oceanography. Plenum Press, New York, New York.

Hoerner, S.F. 1965. Fluid dynamic drag. Hoerner, Bricktown, New Jersey, USA.

Hunter, J.R. 1966. Procedure for analysis of schooling behavior. Journal of the Fisheries Research Board of Canada 23:547-562.

Kalleberg, H. 1958. Observations in a stream tank of territoriality and competition in juvenile salmon and trout (*Salmo salar* L. and *S. trutta* L.). Institute of Freshwater Research Drottningholm Report 39:55-98.

Katz, J., and D. Weihs. 1978. Hydrodynamic propulsion by large amplitude oscillation of an airfoil with chordwise flexibility. Journal of Fluid Mechanics 88:485-497.

Kerr, S.R. 1971. A simulation model of lake trout growth. Journal of the Fisheries Research Board of Canada 28:815-819.

Kliashtorin, L.B. 1973. The swimming energetics and hydronomic characteristics of actively swimming fish. Ekspress Informatsiya. Promyslovaya Okeanologiya i Podvodnaya Tehknika: Commercial Oceanology and Underwater Technology 6:1-19. Translated by the Translation Bureau of Canada.

Lighthill, M.J. 1969. Hydromechanics of aquatic animal propulsion. Annual Review of Fluid Mechanics 1:413-446.

Lighthill, M.J. 1971. Large-amplitude elongated-body theory of fish locomotion. Proceedings of the Royal Society of London B 179:125-138.

Lighthill, M.J. 1975. Biofluid hydrodynamics. Society for Industrial Applied Mathematics, New York, New York, USA.

Magnuson, J.J. 1978. Locomotion by scombrid fishes: hydromechanics, morphology and behavior. Pages 230-313 *in* W.S. Hoar and D.J.

Randall, editors. Fish physiology, volume 7. Academic Press, New York, New York, USA.

Ogilvy, C.S., and A.B. Dubois. 1981. The hydrodynamic drag of swimming bluefish (*Pomatomus saltatrix*) in different intensities of turbulence. Variation with changes of buoyancy. Journal of Experimental Biology 92:67-85.

Olla, B.L., C.E. Samet, and A.L. Studholme. 1972. Activity and feeding behavior of the summer flounder (*Paralichthys dentatus*) under controlled laboratory conditions. US National Marine Fisheries Service Fishery Bulletin 70:1127-1136.

Osborne, M.F.M. 1961. The hydrodynamical performance of migratory salmon. Journal of Experimental Biology 38:365-390.

Partridge, B.L., and T.J. Pitcher. 1979. Evidence against a hydrodynamic function for fish schools. Nature (London) 279:418-419.

Peterson, C.H. 1976. Cruising speeds during migration of the striped mullet (*Mugil cephalus* L.): an evolutionary response to predation? Evolution 30:393-396.

Shaw, E. 1978. Schooling fishes. American Scientist 66:166-175.

Trump, C.L., and W.C. Leggett. 1980. Optimum swimming speeds in fish: the problem of currents. Canadian Journal of Fisheries and Aquatic Science 37:1086-1092.

Veen, J.F. de. 1967. On the phenomenon of soles (*Solea solea* L.) swimming at the surface. Journal du Conseil Conseil international pour l'Exploration de la Mer 31:207-236.

Videler, J.J., and C.S. Wardle. 1978. New kinematic data from high speed cine film recordings of swimming cod (*Gadus morhua*). Netherlands Journal of Zoology 28:465-484.

Videler, J.J., and D. Weihs. 1982. Energetic advantages of burst-and-coast swimming of fish at high speeds. Journal of Experimental Biology 97:169-178.

Ware, D.M. 1975. Growth, metabolism, and optimal swimming speed of a pelagic fish. Journal of the Fisheries Research Board of Canada 32:33-41.

Ware, D.M. 1978. Bioenergetics of pelagic fish: theoretical change in swimming speed and ration with body size. Journal of the Fisheries Research Board of Canada 35:220-228.

Webb, P.W. 1971a. The swimming energetics of trout I. Thrust and power output at cruising speeds. Journal of Experimental Biology 55:489-520.

Webb, P.W. 1971b. The swimming energetics of trout II. Oxygen consumption and swimming efficiency. Journal of Experimental Biology 55:521-540.

Webb, P.W. 1975. Hydrodynamics and energetics of fish propulsion. Bulletin of the Fisheries Research Board of Canada 190:1-159.

Weihs, D. 1973a. Optimal fish cruising speed. Nature (London) 245:48-50.

Weihs, D. 1973b. Mechanically efficient swimming techniques for fish with negative buoyancy. Journal of Marine Research 31:194-209.

Weihs, D. 1973c. Hydromechanics of fish schooling. Nature (London) 241:290-291.

Weihs, D. 1974. Energetic advantages of burst swimming of fish. Journal of Theoretical Biology 48:215-229.

Weihs, D. 1975. Some hydrodynamical aspects of fish schooling. Pages 703-718 *in* T.Y. Wu, C.J. Brokaw, and C. Brennen, editors. Swimming and flying in nature, volume 2. Plenum Press, New York, New York, USA.

Weihs, D. 1977. Effects of size on sustained swimming speeds of aquatic organisms. Pages 333-338 *in* T.J. Pedley, editor. Scale effects in animal locomotion. Academic Press, New York, New York, USA.

Weihs, D. 1978. Tidal stream transport as an efficient method for migration. Journal du Conseil Conseil international pour l'Exploration de la Mer 38:92-99.

Weihs, D., R.S. Keyes, and D.M. Stalls. 1981. Voluntary swimming speeds of two species of carcharhinid sharks. Copeia 1981:219-222.

Weihs, D., and P.W. Webb. in press. Optimization of locomotion. *in* P.W. Webb and D. Weihs, editors. Fish biomechanics. Praeger, New York, New York, USA.

Winberg, G.G. 1956. Rate of metabolism and food requirements of fishes. (Inten'sivnost obema i pishchevye potrebnosti ryb. Nauchnye Trudy Belorusskova Gosudarstvennovo Universiteta imeni V.I. Lenina, Minsk, USSR.) Fisheries Research Board of Canada Translation Series Number 194.

Yates, G. in press. Body and caudal fin propulsion. *in* P.W. Webb and D. Weihs, editors. Fish biomechanics. Praeger, New York, New York, USA.

Zuyer, G.V., and V.V. Belyayev. 1970. An experimental study of the swimming of fish in groups as exemplified by the horsemackerel (*Trachurus mediterraneus ponticus* Aleev). Journal of Ichthyology 10:545-549.

APPENDIX: NOMENCLATURE.

b - Amplitude (maximal) of sideways oscillation of caudal fin.

C_D, C_D - Drag coefficient in direction parallel to and normal to the fish longitudinal axis respectively. This nondimensional coefficient is obtained by dividing the drag force by the dynamic pressure and the fish volume to the 2/3 power.

C_D - Reduced drag coefficient, independent of velocity.

C_L, C_T - Lift and thrust coefficients, obtained in same manner as the drag coefficient.

D - Drag force (MLT^{-2}), in the direction of motion and in direction perpendicular to motion, respectively.

E - Total rate of energy expenditure (ML^2T^{-3}).

L - Lift force (MLT^{-2}).

L_f - Fork length (L).

m - mass (M).

M - Resting (standard) metabolic rate (ML^2T^{-3}).

P - Rate of energy expenditure for propulsive purposes (ML^2T^{-3}).

R_e - Reynolds number defined by characteristic length multiplied by speed and divided by kinematic viscosity.

T - Thrust (MLT^{-2}).

T_a - Thrust available for sustained swimming (MLT^{-2}).

U - Relative speed between fish and water, along axis (LT^{-1}).

U_o - Optimum swimming speed (LT^{-1}).

V - Fish volume (L^3).

W - Submerged weight (MLT^{-2}).

x, z - Coordinates parallel, and normal to the longitudinal axis of the fish, respectively (L).

$\ddot{x}, \ddot{z}$ - Acceleration in x, z directions respectively (LT^{-2}).

α, β - Empirical constants in equation relating efficiency and swimming speed.

γ - Angle of motion relative to horizon.

η - Efficiency ($o<\eta<1$).

δ - Water density.

ζ - Ratio of drag when swimming actively to gliding at the same speed.

MIGRATION AND NAVIGATION IN BIRDS: A PRESENT-STATE SURVEY WITH SOME DIGRESSIONS TO RELATED FISH BEHAVIOUR

Hans G. Wallraff

Max-Planck-Institut für Verhaltensphysiologie
D-8131 Seewiesen
Federal Republic of Germany

ABSTRACT

This paper reviews various long-distance orientation and navigation mechanisms believed responsible for bird migration patterns observed in nature. Birds appear to have developed spontaneously intended directions of migration which are population-specific and permit naive juveniles to fly particular compass directions and distances. Three types of compass mechanisms are known to exist in birds, the sun compass, the star compass, and the magnetic compass. Once birds acquire experience of a particular area, they are able to return to that area from unknown sites and thus display true goal orientation. Based on experiments with homing pigeons, the theoretical principles of goal orientation are considered and it is demonstrated that pigeons use a type of "map-and-compass" mechanism. Different types of maps are discussed, in particular a "mosaic map" and a "gradient map." Evidence is presented that atmospheric odours constitute a substantial basis of pigeon homing. They are assumed to form long-range gradients and thus navigational coordinates. It is concluded that, as both birds and fishes are highly mobile in three dimensions, similar mechanisms of migratory orientation may be used by both classes in cases where birds and fish refer to similar environmental clues.

INTRODUCTION

In some ways birds and fishes are more closely related to each other than to the rest of the vertebrates. Most representatives of both classes are highly mobile in three dimensions and thus to a high degree independent of structures of the solid ground. Mobility allows large living areas not only of species, but also of individuals, and these large areas are covered during extended migrations. As the migrations do not primarily consist of random dispersal, but of well-organized movements to and from population-specifically predetermined geographic regions often hundreds and thousands of kilometers distant from each other, they demand and depend on appropriate capabilities to orientate and navigate over such long distances. It is reasonable to ask whether both groups of animals may have developed similar or equal orientational mechanisms. If so, researchers investigating one group might be able to profit from progress achieved by experiments dealing with the other group.

In this volume the birds are guests, and their related behaviour has to be summarized within one contribution. To make it not too cursory I will focus on only some aspects of bird migration and neglect others which are, however, important for an understanding of the whole complex as well. I will deal very little with ecological adaptations and entirely omit the whole field of physiological control of migration as well as related problems of energetics. Instead, after a brief look at migration patterns as they occur in nature, I will concentrate on the various mechanisms involved in long-distance orientation and navigation. The resulting survey may enable investigators of fish migration and orientation to evaluate whether our knowledge about related bird behaviour may elucidate their own field, or whether they can ignore it. Even then, a glance in the neighbour's garden may not be without any interest.

PATTERNS OF BIRD MIGRATION

Our knowledge about long-range routes and destination areas of bird migrants results primarily from ringing of free-living individuals. About 80 years of ringing has yielded considerable amounts of data on the movements of a great variety of species (e.g., Zink 1973-1982). From these data the following basic conclusions can be drawn:

1. Offspring of a particular population migrate, in their first autumn of life, in similar directions. Birds belonging to different populations of the same species may select different directions. The same is true for birds of different species originating from the same area. (Nevertheless, the majority of species, breeding in a given area, may prefer similar directions. Birds from central Europe, for instance, migrate preferably toward the southwest.)

2. Flight routes between breeding and wintering grounds may follow, by and large, a straight beeline course, yet in other cases, they may include detours and thus some directional changes. Usually detours appear in some way adapted to large-scale geomorphological features.

3. Autumn and spring migrations do not always follow the same routes.

4. Distances of travel vary enormously from small-scale vagrancy to global migrations connecting the Arctic with the Antarctic.

Ringing reveals rough travel routes and winter quarters of individuals of known populations. To elucidate which way actual migration takes place, however, more direct observation is required. The most important methods used for this purpose are: 1) observation, count and catch of grounded migrants; 2) visual observation of actual migration during day and night (nocturnal observation is conducted by telescopic monitoring of the moon's disc--"moon-watching" --or of a vertically-directed light beam--"ceilometer"); and 3) observation by various kinds of radar techniques. (For advantages, disadvantages, and biases connected with the different methods, and for related literature, see Richardson 1978.)

Many species whose normal activity is during daytime migrate at night. Nocturnal observation methods, unfortunately, are mostly unable to determine the species involved, and appropriate guesses can be made in only a few cases.

Migrating birds respond to spatial as well as temporal features of their environment. Interrelations are complex, more or less species-specific, and rarely follow an all-or-nothing scheme. Some general tendencies, nevertheless, can be extracted from the vast amount of literature. On the one hand, many birds surmount extended unfavourable areas as high mountains, seas and deserts, sometimes performing nonstop flights of 50 hours and more (e.g., Williams et al. 1977, Richardson 1980). On the other hand, flight courses are often influenced by rather small-scale topographical features which act as so-called leading lines, yet should often better be designated as either deflection lines or attraction lines, depending on whether they act as barriers deflecting the birds from their originally intended direction, or whether they attract birds to approach favoured habitats or even to use those features as aids for orientation. Whether mountain ridges, coasts or valleys influence flight directions depends on the species as well as on many concomitant conditions such as angular difference between intended direction and interfering structure, size and shape of this structure, height of flight, wind, cloud cover, visibility, etc. (e.g., van Dobben 1955; Mueller and Berger 1967; Schüz 1971). Naturally, topographical influences are stronger in daytime than at night, yet nocturnal migration often also appears affected (e.g., Wallraff and Kiepenheuer 1963; Bruderer 1978, 1982; Bingman et al. 1982).

Despite many interrelations between geomorphological structures

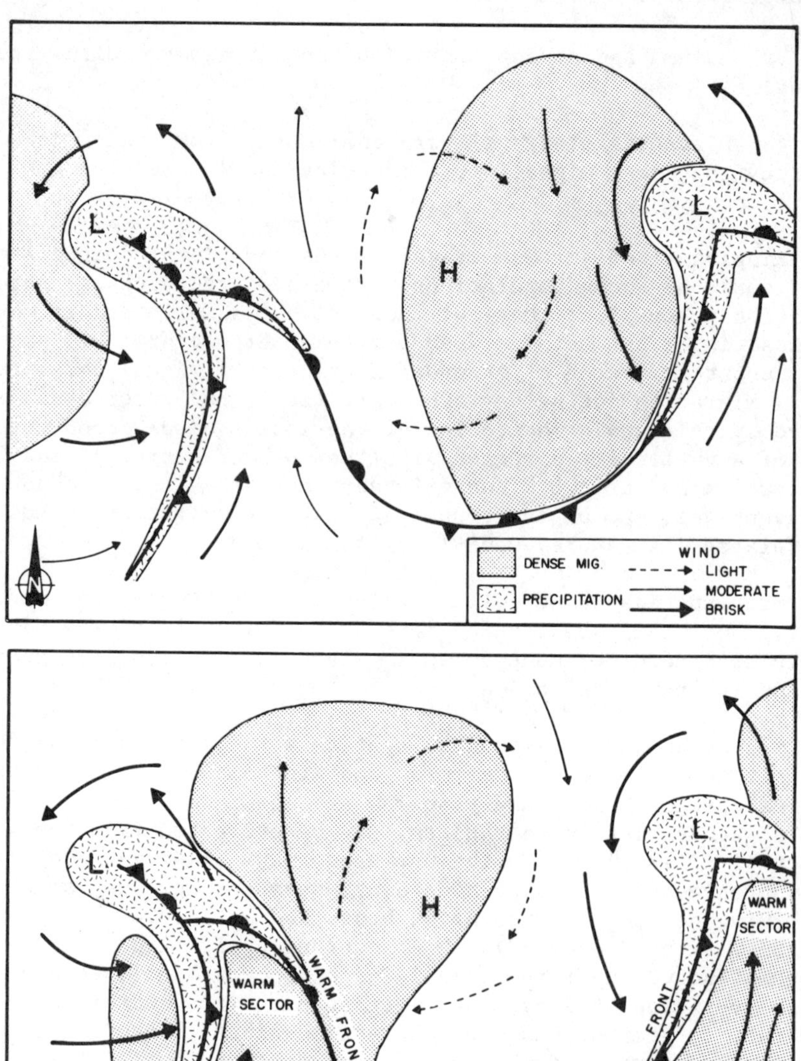

Figure 1. Typical synoptic weather situation at northern temperature latitudes with high (H) and low (L) pressure systems, fronts, precipitation and winds. Stipplings show parts of pressure systems where peak migrations tend to occur in autumn (top) and spring (bottom). From Richardson (1978).

and bird-migration routes (large-scale adaptations as well as small-scale responsiveness), influences are always more or less relative, since for birds, in contrast to fishes, absolutely unsurmountable barriers practically do not exist.

In principle the same is true with respect to temporal variations of environmental conditions. Bird migration takes place in almost all kinds of weather, but not always to the same extent and in the same way. Peak migration is generally correlated with following winds. If inspected in more detail, however, the field of interrelations between wind and bird migration turns out to be quite complex (for theoretical aspects, see Alerstam 1979). Wind is part of a general weather situation, and Figure 1 shows that peaks of migration appear differently adapted in autumn and spring to the high and low pressure systems as they typically occur in temperate northern latitudes. All the meteorological parameters are in some way correlated with each other, and in most cases it is difficult to find out to which of the components the birds directly respond (Richardson 1978). As it is interesting in the context of orientation, it should be mentioned that, on average, fewer birds migrate under overcast than under clear skies. Yet even under dense cloud cover they are commonly well oriented. However, without any useful visual information available, e.g. when flying within clouds, migrating birds appear poorly or inappropriately oriented (Able 1982a,b).

BIRD ORIENTATION AND NAVIGATION

Migratory Orientation: A Complex System

If we seek the variables controlling a bird's course that enable it to follow the population-specific paths which had been followed by its ancestors, we have to seek two sorts of variables. There must be some endogenous intention to respond to environmental features in a particular way, and there must be such features suited to release meaningful responses. We may first inquire about the endogenous aspect and thus consider intended directions, etc., without specifying with which physical clues they might be associated.

Direct observation as well as ringing data leave no doubt that birds are able to select a particular compass direction and to keep it over considerable distances. As it is unlikely that juveniles of small nocturnal migrants remain in sufficient social contact with adult conspecifics, the intended direction of naive birds appears to be phylogenetically transmitted in a population-specific way. That this is true even in a large species migrating in flocks was demonstrated experimentally some fifty years ago (Fig. 2). This example of the white stork is particularly remarkable, as Schüz

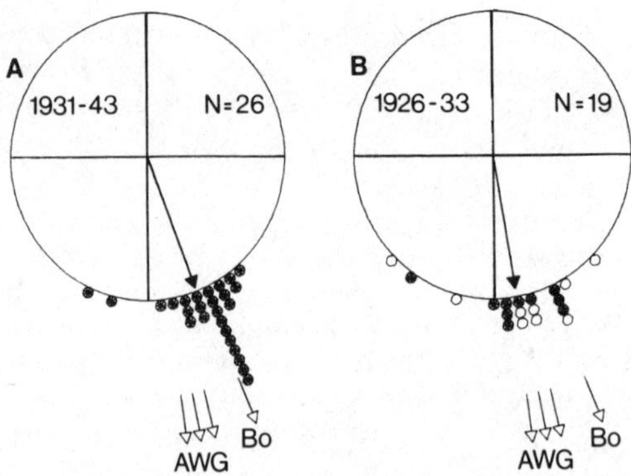

Figure 2. Directions of autumn recoveries of juvenile white storks (*Ciconia ciconia*) hatched in East Prussia and found within Europe at distances of more than 100 km. A. Normally migrating birds. B. Birds detained until after departure of wild-living conspecifics; filled dots, released in East Prussia; open dots, released in western Germany (from where endemic storks migrate southwest). Resulting mean vectors are given as arrows (maximum length 1 = radius). Peripheral arrows indicate directions from home area to African wintering grounds (AWG) and to the first destination, the Bosporus (Bo). B after Thienemann (1931) and Schüz (1949); from Wallraff (1977).

(1950) also demonstrated that juveniles of the same population are deflected from their endogenously intended direction when they start migrating together with conspecifics of another population preferring another direction. Hence the directions actually chosen may depend on social influences as well as on endogenous tendencies. Traditional flight routes are rather common in larger birds as in waterfowl (e.g., Sterbetz and Szijj 1968; Bellrose and Crompton 1970).

Methodologically it was an important step forward when Gustav Kramer (1949) started to investigate migratory orientation in the laboratory. He made use of the migratory restlessness (Zugunruhe) exhibited by birds of many species in autumn and spring, even under cage conditions. If the birds are kept in radially symmetrical cages, one can observe or automatically record whether their locomotory activity is oriented in some way. Various types of such "Kramer cages" have been in use, most successfully the octagonal "Merkel cage" with eight radial perches (Merkel and Fromme 1958) and the "Emlen funnel" which by its shape compels the bird to start its hops always from the centre (Emlen and Emlen 1966).

Passerine birds tested in such cages or funnels usually prefer directions corresponding, by and large, to those flown by their free-living conspecifics (e.g., Emlen 1967, 1975; Wiltschko 1968). If they are raised, kept and tested without useful visual cues, however, this correspondence is not always impressive, either due to a large angular scatter (Fig. 3A,B) or to some deviation from the expected direction (Fig. 3C) or to a bipolar distribution of directional preferences (Fig. 3D). Temporal changes in average bearings (Fig. 3A,B) have been interpreted as reflecting a temporal

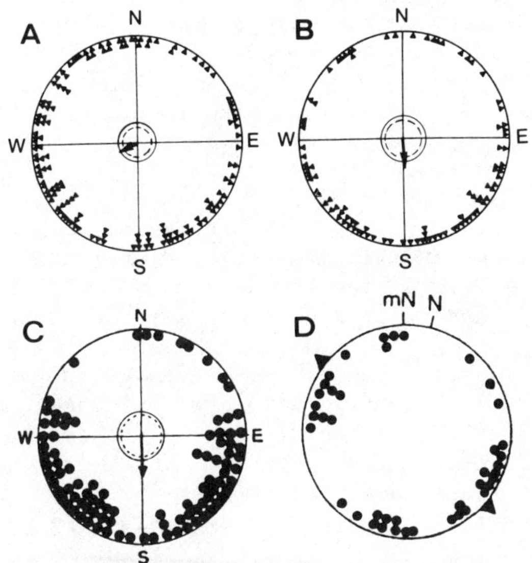

Figure 3. Bearings of Zugunruhe in Merkel cages (A,B) and Emlen funnels (C,D). Each symbol indicates the bearing of the mean vector calculated from the activity of one bird during one night. Arrows give mean vectors calculated from the distributions shown (maximum length 1 = radius; small circles refer to significance levels P = 0.05 and 0.01 according to the Rayleigh test); in D, two calculated modes are given. mN = magnetic north (in A-C practically identical with geographic north). All data are from birds raised in captivity without view of the sky and tested in their first autumn without visual cues; birds lived and were tested under natural magnetic conditions. A,B. Garden warblers (*Sylvia communis*) tested in August and September (A) and in October to December (B) (from Wiltschko et al. 1980); expected directions are SW and about SSE, respectively. C. Pied flycatchers (*Ficedula hypoleuca*) tested 21 August to 15 October (from Beck and Wiltschko 1982); the direction expected is SW. D. Savannah sparrows (*Passerculus sandwichensis*) (from Bingman 1981).

program controlling the directional shift garden warblers have to perform when continuing their flight from the Iberian peninsula to Africa (Gwinner and Wiltschko 1978), yet in captivity, temporal shifts have also been observed with the sign opposed to that expected (Beck and Wiltschko 1982).

In conclusion we may state that during evolution birds have developed spontaneously intended directions of migration which are specific for particular populations (see also Bellrose 1958a; Perdeck 1958). Indications that these directions may not be static, but vary systematically according to an endogenous temporal program, require confirmation. (In some ways, analogous results were published earlier by Groot [1965] for sockeye salmon. Yet coincidence between experimental data and open-water behaviour remained questionable [Wallraff 1966b].) Nevertheless a temporal program may presently be the most plausible explanation of the directional shifts occurring in natural migration routes. In some cases shifts may not actively be performed by the birds, but result from more or less constant headings in conjunction with position-dependent wind drift (Williams et al. 1977; Williams and Williams 1978; Richardson 1980). Other possibilities have been listed and discussed (Wallraff 1977), but supporting experimental evidence is weak or non-existent.

Much better founded than time-dependent directional shifts, and supported by a large body of experimental evidence, is Gwinner's (1968, 1972, 1977, 1981) hypothesis that distances to be travelled result from an endogenous time program. Under cage conditions phases of Zugunruhe last longer and reach higher levels of intensity in long-distance than in short-distance migrants (Fig. 4). Gwinner translated cage activity to distances of flight and found fairly good coincidence with the distances covered by the respective species in nature. Zugunruhe, like other functions (fat deposition, molt, etc.), has been shown to be part of an endogenous circannual periodicity which is synchronized with the physical annual cycle primarily by photoperiodic events, but continues freerunning under constant conditions (cf. Gwinner 1981).

Endogenous programs, of course, can supply a bird with only rough prescriptions of what to do and how to respond to the environment. The bird's behaviour is modified by various exogenous influences. There is no doubt that the birds are affected by ecological conditions, and certainly they do not finish their journey without having reached sufficiently favourable resting and feeding grounds (cf. Perdeck 1964). Within the predetermined schedule the birds have opportunity to explore the areas they are visiting, hence to fit Baker's (1978, 1982) "exploration model" (for what it may be worth), and potentially to make use of their experience in later years (see, e.g., Bellrose and Crompton 1970). One must bear in mind, however, that occasions to explore detailed features are restricted in long-distance migrants which cover hundreds or

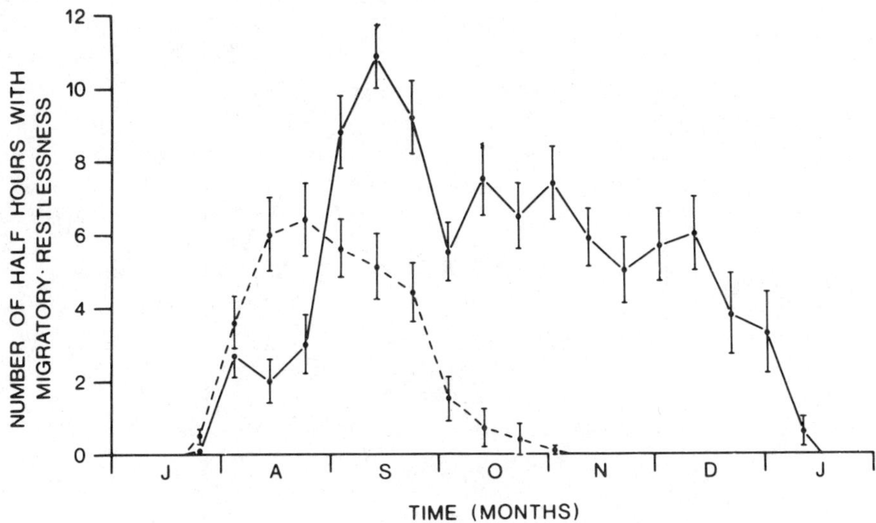

Figure 4. Migratory restlessness of blackcap warblers (*Sylvia atricapilla*) from Finland (solid line) and France (broken line) kept under identical constant light-dark and temperature conditions. Means and standard errors of means per 10-day period. Finnish populations migrate over quite long distances, whereas French blackcaps are short-distance, partial migrants. From Berthold (1977).

thousands of kilometres by non-stop flights, often during nighttime and partly over oceans or deserts.

So far we have focussed on young birds migrating for the first time in their lives and thus unfamiliar with all the areas they approach. They appear to reach their winter quarters primarily by means of a bearing-and-distance mechanism ("vector orientation") with the distance given in measures of time. However, from their first return flight onwards their orientational performance may be improved. When Perdeck (1958) displaced juvenile starlings perpendicularly to their migration route, they continued flying their normal compass course, yet adults tended to approach the normal wintering grounds where they earlier had spent at least one winter (Fig. 5). From these experiments we may conclude that phylogenetic programming was able to provide the birds with intentions to fly particular compass directions and distances, but not with information defining the goal area itself. Once a bird has been living for a while in a particular area, however, it is able to return to this area even from unknown sites where it never had been before. This capability is designated as goal orientation or "true" navigation (in contrast to sole compass orientation), and it deserves its own section (see below).

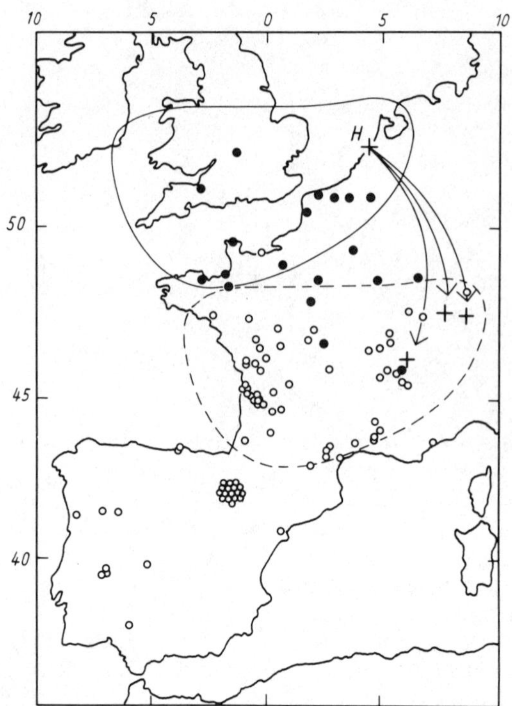

Figure 5. Recovery sites of starlings (*Sternus vulgaris*) caught during migration in Holland (H) and displaced to Switzerland (three release sites indicated as crosses). Their approximate normal wintering range is circumscribed by a solid line, an accordingly displaced range by a broken line. Symbols indicate recovery sites of adults (filled dots) and juveniles (open dots) in subsequent winter months. Adults tend toward normal winter range, juveniles maintain normal compass direction. Modified from Perdeck (1958).

This section cannot be closed without mention that an intended direction, independently of whether its origin is endogenous or exogenous, has to be defined as an angle to some physical feature of the environment. "Compass direction" is an abstract term; in reality it is given, according to our present knowledge, as an angle to some astronomical clue(s) or to the geomagnetic field. Both sorts of clues are utilized by birds, but it is not yet clear whether one of them is primary in the sense that the orientational system is genetically encoded with reference to this one sort and only secondarily calibrated to the other. There are indications for the stars playing a primary role (Emlen 1970) as well as for the geomagnetic field (Wiltschko and Wiltschko 1975, 1976), and it seems possible that different species react differently or that none of the clues is more "primary" than the other (cf. Wiltschko 1982). The whole field of interrelations between different types of compasses, as investigated by cage experiments, has become rather puzzling in recent years and requires clarification.

I will not discuss all the variables that may or do influence flight directions of migratory birds nor all aspects of the integration of multiple orientational clues (cf. Emlen 1975, 1980; Able 1980), but will only consider the three basic types of compass mechanisms which are known to be used by birds.

Mechanisms of Compass Orientation

The sun compass. The time-compensated sun compass is a well-known and widely distributed orientational mechanism whose function has been described many times (e.g., Kramer 1951, 1952, 1953a; Hoffmann 1960, 1965; Matthews 1968; Emlen 1975; Schmidt-Koenig 1979; R. Wiltschko 1980, 1981; Wallraff 1981a). In order to compensate for the sun's daily movement, this type of orientation is closely linked with the animals' circadian clock.

There are some indications that the birds' sun compass might differ from that of fishes in two ways. Pigeons apparently have to learn to interpret the sun's movement individually by observation of its whole daily path, at least in the context of homing-flights (R. Wiltschko and W. Wiltschko 1980), whereas fishes are reported to respond spontaneously in an appropriate manner (Braemer 1960). Sun-compass orientation appears to be independent of sun altitude in birds (e.g., Hoffmann 1960; R. Wiltschko 1980), but not in fishes (Schwassmann and Hasler 1964). However, since all the related experiments were conducted with few species and often with different methods, general conclusions regarding consistent differences between classes of animals cannot be drawn.

Little is known about the application of the sun compass in the birds' natural lives. There is no doubt that it is incorporated in the homing system of pigeons (e.g., Schmidt-Koenig 1958, 1979; Keeton 1974; Wallraff 1974) and that it is used in initial orientation by displaced ducks (Bellrose 1958b; Matthews 1963b). There are indications in only one species that it is also used during daytime migration (Kramer 1951; W. Wiltschko 1980). It is still unclear to what extent it is used in nature (see also R. Wiltschko 1981).

Oddly enough, use of the sun appears better demonstrated now for nocturnal than for diurnal migration. The direction of sunset has been shown to influence orientation not only during twilight, but also in the subsequent night (Moore 1980). Sunset orientation is possibly not part of the time-compensated sun compass, but fixed-angle menotaxis independent of circadian rhythmicity. Recent

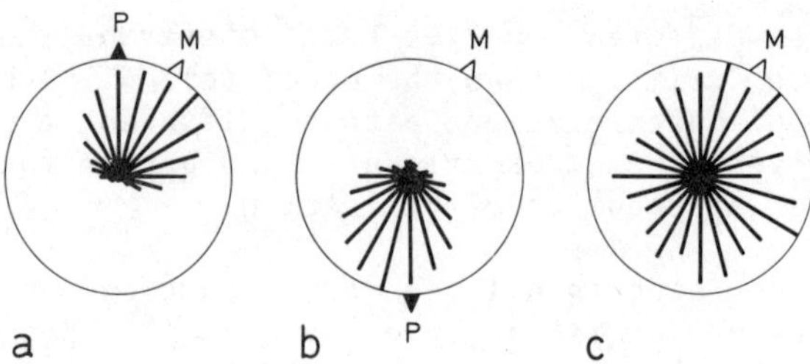

Figure 6. Migratory activity of one indigo bunting (*Passerina cyanea*) in a planetarium in spring. Radii indicate relative frequency of activity toward the respective direction in an Emlen funnel. P, north of planetarium sky; M, magnetic north. a. P coincides with geographic north; b. P reversed; c. Artificial stars shut off. After Emlen (1967).

experiments by Able (1982c) open the possibility that skylight-polarisation patterns rather than the setting sun itself are used as orientational clues.

The star compass. Use of natural as well as artificial stellar skies in avian migratory orientation has been demonstrated many times (e.g., Sauer 1957; Emlen 1967, 1970); an example is shown in Fig. 6. In contrast to sun orientation, stellar orientation does not necessarily require knowledge of time of day, and it has in fact been shown to operate independently of the circadian clock (Matthews 1963b; Emlen 1967; Wallraff 1968). There are, however, some results indicating dependence on the phase relation between "sky time" and "bird time" (Sauer 1957; Sauer and Sauer 1960; Liepa 1978; Vilks et al. in press). The discrepancies may result from the fact that the moving stellar sky can be evaluated in different ways, and the ways birds respond to an unnatural situation may depend on the particular circumstances of testing.

Most of the related experiments support the conclusion that birds recognize stellar configurations by means of some sort of pattern recognition (Emlen 1967, 1970; Wallraff 1972a), but Vilks et al. (in press) favour orientation with respect to the resulting mean vector of light distribution.

It seems extremely unlikely that any birds possess an "inherited star map" (as Sauer 1957 assumed; but see Emlen 1970). It has been shown that ducks are provided with a great capacity in learning, distinguishing and memorizing complex stellar patterns (Wallraff 1969, 1972a). Whether the birds calibrate these patterns to earth-bound compass bearings by observation of celestial rotation (Emlen 1970) or via the magnetic compass (W. Wiltschko and R. Wiltschko

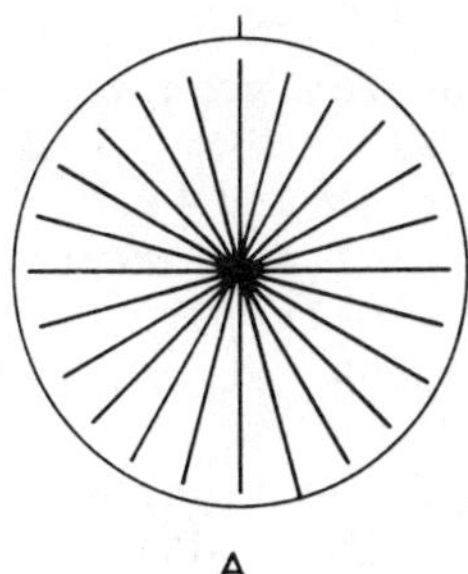

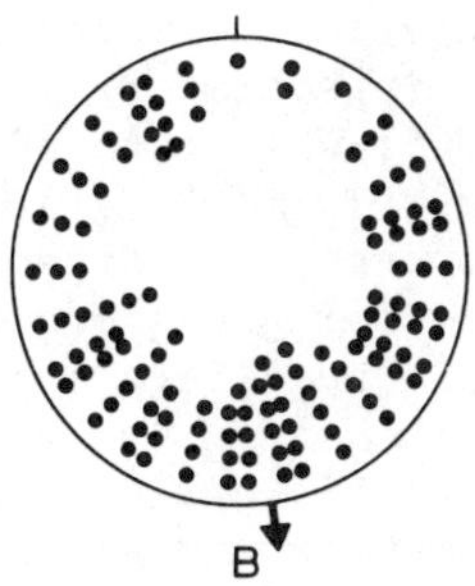

Figure 7. Directional distribution of migratory activity of indigo buntings during 122 bird-nights in Emlen funnels without visual clues (natural magnetic field). A. Frequency distribution of all activities per 15° sector; B. Mean bearings calculated from each of the individual bird-nights (mean of these means indicated by arrow; vector length ca. 0.25; $P<0.001$ by Rayleigh test). From Emlen (1975).

1975, 1976), or whether species-specific differences exist, is not yet clear.

The magnetic compass. For more than a decade, starting with Merkel and Fromme (1958), it was debated whether birds are able to select their migratory direction even in the absence of visual clues. Finally, it was concluded that they are able to do so, but it also became understandable why this had been doubted for such a long time. Directional preferences of Zugunruhe activity under such conditions are so weak that all the activity traces of one or several nights pooled result in an almost uniform angular distribution (Fig. 7A). However, consistent directional tendencies may become visible, if mean bearings calculated from the data of each individual bird-night are treated as units. These means are not randomly distributed, as they should be in the case of entirely disoriented behaviour, but reveal directional preferences which roughly coincide with the seasonally appropriate direction of migration (Fig. 7B). If directional preferences are hidden behind so much noise, one may expect that it depends on seemingly irrelevant details of the experimental design whether they become observable at all (cf. Wallraff 1972b). With visual clues (stars or star-like light spots) available the birds' movements in Merkel cages or Emlen funnels are much more concentrated in direction (cf. Fig. 6). The difference, however, appears to concern the ability to maintain a particular bearing over a length of time rather than the ability to select the bearing (R. Wiltschko and W. Wiltschko 1978).

Today it seems sufficiently established that this kind of

non-visual orientation is magnetic orientation, although almost all the positive findings came from only one research group (e.g., Merkel and Wiltschko 1965; Wiltschko 1968; W. Wiltschko and R. Wiltschko 1972; Emlen et al. 1976; Bingman 1981; Beck and Wiltschko 1982). The findings are, however, supported by data of a different kind obtained with free-flying homing pigeons (Keeton 1971, 1972; Walcott and Green 1974; Visalberghi and Alleva 1979; Wiltschko et al. 1981).

Wiltschko and his colleagues reveal two important characteristics of the magnetic compass of passerine birds: 1) its range of operation is restricted to particular levels of intensity, yet can be adapted to other levels by exposure of the birds to these conditions for several days (Wiltschko 1972, 1978); and 2) reversal of either the horizontal or the vertical component of the magnetic field reverses the direction preferred by the birds, whereas complete reversal of the whole magnetic vector is ineffective (W. Wiltschko and R. Wiltschko 1972; Beck and Wiltschko 1982). These results lead to the conclusion that the birds are unable to distinguish between the poles of the magnetic vector but use the angle of inclination for making magnetic information unambiguous (Fig. 8). Consequently, this kind of compass is ambiguous when the magnetic field lines are horizontal, as they are at the magnetic equator. Furthermore, it poses difficulties for transequatorial migrants, as an unchanged response would create reversed migration in the other hemisphere.

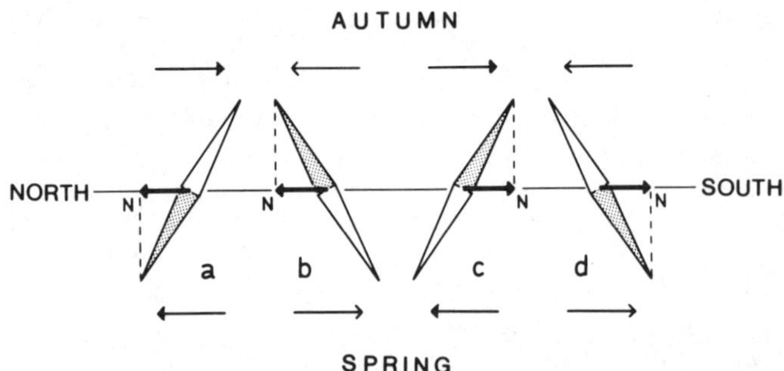

Figure 8. Scheme of the magnetic inclination compass according to Wiltschko and Wiltschko (1972). Compass needles are thought to be free to rotate in a vertical N-S plane; north-seeking pole dotted, heavy arrow indicating horizontal component of magnetic field vector. Top and bottom arrows show flight directions of birds in autumn and spring. a. Natural condition in median northern latitudes, birds fly normally; b. Vertical field component reversed, birds fly in opposite direction; c. Complete field vector reversed, birds fly normally; d. Horizontal field component reversed, birds fly oppositely.

Still a complete "black box" is the mechanism of magnetic perception in birds. Several hypotheses about various kinds of elementary physical processes potentially producing respective sensitivities have been published (e.g., Barnothy 1964; Leask 1977; Bamberger et al. 1978; Kirschvink and Gould 1981), but none of them has left the stage of academic speculation. Magnetite found in various animal tissues (e.g., Walcott and Walcott 1982) has not been shown thus far to be involved in magnetic field detection (but might well be). Preconditions for a mechanism based on electromagnetic induction, as described for elasmobranch fishes (Kalmijn 1978), appear more difficult for terrestrial animals than for marine animals, yet this does not exclude a different type of induction mechanism in birds. Attempts have been made to explain, by means of the proposed characteristics of the receptor system, the peculiarities of magnetic orientation, such as the enormous directional scatter in cage studies, the involvement of the inclination angle in horizontal orientation, and the apparent necessity to use visual clues in addition to the magnetic compass (Wallraff 1978b). These attempts are as speculative as all the other considerations in this field.

Mechanisms of Goal Orientation

This section deals almost exclusively with a domestic animal, the homing pigeon. There is so far no indication that this model bird uses a navigational mechanism different from that of other birds. Because of practical reasons, it is much better suited for experimental work than any species in the wild. Although pigeon homing does not involve a true migration, it may well be of interest with respect to both bird and fish migration. It is certainly true that goal-oriented navigation is an integral part of many migrations in both animal classes, yet it is difficult to investigate in its natural context. Thus it makes sense to first investigate homing behaviour as an isolated phenomenon. For that purpose the homing pigeon may be suited as the most closely related paradigmatic animal even with respect to fish navigation in open oceans or lakes.

Theoretical principles. Six theoretically possible methods of goal orientation may be considered: 1) animals may have direct sensory contact to the goal area or to stimuli emanating from there; 2) animals may find the goal by use of some random-search strategy; 3) animals may track their own traces (or those of colony members, etc.) coming from the goal, back to the goal; 4) animals may integrate the path they follow on leaving the goal and then return by means of dead reckoning; 5) animals may find their goal by use of some cognitive "map" based on site-specific features (landmarks) they learn by experience; and 6) animals may orientate by means of a bicoordinate or multicoordinate navigation system. None of these possibilities needs be realized in pure form; intermediate as well as hybrid methods are conceivable.

As pigeons orientate homeward over several hundreds of kilometres, direct sensory contact (item 1) can scarcely be assumed a realistic possibility. More specific findings show that it is also not primarily responsible for homing over shorter distances (see below, e.g. clock-shift experiments). Random search (item 2) cannot be the sole principle of homing, as the pigeons can be shown to be systematically goal-oriented from the beginning. Item 3 can be excluded because pigeons displaced by car do not leave traces; moreover, they do not follow their outward route during the return trip.

To be considered more seriously is the fourth possibility. Homing by path integration is widely distributed in arthropods (e.g., von Frisch 1967; Wehner 1982), but has also been shown in mammals and birds walking over distances from less than a metre to several hundreds of metres (Mittelstaedt and Mittelstaedt 1982; von Saint Paul 1982). In pigeons passively displaced over longer distances some effects of conditions during transportation to the release site have been observed (e.g., Papi et al. 1978a; Kiepenheuer 1980; W. Wiltschko and R. Wiltschko 1981; Baldaccini et al. 1982; Benvenuti et al. 1982). These findings, which almost exclusively concern short-term initial orientation, show no evidence of a path-integration system being at work. Severe interference with any kind of stimulus perception during displacement had no or very slight effects on subsequent initial orientation as well as return speeds and hence proved the navigational system not to depend on information gained during the outward journey (Wallraff 1980b; Wallraff et al. 1980). Assumptions that very young pigeons nevertheless might use a path-integration system (W. Wiltschko and R. Wiltschko 1982) are not based on conclusive experimental evidence. It should be emphasized, however, that from the existing data we are not allowed to conclude that some sort of path integration does not occur in free-living birds. It may well be involved when birds are not displaced passively, but perform active flights outwards and back. Yet we can conclude that pigeons are able to home without a dead-reckoning system being at work.

Hence we remain with the two last-mentioned possibilities (items 5 and 6) and thus with the assumption that pigeons are provided with some sort of map which enables localization of the site of momentary stay in relation to the home site. The two types of maps have been designated as "mosaic map" and "gradient map," respectively (Wallraff 1974, in press b; W. Wiltschko and R. Wiltschko 1978, 1982; see also "familiar area map" and "grid map" in Baker 1978, 1982). The first type (Fig. 9A,B) depends on individual experience of the spatial distribution of particular features. It is hardly conceivable to extend much beyond the range of an animal's preceding activities. The second type of map, in contrast, is thought to be based on at least two large fields of gradients which can be used to define the location of each site in terms of coordinates (Fig. 9C,D). Since pigeons are able to orientate homeward from sites far beyond the

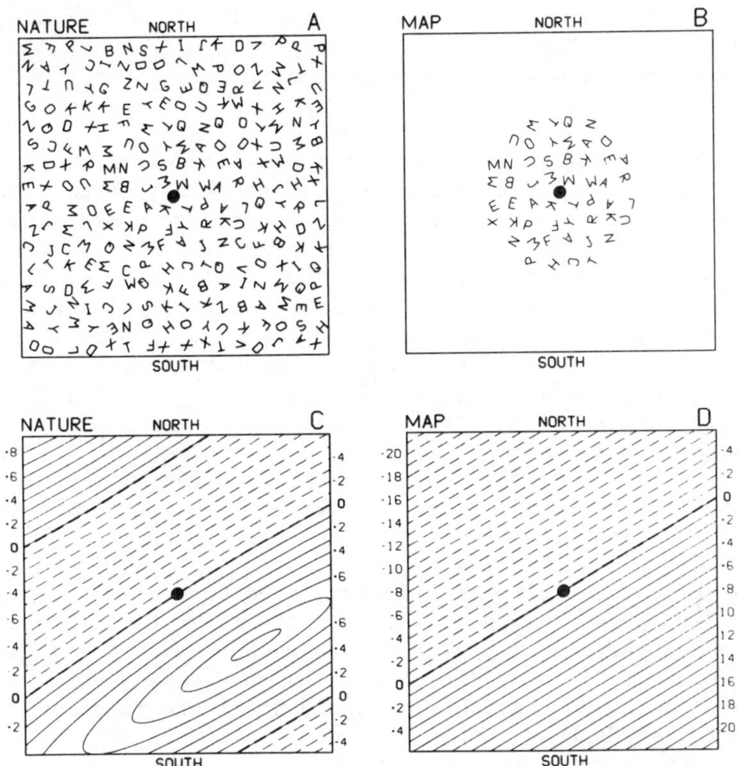

Figure 9. A. Mosaic of landmarks, symbolized by letters, surrounding an animal's home site (central dot). B. The corresponding mosaic map which is limited in extension in dependence of the animal's experience. C. Iso-lines (arbitrary units) of a field of gradients. The line running through the animal's home site is designated as 0, higher values indicated by solid and lower values by broken lines. D. The animal's corresponding gradient map as established by extrapolation of home-site conditions. Notice that in C and D, for the sake of simplicity, only one coordinate is shown. For complete site localization, at least two sorts of gradient fields are required with the gradients intersecting at sufficiently large angles. The sections shown are thought to be of different size; sides of the square in A,B may be in the order of maximally a few hundred, those in C,D of a few thousand, kilometres. Hence the map area in B may roughly coincide in size with the central dot in C and D. From Wallraff (in press b).

range they had ever visited before (see below), it is reasonable to hypothesize that they possess some sort of gradient map, possibly in conjunction with a mosaic map used at short distances.

If this conclusion is justified, we must further conclude that pigeons at the home site are able not only to determine the local scalar values of (at least) two coordinates but also to develop some idea of the compass directions of the respective coordinates (= gradients). Otherwise, when released at a distant site they would have to scan the field in order to find this direction. They do not do this, usually deciding immediately for a particular course. Furthermore, they are systematically deflected from this course if their sun compass is deflected due to an experimental clock shift (Schmidt-Koenig 1958, 1979; see also Keeton 1974; Wallraff 1974). All investigators agree therefore that pigeon navigation involves two steps according to Kramer's (1953b) "map-and-compass" concept. In a first step the bird localizes, by means of a map, its position in relation to home, and in a second step it determines, by means of a compass, the direction concluded to be appropriate for flight.

Basic features of pigeon homing. The majority of a group of pigeons displaced from their loft for the first time in their lives and released individually at a site somewhere abroad typically start flying in about the same direction. In most cases, this direction is neither directly homeward nor independent of the relation to home. Usually the birds tend to fly in a certain compass direction. This tendency can be stronger, equally strong or somewhat weaker than the tendency to fly toward home (for examples see Fig. 10). The strength relation between the two tendencies varies temporally, from loft site to loft site, and in dependence on individual homing experience, yet their basic existence is typical for initial orientation data in pigeons (e.g., Wallraff 1967, 1978a, 1982a, in press a; Windsor 1975). If such data from several release sites symmetrically distributed around home are combined, they reveal a homeward pointing component as well as a generally preferred compass direction (PCD) which may be different for pigeons settled at different home sites (lower diagrams in Fig. 10).

It is necessary to keep these general rules in mind in order to interpret the individual "release-site bias" of initial bearings (cf. Keeton 1974, 1980) at a given location. It must not automatically be interpreted as reflecting some site-specific physical peculiarity of the environment. In some cases such a peculiarity may be involved, but it can only be recognized if the general directional pattern of initial orientation around the respective loft is known and adequately considered. As long as a particular release-site bias can be explained as resulting from the relation between the direction toward home and the loft-specific PCD, local anomalies of navigational clues need not be presupposed (cf. Wallraff 1980a, 1982a, in press a).

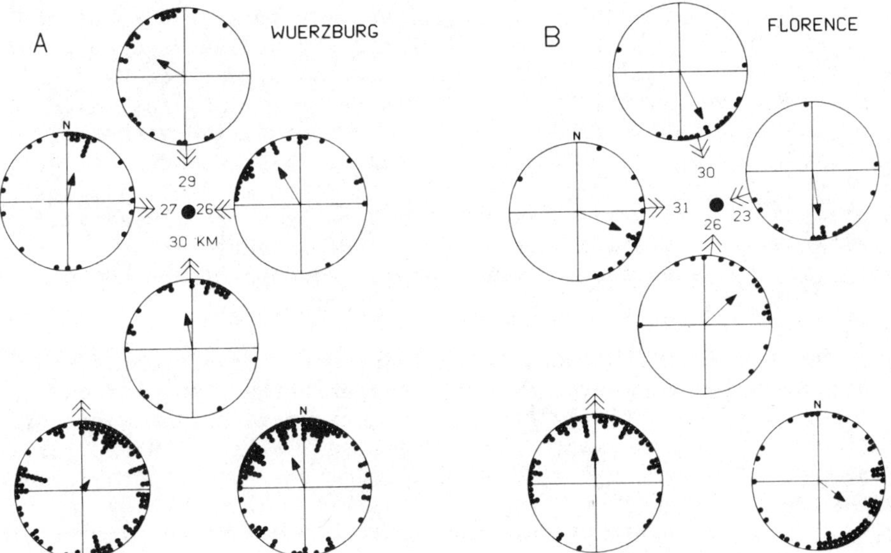

Figure 10. Initial orientation of inexperienced pigeons from lofts near Würzburg, Germany (A), and Florence, Italy (B). Each dot represents the vanishing bearing (field-glass observation) of one bird. Resulting mean vectors are shown as arrows (maximum length 1 = radius). Homeward directions (double-headed arrows) and distances in km (numbers) indicated. Data are shown for the 2 x 4 individual release sites as well as pooled with respect to home (lower left) and north (lower right). From Foà et al. (1982).

The PCDs in homing pigeons are probably homologous to similar directional preferences observed in ducks and other wild birds (Griffin and Goldsmith 1955; Bellrose 1958b; Matthews 1961, 1963a,b). There are reasons to assume that they do not reflect "nonsense orientation" (as they are termed by Matthews), but result from the particular navigational strategies used by the pigeons and probably by other birds as well (for detailed considerations, see Wallraff 1974, 1980a).

The sensory basis of pigeon navigation. Over many years the deductive approach was the most common way to search for environmental features that might provide the birds with sufficient navigational information. Researchers considered the various potentially useful physical parameters and tested whether birds make adequate use of them. Few suitable parameters are known, however, and most of them were favoured for some time by one or the other author. So far none of them can hold its favoured position. This is true for astronomical clues such as the sun (Matthews 1953, 1968; but see, e.g., Kramer 1961; Keeton 1974; Wallraff 1974; Schmidt-Koenig 1979) or the stars (Sauer 1957; Sauer and Sauer 1960; but see Wallraff 1960; Emlen 1967, 1975), and is also true for geophysical grids built up by the geomagnetic field, possibly in conjunction with the Coriolis force or other variables (Yeagley 1947; Gould 1982; Walcott 1982; but see Kramer 1948; Griffin 1952; Matthews 1968; Wallraff in press a). Similarly, hypotheses proposing inertial navigation (Barlow 1964), operating more or less independently of environmental clues, are not supported by experimental evidence (see above, *path integration*).

The inductive approach, being independent of anticipated plausibility, proved more fruitful. Experiments with pigeons confined in differentially shielded aviaries led to the conclusion that some atmospheric factors are involved (Wallraff 1966a, 1970). This conclusion remained rather vague, however, until a real breakthrough was achieved when Papi et al. (1971, 1972; for review see Papi 1976, 1982a) were able to show that atmospheric odours constitute a substantial basis of pigeon navigation. Homeward orientation of pigeons deprived of olfaction is highly and significantly reduced as compared with that of control birds. This can be seen in initial orientation (Fig. 11), in the distributions of recoveries (Fig. 12), and in homing success (see legend to Fig. 12). Disorientation can also be obtained by elimination of the somewhat larger molecules from the ambient air the pigeons breathe (Fig. 13).

It is not surprising that the idea of olfactory navigation in birds was generally received with scepticism. It was reasonable to try to explain the experimental findings on the basis of some nonspecific distraction or reduced motivation. However, attempts of this kind (e.g., Keeton et al. 1977; Keeton and Brown in Papi et al. 1978b; Keeton 1980) could not stand up to the increasing amount of favourable results collected over more than a decade. Claimed discrepancies between experimental results have been shown not to exist in reality (Wallraff 1980c, in press a; Papi 1982a). In his recent review Gould (1982) could uphold doubts about the relevance of olfaction in pigeon homing only by inaccurate representation or neglect of essential empirical findings (see Papi 1982b; Wallraff 1982b, in press a). At the moment I do not see any fact or serious argument contradicting the conclusion that olfaction is a substantial component of pigeon navigation which mediates the flow of information that is necessary to orientate the course homeward from distant unfamiliar areas (for extensive discussion of the matter, see Wallraff in press a). Experiments with two additional avian species indicate that the conclusion need not be restricted to pigeons (Fiaschi et al. 1974; Wallraff and Hund 1982).

Pigeons have been shown to navigate adequately by means of local olfactory stimuli at distances from home as large as 200, 300 and 500

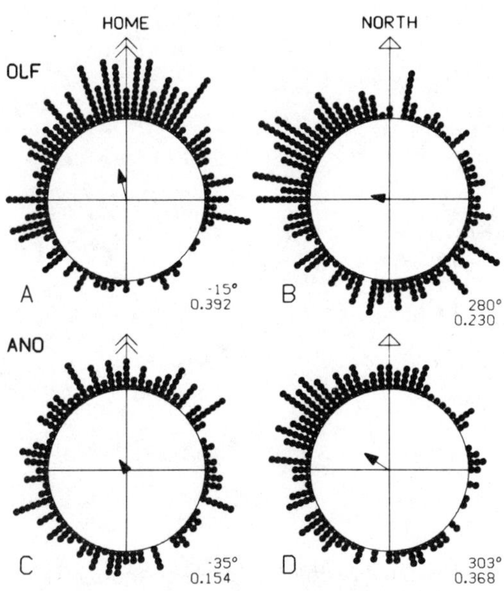

Figure 11. Initial orientation of inexperienced Würzburg pigeons, some with olfactory apparatus unaffected (OLF) and some made anosmic by bilateral olfactory nerve section (ANO). Vanishing bearings from eight centrally symmetrical release sites at distances of 180 km (6) and 300 km (2) pooled with respect to home (left) and north (right). Resulting mean vectors are shown as arrows and given numerically. Difference between A and C distribution is highly significant (P<0.0001; for further statistics see Wallraff, in press a: Fig. 2). Data from Wallraff (1980c, 1981b) and Wallraff and Foà (1982).

km. Smelling during transportation to those remote sites is not necessary (Wallraff 1980c, 1981b; Wallraff and Foà 1982). It has also been shown, however, that the navigational range is not unlimited in its spatial extension. Pigeons released at a particular site 700 km distant from their home loft were not homeward oriented unless they had opportunity to smell during the journey to this site (Ioalé et al. in press). Yet this distance is not always beyond the navigational range.

The functional system of pigeon navigation. Intuitively, nobody can imagine that odours might help a pigeon in finding its home over hundreds of kilometres. Nevertheless, empirical evidence leads to the conclusion that substances dispersed in the atmosphere contain suitable information to make this performance possible. Although at present it is unknown how a system based on this kind of information operates, some indications exist.

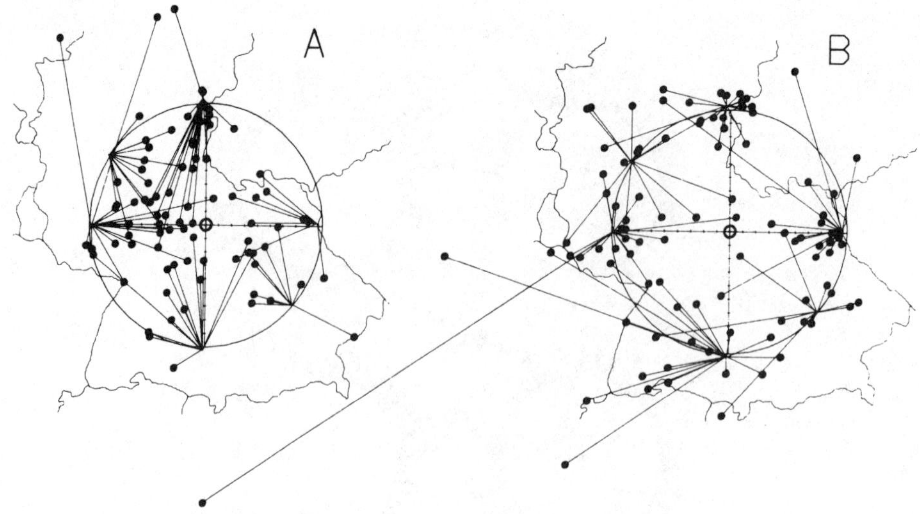

Figure 12. Maps showing recovery sites--connected with the respective release sites--of inexperienced pigeons displaced to six places about 180 km (radius of circle) distant from the home loft near Würzburg (centre). A. Control birds; B. Anosmic birds (bilateral olfactory nerve section). The borders of the Federal Republic of Germany are indicated. Not included in the figure are those pigeons which directly returned to the loft; they are 24% of the controls released, but none of the experimentals. After Wallraff (1980c, 1981b).

Homeward orientation of pigeons does not only depend on an intact olfactory apparatus and availability of ambient air at the release site, but also on preceding long-term access to natural wind conditions at the home site. This has been shown by use of pigeons settled in aviaries with various types of lateral shielding, partly with some additional arrangement. The most important types of experiments are shown in Fig. 14 and summarized as follows:

A. Complete screening with glass walls highly reduces homeward orientation, whereas visual shielding with permission of airflow does not (Wallraff 1966a, 1970, 1979).

B. Two-sided shieldings with an open "air corridor" along one axis not only led to some reduction of homeward orientation when the pigeons were released on an axis perpendicular to this corridor (Ioalé 1982), but also to a not yet understood reversal of the preferred compass direction (Wallraff 1979).

C. Clockwise or counterclockwise deflection of the wind caused

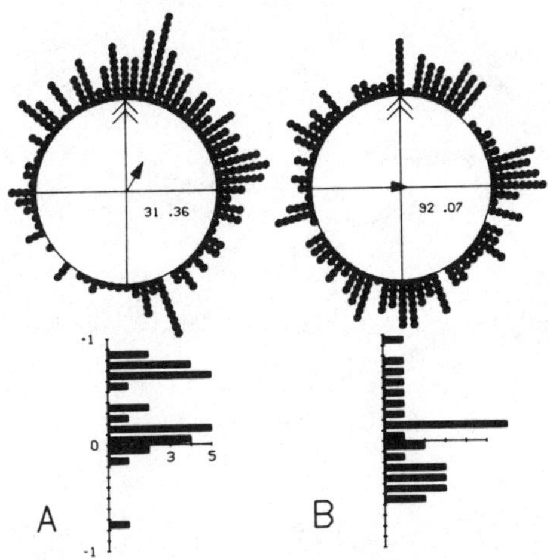

Figure 13. Effect of filtration of ambient air on initial orientation. During transportation and until individual release, pigeons were kept in airtight containers and ventilated with either normal (A) or charcoal-filtered (B) environmental air. Between removal from container and release, olfactory epithelia of each pigeon were locally anaesthetized so that after-treatment smelling was impossible for members of both groups. Circular diagrams show pooled vanishing bearings of 28 releases (14 pairs at centrally symmetrical sites); direction (in degrees) and length of mean vectors are given. Home direction points upward. Histograms show mean homeward components (= homeward pointing cosine components of mean vectors) per release. In 25 out of 28 releases the homeward component was lower in experimentals than in controls. Data from Wallraff and Foà (1981) and unpublished.

respective deflections of initial bearings (Baldaccini et al. 1975, 1978; Kiepenheuer 1978; Waldvogel et al. 1978). Although some findings complicating the matter were published recently (Kiepenheuer 1982; Phillips and Waldvogel 1982; Waldvogel and Phillips 1982), the basic conclusions from experiments with pigeons permanently living in deflector cages have so far not been shown to be invalid.

D. Reversal of natural winds led to initial bearings pointing preferably away from home (Ioalé et al. 1978; Ioalé 1980).

Together with the olfactory deprivation experiments these findings make sense if one assumes that the pigeons associate, at the home site, particular odorous situations with particular wind situations. They are thought to develop, by this way, an "olfactory

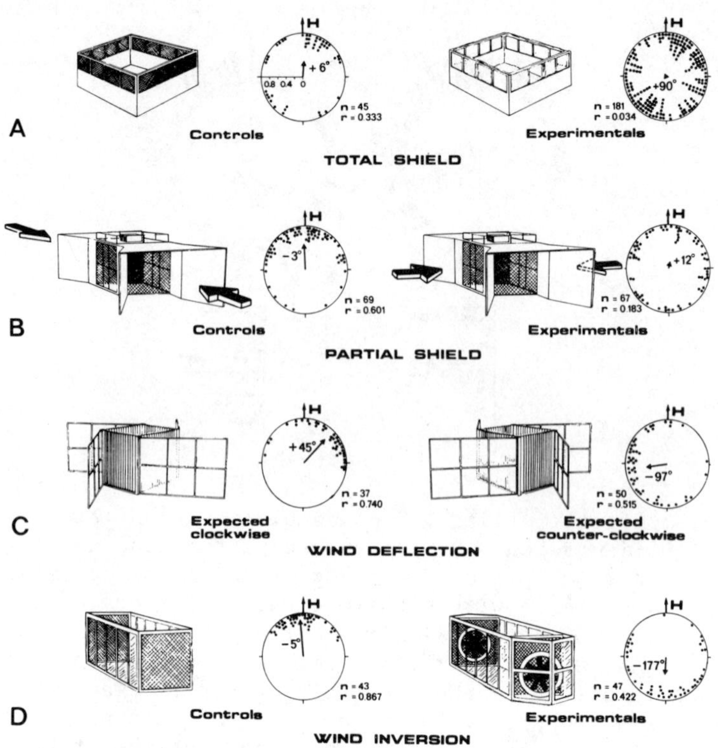

Figure 14. Summary of experiments varying wind direction at the home site. Aviary types are shown together with pooled initial-orientation data (H = home direction). A. Wire mesh vs. glass walls; B. Corridors open on either N-S or E-W axis (controls released at sites in prolongation of the respective axis, experimentals perpendicular to it); C. Clockwise vs. counterclockwise wind deflection; D. Corridor open to natural wind flow vs. wind reversed by fans. See text for explanation. From Papi (1982a).

map." This might be, as Papi (1976; Papi et al. 1972) assumes, a "mosaic map" (cf. Fig. 9B) which extends, due to odours carried by winds, to an area substantially larger than the home range covered by the pigeons' spontaneous flights. More probably, in my view, the pigeons are provided with a "gradient map" (cf. Fig. 9D) which seems better suited to cover a radius of some 500 km from home and which does not require highly developed olfactory pattern recognition. (For a more detailed discussion of the alternative hypotheses, see Wallraff 1980c; for the last-mentioned aspect, see also Wallraff and Hund 1982). Then, however, we would have to predict the existence of accordingly large fields of olfactory gradients which must be reliable enough to be evaluated in an appropriate manner despite the fact that meteorological conditions in the respective areas are not at all stable. I admit that this sounds unlikely, yet we must consider that pigeon homing behaviour is temporally variable as well, and some degree of compensation for environmental fluctuations might be performed by the navigational system (cf. Wallraff 1974).

CONCLUSION: BIRDS AND FISHES

Many aspects of migration and orientation in birds and fishes are directly comparable, yet only a few shall briefly be considered.

As a rule both animal groups adapt their migratory movements to the movements of the medium in which they live. Generally, adaptation appears closer in fishes than in birds, and this makes sense, because underwater conditions usually are less variable than atmospheric conditions. Water currents must be taken as unavoidable facts, yet due to their relative stability they can also be utilized as trustworthy guideways. Atmospheric currents are often less reliable, but also more avoidable, such that bird migration systems involve, by and large, somewhat more flexibility and independence of the medium. In birds entirely passive drift does not and cannot occur as birds cannot float in the air without horizontal movement with respect to the air. Nevertheless, wind drift and its possible utilization or compensation is one of the major problems in the study of bird migration.

Basic problems of compass orientation are similar for both groups of animals with the exception of stellar orientation, which never has been described in fishes and may be problematic underwater. In fact similar cage or tank experiments utilizing Zugunruhe, as well as training experiments, have been conducted or are ongoing in both groups. (Literature concerning fish orientation may be found elsewhere in this volume.)

When olfaction unexpectedly came onto the scene in bird navigation, it was immediately reminiscent of salmon migrations guided by olfactory stimuli. As far as both systems are known, however, they appear to operate differently. The salmon try to follow an olfactory gradient toward the source of the odorous signals. This is facilitated in the one-dimensional anadromous migrations where the gradient coincides with the upstream direction, but is also possible in open waters, provided the goal is in the centre of the gradient field. There are some indications of analogous behaviours in birds: petrels appear to find their nesting burrows as well as their breeding islands, over shorter distances, by following an olfactory gradient upwind (Grubb 1974, 1979). Long-distance navigation of pigeons, however, involves immediate homeward orientation even with winds blowing away from the goal, and other findings are also incompatible with the idea of direct scanning and following the gradient of a home-area smell.

If the conclusion is justified that pigeons make use of a presently unknown navigational system based on olfactory clues, it seems not too unrealistic to assume that a mechanism of similar nature may be used by fishes in their open-ocean migrations as well.

REFERENCES

Able, K.P. 1980. Mechanisms of orientation, navigation, and homing. Pages 283-373 *in* S.A. Gauthreaux, editor. Animal migration, orientation, and navigation. Academic Press, New York, New York, USA.

Able, K.P. 1982a. The effects of overcast skies on the orientation of free-flying nocturnal migrants. Pages 38-49 *in* F. Papi and H. G. Wallraff, editors. Avian navigation. Springer-Verlag, Berlin, Federal Republic of Germany.

Able, K.P. 1982b. Field studies of avian nocturnal migratory orientation. I. Interaction of sun, wind and stars as directional cues. Animal Behaviour 30:761-767.

Able, K.P. 1982c. Skylight polarization patterns at dusk influence migratory orientation in birds. Nature (London) 299:550-551.

Alerstam, T. 1979. Optimal use of wind by migrating birds: combined drift and overcompensation. Journal of Theoretical Biology 79: 341-353.

Baker, R.R. 1978. The evolutionary ecology of animal migration. Hodder & Stoughton, London, England.

Baker, R.R. 1982. Migration: Paths through time and space. Hodder & Stoughton, London, England.

Baldaccini, N.E., S. Benvenuti, V. Fiaschi, and F. Papi. 1975. Pigeon navigation: Effects of wind deflection at home cage on homing behaviour. Journal of Comparative Physiology A. Sensory, Neural, and Behavioral Physiology 99A:177-186.

Baldaccini, N.E., S. Benvenuti, V. Fiaschi, P. Ioalé, and F. Papi. 1978. Investigation of pigeon homing by means of "deflector cages." Pages 78-91 *in* K. Schmidt-Koenig and W.T. Keeton, editors. Animal migration, navigation, and homing. Springer-Verlag, Berlin, Federal Republic of Germany.

Baldaccini, N.E., S. Benvenuti, V. Fiaschi, P. Ioalé, and F. Papi. 1982. Pigeon orientation: Experiments on the role of olfactory stimuli perceived during the outward journey. Pages 160-169 *in* F. Papi and H.G. Wallraff, editors. Avian navigation. Springer-Verlag, Berlin, Federal Republic of Germany.

Bamberger, S., G. Valet, F. Storch, and G. Ruhenstroth-Bauer. 1978. Electromagnetically induced fluid streaming as a possible mechanism of the biomagnetic orientation of organisms. Zeitschrift für Naturforschung 33c:159-160.

Barlow, J.S. 1964. Inertial navigation as a basis for animal navigation. Journal of Theoretical Biology 6:76-117.

Barnothy, J.M. 1964. Proposed mechanisms for the navigation of migrating birds. Pages 287-293 *in* M.F. Barnothy, editor. Biological effects of magnetic fields. Plenum Press, New York, New York, USA.

Beck, W., and W. Wiltschko. 1982. The magnetic field as a reference system for genetically encoded migratory direction in pied flycatchers (*Ficedula hypoleuca* Pallas). Zeitschrift für Tierpsychologie 60:41-46.

Bellrose, F.C. 1958a. The orientation of displaced waterfowl in migration. Wilson Bulletin 70:20-40.

Bellrose, F.C. 1958b. Celestial orientation by wild mallards. Bird-Banding 29:75-90.

Bellrose, F.C., and R.D. Crompton. 1970. Migrational behavior of mallards and black ducks as determined from banding. Illinois Natural History Survey Bulletin 30:167-234.

Benvenuti, S., N.E. Baldaccini, and P. Ioalé. 1982. Pigeon homing: Effect of altered magnetic field during displacement on initial orientation. Pages 140-148 *in* F. Papi and H.G. Wallraff, editors. Avian navigation. Springer-Verlag, Berlin, Federal Republic of Germany.

Berthold, P. 1977. The "Vektoren-Navigations-Hypothese": Its validity for different populations of the same species. Naturwissenschaften 64:389.

Bingman, V.P. 1981. Savannah sparrows have a magnetic compass. Animal Behaviour 29:962-963.

Bingman, V.P., K.P. Able, and F. Kerlinger. 1982. Wind drift, compensation, and the use of landmarks by nocturnal bird migrants. Animal Behaviour 30:768-773.

Braemer, W. 1960. A critical review of the sun-azimuth hypothesis. Cold Spring Harbor Symposia on Quantitative Biology 25:413-427.

Bruderer. B. 1978. Effects of Alpine topography and winds on migrating birds. Pages 252-265 *in* K. Schmidt-Koenig and W.T. Keeton, editors. Animal migration, navigation, and homing. Springer-Verlag, Berlin, Federal Republic of Germany.

Bruderer, B. 1982. Do migrating birds fly along straight lines? Pages 3-14 *in* F. Papi and H.G. Wallraff, editors. Avian navigation. Springer-Verlag, Berlin, Federal Republic of Germany.

Dobben, W.H. van. 1955. Nature and strength of the attraction exerted by leading lines. Pages 165-166 *in* A. Portmann and E. Sutter, editors. Acta XI Congressus Internationalis Ornithologici. Birkhäuser, Basel, Switzerland.

Emlen, S.T. 1967. Migratory orientation in the indigo bunting, *Passerina cyanea*. Auk 84:309-342 and 463-489.

Emlen, S.T. 1970. Celestial rotation: its importance in the development of migratory orientation. Science (Washington DC) 170:1198-1201.

Emlen, S.T. 1975. Migration: orientation and navigation. Pages 129-219 *in* D.S. Farner and J.R. King, editors. Avian biology. volume 5. Academic Press, New York, New York, USA.

Emlen, S.T. 1980. Decision making by nocturnal bird migrants: The integration of multiple cues. Pages 553-560 *in* R. Nöhring, editor. Acta XVII Congressus Internationalis Ornithologici. Deutsche Ornithologen-Gesellschaft, Berlin, Federal Republic of Germany.

Emlen, S.T., and J.T. Emlen. 1966. A technique for recording migratory orientation of captive birds. Auk 83:361-367.

Emlen, S.T., W. Wiltschko, N.J. Demong, R. Wiltschko, and S. Bergman. 1976. Magnetic direction finding: Evidence for its use in migratory indigo buntings. Science (Washington DC) 193:505-508.

Fiaschi, V., A. Farina, and P. Ioalé. 1974. Homing experiments on swifts *Apus apus* (L.) deprived of olfactory perception. Monitore Zoologico Italiano (N.S.) 8:235-244.

Foà, A., H.G. Wallraff, P. Ioale, and S. Benvenuti. 1982. Comparative investigations of pigeon homing in Germany and Italy. Pages 232-238 *in* F. Papi and H.G. Wallraff, editors. Avian navigation. Springer-Verlag, Berlin, Federal Republic of Germany.

Frisch, K. von. 1967. The dance language and orientation of bees. Harvard University Press, Cambridge, Massachusetts, USA.

Gould, J.L. 1982. The map sense of pigeons. Nature (London) 296:205-211.

Griffin, D.R. 1952. Bird navigation. Biological Reviews 27:359-393.

Griffin, D.R., and T.H. Goldsmith. 1955. Initial flight directions of homing birds. Biological Bulletin 108:264-276.

Groot, C. 1965. On the orientation of young sockeye salmon (*Oncorhynchus nerka*) during their seaward migration out of lakes. Behaviour, Supplement 14:1-198.

Grubb, T.C. 1974. Olfactory navigation to the nesting burrow in Leach's petrel (*Oceanodroma leucorrhea*). Animal Behaviour 22: 192-202.

Grubb, T.C. 1979. Olfactory guidance of Leach's storm petrel to the breeding island. Wilson Bulletin 91:141-143.

Gwinner, E. 1968. Artspezifische Muster der Zugunruhe bei Laubsängern und ihre mögliche Bedeutung für die Beendigung des Zuges im Winterquartier. Zeitschrift für Tierpsychologie 25:843-853.

Gwinner, E. 1972. Endogenous timing factors in bird migration. Pages 321-338 *in* S.R. Galler et al., editors. Animal orientation and navigation. National Aeronautics and Space Administration (SP-262), Washington, District of Columbia, USA.

Gwinner, E. 1977. Circannual rhythms in bird migration. Annual Reviews of Ecology and Systematics 8:381-405.

Gwinner, E. 1981. Circannual systems. Pages 391-410 *in* J. Aschoff, editor. Handbook of behavioral neurobiology, volume 4: Biological rhythms. Plenum Press, New York, New York, USA.

Gwinner, E., and W. Wiltschko. 1978. Endogenously controlled changes in migratory direction of the garden warbler, *Sylvia borin*. Journal of Comparative Physiology A. Sensory, Neural, and Behavioral Physiology 125A:267-273.

Hoffmann, K. 1960. Experimental manipulation of the orientational clock in birds. Cold Spring Harbor Symposia on Quantitative Biology 25:379-387.

Hoffmann, K. 1965. Clock-mechanisms in celestial orientation of animals. Pages 426-441 *in* J. Aschoff, editor. Circadian clocks. North-Holland Publishing Company, Amsterdam, Holland.

Ioalé, P. 1980. Further investigation on homing behaviour of pigeons subjected to reversing of wind directions at the loft. Monitore Zoologico Italiano (N.S.) 14:77-87.

Ioalé, P. 1982. Pigeon homing: Effects of differential shielding of home cages. Pages 170-178 *in* F. Papi and H.G. Wallraff, editors. Avian navigation. Springer-Verlag, Berlin, Federal Republic of Germany.

Ioalé, P., F. Papi, V. Fiaschi, and N.E. Baldaccini. 1978. Pigeon navigation: Effects upon homing behaviour by reversing wind direction at the loft. Journal of Comparative Physiology A. Sensory, Neural, and Behavioral Physiology 128A:285-295.

Ioalé, P., H.G. Wallraff, F. Papi, and A. Foà. In press. Long-distance releases to determine the spatial range of pigeon navigation. Comparative Biochemistry and Physiology A.

Kalmijn, A.J. 1978. Experimental evidence of geomagnetic orientation in elasmobranch fishes. Pages 347-353 *in* K. Schmidt-Koenig and W.T. Keeton, editors. Animal migration, navigation, and homing. Springer-Verlag, Berlin, Federal Republic of Germany.

Keeton, W.T. 1971. Magnets interfere with pigeon homing. Proceedings of the National Academy of Sciences of the United States of America 68:102-106.

Keeton, W.T. 1972. Effects of magnets on pigeon homing. Pages 579-594 *in* S.R. Galler et al., editors. Animal orientation and navigation. National Aeronautics and Space Administration (SP-262), Washington, District of Columbia, USA.

Keeton, W.T. 1974. The orientational and navigational basis of homing in birds. Advances in the Study of Behavior 5:47-132.

Keeton, W.T. 1980. Avian orientation and navigation: New developments in an old mystery. Pages 137-157 *in* R. Nöhring, editor. Acta XVII Congressus Internationalis Ornithologici. Deutsche Ornithologen-Gesellschaft, Berlin, Federal Republic of Germany.

Keeton, W.T., M.L. Kreithen, and K.L. Hermayer. 1977. Orientation by pigeons deprived of olfaction by nasal tubes. Journal of Comparative Physiology A. Sensory, Neural, and Behavioral Physiology 114A:289-299.

Kiepenheuer, J. 1978. Pigeon homing: A repetition of the deflector loft experiment. Behavioral Ecology and Sociobiology 3:393-395.

Kiepenheuer, J. 1980. The importance of outward journey information in the process of pigeon homing. Pages 593-598 *in* R. Nöhring, editor. Acta XVII Congressus Internationalis Ornithologici. Deutsche Ornithologen-Gesellschaft, Berlin, Federal Republic of Germany.

Kiepenheuer, J. 1982. Pigeon orientation: A preliminary evaluation of factors involved or not involved in the deflector loft effect. Pages 203-210 *in* F. Papi and H.G. Wallraff, editors. Avian navigation. Springer-Verlag, Berlin, Federal Republic of Germany.

Kirschvink, J.L., and J.L. Gould. 1981. Biogenic magnetite as a basis for magnetic field detection in animals. BioSystems 13:181-201.

Kramer, G. 1948. Neue Beiträge zur Frage der Fernorientierung der Vögel. Ornithologische Berichte 1:228-238.

Kramer, G. 1949. Über Richtungstendenzen bei der nächtlichen Zugunruhe gekäfigter Vögel. Pages 269-283 *in* E. Mayr and E. Schüz, editors. Ornithologie als biologische Wissenschaft. Winter, Heidelberg, Federal Republic of Germany.

Kramer, G. 1951. Eine neue Methode zur Erforschung der Zugorientierung und die bisher damit erzielten Ergebnisse. Pages 269-280 *in* S. Hörstadius, editor. Proceedings of the Xth International Ornithological Congress. Almqvist & Wiksell, Uppsala, Sweden.

Kramer, G. 1952. Experiments on bird orientation. Ibis 94:265-285.

Kramer, G. 1953a. Die Sonnenorientierung der Vögel. Verhandlungen der Deutschen Zoologischen Gesellschaft Freiburg 1952:72-84.

Kramer, G. 1953b. Wird die Sonnenhöhe bei der Heimfindeorientierung verwertet? Journal für Ornithologie 94:201-219.

Kramer, G. 1961. Long-distance orientation. Pages 341-371 *in* A.J. Marshall, editor. Biology and comparative physiology of birds, volume 2. Academic Press, New York, New York, USA.

Leask, M.J.M. 1977. A physicochemical mechanism for magnetic field detection by migratory birds and homing pigeons. Nature (London) 267:144-145.

Liepa, V. 1978. Orientation of European robin in circular cages under artificial light clues. Pages 77-179 *in* H. Mihelsons, P. Blums and J. Baumanis, editors. *Orientatsija ptits*, Bird orientation. Zinatne, Riga, USSR (in Russian).

Matthews, G.V.T. 1953. Sun navigation in homing pigeons. Journal of Experimental Biology 30:243-267.

Matthews, G.V.T. 1961. "Nonsense" orientation in mallard *Anas platyrhynchos* and its relation to experiments on bird navigation. Ibis 103a:211-230.

Matthews, G.V.T. 1963a. "Nonsense" orientation as a population variant. Ibis 105:185-197.

Matthews, G.V.T. 1963b. The astronomical bases of "nonsense" orientation. Proceedings of the XIIIth International Ornithological Congress 415-429.

Matthews, G.V.T. 1968. Bird navigation. 2nd edition. Cambridge University Press, Cambridge, England.

Merkel, F.W., and H.G. Fromme. 1958. Untersuchungen über das Orientierungsvermögen nächtlich ziehender Rotkehlchen (*Erithacus rubecula*). Naturwissenschaften 45:499-500.

Merkel, F.W., and W. Wiltschko. 1965. Magnetismus und Richtungsfinden zugunruhiger Rotkehlchen (*Erithacus rubecula*). Vogelwarte 23: 71-77.

Mittelstaedt, H., and M.-L. Mittelstaedt. 1982. Homing by path integration. Pages 290-297 *in* F. Papi and H.G. Wallraff, editors. Avian navigation. Springer-Verlag, Berlin, Federal Republic of Germany.

Moore, F.R. 1980. Solar cues in the migratory orientation of the savannah sparrow (*Passerculus sandwichensis*). Animal Behaviour 28:684-704.

Mueller, H.C., and D.D. Berger. 1967. Wind drift, leading lines, and diurnal migration. Wilson Bulletin 79:50-63.

Papi, F. 1976. The olfactory navigation system of the homing pigeon. Verhandlungen der Deutschen Zoologischen Gesellschaft 69: 184-205.

Papi, F. 1982a. Olfaction and homing in pigeons: Ten years of experiments. Pages 149-159 *in* F. Papi and H.G. Wallraff, editors. Avian navigation. Springer-Verlag, Berlin, Federal Republic of Germany.

Papi, F. 1982b. The homing mechanism of pigeons. Nature (London) 300:293-294.

Papi, F., L. Fiore, V. Fiaschi, and S. Benvenuti. 1971. The influence of olfactory nerve section on the homing capacity of carrier pigeons. Monitore Zoologico Italiano (N.S.) 5:265-267.

Papi, F., L. Fiore, V. Fiaschi, and S. Benvenuti. 1972. Olfaction and homing in pigeons. Monitore Zoologico Italiano (N.S.) 6:85-95.

Papi, F., P. Ioalé, V. Fiaschi, S. Benvenuti, and N.E. Baldaccini. 1978a. Pigeon homing: Cues detected during the outward journey influence initial orientation. Pages 65-77 *in* K. Schmidt-Koenig and W.T. Keeton, editors. Animal migration, navigation, and homing. Springer-Verlag, Berling, Federal Republic of Germany.

Perdeck, A.C. 1958. Two types of orientation in migrating starlings, *Sturnus vulgaris* L., and chaffinches, *Fringilla coelebs* L., as revealed by displacement experiments. Ardea 46:1-37.

Perdeck, A.C. 1964. An experiment on the ending of autumn migration in starlings. Ardea 52:133-139.

Phillips, J.B., and J.A. Waldvogel. 1982. Reflected light cues generate the short-term deflector-loft effect. Pages 190-202 *in* F. Papi and H.G. Wallraff, editors. Avian navigation. Springer-Verlag, Berlin, Federal Republic of Germany.

Richardson, W.J. 1978. Timing and amount of bird migration in relation to weather: a review. Oikos 30:224-272.

Richardson, W.J. 1980. Autumn landbird migration over the western Atlantic Ocean as evident from radar. Pages 501-506 *in* R. Nöhring, editor. Acta XVII Congressus Internationalis Ornithologici. Deutsche Ornithologen-Gesellschaft, Berlin, Federal Republic of Germany.

Saint Paul, U. von. 1982. Do geese use path integration for walking home? Pages 298-307 *in* F. Papi and H.G. Wallraff, editors. Avian navigation. Springer-Verlag, Berlin, Federal Republic of Germany.

Sauer, F. 1957. Die Sternenorientierung nächtlich ziehender Grasmücken (*Sylvia atricapilla, borin* und *curruca*). Zeitschrift für Tierpsychologie 14:29-70.

Sauer, E.G.F., and E.M. Sauer. 1960. Star navigation of nocturnal

migrating birds: The 1958 planetarium experiments. Cold Spring Harbor Symposia on Quantitative Biology 25:463-473.
Schmidt-Koenig, K. 1958. Experimentelle Einflussnahme auf die 24-Stunden-Periodik bei Brieftauben und deren Auswirkungen unter besonderer Berücksichtigung des Heimfindevermögens. Zeitschrift für Tierpsychologie 15:301-331.
Schmidt-Koenig, K. 1979. Avian orientation and navigation. Academic Press, London, England.
Schüz, E. 1949. Die Spät-Auflassung ostpreussischer Jungstörche in West-Deutschland durch die Vogelwarte Rossitten 1933. Vogelwarte 15:63-78.
Schüz, E. 1950. Die Frühauflassung ostpreussischer Jungstörche in West-Deutschland durch die Vogelwarte Rossitten 1933-1936. Bonner zoologische Beitrage 1:239-253.
Schüz, E. 1971. Grundriss der Vogelzugskunde. Parey, Berlin, Federal Republic of Germany.
Schwassmann, H.O., and A.D. Hasler. 1964. The role of the sun's altitude in sun orientation of fish. Physiological Zoology 37: 163-178.
Sterbetz, I., and J. Szijj. 1968. Das Zugverhalten der Rothalsgans (*Branta ruficollis*) in Europa. Vogelwarte 24:266-277.
Thienemann, J. 1931. Vom Vogelzuge in Rossitten. Neumann Verlag, Neudamm, German Democratic Republic.
Vilks, I., Y. Katz, and V. Liepa. In press. The role of artificial light sources for the visual orientation of European robins (*Erithacus rubecula*). Acta XVIII Congressus Internationalis Ornithologici, Moscow, USSR.
Visalberghi, E., and F. Alleva. 1979. Magnetic influences on pigeon homing. Biological Bulletin 156:246-256.
Walcott, B., and C. Walcott. 1982. A search for magnetic field receptors in animals. Pages 338-343 *in* F. Papi and H.G. Wallraff, editors. Avian navigation. Springer-Verlag, Berlin, Federal Republic of Germany.
Walcott, C. 1982. Is there evidence for a magnetic map in homing pigeons? Pages 99-108 *in* F. Papi and H.G. Wallraff, editors. Avian navigation. Springer-Verlag, Berlin, Federal Republic of Germany.
Walcott, C., and R.P. Green. 1974. Orientation of homing pigeons altered by a change in the direction of an applied magnetic field. Science (Washington DC) 184:180-182.
Waldvogel, J.A., and J.B. Phillips. 1982. Pigeon homing: New experiments involving permanent-resident deflector-loft birds. Pages 179-189 *in* F. Papi and H.G. Wallraff, editors. Avian navigation. Springer-Verlag, Berlin, Federal Republic of Germany.
Waldvogel, J.A., S. Benvenuti, W.T. Keeton, and F. Papi. 1978. Homing pigeon orientation influenced by deflected winds at home loft. Journal of Comparative Physiology A. Sensory, Neural, and Behavioral Physiology 128A:297-301.
Wallraff, H.G. 1960. Does celestial navigation exist in animals?

Cold Spring Harbor Symposia on Quantitative Biology 25:451-461.
Wallraff, H.G. 1966a. Über die Heimfindeleistungen von Brieftauben nach Haltung in verschiedenartig abgeschirmten Volieren. Zeitschrift für vergleichende Physiologie 52:215-259.
Wallraff, H.G. 1966b. (Review of Groot, 1965) Berichte über die wissenschaftliche Biologie 265:485-487.
Wallraff, H.G. 1967. The present status of our knowledge about pigeon homing. Pages 331-358 *in* D.W. Snow, editor. Proceedings of the XIV International Ornithological Congress. Blackwell, Oxford, England.
Wallraff, H.G. 1968. Direction training of birds under a planetarium sky. Naturwissenschaften 55:235-236.
Wallraff, H.G. 1969. Über das Orientierungsvermögen von Vögeln unter natürlichen und künstlichen Sternenmustern. Dressurversuche mit Stockenten. Zoologischer Anzeiger, Supplement 32 (Verhandlungen der Deutschen Zoologischen Gesellschaft):348-357.
Wallraff, H.G. 1970. Weitere Volierenversuche mit Brieftauben: Wahrscheinlicher Einfluss dynamischer Faktoren der Atmosphäre auf die Orientierung. Zeitschrift für vergleichende Physiologie 68:182-201.
Wallraff, H.G. 1972a. An approach toward an analysis of the pattern recognition involved in the stellar orientation of birds. Pages 211-222 *in* J.R. Galler et al., editors. Animal orientation and navigation. National Aeronautics and Space Administration (SP-262), Washington, District of Columbia, USA.
Wallraff, H.G. 1972b. Nicht-visuelle Orientierung zugunruhiger Rotkehlchen (*Erithacus rebecula*). Zeitschrift für Tierpsychologie 30:374-382.
Wallraff, H.G. 1974. Das Navigationssystem der Vögel. Oldenbourg, München, Federal Republic of Germany.
Wallraff, H.G. 1977. Selected aspects of migratory orientation in birds. Vogelwarte 29 Sonderheft:64-76.
Wallraff, H.G. 1978a. Preferred compass directions in initial orientation of homing pigeons. Pages 171-183 *in* K. Schmidt-Koenig and W.T. Keeton, editors. Animal migration, navigation, and homing. Springer-Verlag, Berlin, Federal Republic of Germany.
Wallraff, H.G. 1978b. Proposed principles of magnetic field perception in birds. Oikos 30:188-194.
Wallraff, H.G. 1979. Goal-oriented and compass-oriented movements of displaced homing pigeons after confinement in differentially shielded aviaries. Behavioral Ecology and Sociobiology 5: 201-225.
Wallraff, H.G. 1980a. Homing strategy of pigeons and implications for the analysis of their initial orientation. Pages 604-608 *in* R. Nöhring, editor. Acta XVII Congressus Internationalis Ornithologici. Deutsche Ornithologen-Gesellschaft, Berlin, Federal Republic of Germany.
Wallraff, H.G. 1980b. Does pigeon homing depend on stimuli perceived during displacement? I. Experiments in Germany. Journal of

Comparative Physiology A. Sensory, Neural, and Behavioral Physiology 139A:193-201.

Wallraff, H.G. 1980c. Olfaction and homing in pigeons: Nerve-section experiments, critique, hypotheses. Journal of Comparative Physiology A. Sensory, Neural, and Behavioral Physiology 139A:209-224.

Wallraff, H.G. 1981a. Clock-controlled orientation in space. Pages 299-309 *in* J. Aschoff, editor. Handbook of Behavioral Neurobiology, volume 4: Biological rhythms. Plenum Press, New York, New York, USA.

Wallraff, H.G. 1981b. The olfactory component of pigeon navigation: Steps of analysis. Journal of Comparative Physiology A. Sensory, Neural, and Behavioral Physiology 143A:411-422.

Wallraff, H.G. 1982a. Homing to Würzburg: An interim report on long-term analyses of pigeon navigation. Pages 211-221 *in* F. Papi and H.G. Wallraff, editors. Avian navigation. Springer-Verlag, Berlin, Federal Republic of Germany.

Wallraff, H.G. 1982b. The homing mechanism of pigeons. Nature (London) 300:293.

Wallraff, H.G. In press a. Relevance of atmospheric odours and geomagnetic field to pigeon navigation: What is the 'map' basis? Comparative Biochemistry and Physiology A.

Wallraff, H.G. In press b. Theoretical aspects of avian navigation. Acta XVIII Congressus Internationalis Ornithologici, Moscow, USSR

Wallraff, H.G., and A. Foà. 1981. Pigeon navigation: Charcoal filter removes relevant information from environmental air. Behavioral Ecology and Sociobiology 9:67-77.

Wallraff, H.G., and A. Foà. 1982. The roles of olfaction and magnetism in pigeon homing. Naturwissenschaften 69:504-505.

Wallraff, H.G., and K. Hund. 1982. Homing experiments with starlings (*Sturnus vulgaris*) subjected to olfactory nerve section. Pages 313-318 *in* F. Papi and H.G. Wallraff, editors. Avian navigation. Springer-Verlag, Berlin, Federal Republic of Germany.

Wallraff, H.G., and J. Kiepenheuer. 1963. Migración y orientación en aves: Observaciones en otoño en el Sur-Oeste de Europa. Ardeola (Madrid) 8:19-40.

Wallraff, H.G., A. Foà, and P. Ioalé. 1980. Does pigeon homing depend on stimuli perceived during displacement? II. Experiments in Italy. Journal of Comparative Physiology A. Sensory, Neural, and Behavioral Physiology 139A:203-208.

Wehner, R. 1982. Himmelsnavigation bei Insekten. Orell Füssli, Zürich, Switzerland. (Neujahrsblatt der Naturforschenden Gesellschaft in Zurich 184:1-132).

Williams, T.C., and J.M. Williams. 1978. Orientation of transatlantic migrants. Pages 239-251 *in* K. Schmidt-Koenig and W.T. Keeton, editors. Animal migration, navigation, and homing. Springer-Verlag, Berlin, Federal Republic of Germany.

Williams, T.C., J.M. Williams, L.C. Ireland, and J.M. Teal. 1977. Autumnal bird migration over the western North Atlantic Ocean. American Birds 31:251-267.

Wiltschko, R. 1980. Die Sonnenorientierung der Vögel. I. Die Rolle der Sonne im Orientierungssystem und die Funktionsweise des Sonnenkompass. Journal für Ornithologie 121:121-143.

Wiltschko, R. 1981. Die Sonnenorientierung der Vögel. II. Entwicklung des Sonnenkompass und sein Stellenwert im Orientierungssystem. Journal für Ornithologie 122:1-22.

Wiltschko, R., and W. Wiltschko. 1978. Relative importance of stars and the magnetic field for the accuracy of orientation in night-migrating birds. Oikos 30:195-206.

Wiltschko, R., and W. Wiltschko. 1980. The process of learning sun compass orientation in young homing pigeons. Naturwissenschaften 67:512-513.

Wiltschko, R., D. Nohr, and W. Wiltschko. 1981. Pigeons with a deficient sun compass use the magnetic compass. Science (Washington DC) 214:343-345.

Wiltschko, W. 1968. Über den Einfluss statischer Magnetfelder auf die Zugorientierung der Rotkehlchen (*Erithacus rubecula*). Zeitschrift für Tierpsychologie 25:537-558.

Wiltschko, W. 1972. The influence of magnetic total intensity and inclination on directions preferred by migrating European robins (*Erithacus rubecula*). Pages 569-578 *in* S.R. Galler et al., editors. Animal orientation and navigation. National Aeronautics and Space Administration (SP-262), Washington, District of Columbia, USA.

Wiltschko, W. 1978. Further analysis of the magnetic compass of migratory birds. Pages 302-310 *in* K. Schmidt-Koenig and W.T. Keeton, editors. Animal migration, navigation, and homing. Springer-Verlag, Berlin, Federal Republic of Germany.

Wiltschko, W. 1980. The relative importance and integration of different directional cues during ontogeny. Pages 561-565 *in* R. Nöhring, editor. Acta XVII Congressus Internationalis Ornithologici. Deutsche Ornithologen-Gesellschaft, Berlin, Federal Republic of Germany.

Wiltschko, W. 1982. The migratory orientation of garden warblers, *Sylvia borin*. Pages 50-58 *in* F. Papi and H.G. Wallraff, editors. Avian navigation. Springer-Verlag, Berlin, Federal Republic of Germany.

Wiltschko, W., and R. Wiltschko. 1972. Magnetic compass of European robins. Science (Washington DC) 176:62-64.

Wiltschko, W., and R. Wiltschko. 1975. The interaction of stars and magnetic field in the orientation system of night migrating birds. Zeitschrift für Tierpsychologie 37:337-355 and 39:265-282.

Wiltschko, W., and R. Wiltschko. 1976. Interrelation of magnetic compass and star orientation in night-migrating birds. Journal of Comparative Physiology A. Sensory, Neural, and Behavioral Physiology 109A:91-99.

Wiltschko, W., and R. Wiltschko. 1978. A theoretical model of migratory orientation and homing in birds. Oikos 30:177-187.

Wiltschko, W., and R. Wiltschko. 1981. Disorientation of inexperienced young pigeons after transportation in total darkness. Nature (London) 291:433-434.

Wiltschko, W., and R. Wiltschko. 1982. The role of outward journey information in the orientation of homing pigeons. Pages 239-252 *in* F. Papi and H.G. Wallraff, editors. Avian navigation. Springer-Verlag, Berlin, Federal Republic of Germany.

Wiltschko, W., E. Gwinner, and R. Wiltschko. 1980. The effect of celestial cues on the ontogeny of non-visual orientation in the garden warbler (*Sylvia borin*). Zeitschrift für Tierpsychologie 53:1-8.

Windsor, D.M. 1975. Regional expression of directional preferences by experienced homing pigeons. Animal Behaviour 23:335-343.

Yeagley, H.L. 1947. A preliminary study of a physical basis of bird navigation. Journal of Applied Physics 18:1035-1063.

Zink, G. 1973, 1975, 1982. Der Zug europäischer Singvögel (1.-3. Lieferung). Vogelzug-Verlag, Möggingen, Federal Republic of Germany.

FISH MIGRATION STUDIES: FUTURE DIRECTIONS

James D. McCleave

Department of Zoology and
Migratory Fish Research Institute
University of Maine
Orono, Maine 04469 USA

F.R. Harden Jones

Fisheries Laboratory
Ministry of Agriculture, Fisheries and Food
Lowestoft, Suffolk NR33 OHT England

W.C. Leggett

Department of Biology
McGill University
1205 Avenue Docteur Penfield
Montreal, Quebec H3A 1B1 Canada

T.G. Northcote

Institute of Animal Resource Ecology
University of British Columbia
Vancouver, British Columbia V6T 1W5 Canada

ABSTRACT

General recommendations for future research arising out of a NATO Advanced Research Institute "Mechanisms of Migration in Fishes" relate to approaches and methods as well as examination of hypotheses. Whenever possible experimental designs should be couched in terms of testable hypotheses in conceptually complete frameworks. The spatial and temporal scale of observations of environmental factors should be appropriate to the scale of fish movements being observed. Modeling, rigorous statistical analysis, and careful choice of experimental techniques should become routine parts of future research. Development and use of sophisticated remote-sensing techniques for environmental and biological observations are to be welcomed, as are imaginative new uses of simple manipulative or experimental techniques.

Greater understanding of the learning process relative to migratory mechanisms is needed. Definitive work to document the use in migration of various orienting mechanisms is lacking, and the extent to which a modest directional bias can account for migration needs to be investigated. Little is known about redundancy or hierarchical organization of orienting mechanisms in fish migration. Likewise, little is known of the ontogenetic and cyclic changes in sensory systems that occur in migratory species.

The extent to which true passive drift can account for migration is largely undocumented. Selective drift is an important process, especially in tidal waters, but the cues which trigger the selective movements have not been identified. The ontogeny of locomotor performance and orienting ability needs study in relation to the interplay of transport by water currents and by active swimming.

INTRODUCTION

One aim of this conference[1] was to make recommendations for future research in fish migration. This aim has been met, at least in part, in two ways. First, many of the authors in this volume have proposed specific, testable hypotheses to advance knowledge in a particular line of research. Their papers should be read in that regard. Second, a more general set of recommendations arose from panel discussions on open-ocean, coastal, and riverine migration held on the last day of the conference. Many of these general recommendations came from more than one panel in similar form. Here we present the recommendations in an integrated form.[2]

---

[1] A NATO Advanced Research Institute "Mechanisms of Migration in Fishes," held 13-17 December 1982 in Acquafredda di Maratea, Italy.

[2] While we accept the responsibility for erroneous interpretation of the panel discussions, we gratefully acknowledge that most of the ideas herein were freely given by all 43 participants at the conference. Further use of the term "we" in this paper is construed in this universal sense.

EXPERIMENTAL DESIGN

There is a strong conviction that more attention must be given to the clear definition of specific key questions, followed by careful design of experimental approaches. For some species at least, there is now a sufficient body of knowledge on their migrations to obviate the need for data gathering for data gathering's sake. While specific points may need further factual support, this should not prevent us from using what is already known in formulating testable hypotheses. Hypotheses should, ideally, be formulated in a conceptually complete framework (see van der Steen 1984--this volume). Then appropriate experimental designs and experimental fishes can be chosen. Adherence to this recommendation will yield both advances in our understanding of fish migrations and significant reductions in the cost of this understanding. Quinn's (1984a,b--this volume) papers seem instructive in this regard.

For some species and some geographic areas there is still a need to conduct basic descriptive studies before significant testable hypotheses and critical experiments can be developed. An outstanding example is provided by the highly migratory characins and catfishes of South American rivers (Northcote 1984--this volume). However, even these studies can be conducted in the best hypothetical framework available without violating the tenets of the preceding paragraph.

THE SCALE OF OBSERVATIONS

Studies of migrations of fishes cover a range of time and distance scales from short (hours and tens of meters), through medium (days and tens of kilometers), to long (months to years and hundreds to thousands of kilometers). There is a great need--pervading most of our migration research--for a more fundamental understanding of the relationship between physical and biological features of the environment and observed behavior. This understanding will be achieved only if observations of both the physical and biological environment and of migratory behavior are taken at scales of time and space which are relevant to the processes under investigation. Studies to date have generally been negligent in this regard, but some notable exceptions are Laurs and Lynn (1977), Fortier and Leggett (1982, 1983) and Arnold and Cook (1984--this volume).

METHODS

Data Collection

Environmental data (such as water current velocity, temperature and salinity) and biological observations (such as depth of fish, swimming speed, direction of travel) should be taken, then, on appropriate time and space scales. On the short scale this leads to the need for detailed physical and chemical information at the time and position at which the biological observations are made. Westerberg (1984--this volume) makes us aware of the richness of environmental information available on this short scale. Environmental information on this scale is important in rivers, coastal waters and even open oceans (e.g. see Johnsen 1984--this volume; Dodson and Dohse 1984--this volume; Cook 1984--this volume, respectively). The requirement for detail relaxes as the resolution of the biological requirement decreases; on the macroscale we need not "...keep track of every dart and turn of a fish" (DeAngelis and Yeh 1984--this volume), although long-range migrations may involve a more or less continuous process of decision making by the fish. The time scales on which environmental information is important may be longer than heretofore realized (Mysak 1984--this volume).

Statistics

It is important that the appropriate statistical support be available both in the planning and execution of research on fish migrations. It is helpful to make the distinction among statistics relating to experimental design and data collection, descriptive statistics and inferential statistics. Statistical help should be actively sought and heeded. Lack of statistical rigor may be one of the greatest weaknesses in previous work. Statistical research per se is also needed, as despite substantial recent development (e.g. Batschelet 1981), we still must on occasion simply throw up our hands (e.g. McCleave and Kleckner 1982).

Models

Models can be helpful at most stages of research. It is recommended that explicit models be constructed, as a matter of course, to provide the framework for synthesis and evaluation of ideas about migratory mechanisms. Such models offer the promise of a cost-effective basis for screening hypotheses (Leggett 1984--this volume; Neill 1984--this volume). Those that are likely to fail owing to inconsistencies and insufficiencies can be eliminated from further consideration. Hypotheses that remain may be sharpened further by the modeling process. Innovative models may even suggest migratory mechanisms that are, at present, entirely unsuspected (Leggett 1984--this volume). It is important, when appropriate, that the results derived from models should be subjected to no less statistical rigor than the results of field or laboratory experiments.

Examples of several modeling approaches were presented at the conference, e.g. by Balchen (1979), Arnold and Cook (1984--this volume), Cook (1984--this volume), DeAngelis and Yeh (1984--this volume), Dodson and Dohse (1984--this volume), Power (1984--this volume).

Techniques

Although there is an obvious need to stress experimental studies, we must clearly understand how our experimental approaches, techniques and equipment may limit or bias the responses which the animals show. For example, there are serious limitations with arenas (Quinn 1984a--this volume) and Y-mazes; slight changes in design of an apparatus can greatly alter results of behavioral experiments (Bitterman 1984--this volume).

While modern, sophisticated techniques such as satellite imagery (Lynn 1984--this volume) should be exploited to a greater degree, there remains a great need for simple manipulative and experimental approaches, such as transplantation experiments (Brannon 1984--this volume; Ogden and Quinn 1984--this volume), imaginatively applied to questions of migration.

Attention should be given to the development of techniques which would allow the movement of several fish to be tracked simultaneously in the sea or in coastal waters. Also seriously needed are techniques which would allow tracking time to be extended from the relatively short scale now possible to the medium or long time scales. These developments would allow more detailed observations of behavior to be extended into the time scale normally associated with changes of fish distribution.

Incorporation of sensors for well-chosen biological and environmental factors into telemetry tags would provide data collected at the point of the fish, so important on the shorter time and space scales (see above). While a considerable effort has been expended in developing telemetering tags for temperature, pressure, salinity, heart-rate, etc., only temperature transmitters have received extensive use in the field. Further development and use should be encouraged. That such development is important is shown by the fact that a relatively crude compass-direction transmitter is already allowing a new look at the possibility of inertial mechanisms in short-term movements of fish (Harden Jones 1984a--this volume).

LEARNING

There is a strong belief that we need a greater understanding of the importance of learning processes in fish migration. Among the specific requirements identified are a greater knowledge of the importance of developmental sequences in learning and in sensitivity to particular stimuli which may provide directional clues. Although "imprinting" is an explicit learning process in the parent-stream odor hypothesis of salmonid homing (e.g. Hara et al. 1984--this volume; Johnsen 1984--this volume), the process which occurs is poorly known, when and how rapidly it occurs needs clarification, and how it is affected by the physiological state of the fish awaits investigation. Associative learning is the integral feature of one hypothesis of orientation in tidal waters (Dodson and Dohse 1984--this volume). The learning components of most of our hypotheses concerning migration are perhaps the least investigated ones.

Directly linked to the question of learning is the question of whether a time-distance sense exists in fishes (Ogden and Quinn 1984--this volume), whether it is universal, and given its existence, the time and distance scales beyond which it breaks down.

ORIENTING MECHANISMS

There is a pressing need for definitive work to demonstrate the existence and role of various guiding, orienting and navigational mechanisms used by fishes to achieve directional bias and movement. Both Harden Jones (1984b--this volume) and Leggett (1984--this volume) have warned us not to be overly zealous in a search for precise mechanisms of orientation and navigation and not to ignore the behavior of fish that do not perform to the specifications of our hypotheses. The degree to which a modest directional bias can account for observed migrations should be examined.

Given the redundancy which has become evident in the orienting ability of birds (see Wallraff 1984--this volume) and the considerable array of potential orienting mechanisms suggested to date for fishes, research should be directed not only to the identification of these mechanisms but also to a study of the temporal sequence of their use. Included in this category are inertial navigation (Harden Jones 1984a--this volume), sun orientation, geomagnetic (Quinn 1984a--this volume; Walker 1984--this volume) and geoelectric orientation, polarized-light orientation, chemical (Hara et al. 1984--this volume; Johnsen 1984--this volume) and visual (and auditory?) imprinting, and behavioral regulation of components of the environment (Neill 1984--this volume). There is a wide gap, except with regard to olfactory orientation of homing salmonids, between laboratory studies of a sensory capability and field demonstration of use of that capability in migration.

Despite, or perhaps because of, the extensive work on olfactory-mediated homing in response to home-stream odors or pheromones in the last 30 years, many questions remain open. The behavioral studies implicating olfaction in homing have seemingly built a convincing case (Johnsen 1984--this volume). But many of the behavioral studies linking homing with a parent-stream odor have depended on the synthetic chemical morpholine, whose effectiveness as an olfactory stimulant is still in dispute (Hara et al. 1984--this volume). Furthermore, the high sensitivity of the gustatory sense, which has been little studied to date, makes it a contender for a role in chemical orientation (Hara et al. 1984--this volume). Participation by natural-products chemists in future studies might focus our attention back on the bouquet of natural waters.

For most postulated sensory mechanisms used in migration the ontogenetic and cyclic changes in the sensory systems are unknown, at least relative to migration. Research needs to be focused on these developmental and physiological (hormonal) changes in sensory mechanisms. More specifically we require a greater understanding of: 1) changes in sensitivity occurring at certain life history stages (e.g. nonmigratory parr-migratory smolts of salmonids [Thorpe 1984--this volume]; nonmigratory-migratory adult anguillids [Pankhurst 1984--this volume]); 2) changes in capability under specific stimulus conditions (e.g. freshwater-saltwater transition during migration); and 3) daily and seasonal changes in sensitivity.

PASSIVE AND SELECTIVE DRIFT

Despite our general assumption that many legs of migration, especially by larval fishes, are accomplished by passive drift, there are few good documentations for oceans (John 1984--this volume) or rivers (Northcote 1984--this volume). Cooperative research involving both biological and physical oceanographers and limnologists is needed to determine the extent to which migrations may be explained by passive drift.

Given the importance of behavioral interaction of migrating fishes with tidal and residual currents in coastal waters emphasized at the conference (McCleave and Kleckner 1982; Fortier and Leggett 1983; Arnold and Cook 1984--this volume; Dodson and Dohse 1984--this volume; Miller et al. 1984--this volume), we stress the importance of describing the cues which trigger the vertical (or other) movements and the clues for vertical placement of fishes whose migrations occur by transport in tidal or residual streams. There is a great need to explain these behavioral responses in terms of sensory physiology and endocrinology.

Related to the question of passive and selective drift is the need to know: 1) the ontogeny of locomotor performance in migratory fishes; and 2) the precision with which fish can hold a directional heading over time. These determinations are pertinent to the extent to which fish migrations may be independent of the currents in the environment.

Because of the importance of energetics in the migrations of fishes in general, and the considerable energy savings that may be achieved by passive, or partially passive, transport (Weihs 1984--this volume), further studies into the energetic consequences of the partitioning of time during migration between passive transport and active, oriented movement are strongly recommended. One conceptual approach might be based upon the assumption that energetic efficiency of transport (migration) increases exponentially for active, oriented movement and decreases exponentially for passive (or selective passive) transport as the ratio of swimming speed to current speed increases.

REFERENCES[1]

Arnold, G.P., and P.H. Cook. 1984. Fish migration by selective tidal stream transport: First results with a computer simulation model for the European continental shelf. Pages 227-261 *in* this volume.

Balchen, J.G. 1979. Modeling, prediction, and control of fish behavior. Pages 99-146 *in* C.T. Leondes, editor. Control and dynamic systems, volume 15. Academic Press, New York, USA.

Batschelet, E. 1981. Circular statistics in biology. Academic Press, London, England.

Bitterman, M.E. 1984. Migration and learning in fishes. Pages 397-420 *in* this volume.

Brannon, E.L. 1984. Influence of stock origin on homing behavior of salmon. Pages 103-111 *in* this volume.

Cook, P.H. 1984. Directional information from surface swell: Some possibilities. Pages 79-101 *in* this volume.

DeAngelis, D.L., and G.T. Yeh. 1984. An introduction to modeling fish migratory behavior. Pages 445-469 *in* this volume.

Dodson, J.J., and L.A. Dohse. 1984. A model of olfactory-mediated conditioning of directional bias in fish migrating in reversing tidal currents based on the homing migration of American shad (*Alosa sapidissima*). Pages 263-281 *in* this volume.

Fortier, L., and W.C. Leggett. 1982. Fickian transport and the dispersal of fish larvae in estuaries. Canadian Journal of Fisheries and Aquatic Sciences 39:1150-1163.

Fortier, L., and W.C. Leggett. 1983. Vertical migrations and transport of larval fish in a partially mixed estuary. Canadian Journal of Fisheries and Aquatic Sciences 40:1543-1555.

Hara, T.J., S. Macdonald, R.E. Evans, T. Marui, and S. Arai. 1984. Morpholine, bile acids and skin mucus as possible chemical cues in salmonid homing: electrophysiological re-evaluation. Pages 363-378 *in* this volume.

---

[1] All references to this volume are:
McCleave, J.D., G.P. Arnold, J.J. Dodson, and W.H. Neill, editors. Mechanisms of migration in fishes. Plenum Press, New York, New York, USA.

Harden Jones, F.R. 1984a. Could fish use inertial clues when on migration? Pages 67-78 *in* this volume.

Harden Jones, F.R. 1984b. A view from the ocean. Pages 1-26 *in* this volume.

John, H-Ch. 1984. Drift of larval fishes in the ocean: results and problems from previous studies and a proposed field experiment. Pages 39-59 *in* this volume.

Johnsen, P.B. 1984. Establishing the physiological and behavioral determinants of chemosensory orientation. Pages 379-385 *in* this volume.

Laurs, R.M., and R.J. Lynn. 1977. Seasonal migration of North Pacific albacore, *Thunnus alalunga*, into north American coastal waters: Distribution, relative abundance and association with Transition Zone waters. US National Marine Fisheries Service Fishery Bulletin 75:795-822.

Leggett, W.C. 1984. Fish migrations in coastal and estuarine environments: a call for new approaches to the study of an old problem. Pages 159-178 *in* this volume.

Lynn, R. 1984. Aspects of physical oceanography as they relate to the migration of fish. Pages 471-486 *in* this volume.

McCleave, J.D., and R.C. Kleckner. 1982. Selective tidal stream transport in the estuarine migration of glass eels of the American eel (*Anguilla rostrata*). Journal du Conseil Conseil international pour l'Exploration de la Mer 40:262-271.

Miller, J.M., J.P. Reed, and L.J. Pietrafesa. 1984. Patterns, mechanisms and approaches to the study of migrations of estuarine-dependent fish larvae and juveniles. Pages 209-225 *in* this volume.

Mysak, L.A. 1984. Large-scale interannual fluctuations in ocean parameters and their influence on fish populations. Pages 205-207 *in* this volume.

Northcote, T.G. 1984. Mechanisms of fish migration in rivers. Pages 317-355 *in* this volume.

Neill, W.H. 1984. Behavioral enviroregulation's role in fish migration. Pages 61-66 *in* this volume.

Ogden, J.C., and T.P. Quinn. 1984. Migration in coral reef fishes: ecological significance and orientation mechanisms. Pages 293-308 *in* this volume.

Pankhurst, N.W. 1984. Artificial maturation as a technique for investigating adaptations for migration in the European eel *Anguilla anguilla* (L.). Pages 143-157 *in* this volume.

Power, J.H. 1984. Advection, diffusion, and larval fish drift migrations. Pages 27-37 *in* this volume.

Quinn, T.P. 1984a. An experimental approach to fish compass and map orientation. Pages 113-123 *in* this volume.

Quinn, T.P. 1984b. Homing and straying in Pacific salmon. Pages 357-362 *in* this volume.

Steen, W.J. van der. 1984. Methodological aspects of migration and orientation in fishes. Pages 421-444 *in* this volume.

Thorpe, J.E. 1984. Downstream movements of juvenile salmonids: a forward speculative view. Pages 387-396 *in* this volume.

Walker, M.M. 1984. Magnetic sensitivity and its possible physical basis in the yellowfin tuna, *Thunnus albacares*. Pages 125-141 *in* this volume.

Wallraff, H-G. 1984. Migration and navigation in birds: a present-state survey with some digressions to related fish behavior. Pages 509-544 *in* this volume.

Weihs, D. 1984. Bioenergetic considerations in fish migration. Pages 487-508 *in* this volume.

Westerberg, H. 1984. The orientation of fish and the vertical stratification at fine- and microstructure scales. Pages 179-203 *in* this volume.

GEOGRAPHIC AND HYDROGRAPHIC INDEX

In the Geographic and Hydrographic Index detailed feature names are located by general area according to the following key:

AF	Africa	LAB	Labrador, Canada
AK	Alaska, USA	LS	Labrador Sea
AOn	Atlantic Ocean, North	MAUR	Mauritania
AOs	Atlantic Ocean, South	ME	Maine, USA
AUST	Australia	MEX	Mexico
BC	British Columbia, Canada	NC	North Carolina, USA
BS	Baltic Sea	NETH	Netherlands, The
BZ	Brazil	NEWF	Newfoundland, Canada
CA	California, USA	NOR	Norway
CAN	Canada	NS	North Sea
CH	Chile	NY	New York, USA
CS	Celtic Sea	OR	Oregon, USA
CT	Connecticut, USA	POn	Pacific Ocean, North
DEN	Denmark	POs	Pacific Ocean, South
ENG	England	PORT	Portugal
ENG CHAN	English Channel	RI	Rhode Island, USA
FL	Florida, USA	SA	South Africa
FR	France	SCOT	Scotland
FRG	Federal Republic of Germany	SHET	Shetland Islands, Scotland
GBR	Great Britain	TUR	Turkey
GREEN	Greenland	USA	United States of America
IO	Indian Ocean	USSR	Union of Soviet Socialist Republics
JAP	Japan	WA	Washington, USA
		VA	Virginia, USA

TAXONOMIC INDEX

SUBJECT INDEX